Fuzzy Sets in Decision Analysis, Operations Research and Statistics

THE HANDBOOKS OF FUZZY SETS SERIES

Series Editors

Didier Dubois and Henri Prade
IRIT, Université Paul Sabatier, Toulouse, France

FUZZY SETS IN DECISION ANALYSIS, OPERATIONS RESEARCH AND STATISTICS

Edited by
Roman SŁOWIŃSKI
Poznań University of Technology

SPRINGER SCIENCE+BUSINESS MEDIA, LLC

Library of Congress Cataloging-in-Publication Data

Fuzzy sets in decision analysis, operations research, and statistics /
edited by Roman Słowiński.
p. cm. -- (The handbooks of fuzzy sets series)
Includes bibliographical references and index.
ISBN 978-1-4613-7583-8 ISBN 978-1-4615-5645-9 (eBook)
DOI 10.1007/978-1-4615-5645-9
1. Decision-making. 2. Fuzzy sets. 3. Mathematical optimization.
I. Słowiński, Roman. II. Series.
T57.95.F894 1998
658.4'03--dc21 98-3700
CIP

Originally published by Kluwer Academic Publishers in 1998
Softcover reprint of the hardcover 1st edition 1998

Contents

Part II MATHEMATICAL PROGRAMMING

Series Foreword

Fuzzy sets were introduced in 1965 by Lotfi Zadeh with a view to reconcile mathematical modeling and human knowledge in the engineering sciences. Since then, a considerable body of literature has blossomed around the concept of fuzzy sets in an incredibly wide range of areas, from mathematics and logics to traditional and advanced engineering methodologies (from civil engineering to computational intelligence). Applications are found in many contexts, from medicine to finance, from human factors to consumer products, from vehicle control to computational linguistics, and so on. Fuzzy logic is now currently used in the industrial practice of advanced information technology.

As a consequence of this trend, the number of conferences and publications on fuzzy logic has grown exponentially, and it becomes very difficult for students, newcomers, and even scientists already familiar with some aspects of fuzzy sets, to find their way in the maze of fuzzy papers. Notwithstanding circumstancial edited volumes, numerous fuzzy books have appeared, but, if we except very few comprehensive balanced textbooks, they are either very specialized monographs, or remain at a rather superficial level. Some are even misleading, conveying more ideology and unsustained claims than actual scientific contents.

What is missing is an organized set of detailed guidebooks to the relevant literature, that help the students and the newcoming scientist, having some preliminary knowledge of fuzzy sets, get deeper in the field without wasting time, by being guided right away in the heart of the literature relevant for her or his purpose. The ambition of the HANDBOOKS OF FUZZY SETS is to address this need. It will offer, in the compass of several volumes, a full picture of the current state of the art, in terms of the basic concepts, the mathematical developments, and the engineering methodologies that exploit the concept of fuzzy sets.

This collection will propose a series of volumes that aim at becoming a useful source of reference for all those, from graduate students to senior researchers, from pure mathematicians to industrial information engineers as well as life, human and social sciences scholars, interested in or working with fuzzy sets. The original feature of these volumes is that each chapter is written by one or several experts in the concerned topic. It provides introduction to the topic, outlines its development, presents the major results, and supplies an extensive bibliography for further reading.

The core set of volumes are respectively devoted to fundamentals of fuzzy set, mathematics of fuzzy sets, approximate reasoning and information systems, fuzzy models for pattern recognition and image processing, fuzzy sets in decision analysis, operations research and statistics, fuzzy systems modeling and control, and a guide to practical applications of fuzzy technologies.

Didier DUBOIS Henri PRADE
Toulouse

Preface

At the end of sixties fuzzy sets were proposed by Bellman and Zadeh as a natural setting for the representation and aggregation of ill-defined goals in the scope of multiple-stage decision making. Earlier, in the late sixties, Zadeh had also connected the notions of fuzzy sets to probability theory, showing their orthogonality in concern. These seminal proposals have given birth to a coherent body of literature pertaining to computer-based decision support, including the handling of various forms of uncertainty by mixing fuzzy sets and statistics. Fuzzy sets and fuzzy relations have now become familiar tools for representing criteria, preferences and constraints in decision problems. The present volume is the first one in the HANDBOOK OF FUZZY SETS Series and covers the impact of fuzzy set theory in *Decision Analysis, Operations Research and Statistics.*

It is worth recalling that the use of fuzzy sets has been associated with different semantics. One can distinguish three basic semantics. The first semantics, and historically the oldest one, is the expression of similarity or closeness. This understanding of fuzzy sets is used in clustering, recognition and classification.

The second semantics of fuzzy sets is related to the representation of incomplete or vague states of information under the form of possibility distributions. This view of fuzzy sets enables representation of imprecise or uncertain information in mathematical models of decision problems considered in operations research. In models formulated in terms of mathematical programming, the imprecision and uncertainty of information (data) is taken into account through the use of fuzzy numbers or fuzzy intervals instead of crisp coefficients. It involves fuzzy arithmetic and other mathematical operations on fuzzy numbers which are defined with respect to the famous Zadeh's extension principle.

The third semantics of fuzzy sets is useful to express preferences concerning satisfaction of flexible constraints and/or attainment of goals. This is especially important for exploiting information in decision making. The gradedness introduced by fuzzy sets refines the simple binary distinction made by ordinary constraints. It also refines the crisp specification of goals and "all-or-nothing" decisions. Constraint satisfaction algorithms, optimization techniques and multi-criteria decision analysis are typically involving flexible requirements which can be represented by fuzzy relations.

The uncertainty and the preference semantics are often encountered together. It is typical for decision analysis and operations research where, in order to deal with both uncertain data and flexible requirements, one can use a fuzzy set representation. These representations are comprehensively described in the present volume on the use of *Fuzzy Sets in Decision Analysis, Operations Research and Statistics.*

The book includes thirteen chapters divided into four parts. They are devoted to the use of fuzzy sets in:

- Decision Making,
- Mathematical Programming,
- Statistics and Data Analysis, and
- Reliability, Maintenance and Replacement.

The skeleton of the book has resulted from multiple discussions between the editor of the volume, the editors of the whole series and their advisory board of experts. Then, researchers well-known for achievements in the fields covered by particular chapters, were invited by the editor to write the first draft. These drafts were submitted for review to other experts who sometimes became co-authors because of the extent of their contribution to the second draft. Finally, the thirteen chapters have been signed by twenty five authors from twelve countries and four continents. This enhances the international character of the project and gives an idea how intensive was the correspondence between the editor, the authors and the reviewers during the three-year period of cooperation.

The contents of particular parts is the following.

In chapter 1, P. Perny and M. Roubens survey fuzzy preference models for single and multiple criteria decision aiding. The first part begins with the definition of classical relational preference structures from a weak preference relation and proceeds to the extension of these structures to the fuzzy case. The second part is devoted to construction of the preference structures from criterion values in single or multiple criteria decision problems. It is shown first how crisp preference relations are constructed from criterion values, and then, how fuzzy preference relations can be constructed from the same criterion values. The latter case is finally extended to fuzzy criterion values.

In chapter 2, M. Grabisch, S.A. Orlovski and R.R. Yager are addressing the problem of aggregating numerical preferences. The first part of this chapter deals with the aggregation of criteria into a single criterion. Properties which are suitable for this kind of aggregation are presented together with the most common aggregating operators, in particular, the ordered weighted averaging operators and fuzzy integrals. The second part is devoted to aggregation of fuzzy relational preferences in a multiattribute context using two different approaches: the Pareto principle and the agreement-discordance (or concordance-discordance) principle being the key concept of the outranking methods for multiple criteria decision aiding.

In chapter 3, J. Fodor, S.A. Orlovski, P. Perny and M. Roubens review several approaches to elaboration of a recommendation from a fuzzy relational preference model in multiple criteria choice, ranking and sorting. As the overall fuzzy preference model is, in general, neither transitive nor complete, its exploitation in view of finding a recommendation is not an easy task. The chapter starts with the choice based on a fuzzy preference relation; fuzzy non-dominated set of alternatives and fuzzy choice rules, like the core, are characterized. Then, several choice functions based on scoring functions, covering relations and kernels are presented. Ranking and sorting procedures from fuzzy relations are surveyed in the final part.

In chapter 4, J. Kacprzyk and H. Nurmi describe main developments in group decision making under fuzziness. The key question addressed in this chapter is how to find an alternative or a set of alternatives, from among the feasible ones, which best reflects the preferences of a group of individuals as a whole. Procedures used for answering this question are based on fuzzy representation of individual preferences and on fuzzy majority equated with a fuzzy linguistic quantifier, like most, almost all, etc. The fuzzy majority is determined via fuzzy logic based calculus of linguistically quantified statements or via the ordered weighted averaging operators. Another question discussed in this chapter is how to define a degree of consensus in the group under individual fuzzy preference relations and a fuzzy majority.

In chapter 5, A. Billot presents elements of fuzzy game theory. In the first part the intuitive relationships between games and fuzziness are characterized. Then, some arguments to justify the two main concepts of fuzzy information and fuzzy coalition are proposed on which cooperative and noncooperative analyses are founded. In the main part, the noncooperative and the cooperative fuzzy game theory are surveyed.

In chapter 6, which is the first chapter of Part II, H. Rommelfanger and R. Słowiński present an overview of methods for solving fuzzy linear programming (FLP) problems with single or multiple objective functions. The case of FLP with soft goals and constraints (called flexible programming) is distinguished from the case of FLP with fuzzy coefficients. They are associated with different semantics of fuzzy sets mentioned above, but these semantics can also appear together in a common model of FLP. Essential questions connected with FLP are discussed: modeling of fuzzy data, extended addition for aggregating fuzzy objectives and left-hand sides of fuzzy constraints, inequality relations between fuzzy sets in constraints, treatment of fuzzy objectives and computation of a compromise solution in the multi-objective case. A survey of applications of FLP and of related bibliography is given.

In chapter 7, M. Sakawa generalizes the considerations of the previous chapter to fuzzy nonlinear programming (FNLP). The distinction between flexible programming and FNLP with fuzzy coefficients is maintained. Special attention is paid to the definition of a generalized Pareto optimality in both distinguished cases, as well as in the case where the two semantics are combined within fuzzy multi-objective nonlinear programming with fuzzy coefficients. Another crucial point carefully discussed is an interactive search for a compromise solution in the set of generalized Pareto optimal solutions.

In chapter 8, S. Chanas and D. Kuchta survey various applications of fuzzy sets in discrete optimization problems representing traditional operations research models. Among these problems, the best known are connected with network optimization and scheduling; their fuzzy counterparts have an important place in this chapter. In particular, the survey includes: fuzzy optimization on graphs, transportation and assignment with fuzzy coefficients, project network scheduling with fuzzy time parameters, including fuzzy CPM, fuzzy PERT and fuzzy scheduling on machines, fuzzy set covering, fuzzy knapsack problem and the general fuzzy 0-1 linear programming problem.

In chapter 9, A.O. Esogbue and J. Kacprzyk provide a review of main developments within fuzzy dynamic programming for dealing with optimal multistage decision making where constraints on decisions made and goals on states attained are fuzzy. The cases of deterministic, stochastic and fuzzy state transitions are discussed for different definitions on termination time: fixed and specified, implicitly given, fuzzy and infinite. A brief review of applications of fuzzy dynamic programming is given.

In chapter 10, opening Part III, J. Gebhardt, M.A. Gil and R. Kruse describe existing approaches to fuzzy statistical data analysis. The chapter starts with a discussion of fuzziness in random experiments. Concepts combining basic notions of fuzzy sets and probability theory are defined. Fuzzy-valued and real-valued statistics with fuzzy experimental data are characterized with an emphasis on parameter estimation and hypothesis testing. After the problems of statistical inference, statistical decision making with fuzzy data is discussed, involving fuzzy utility within fuzzy Bayesian framework.

In chapter 11, P. Diamond and H. Tanaka make an overview of regression analysis with fuzzy data. All methods described fit a fuzzy linear model to the given data, for both crisp input - fuzzy output and fuzzy input - fuzzy output. The fitting methods fall into two distinct groups of techniques: possibilistic regression and least squares regression, based on possibilistic distributions and on the notion of a metric distance between fuzzy sets, respectively.

In chapter 12, belonging to Part IV, E. Kerre, T. Onisawa, B. Cappelle and I. Gazdik sketch the progress made in the relatively young discipline of system reliability engineering in fuzzy environment. In fuzzy reliability analysis, the transition from a success state to a failure state of an object is not abrupt but changes gradually. First, a historical perspective of reliability engineering is given and some questions of classical reliability are reviewed. The probabilistic approach to system reliability analysis is characterized from the fuzzy set-theoretic viewpoint. Non-probabilistic measures of reliability allow natural language expressions of expert's subjective judgments.

In chapter 13, A.O. Esogbue and W.E. Hearnes II describe the use of fuzzy sets and possibility theory to model uncertainty in replacement decisions via fuzzy variables and fuzzy numbers. In particular, a fuzzy set approach to economic life of an asset calculation as well as to finite horizon single asset replacement problem with multiple challengers is discussed. The algorithms used to determine the optimal replacement policy incorporate fuzzy arithmetic, dynamic programming with fuzzy rewards, the vertex method and various ranking methods for fuzzy numbers.

The volume editor wishes to acknowledge the harmonious cooperation with and between the authors which finally results in this original guidebook on the use of *Fuzzy Sets in Decision Analysis, Operations Research and Statistics.* Special thanks are due to Dr. Barbara Wołyńska for her patience in preparation of the camera-ready manuscript.

Roman SŁOWIŃSKI

Contributing Authors

Antoine Billot
Department of Economics
University of Paris 2 Panthéon-Assas,
92, rue d'Assas, 75006 Paris, France
e-mail: Antoine.Billot@ens.u-paris2.fr

Bart Cappelle
Abnamrobank NV Belgian Branch
Regentlaan 53,
B-1000 Brussels, Belgium
e-mail: bart.cappelle@abnamro.com

Stefan Chanas
Institute of Industrial Engineering and Management
Wrocław University of Technology
Wybrzeze Wyspianskiego 27, 50-372 Wrocław, Poland
e-mail: chanas@ioz.pwr.wroc.pl

Phil Diamond
Department of Mathematics
University of Queensland
St. Lucia, QLD 4067, Australia
e-mail: pmd@maths.uq.oz.au

Augustine O. Esogbue
School of Industrial and Systems Engineering
Georgia Institute of Technology
Atlanta, GA 30332-0205, U.S.A.
e-mail: aesogbue@isye.gatech.edu

János Fodor
Department of Mathematics, Institute of Mathematics and Computer Science
University of Agricultural Sciences (GATE)
Páter Károly u.1. H-2103 Gödöllö , Hungary
e-mail: jfodor@cs.elte.hu, jfodor@mszi.gau.hu

Igor Gazdik
IG Innovation
Elinsborgsbacken 23,
S-16364 Spanga, Sweden
e-mail: igor.gazdik@mailbox.swipnet.se

Joerg Gebhardt
Department of Mathematics and Computer Science
University of Braunschweig
Bueltenweg 74-75, D-38106 Braunschweig, Germany
e-mail: gebhardt@ibr.cs.tu-bs.de

Maria Angeles Gil
Department of Statistics OR and DM
University of Oviedo, C/Calvo Sotelo, s/n
E-33007 Oviedo, Spain
e-mail: angeles@pinon.ccu.uniovi.es

Michel Grabisch
Thomson-CSF, Corporate Research Laboratory
Domaine de Corbeville
91404 Orsay, France
e-mail: grabisch@thomson-lcr.fr

Warren E. Hearnes II
School of Industrial and Systems Engineering
Georgia Institute of Technology
Atlanta, GA 30332-0205, U.S.A.
e-mail: wp46268@west-point.org

Janusz Kacprzyk
Systems Research Institute
Polish Academy of Sciences
ul. Newelska 6, 01–447 Warsaw, Poland
e-mail: kacprzyk@ibspan.waw.pl

Etienne Kerre
Fuzziness and Uncertainty Modeling
Applied Mathematics and Computer Science, University Gent
Krijgslaan 281 - S9, B-9000 Gent, Belgium
e-mail: etienne.kerre@rug.ac.be

Rudolf Kruse
Department of Computer Science
Otto-von-Guericke University Magdeburg
Universitaetsplatz 2, D-39106 Magdeburg, Germany
e-mail: kruse@iik.cs.uni-magdeburg.de

Dorota Kuchta
Institute of Industrial Engineering and Management
Wrocław University of Technology
Wybrzeze Wyspiańskiego 27, 50-372 Wrocław, Poland
e-mail: kuchta@ioz.pwr.wroc.pl

Hannu Nurmi
Department of Political Science
University of Turku
20500 Turku, Finland
e-mail: hnurmi@sara.cc.utu.fi

Takehisa Onisawa
Institute of Engineering Mechanics, College of Engineering Systems
University of Tsukuba
1-1-1, Tennodai, Tsukuba 305-8573, Japan
e-mail: onisawa@esys.tsukuba.ac.jp

Sergei A. Orlovski
Computing Center
Russian Academy of Sciences
Vavilova 40, 117333 Moscow, Russia
e-mail: orlov@ccas.ru

Patrice Perny
Université Pierre et Marie Curie
LIP6, Box 169, 4, Place Jussieu
F-75252 Paris Cedex 05, France
e-mail: Patrice.Perny@lip6.fr

Heinrich J. Rommelfanger
Institute of Statistics and Mathematics
J.W. Goethe-University Frankfurt am Main
Mertonstrasse 17-23, D-60054 Frankfurt am Main, Germany
e-mail: rommelfanger@wiwi.uni-frankfurt.de

Marc Roubens
University of Liège
Institute of Mathematics
15, avenue des Tilleuls, B-4000 Liège, Belgium
e-mail: roubens@math.ulg.ac.be

Masatoshi Sakawa
Department of Industrial and Systems Engineering
Faculty of Engineering, Hiroshima University
Kagamiyama 1-4-1, Higashi-Hiroshima, 739 Japan
e-mail: sakawa@msl.sys.hiroshima-u.ac.jp

Roman Słowiński
Institute of Computing Science
Poznań University of Technology
Piotrowo 3a, 60-965 Poznań, Poland
e-mail: slowinsk@sol.put.poznan.pl

Hideo Tanaka
University of Osaka Prefecture
College of Engineering
1-1, Gakuen-cho, Sakai, Osaka 593, Japan
e-mail: hideot@center.osakafu-u.ac.jp

Ronald R. Yager
Machine Intelligence Institute
Iona College
New Rochelle, NY 10801,U.S.A.
e-mail: yager@panix.com, RYager@iona.edu

I Decision Making

1 FUZZY PREFERENCE MODELING

Patrice Perny
Marc Roubens

Abstract: We survey of classical and fuzzy relational preference models for single or multiple criteria decision aid (MCDA). After a brief introduction to preference modeling for decision aid (§ 1.1), the chapter is divided in two complementary parts. The first part (§ 1.2) deals with the definition of classical preference structures from a weak preference relation R and with the extension of these structures to the fuzzy case. A particular attention is given to the definition of fuzzy strict preference (P), indifference (I) and incomparability (J) relations from a fuzzy weak preference relation R. The second part (§ 1.3) is devoted to the construction of the preference structures in single or multiple criteria decision problems. After presenting the construction of crisp preference relations from criterion values, we show the specific interest of fuzzy sets in this context. Then we elaborate fuzzy preference structures from criterion values and indicate some extensions of these models when criterion values are imperfectly known.

1.1 PREFERENCE MODELING AND DECISION AID

Most of the real-world decision problems take place in a complex environment where conflicting systems of logic, uncertain and imprecise knowledge and possibly vague preferences have to be considered. To face such complexity, preference modeling requires the use of specific tools, techniques and concepts which allow the available information to be represented with the appropriate granularity. The aim of this chapter is precisely to provide some material based on multi-valued logic and fuzzy set theory for the building of preference models.

In real decision situations, we must keep in mind that the analyst has to provide a crisp and intelligible recommendation, even if preference information is given with vagueness, imprecision or uncertainty. To achieve this goal, one often requires the use of a procedure aiding to draw a nonfuzzy conclusion from fuzzy information. As suggested in Perny and Roy (1992), one can distinguish two parallel approches to carry out this critical step.

- *The Conventional Approach: "Dissolving fuzziness then processing nonfuzzy data"* This approach is founded on the construction of a nonfuzzy model of preferences from the initial information which is often imprecise or uncertain (fuzziness dissolution). This information is then analysed and exploited within a decision aid system. The system provides crisp outputs leading to clearcut recommendations (crisp data processing). However, the fuzziness dissolution performed at an early stage of the overall decision process is not free of arbitrariness. The construction of crisp preference models necessarily implies many simplifications and distortions of the initial information. Hence, the precise and discriminating nature of the resulting outputs is often illusive. Sometimes, to cope with this difficulty, a sensitivity analysis is implemented a posteriori. Such an analysis consist of considering many different sets of data which must be separately processed. The different resulting outputs are compared and synthetised in order to obtain the final recommendation.

- *The Fuzzy Approach: "Processing fuzzy data then dissolving fuzziness"* This approach aims at dissolving fuzziness at a later stage of the overall decision process. A fuzzy model of preferences is elaborated from the initial information and this model is really processed as fuzzy information. Usually, the outputs of the decision system are fuzzy and lead to derive more nuanced recommendations. If necessary, a fuzziness dissolution step is performed. At this stage, the user has the possibility to obtain several sets of crisp outputs allowing clearcut recommendations to be derived. By cutting the information so as to keep only conclusions whose confidence degree exceeds a certain threshold, one gets a very reliable information. When this cut is not sufficient to derive non-trivial conclusions, one can get more discriminating results by weakening the threshold. It allows a compromise between reliability and discrimination to be obtained before presenting the conclusions. This flexibility is a typical feature of the fuzzy approach.

The main difference between the two approches lies in the position of the fuzziness dissolution step within the decision aid system. In the fuzzy approach, this step which often induces weakening and distortion of information stands after (instead of before) the analysis and exploitation step within the overall decision aid procedure. This increases the reliability and flexibility of the model but disturbs the simplicity of the subsequent analysis and exploitation step.

The first part of this book provides an illustration of the fuzzy approach for preference modeling, in the context of single and multiple Criteria Decision Aid (MCDA). This first chapter is divided in two parts. The first section is devoted to preference structures. We first recall some classical notions in crisp and fuzzy preference modeling and the properties of preference structures are investigated (§ 1.2). The second part of the chapter deals with the elaboration of preference structures in single or multicriteria decision problems (§ 1.3). We discuss the construction of crisp and fuzzy preference relations from one or several criteria. Aggregation and exploitation of these relations are discussed in chapters 2–4 of this book.

1.2 PREFERENCE STRUCTURES

This section defines preference structures using classic and many-valued logics. We first recall some basic notions about the classical language used to model preferences. Some classical preference structures based on crisp relations are presented with their properties. Then, the standard properties of preference relations are extended using many-valued logics and we investigate the definition of fuzzy preference structures using De Morgan triples $< T, S, n >$. In this direction, most of the results concentrate on obtaining as many conditions that are valid in the classical language as possible. The framework is determined by the seminal paper of Orlovski (1978) which starts his analysis with a reflexive fuzzy preference relation R and defines an indifference (or similarity) relation I and a strict preference relation P with the use of Lukasiewicz and min t-norms. Moreover, Alsina (1985) formulates an "impossibility theorem" that proves the non existence of a single, consistent many-valued logic as the logic of preference. Extensions to many-valued languages are described on the basis of results obtained by Ovchinnikov and Roubens (1992), Fodor (1992), Fodor and Roubens (1994b), Perny and Roy(1992), Perny(1995). An important step consists in the introduction of an incomparability relation J. Final conditions for transitivity of (P, I, J) are given.

1.2.1 Basic Notions of Crisp Preference Modeling

In the classical theory of preference modeling (see Doignon *et al.* (1986), Fishburn (1970, 1985), Monjardet (1978), Pirlot and Vincke (1997), Roberts (1979), Roubens and Vincke (1985), Roy (1996), Vincke (1992), one usually assumes a finite set A of alternatives and a weak preference relation R defined on it. The crisp preference relation R corresponds to a mapping $R : A \times A \to \{0, 1\}$ where $R(a, b) = 1$ denotes aRb and corresponds to the assertion "alternative a is not worse than alternative b". Relation R is usually called a *weak preference relation* or *outranking relation* (see *e.g.* Roy(1996)). Some other preference relations can be considered, depending on the particular decision problem to deal with (see *e.g.* Perny and Roy (1992):

- the *strict preference* relation is used to represent strict (asymmetric) preference situations. The preference of a over b usually corresponds both to the presence of arguments strong enough to justify the assertion aRb and the absence of similar arguments to justify the reverse assertion bRa.

- the *indifference* relation is used to represent the similarity between two alternatives. Indifference for a pair (a, b) usually corresponds to the presence of arguments strong enough to justify both the assertion aRb and the assertion bRa.
- the *incomparability* relation is used to decribe partial preference structures where some alternatives cannot be compared in terms of preference or indifference. Incomparability for a pair (a, b) usually corresponds to the absence of arguments strong enough to justify at least one of the two assertions aRb and bRa.

Formally, to any weak preference relation R, one can associate three preference relations P, I, J, defined as follows:

- the strict preference relation P :

  $P(a,b) = 1$ (aPb) if and only if $R(a,b) = 1$ (aRb) and $R(b,a) = 0$ (not bRa)

- the indifference relation I :

  $I(a,b) = 1$ (aIb) if and only if $R(a,b) = 1$ (aRb) and $R(b,a) = 1$ (bRa)

- the incomparability relation J :

  $J(a,b) = 1$ (aJb) if and only if $R(a,b) = 0$ (not aRb) and $R(b,a) = 0$ (not bRa)

Using set-theoretical operators, the previous linguistic operators (... is better than ..., ... is indifferent to ..., ... is incomparable to ...) can be translated as :

$$P = R \cap R^d \quad (1)$$
$$I = R \cap R^{-1} \quad (2)$$
$$J = R^c \cap R^d \quad (3)$$

where R^{-1}, R^c, R^d represent respectively the inverse, the complementary and the dual relations ($R^{-1}(a,b) = R(b,a)$, $R^c(a,b) = 1 - R(a,b)$ and $R^d(a,b) = 1 - R(b,a), \forall a, b \in A$)

The triple (P, I, J) forms a preference structure which components are linked together according to :

$$P \cup I = R \quad (4)$$
$$P \cup J = R^d \quad (5)$$
$$P \cap P^{-1} = \emptyset \quad (6)$$
$$I = I^{-1} \quad (7)$$

$$J = J^{-1} \tag{8}$$
$$P \cap I = \emptyset \tag{9}$$
$$P \cap J = \emptyset \tag{10}$$
$$I \cap J = \emptyset \tag{11}$$
$$P \cup I \cup J \cup P^{-1} = A \times A \tag{12}$$
$$P \cup I \cup P^{-1} = R \cup R^{-1} \tag{13}$$

a) Classical properties of preference relations

A binary relation R is said to be:

Reflexive iff $\forall a \in A, aRa$
Complete iff $\forall a, b \in A, aRb$ or bRa
Symmetric iff $\forall a, b \in A, aRb \Rightarrow bRa$
Antisymmetric iff $\forall a \in A, \forall b \in A - \{a\}, not(aRb$ and $bRa)$
Asymmetric iff $\forall a, b \in A, not(aRb$ and $bRa)$
Transitive iff $\forall a, b \in A, (aRb$ and $bRc) \Rightarrow aRc$
Semi-transitive iff $\forall a, b, c, d \in A, (aRb$ and $bRc) \Rightarrow (aRd$ or $dRc)$
Ferrers iff $\forall a, b, c, d \in A, (aRb$ and $cRd) \Rightarrow (aRd$ or $cRb)$

By definition, for any relation R, the associated relation P is asymmetric (see equation (6)) whereas relations I and J are symmetric (see equations (7) and (8)). From these definitions, many preference structures of type (P, I, J) can be defined, depending on the properties of relation R.

b) Particular preference structures

The main classical structures are:

- *The total preorder:* this is a weak-preference relation R which is complete and transitive. The associated preference structure is a triple (P, I, J) where P and I are transitive and J is empty (there is no incomparability). When R is antisymmetric, I is restricted to pairs of type (a, a) and the structure is called a *total order*. When R is asymmetric, I is empty and the structure is called a *strict total order*.

- *The partial preorder:* this is a weak-preference relation R which is reflexive and transitive. The associated preference structure is a triple (P, I, J) where P and I are transitive and J is not empty. Moreover $(aPb$ and $bIc) \Rightarrow aPc$ and $(aIb$ and $bPc) \Rightarrow aPc$. When R is antisymmetric, I is restricted to pairs of type (a, a) and the structure is called a *partial order*. When R is asymmetric, I is empty and the structure is called a *strict partial order*.

- *The semiorder:* this is a weak-preference relation R which is complete, Ferrers and semi-transitive. The associated preference structure is a triple (P, I, J) where P is transitive (but not necessarily I) and J is empty. Moreover, with this structure we have $(aPb, bIc, cPd) \Rightarrow aPd$ and $(aPb, bPc, aId) \Rightarrow dPc$.

- *The interval order:* this is a weak-preference relation R which is complete and Ferrers. The associated preference structure is also a triple (P, I, J) where P is transitive (but not necessarily I) and J is empty. Nevertheless, in this case, the only relation linking P and I is $(aPb, bIc, cPd) \Rightarrow aPd$.

Several other structures may be considered. For a survey about these structures, see *e.g.* Vincke (1992). We now translate the above scheme into fuzzy binary preference relations :

$$R : A \times A \to [0, 1].$$

1.2.2 Fuzzy Preference Modeling

In this section, we consider the case where preferences between alternatives are described by a fuzzy preference relation R such that the value $R(a, b)$, between 0 and 1, corresponds to the degree to which "alternative a is not worse than alternative b" is true.

In order to extend preference structures in this context, it is necessary to extend properties of binary relations to the fuzzy case and to define fuzzy preference structures including valued strict preferences, valued indifference and valued incomparability relation.

a) Properties of fuzzy preference relations

The properties introduced above in the crisp case can easily be extended to the fuzzy case by replacing the classical predicate logic by a many-valued logic admitting any possible truth value within the unit interval. In this context, an interpretation is a mapping ν which associates a truth value $\nu(F) \in [0, 1]$ to any well-formed formula of the predicate calculus. The interpretation of any complex formula composed of atomic predicates $R(x, y)$, $P(x, y)$, $I(x, y)$, $J(x, y)$ is defined from a De Morgan triple $< T, S, n >$ by the following rules (see Perny (1995)):

$$\begin{aligned}
\nu(\text{not } F) &= n\nu(F) && (14)\\
\nu(F_1 \text{ and } F_2) &= T(\nu(F_1), \nu(F_2)) && (15)\\
\nu(F_1 \text{ or } F_2) &= S(\nu(F_1), \nu(F_2)) && (16)\\
\nu(F_1 \Rightarrow F_2) &= I_T(\nu(F_1), \nu(F_2)) && (17)\\
\nu(F_1 \Leftrightarrow F_2) &= E_T(\nu(F_1), \nu(F_2)) && (18)\\
\nu(\forall a F) &= \inf_{a \in A}\{\nu(F)\} && (19)\\
\nu(\exists a F) &= \sup_{a \in A}\{\nu(F)\} && (20)
\end{aligned}$$

where T is a t-norm, n a strict negation, S a t-conorm, I_T an implication operator and E_T an equivalence operator defined from T by:

$$S(x,y) = nT(nx,ny) \quad (21)$$
$$I_T(x,y) = \sup\{z \in [0,1] : T(x,z) \leq y\} \quad (22)$$
$$E_T(x,y) = T(I_T(x,y), I_T(y,x)) \quad (23)$$

For the sake of simplicity quantities of type $\nu(R(a,b))$ are denoted $R(a,b)$. As shown in Perny (1995), this interpretation system allow any classical property to be extended. For instance, generalisations of transitivity are obtained by setting to 1 ("true") the interpretation of the classical formula for transitivity as shown by the following equations.

$$\nu[(\forall a, \forall b, \forall c, (R(a,b) \text{ and } R(b,c)) \Rightarrow R(a,c)] = 1$$
$$\forall a, \forall b, \forall c,\ \nu[(R(a,b) \text{ and } R(b,c)) \Rightarrow R(a,c)] = 1 \quad \text{(by (19))}$$
$$\forall a, \forall b, \forall c,\ I_T(\nu[R(a,b) \text{ and } R(b,c)], R(a,c)) = 1 \quad \text{(by (17))}$$
$$\forall a, \forall b, \forall c,\ \nu[R(a,b) \text{ and } R(b,c)] \leq R(a,c) \quad \text{(by (22))}$$
$$\forall a, \forall b, \forall c,\ T(R(a,b), R(b,c)) \leq R(a,c) \quad \text{(by (15))}$$

and we get a new property known as T-transitivity. Hence, the multiplicity of possible choices for T induce a entire family of transitivity properties for fuzzy relations. Within this family, the strongest version of transitivity is obtained for the *min* t-norm. This (min-)transitivity is often considered as the more natural extension of transitivity for fuzzy preferences because it keeps a strong relationship with its classical counterpart for ordinary relations. It is indeed easy to show that a fuzzy relation R is min-transitive if and only if the α-cuts of R (crisp relations derived from R by keeping only the ordered pairs (a,b) such that $R(a,b) > \alpha$) are transitive for all $\alpha \in [0,1)$. This result also holds for any "min-property" derived from classical reflexivity, completeness, symmetry, asymmetry, semi-transitivity, Ferrers. Therefore, in the sequel, we shall use the following definitions derived by interpreting classical definitions, using equations (14–23) with $T(x,y) = \min(x,y)$ and $n(x) = 1 - x$.

A fuzzy relation R is said to be:

Reflexive iff $\forall a \in A,\ R(a,a) = 1$
Complete iff $\forall a, b \in A,\ \max(R(a,b), R(b,a)) = 1$
Symmetric iff $\forall a, b \in A,\ R(a,b) = R(b,a)$
Antisymmetric iff $\forall a \in A, \forall b \in A - \{a\},\ \min(R(a,b), R(b,a)) = 0$
Asymmetric iff $\forall a, b \in A,\ \min(R(a,b), R(b,a)) = 0$
Transitive iff $\forall a, b \in A,\ \min(R(a,b), R(b,c)) \leq R(a,c)$
Semi-transitive iff $\forall a, b, c, d \in A,\ \min(R(a,b), R(b,c)) \leq \max(R(a,d), R(d,c))$
Ferrers iff $\forall a, b, c, d \in A,\ \min(R(a,b), R(c,d)) \leq \max(R(a,d), R(c,b))$

b) Fuzzy preference structures

From the properties listed above, we can derive the following definitions, as natural extensions of those introduced for crisp preferences:

- A *fuzzy total preorder* is a fuzzy relation R which is complete and transitive. When R is antisymmetric this structure is called a *fuzzy total order*. When R is asymmetric this structure is called a *fuzzy strict total order*.
- A *fuzzy partial preorder* is a fuzzy relation R which is reflexive and transitive. When R is antisymmetric this structure is called a *fuzzy partial order*. When R is asymmetric this structure is called a *fuzzy strict partial order*.
- A *fuzzy semiorder* is a fuzzy relation R which is complete, Ferrers and semi-transitive.
- A *fuzzy interval order* is a fuzzy relation R which is complete and Ferrers.

We examine now the fuzzy preference structures of type (P, I, J) that may be derived from R. The problem of defining valued strict preference, valued indifference and valued incomparability has been investigated by Fodor (1991, 1992), Fodor and Roubens (1991, 1994a, 1994b), Ovchinnikov and Roubens (1991, 1992), Perny and Roy (1992)). The idea is trying to translate relations (1) - (13) in terms of fuzzy logical operations defined from a De Morgan triple $< T, S, n >$

If we try to combine (1), (2) and (4), denoting $R(a, b) = x$, $R(b, a) = y$,

$$P(a,b) = T[R(a,b), nR(b,a)] = T[x, ny]$$
$$I(a,b) = T[R(a,b), R(b,a)] = T[x, y]$$

$$S[T(x, ny), T(x,y)] = x$$

we immediately obtain a contradiction (see Alsina 1985, Fodor and Roubens 1994a). Therefore, we intend to find alternative definitions of P, I, J that would preserve (4) - (13).

In different papers, Fodor, Ovchinnikov and Roubens have considered three general axioms, namely Independence of Irrelevant Alternatives (IA), Positive Association (PA) and Symmetry (SY) to define (P, I, J) :

(IA) : for any two alternatives $a, b \in A$, the values $P(a, b)$, $I(a, b)$ $J(a, b)$ depend only upon the values $R(a, b)$ and $R(b, a)$.

As a consequence, it is assumed the existence of three functions

$$p, i, j : [0,1] \times [0,1] \to [0,1]$$

such that

$$P(a,b) = p(R(a,b), R(b,a)) \quad (24)$$
$$I(a,b) = i(R(a,b), R(b,a)) \quad (25)$$
$$J(a,b) = j(R(a,b), R(b,a)) \quad (26)$$

This standard assumption might be criticized in the sense that preferences between two alternatives might depend on their respective performance vis à vis other alternatives (see Tversky and Simonson, 1993).

(PA) : $p(x, y)$ is nondecreasing with respect to its first argument and nonincreasing with respect to its second argument,

$i(x, y)$ is nondecreasing with respect to both arguments and

$j(x, y)$ is nonincreasing with respect to both arguments.

(SY) : $i(x, y) = i(y, x)$ and $j(x, y) = j(y, x)$.

In order to preserve the valued extension of (4) and (5), we link (p, i, j) in terms of the following functional equations

$$S\,[p(x, y), i(x, y)] = x \tag{27}$$
$$S\,[p(x, y), j(x, y)] = ny \tag{28}$$

The valued relation R is a disjunction of the strict preference and the indifference and the remaining of the valued relation R should correspond either to the opposite valued strict preference either to the incomparability.
Relations (27) and (28) can be rewritten in the following form :

$$\begin{aligned} P \cup_S I &= R \\ P \cup_S J &= nR^{-1} \end{aligned}$$

Fodor and Roubens (1991, 1994a, 1994b) could find the following properties of $\langle p, i, j, T, S, n\rangle$ under assumptions (24) – (26) and (PA)–(SY) properties :

$$\begin{aligned} i(0, y) &= p(0, y) = 0, \quad p(x, 0) = x, \quad j(0, y) = ny \\ i(x, 1) &= x, \quad j(x, 1) = p(x, 1) = 0, \quad p(1, y) = ny \\ T(x, y) &= \varphi^{-1}\left\{\max\{\varphi(x) + \varphi(y) - 1, 0\}\right\} = W_\varphi(x, y) \end{aligned}$$

where φ represents an automorphism of the unit interval.

$$n(x) = \varphi^{-1}[1 - \varphi(x)] = N_\varphi(x)$$

$$W_\varphi(x, N_\varphi(y)) \leq p(x, y) \leq \min\{x, N_\varphi(y)\} \tag{29}$$
$$W_\varphi(x, y) \leq i(x, y) \leq \min\{x, y\} \tag{30}$$
$$W_\varphi(N_\varphi(x), N_\varphi(y)) \leq j(x, y) \leq \min\{N_\varphi(x), N_\varphi(y)\} \tag{31}$$

(29)–(31) can be rewritten as

$$\begin{aligned} R \cap_{W_\varphi} N_\varphi R^{-1} &\subseteq P \subseteq R \cap_{\min} N_\varphi R^{-1} \\ R \cap_{W_\varphi} R^{-1} &\subseteq I \subseteq R \cap_{\min} R^{-1} \\ N_\varphi R \cap_{W_\varphi} N_\varphi R^{-1} &\subseteq J \subseteq N_\varphi R \cap_{\min} N_\varphi R^{-1} \end{aligned}$$

$$W_\varphi\{p(x,y),p(y,x)\} = 0 \tag{32}$$

and P is W_φ-asymmetric : $P \cap_{W_\varphi} P^{-1} = \phi$

$$W_\varphi\{p(x,y),i(x,y)\} = 0 \tag{33}$$
$$W_\varphi\{p(x,y),j(x,y)\} = 0 \tag{34}$$
$$W_\varphi\{i(x,y),j(x,y)\} = 0 \tag{35}$$

which are the valued translations for (9)–(11) :

$$P \cap_{W_\varphi} I = \emptyset, \quad P \cap_{W_\varphi} J = \emptyset, \quad I \cap_{W_\varphi} J = \emptyset.$$

$$W'_\varphi\{p(x,y),p(y,x),i(x,y),j(x,y)\} = 1$$

which extends formula (12) : $P \cup_{W'_\varphi} P^{-1} \cup_{W'_\varphi} I \cup_{W'_\varphi} J = A \times A$, if

$$S(x,y) = W'_\varphi(x,y) = \varphi^{-1}\{\min\{\varphi(x)+\varphi(y),1\}\}$$

Van de Walle, De Baets and Kerre (see De Baets et al. 1995a, 1995b, 1997, Van de Walle 1996, Van de Walle et al. 1997) have considered a fuzzy preference structure (P, I, J) without reference to the basic binary relation R. Starting with a de Morgan triple $< T, S, n >$ they consider (P, I, J) satisfying the following axioms :

P is T-asymmetric $(P \cap_T P^{-1} = \emptyset)$
I is reflexive and J is irreflexive $(I(a,a) = 1, J(a,a) = 0)$
I and J are symmetric $(I = I^{-1},\ J = J^{-1})$
$P \cap_T I = \emptyset$, $P \cap_T J = \emptyset$, $I \cap_T J = \emptyset$
$P \cup_S P^{-1} \cup_S I \cup_S J = A \times A$.

If T corresponds to a continuous t-norm without zero divisors, it can easily be proved that the structure (P, I, J) must be crisp and only allows for the classical preference structure. To deal with truly fuzzy preferences, one has to turn to t-norms with zero divisors.

If continuous non-Archimedean t-norms with zero divisors are considered, Van de Walle et al. (1997) could prove that the related fuzzy structures are counter-intuitive : the degrees of strict preference, indifference and incomparability in $[0, 1]$ are always bounded from above by a value strictly smaller than 1. This result induces a basic argument for the use of continuous Archimedean t-norms having zero divisors and consequently $< T, S, n >$ is a Lukasiewicz triple $< W_\varphi, W'_\varphi, N_\varphi >$ completely determined by the choice of the automorphism φ to give a Lukasiewicz fuzzy preference structure denoted in a non ambiguous way, $\varphi - FPS$.

1.2.3 Particular Solutions

(i) Suppose first that R is *complete*, we immediately obtain that the system (29)- (31) gives

$$\begin{aligned} P(a,b) &= R^d(a,b) = 1 - R(b,a) \\ I(a,b) &= \min\{R(a,b), R(b,a)\} \\ J(a,b) &= 0 \end{aligned}$$

(ii) It has been proved by Ovchinnikov and Roubens (1992) that the necessary and sufficient condition to obtain the following solution of system (27)–(28) :

$$\begin{aligned} P(a,b) &= W_\varphi\{R(a,b), N_\varphi(R(b,a))\} \\ I(a,b) &= \min\{R(a,b), R(b,a)\} \\ J(a,b) &= \min\{N_\varphi R(a,b), N_\varphi R(b,a)\} \end{aligned}$$

corresponds to the asymmetry of P. In this case, P reaches the lower bound in (29) and I, J are on the upper bounds in (30)– (31).

(iii) Suppose now that we want to express $\langle p, i, j\rangle$ in terms of a common continuous t-norm T_1 to obtain

$$\begin{aligned} P(a,b) &= T_1\{R(a,b), N_\varphi R(a,b)\} \\ I(a,b) &= T_1\{R(a,b), R(b,a)\} \\ J(a,b) &= T_1\{N_\varphi R(a,b), N_\varphi R(b,a)\}. \end{aligned}$$

It has been proved by Alsina (1985) that this kind of relations are solutions for (27)–(28) if and only if

$$T_1(x,y) = \varphi^{-1}[\varphi(x)\varphi(y)] = \pi_\varphi(x,y).$$

Giving up conditions (27)–(28), an alternative option relaxing the continuity requirement for T_1 has been proposed by Perny and Roy (1992). They introduced a construction with $n = N_\varphi$ and T_1 defined as follows:

$$T_1(x,y) = M_\varphi(x,y) = \begin{cases} \min(x,y) & \text{if } \varphi(x) + \varphi(y) > 1 \\ 0 & \text{otherwise} \end{cases}$$

$\{M_\varphi\}$ is a family of non-continuous t-norms introduced in Perny(1992) to modify the ordinal *min* t-norm so as to keep the asymmetry in the definition of the strict valued preference P from the weak-preference relation R. This allows to export transitivity from R to P (see the next subsection).

The result obtained by Alsina, has been extended by Fodor and Roubens (1994a) if p on one side, i, j on the other side should be expressed in terms of two continuous t-norms T_1 and T_2.

(iv)

$$\begin{aligned} P(a,b) &= T_1\{R(a,b), N_\varphi R(a,b)\} \\ I(a,b) &= T_2\{R(a,b), R(b,a)\} \\ J(a,b) &= T_2\{N_\varphi R(a,b), N_\varphi R(b,a)\} \end{aligned}$$

are solutions of (33)–(34) if and only if T_1 and T_2 belong to the Frank family of continuous Archimedean t-norms :

$$\begin{aligned} T_1(x,y) &= \varphi^{-1}(T^s(\varphi(x),\varphi(y))) = T^s_\varphi(x,y) \\ T_2(x,y) &= \varphi^{-1}(T^{1/s}(\varphi(x),\varphi(y))) = T^{1/s}_\varphi(x,y) \end{aligned}$$

where

$$T^s(u,v) = \log_s\left(1 + \frac{(s^u-1)-(s^v-1)}{s-1}\right), \quad s>0,\ s\neq 1.$$

(v) It is theoretically challenging to determine conditions under which P should reach the upper bound of (29) and I, J would be on the lower bound in (30)–(31). These conditions were obtained by Fodor and Roubens (1994a).

The necessary and sufficient condition to obtain

$$\begin{aligned} P(a,b) &= \min\{R(a,b), N_\varphi R(a,b)\} \\ I(a,b) &= W_\varphi\{R(a,b), R(b,a)\} \\ J(a,b) &= W_\varphi\{N_\varphi R(a,b), N_\varphi R(b,a)\} \end{aligned}$$

as solution for (27)–(28) corresponds to

$$W'_\varphi\{p(x,y), p(y,x), i(x,y)\} = W'_\varphi\{R(x,y), R(y,x)\}$$

or

$$P \cup_{W'_\varphi} P^{-1} \cup_{W'_\varphi} I = R \cup_{W'_\varphi} R^{-1},$$

i.e. the valued counterpart of (13).

(vi) It is well-known that a classical preference structure (P,I,J) on A can be reconstructed from its weak preference relation $R = P \cup I$, in the following way:

$$(P,I,J) = (R \cap R^d, R \cap R^{-1}, R^c \cap R^d).$$

Consider now a fuzzy preference structure (P,I,J) on A with fuzzy large preference relation $R = P \cup_{W'_\varphi} I$. Van de Walle et al. (1997) have identified the conditions under which (P,I,J) can be reconstructed from R. As such, they have established a one-to-one relationship between a fuzzy preference structure and its fuzzy large preference relation. The solution was to impose an additional non-trivial characterizing condition, resulting in the following definition of a two-parameter class of fuzzy preference structures. In the definition, $\cap_{T_\varphi^s}$ is denoted shortly as $\cap_\varphi^s$.

Consider a $[0,1]$-automorphism φ and $s \in [0,\infty]$. An (s,φ)-fuzzy preference structure on A is a triplet (P,I,J) of binary fuzzy relations in A that satisfy:

(F1) I is reflexive and J is irreflexive;

(F2) I and J are symmetrical;

(F3) $P \cap_{W_\varphi} I = \emptyset$ and $P \cap_{W_\varphi} J = \emptyset$;

(F4) $N_\varphi(P \cup_{W'_\varphi} I) = P^{-1} \cup_{W'_\varphi} J$;

(F5) for $s \in]0,1[\cup]1,\infty[$: $s^{\varphi(P\cap_\varphi^s P^{-1})} + s^{-\varphi(I\cap_\varphi^{\frac{1}{s}} J)} = 2$,
for $s \in \{0,1,\infty\}$: $P \cap_\varphi^s P^{-1} = I \cap_\varphi^{\frac{1}{s}} J$.

Condition (F5) allows to characterize (s,φ)-fuzzy preference structures (P,I,J) on A. Using the previous definition, Van de Walle et al. (1997) obtain the following result :

Consider a $[0,1]$-automorphism φ and $s \in [0,\infty]$. A triplet (P,I,J) of binary fuzzy relations in A is an (s,φ)-fuzzy preference structure on A if and only if

$$\begin{aligned} P &= (P \cup_{W'_\varphi} I) \cap_\varphi^{\frac{1}{s}} N_\varphi(P \cup_{W'_\varphi} I)^{-1} \\ I &= (P \cup_{W'_\varphi} I) \cap_\varphi^{s} (P \cup_{W'_\varphi} I)^{-1} \\ J &= N_\varphi(P \cup_{W'_\varphi} I) \cap_\varphi^{s} N_\varphi(P \cup_{W'_\varphi} I)^{-1}. \end{aligned}$$

1.2.4 Transitivity of (P,I,J)

Let us consider the triple (P,I,J) together with $(IA), (PA), (SY)$ conditions and asymmetry of P. Suppose that R is a transitive relation. It has been proved by Ovchinnikov and Roubens (1991) that P is a transitive relation.(see also Fodor and Roubens 1994a).

If we restrict (P,I,J) to be a member of the class of models related to equations (29) - (31) with asymmetry condition on P, then we know that (see § 1.2.3 (ii)) :

$$(P,I,J) \equiv (W_\varphi(R, N_\varphi R^{-1}), \min(R,R^{-1}), \min(N_\varphi R, N_\varphi R^{-1}))$$

and it is easy to prove that, R being a transitive relation, P, I and J are transitive relations.

In this framework, the following propositions hold (see *e.g.* Fodor and Roubens (1994a), Perny and Roy (1992))):

(P1) If R is a fuzzy total preorder, then P is a fuzzy strict partial order, I is a fuzzy similarity relation (*i.e.* reflexive, symmetric and transitive) and J is empty.

(P2) If R is a fuzzy semiorder, then P is a fuzzy strict partial order, I is usually not transitive and J is empty.

(P3) If R is a fuzzy interval order, then P is a fuzzy strict partial order, I is usually not transitive and J is empty.

These results cannot be extended to weaker transitivity properties (*e.g.* T–transitivities with $T \neq \min$). As an example (Fodor and Roubens 1994a), W-transitivity on R does not necessarily imply W-transitivity of $P = W(R, 1 - R^{-1}) = \max(R - R^{-1}, 0)$.

We present now the elaboration of preference structures in multcriteria decision problems.

1.3 PREFERENCE MODELING FOR MCDA

1.3.1 Multicriteria Decision Problems

In Multiple Criteria Decision Aid, the decision problem can often be formulated as comparing and/or discriminating between m potential alternatives (*i.e.* variants, projects, candidates) of a set $A = \{a_1, \ldots, a_m\}$, on the basis of one or several criteria. Usually, the decision aid process consists of the following phases:

1. *Structuration:* identify the alternatives and the various criteria (points of views) to be considered in the problem,

2. *Evaluation and comparison:* evaluate the alternatives on each of the n criteria, and possibly, for each criterion j, construct a binary preference relation R_j on A defined from the partial evaluations of alternatives,

3. *Multicriteria aggregation:* aggregate the n criteria or the relations $R_j, j = 1, \ldots, n$, so as to produce an overall preference relation R (see chapter 2),

4. *Exploitation:* identify the set of best alternatives, or rank, or sort, or cluster the alternatives on the basis of relation R (see chapter 3).

One can distinguish two main types of methods to construct an overall preference relation R from n criteria :

- *R results from a scoring*: The preference analysis is conducted on the set A. An overall score is calculated for each alternative on the basis of its partial evaluations on each criterion. This score is considered as an overall criterion allowing the alternatives to be ordered. The most typical example of this approach is ordering alternatives according to the arithmetic mean of their partial evaluations (other aggregative operators may be considered, see chapter 2). In this method, if the score of alternative a is denoted $g(a)$, the overall preference relation is defined by: aRb iff $g(a) \geq g(b)$ (a is at least as good as b) for any pair (a, b). This method is very simple and provides easily exploitable results since, by construction, relation R is a total preorder. But using this method implies to suppose implictly the *complete and transitive* comparability of alternatives, which is a strong assumption. Moreover, in real decision problems, the definition of a score function g allowing meaningful comparisons of alternatives is a complex task, provided the heterogeneous nature of criterion values. These are the main limits of scoring methods.

- *R results from the aggregation of $(R_1, \ldots, R_n)$:* The preference analysis is conducted on the cartesian product $A \times A$. This approach is mainly used in social choice theory (see chapter 4 and Sen (1986), Kelly (1991) and outranking methods (see Roy and Bouyssou (1993) and Perny(1997)). A preference structure is defined on each dimension j from partial evaluations of alternatives. The resulting preference relations are then aggregated to obtain an overall preference relation R. Usually, this relation is constructed so as to reflect the majoritarian preference among the set of criteria (see Perny (1997) and chapter 2). This approach allows a fine and flexible description of preferences without forcing arbitrarily alternatives to be comparable. Another advantage of this approach is that there is no need to define a score mixing possibly heteregeneous evaluations coming from different criteria. Comparisons of partial evaluations are performed dimension by dimension, and the results of these comparisons are then aggregated. The main difficulty with this approach is that intransitivities or cycles can occur in the overall preference relation R and there is no serious solution to avoid this problem. This crucial point as been mentioned by Condorcet (1785) and was definitely proved by Arrow (1959) and Gibbard (1969), in the case of nonfuzzy relations. More recently, Leclerc (1984), Barrett et al. (1990, 1992), Perny (1990, 1992) have provided extensions of these results in the case of fuzzy relations (see also chapter 4). Therefore, overall preference relations derived from an ordinal aggregation are not directly exploitable. This shows the necessity of resorting to an additional procedure to derive a transitive information when a ranking is required (see chapter 3).

Let us detail now the specific interest of fuzzy sets in these methods.

1.3.2 The Use of Fuzzy Sets in the Definition of Criteria

In a real decision problem, many constraints have to be considered. Some of them are hard constraints of yes/no type and they are used to define the admissible list of alternatives. Some other are more flexible and gradual and may be seen as ill-defined objectives. For example, consider a financial constraint of type "$c(a) \leq \alpha$" where $c(x)$ is a cost function defined on the set of alternatives and α is a boundary which cannot be known precisely. This constraint is ill-defined and is probably better represented by an objective function of type $\min c(x), x \in A$. In this example, the set of values corresponding to an "attractive cost" can be caracterised using the function $c(x)$, defining implicitly a fuzzy subset of the monetary scale.

Hence, an objective may be seen as a flexible constraint restricting the set of all possible consequences on a given attribute to a subset qualified as *desirable* by the decision maker (see Bellman and Zadeh (1970), Dubois and Prade (1985), Mudri and Perny (1994)).

Let us consider a given dimension (or attribute) describing the consequences of alternatives on a particular viewpoint, using a scale X (this scale is often the real line but any nominal, ordinal or cardinal scale could be considered). An objective (or fuzzy constraint) is characterised on this dimension by the fuzzy set:

$$G = \{(x, g(x)), x \in X\}$$

The membership function $g : X \rightarrow [0, 1]$ defining this fuzzy set acts as a criterion function or utility function allowing consequences of alternatives to be compared on a particular dimension. As we can see, fuzzy sets allow criterion functions to be defined in many situations. For instance, the decision maker may express his/her preferences on a particular dimension by roughly specifying the minimal level of requirement for a variable, the "ideal value" of an attribute, or an "aspiration level" on this or that dimension. Such information points out a fuzzy region of interest, on a particular dimension, around the ideal value or near the aspiration level (see figures 1 and 2).

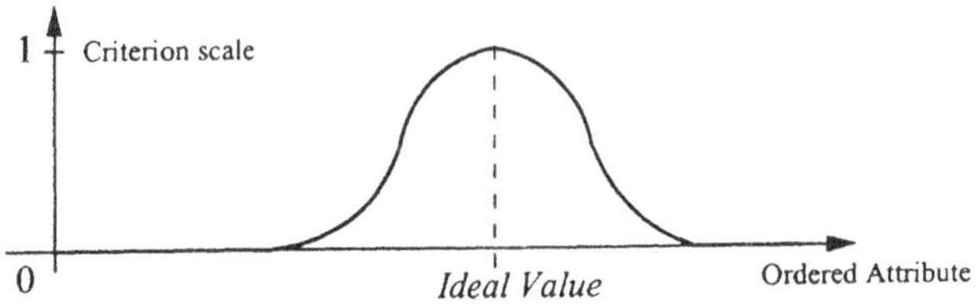

Figure 1 Fuzzy objective defined from a reference value

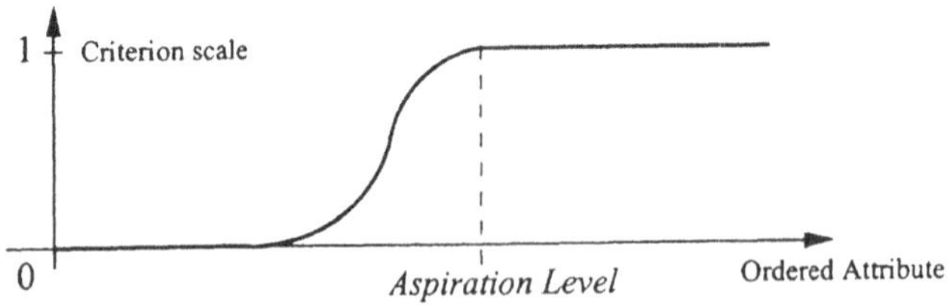

Figure 2 Fuzzy objective defined from an aspiration level

More generally, a *criterion* is a real-valued function g defined on the set of alternatives A, allowing these alternatives to be compared according to the following rule (see Roy (1996)) :

$$g(a) \geq g(b) \;\Rightarrow\; aRb \tag{36}$$

where R is a weak preference (or outranking) relation describing the preference of the decision maker from the particular viewpoint represented by g.

When several possibly conflicting criteria $g_1, \ldots, g_n$ have to be considered in a decision problem, a n-tuple of relations $(R_1, \ldots, R_n)$ is associated to these criteria using (36). Quantities of type $g_j(a), a \in A, j \in \{1, \ldots, n\}$ are called *criterion values* and represent the *partial evaluations* or *scores* of a according to criterion j. They form together the basic preference information, a first step in the elaboration of a more sophisticated preference model for multiple criteria decision aid.

1.3.3 Construction of Preference Relations from Criterion Values

We survey the construction of crisp and fuzzy preference relations from criterion values. In this section, we consider a single criterion denoted g. This criterion may represent one of the g_j's criteria or a synthetical criterion obtained after agregation (scoring approach).

The more conventional way of building a preference model is to interpret function g as a *true-criterion* and thus to consider that a is strictly preferred to b when $g(a)$ exceeds $g(b)$ even by the smallest margin (this amounts to complete equation (36) by susbtituting an equivalence to the implication).

In a real-life problem however, the imprecision and/or uncertainty modifying criterion values and the necessary arbitrariness of the evaluation process imply refusing to assign an absolute discriminating power to criteria. Moreover, even in the case of complete and precise information, a small positive difference of scores does not always justify the preference. For this reason, several methods have been proposed to reduce the possible arbitrariness of traditional highly discriminating technics.

A classical attitude in this framework is to assess discrimination thresholds allowing the distinction between significant and not significant score differences. A typical example of preference model based on the use of a discrimination threshold is given by the standard semi-criterion and the associated semiorder structure (see Luce (1956), Vincke (1980), Roy (1996), Pirlot and Vincke (1997)).

Formally, a (standard) *semi-criterion* is defined by a criterion function g and a threshold function q verifying:

$$g(a) > g(b) \;\Rightarrow\; g(a) + q(g(a)) \geq g(b) + q(g(b)) \tag{37}$$

$$aPb \;\Leftrightarrow\; g(a) - g(b) > q(g(b)) \tag{38}$$

$$aIb \;\Leftrightarrow\; |g(a) - g(b)| \leq \min\{q(g(a)), q(g(b))\} \tag{39}$$

In simple cases, function q is constant and represents the minimal value for a score difference to be significant. Since such a threshold may vary along the criterion scale, q is defined as a function of the position x in this scale. In this case, equation (37) play an important part. It represents a local consistency condition making it impossible to consider the score difference of type $g(c)-g(a)$ as significant when a greater difference of type $g(c) - g(b)$ is not significant (we assume $g(a) > g(b)$).

The preference structure $(P, I, \emptyset)$ obtained by (55- 39) defines a semiorder $R = I \cup P$. When condition (37) is violated, R is an interval order provided $P.I.P \subset P$ (we call it a non standard semi-criterion)

It is worth noting some drawbacks in this modeling technique, specially when the evaluation scale is continuous. Actually, if we recognize a irreducible part of arbitrariness in score evaluation, claiming that we are able to establish a sharp separation between significant and not significant score differences is no longer possible. As an illustration, suppose that two candidates a and b are such that $g(a) - g(b) = q(g(b)) - \epsilon/2$, where ϵ stands for a positive quantity very small compared to $q(g(b))$. If a slightly superior score $(+\epsilon)$ was attached to a, we would obtain $g(a) - g(b) = q(g(b)) + \epsilon/2$. Such a small variation would transform the indifference aIb into the preference aPb.

In order to overcome this difficulty, two discrimination thresholds may be considered so as to define a *pseudo-criterion*. The main idea underlying this concept proposed by Roy and Vincke (1984) is to separate the preference area from the indifference area by inserting a intermediate situation called *weak-preference*. This weakened version of preference must be interpreted as an hesitation between preference and indifference. Formally, two discrimination threshold functions q (indifference threshold) and p (preference threshold) are considered in order to give a meaning to scores differences. Then crisp binary relations I, Q, P called respectively *indifference*, *weak-preference* and *strict-preference* are defined from functions g, q and p using the following conditions (see Roy (1996)):

$$aPb \Leftrightarrow g(a) - g(b) > p(g(b)) \quad (40)$$

$$aQb \Leftrightarrow p(g(b)) \geq g(a) - g(b) > q(g(b)) \quad (41)$$

$$aIb \Leftrightarrow | g(a) - g(b) | \leq \min\{q(g(a)), q(g(b))\} \quad (42)$$

where q and p fulfil condition (37) and the additional consistency condition:

$$\forall a \in A, \ p(a) > q(a) \quad (43)$$

The preference structure (I, Q, P) extends the notion of semiorder and is called a *pseudo-order* (see Roy and Vincke (1987), Vincke (1992), Roy (1996)).

This second model offers new possibilities but unfortunately, it does not really solve the problem of sensititvity illustrated on the semi-criterion. One may indeed readily imagine examples where minor variations of scores lead to significant variations of pseudo-orders, and thus, to various conflicting conclusions. Such undesirable discontinuities make a sensitivity analysis absolutely necessary. However, this important analysis step is quite complex to manage because of the combinatorial nature of the various sets of data. One should indeed combine variations of several thresholds and scores; the number of combinations grows exponentially with the number of criteria and a clear synthesis must be derived from the various conflicting results. All these difficulties will remain as long as we will try to make discrete a continuum of preference situations. The real solution is to define a continuous transition from preference to indifference. This suggests defining fuzzy R, P and I from a criterion g. We now elaborate fuzzy preference structures from criterion values, following the construction proposed in Perny and Roy (1992).

For any function g, a fuzzy outranking relation R can be obtained from a real valued function θ, defined on $\mathbb{R} \times \mathbb{R}$, verifying $R(a,b) = \theta(g(a), g(b))$, for all a, b in A, and such that:

$$\forall y \in X, \quad \theta(x,y) \quad \text{is a nondecreasing function of } x, \tag{44}$$

$$\forall x \in X, \quad \theta(x,y) \quad \text{is a nonincreasing function of } y, \tag{45}$$

$$\forall z \in X, \quad \theta(z,z) \quad = 1. \tag{46}$$

As shown in Perny (1992), relation R defined by (44–46) is a reflexive, complete, semi-transitive and Ferrers fuzzy relation, and thus is a fuzzy semiorder. Every α-cut of R is a crisp semiorder, and these semiorders form together an homogeneous family compatible in the sense of Roberts (1971) with the classical total preorder $\geq$ on scores (for more details see Perny and Roy (1992)). The particular fuzzy preference structure $(P, I, \emptyset)$ associated to R is the same as the one considered in § 1.2.3 (i). P is a fuzzy partial order. I is a fuzzy reflexive symmetric and transitive relation) and thus is a *fuzzy similarity relation* also called "indistinguishability" operator by Valverde (1985).

As an example, the Electre III method proposed by Roy (1978) for the multicriteria ranking of alternatives is based on the construction, on each dimension, of a one-dimensionnal outranking relation R characterised by a function θ defined from thersholds as follows:

$$\theta(x,y) = \frac{p(x) - \min\{y - x, p(x)\}}{p(x) - \min\{y - x, q(x)\}}$$

The resulting fuzzy semiorder structure $(P, I, \emptyset)$ is represented on figure 3. Notice that P is similar to preference relations used in the Promethee methods (see Brans and Vincke (1984)).

When criterion values $g(a)$ and $g(b)$ are perfectly known, the indices $P(a,b)$ and $I(a,b)$ are often interpreted as *intensities* of strict preference and indifference respectively. However, this interpretation is not universally accepted, mainly because the preference $P(a,b) = 1$ is usually reached for a finite and small score difference. If the term "intensity" is used here, it must be distin-

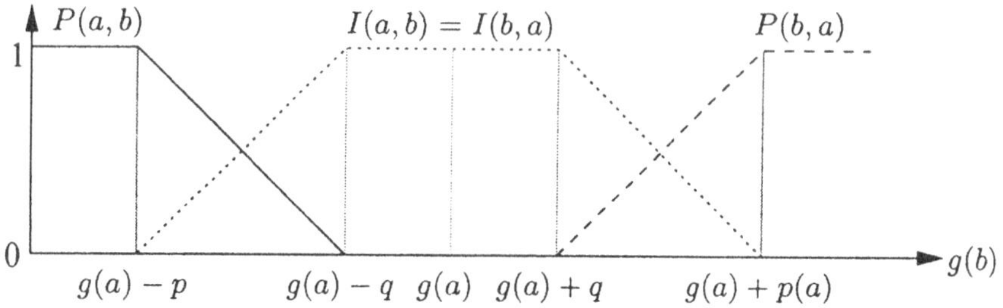

Figure 3 Fuzzy relations P and I defined from criterion g

guished from the classical notion of intensity thought as a difference of utilities (see Fishburn (1970)).

When criterion values are imperfectly known, intervals of type $[g(a) - q(a), g(a)+q(a)]$ represent the range of plausible values for the score $g(a)$, from which fuzzy preferences are defined. In this case, some authors (see *e.g.* Roy (1996)) interpret the indices of type $P(a, b)$ and $I(a, b)$ as the *credibility* of preference and indifference respectively.

1.3.4 The Construction of Preference Relations from Ill-known Consequences

In subsection 1.3.2, a crisp consequence was attached to each alternative, on each dimension. However, in many decision problems, the representation of consequences by crisp evaluations is not appropriate. In some situations indeed, it is not possible to reach a consensus among experts in the estimation of a consequence, and a single value is not sufficient to reflect the diversity of their judgements. In other situations, we are able to estimate the range of possible consequences for each alternative and we want to investigate the possibilities of discrimination between these alternatives knowing they are not completely defined.

In such situations, a natural attitude is representing the ill-known consequences by a possibility distributions (see Dubois and Prade (1985)). The *fuzzy consequence* of alternative a on a given dimension is the fuzzy subset of the evaluation scale X defined by:

$$C_a = \{(x, \pi_{C_a}(x)), x \in X\} \tag{47}$$

where π_{C_a} gives, for any element x of the scale, the possibility degree $\pi_{C_a}(x)$ of the punctual event $C_a = x$, such that :

$$\sup_{x \in X}\{\pi_{C_a}(x)\} = 1 \tag{48}$$

In this context, preference modeling is complicated by the necessity of comparing fuzzy evaluations. We survey now the elaboration of preference structures in this situation (for more details and other approaches, see Baas and Kwakernaak (1977), Dubois and Prade (1983), Buckley (1985), Bortolan and Degani (1985), Roubens and Vincke(1988), Perny and Roy (1992), Fodor and Roubens (1994a)).

a) Definition of consequence/objective compatibility levels

As suggested in Dubois and Prade (1985) and Perny and Mudri (1992), the attractiveness of a fuzzy consequence relatively to a fuzzy objective function may be evaluated as a compatibility between these two fuzzy sets. The compatibility level between a fuzzy consequence C_a and a fuzzy objective G defined on the same scale X can be approximate by the quantities:

$$\Pi(G, C_a) = \sup_{x \in X} \min(g(x), \pi_{C_a}(x)) \qquad (49)$$

$$N(G, C_a) = \inf_{x \in X} \max(g(x), 1 - \pi_{C_a}(x)) \qquad (50)$$

where $\Pi(G, C_a)$ and $N(G, C_a)$ are respectively the possibility and the necessity of the fuzzy set G relatively to C_a. The $\Pi(G, C_a)$ index measures the possibility of the G event relatively to the consequence C_a, in other words the possibility of the statement *"the alternative a fits to the Decision Maker's objective"*. The $N(G, C_a)$ index measures the certitude of the G event relatively to consequence C_a, in other words the degree to which the statement *"the alternative a does not fit to the Decision Maker's objective"* is impossible. Thus, by definition, $N(G, C_a) = 1 - \Pi(G', C_a)$, where $G' = \{(x, 1 - g(x)), x \in X\}$ is the complement of G.

Moreover, from condition (48), we know that $\pi_{C_a}(x_0) = 1$ for some $x_0 \in X$. Hence, we get the following inequalities:

$$\min_{x \in S(C_a)} \{g(x)\} \leq N(G, C_a) \leq g(x_0) \leq \Pi(G, C_a) \leq \max_{x \in S(C_a)} \{g(x)\} \qquad (51)$$

where $S(C_a) = \{x \in X : \pi_{C_a}(x) > 0\}$ is the support of C_a. Therefore, the following inequality holds:

$$\Pi(G, C_a) \geq N(G, C_a) \qquad (52)$$

With such a definition, the typical axiom of possibility measures and the dual axiom for necessity measures hold for any pair (G, G') of fuzzy objectives on the same dimension:

$$\Pi(G \cup G', C_a) = \max\{\Pi(G, C_a), \Pi(G', C_a)\}$$

$$N(G \cap G', C_a) = \min\{N(G, C_a), N(G', C_a)\}$$

However, because G is fuzzy, we have only:

$$max\{\Pi(G, C_a), 1 - N(G, C_a)\} \geq \frac{1}{2}$$

This is weaker than the usual property:

$$max\{\Pi(G,C_a), 1 - N(G,C_a)\} = 1 \tag{53}$$

b) Preference models based on compatibility levels

When a punctual evaluation is required, one can be tempted to aggregate the possibility and necessity of the fuzzy event G in a single value. Thus, we could define the consequence/objective compatibility level by the score:

$$g^{\alpha}(a) = (1-\alpha)\Pi(G,C_a) + \alpha N(G,C_a) \tag{54}$$

where α is a technical parameter allowing to perform a convex combination of the two indices. Choosing $\alpha = 0$ would lead to very optimistic evaluations of compatibilities whereas choosing $\alpha = 1$ would lead to more pessimistic evaluations. Parameter α may be interpreted as a degree of prudence when modulating the confidence we have in our evaluation. Criterion g^{α} allows a total preorder to be defined on A.

However, the real meaning of equation (54) is still to be justified and deriving meaningfull conclusions from the comparison of $g^{\alpha}(a)$ and $g^{\alpha}(b)$ seems difficult. An alternative approach which seems more interesting is to replace the evaluation $g^{\alpha}(a)$ by the interval $[g^-(a), g^+(a)]$ bounded by the following compatibility levels:

$$\begin{aligned} g^-(a) &= N(G,C_a) \\ g^+(a) &= \Pi(G,C_a) \end{aligned}$$

It is worth noting that an interval order structure can be defined from intervals $[g^-(a), g^+(a)]$, $a \in A$ by defining the following relations:

$$aPb \Leftrightarrow g^-(a) > g^+(b) \tag{55}$$

$$aIb \Leftrightarrow [g^-(a), g^+(a)] \cap [g^-(b), g^+(b)] = \emptyset \tag{56}$$

If all the intervals of type $[g^-(a), g^+(a)]$ are of equal length l, this interval order reduces to a semi-order. In this case indeed, the preference structure is representable by a semi-criterion caracterised by the function g^- and the constant indifference threshold l.

The interest of consequence/objective compatibility levels can easily be illustrated by the following, simplified decision problem.

Consider a Professor whose decision problem is to find a suitable place to live with his family. Let us suppose that the "expected duration of the travel from his house to the university" is to be considered. Suppose that an acceptable solution for him should not exceed 35 minutes of travel time and an ideal solution should not exceed 15 minutes. A possible representation of the associated fuzzy objective could be the fuzzy subset of admissible durations defined by:

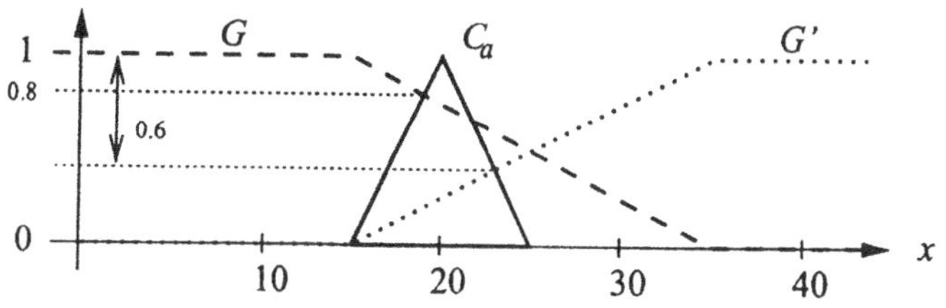

Figure 4 Definition of consequence/objective compatibility

$$\mu_G(x) = \begin{cases} 1 & if \quad 0 \leq x \leq 15 \\ \frac{35-x}{20} & if \quad 15 \leq x \leq 35 \\ 0 & if \quad x \geq 35 \end{cases}$$

For instance, if the expected duration is $x_1 = 30$, the compatibility level is 0.25 whereas it will be 0.75 for a duration $x_2 = 20$. Let us suppose now that the evaluation of the duration of the travel for a location a is about 20 and caracterized by the following possibility distribution:

$$\pi_{C_a}(x) = \begin{cases} 0 & if \quad 0 \leq x \leq 15 \\ \frac{x-15}{5} & if \quad 15 \leq x \leq 20 \\ \frac{25-x}{5} & if \quad 20 \leq x \leq 25 \\ 0 & if \quad x \geq 25 \end{cases}$$

Thus, as shown on figure 4, we get:

$g^+(a) = 0.8, \quad g^-(a) = 0.6$ and $g^{(0.5)}(a) = 0.7$

c) Fuzzy preference relations defined from ill-known criterion values

In the previous paragraph, we have investigated the construction of criterion values and crisp preference structures from ill-known consequences of alternatives. We now elaborate fuzzy preference relations in the same context. Considering a fuzzy objective G caracterised by the criterion function g and an alternative caracterised by its fuzzy consequences C_a, we can define a *fuzzy criterion value* or *fuzzy score* by:

$$\tilde{g}(a) = \{(g(x), \pi_{C_a}(x)), x \in X\} \tag{57}$$

It is worth noting that quantities $g^-(a)$, $g^+(a)$ and $g^\alpha(a)$ are punctual approximations of this fuzzy set.

Suppose that a (crisp or fuzzy) weak preference relation has been defined by $R(a,b) = \theta(g(a), g(b))$ where θ fulfil equations (44–46). Then, knowing the fuzzy sets $\tilde{g}(a)$ for any $a \in A$, we can extend R so as to compare alternatives using the following equations (see Perny (1994)):

$$R^+(a,b) \quad = \sup_{(x,y) \in X \times X} \min\{\theta(g(x), g(y)), \pi_{C_a}(x), \pi_{C_b}(y)\} \tag{58}$$

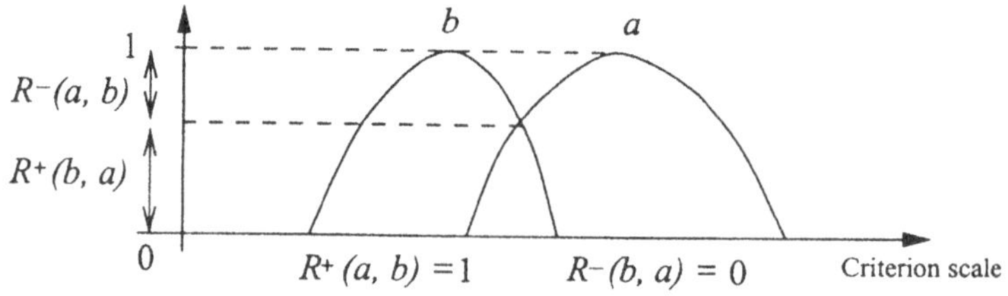

Figure 5 Definition of fuzzy preferences from fuzzy scores

$$R^-(a,b) \quad = \quad \inf_{(x,y)\in X\times X} \max\{\theta(g(x),g(y)), 1-\pi_{C_a}(x), 1-\pi_{C_b}(y)\} \quad (59)$$

In the general case, quantities $R^+(a,b)$ and $R^-(a,b)$ are respectively the possibility and the necessity of the fuzzy relation R relatively to (C_a, C_b). The (R^-, R^+) pair defines an interval valued relation. Moreover, following the idea of equation (54), alternatives could be compared according to the relation:

$$R^\alpha = (1-\alpha)R^+ + \alpha R^-$$

This general construction has been studied by Perny (1994) and some particular results are yet available (many investigations are still to be made on this topic). For example, when $\theta(u,v) \in \{0,1\}$ for all $(u,v) \in I\!R \times I\!R$, R is necessarily a crisp preference relation, and $R^-(a,b) > 0 \Leftrightarrow R^+(a,b) = 1$. In this case, $R^+(a,b)$ and $R^-(a,b)$ measure the possibility and the necessity of the assertion aRb respectively, and for any $\alpha, \beta \in (0,1)$, relations R^α and R^β are identical up to an automorphism of [0, 1].

As an example, suppose that $\theta(u,v) = 1$ if $u \geq v$ and 0 otherwise. In this case R is a total preorder and it has been proved by Roubens and Vincke (1988), Dubois and Prade (1991) that, when scores are fuzzy intervals (see (60) below) on the criterion scale, R^+ is a fuzzy interval order. It is worth noting that the fuzzy strict preference relation associated to R^+ is R^-. Therefore, R^- is a fuzzy partial order. These relations are represented on figure 5

A score is a *fuzzy interval* on the criterion scale $I\!R$ if it is caracterised by a possibility distribution verifying the following condition:

$$\forall u, v, \in I\!R, \forall w \in [u,v], \pi(w) \geq \min\{\pi(u), \pi(v)\} \quad (60)$$

When g is a pseudo-criterion with linear thershold functions q and p (see (40–43)), Fodor and Roubens (1994a) proved that R^- is a fuzzy semi-order, provided the fuzzy score of each alternative $a \in A$ is a fuzzy symmetric interval whose support is $p(a)$ and whose kernel is $q(a)$.

The fuzzy preference relations introduced in this section are well fitted to preference modeling from ill-known evaluations. Here, the valuation of preference or indifference expresses the uncertainty in statements of type $R(a,b)$,

$P(a,b)$, $I(a,b)$ even if these statements are crisp. Like in the previous sections, the use of the fuzzy approach prevents small variations of thresholds and scores from modifying seriously the resulting preference model.

Now, assuming that n criterion functions $g_1, \ldots, g_n$ have to be considered, and having all the marginal weak preference relations R_j, we must determine a global weak-preference relation R with respect to all the objectives. Once more, the subjectivity of the decision maker must be taken into account and we are looking for an operator h such that : $g = h(g_1, \ldots, g_n)$ or alernatively $R = h(R_1, \ldots, R_n)$. This problem is considered in the next chapter of this book.

After the aggregation of preferences, the overall fuzzy preference relation R is usually not complete (excepted in the scoring approach); therefore the preference structure we have to deal with is a *partial* structure of type (P, I, J) where J is not empty. The problem of deriving a clear recommendation from such a structure is considered in chapter 3.

Acknowledgments

We wish to thank D. Dubois, J. Fodor and R. Slowinski for helpful comments on an earlier draft of this text.

References

Alsina C. (1985) On a family of connectives for fuzzy sets, *Fuzzy Sets and Systems* 16, 231–235.

Arrow K.J. (1959) Rational choice functions and orderings, *Econometrica*, 26, 121–127.

Baas and Kwakernaak (1977) Rating and Ranking of multiple aspects using fuzzy sets, Automatica, 12, 47-58.

Barrett, C.R., Pattanaik, P.K. (1990), On the structure of fuzzy welfare functions, Fuzzy Sets and Systems, 19, 1–10.

Barrett, C.R., Pattanaik, P.K., and Salles, M.(1992), Rationality and aggregation of prefererences in an ordinally fuzzy framework, Fuzzy Sets and Systems, 49, 9–13.

Bellman R.E. and Zadeh L.A. (1970) Decision-Making in a Fuzzy Environment, *Management Science*, 17, B141–B164.

Bortolan G. and Degani R. (1985) A review of some methods for ranking fuzzy subsets, *Fuzzy Sets and Systems*, 15, 1-19.

Brans J.P. and Vincke Ph. (1984) A preference ranking organization method, *Management Science*, 31, 647–656.

Buckley J.J. (1985) Ranking alternatives using fuzzy numbers, *Fuzzy sets and systems*, 15, 21-31.

Condorcet M.J.A.M. Caritat, Marquis de, (1785) *Essai sur l'application de l'analyse à la probabilité des decisions rendues à la pluralité des voix* (Imprimerie Royale, Paris).

De Baets B., Van de Walle B. and Kerre E. (1995a) Fuzzy preference structures and their characterization, *The Journal of Fuzzy Mathematics*, 3, 373–381.

De Baets B., Van de Walle B. and Kerre E. (1995b) Fuzzy preference structures without incomparability, *Fuzzy Sets and Systems*, 76, 333–348.

De Baets B., Van de Walle B. and Kerre E. (1997) A plea for the use of Lukasiewicz triplets in fuzzy preference structures. Part 2: The identity case, *Fuzzy Sets and Systems*, in press.

Doignon J.P., Monjardet B., Roubens M., Vincke Ph. (1986) Biorders families, valued relations and preference modeling, *Journal of Mathematical Psychology*, 30, 4, 435–480.

Dubois and Prade (1983) Ranking fuzzy numbers in the setting of possibility theory, in *Fuzzy sets: Theory and application to policy analysis and information systems*, PP. Wang, S.K. Chang (eds.), Plenum Press, 59–75.

Dubois D. and Prade H. (1984) Criteria aggregation and ranking of alternatives in a framework of fuzzy set theory, in : Zimmermann H.J., Zadeh L.A., Gaines B.R. (Eds), *Fuzzy Sets and Decision Analysis*, TIMS Series in Management Sciences, North Holland, 20, 209–240.

Dubois D. and Prade H. (1985) Théorie des possibilités et application à la représentation des connaissances en informatique, Masson, Paris.

Dubois D. and Prade H. (1986) Weighted minimum and maximum operations in fuzzy set theory, *Information Science*, 39, 205–210.

Dubois D. and Prade (1991) On the ranking of ill-known values in possibility theory, *Fuzzy Sets and Systems*, 43, 311-317.

Fishburn P.C. (1970) *Utility theory for decision-making*, Wiley, New-York.

Fishburn P.C. (1985) Interval orders and interval graphs, Wiley, New York.

Fodor J. (1991) Strict preference relations based on weak *t*-norms, *Fuzzy Sets and Systems* 43, 324–336.

Fodor J. (1992) An axiomatic approach to fuzzy preference modeling, *Fuzzy Sets and Systems* 52, 47–52.

Fodor J. and Roubens M. (1991) Fuzzy preference modeling – an overview, *Annales Univ. Sci. Budapest, Sectio Computatorica* XII 93–100.

Fodor J. and Roubens M. (1994a) *Fuzzy preference modeling and multicriteria decision support*,(Kluwer Academic Pub., Dordrecht).

Fodor J. and Roubens M. (1994b) Valued preference structures, *European Journal of Operational Research* 79, 277–286.

Gibbard A. (1969) Intransitive Social Indifference and the Arrow Dilemma, Mimeographed (re. Sen (1986)).

Kelly J.S. (1991), Social Choice Bibliography, *Social Choice and Welfare* **8**, 97–169.

Leclerc B. (1984) Efficient and binary consensus functions on transitively valued relations, *Mathematical Social Sciences*, 8, 45–61.

Luce D. (1956) Semiorders and a theory of utility discrimination, *Econometrica*, 24, 147–191.

Monjardet B. (1978) Axiomatique et proprit des quasi-ordres, *Mathématiques et Sciences Humaines*, 63, 51–82.

Mudri L. and Perny P. (1994) An approach to design support using fuzzy models of architectural objects, in : J.S. Gero and F. Sudweeks (Eds), *Artificial Intelligence in Design'94*,(Kluwer Academic Publisher, Dordrecht), 697–714.

Orlovski S. (1978) Decision making with a fuzzy preference relation, *Fuzzy Sets and Systems* 1, 155–167.

Ovchinnikov S. and Roubens M. (1991) On strict preference relations, *Fuzzy Sets and Systems* 43, 319–326.

Ovchinnikov S. and Roubens M. (1992) On fuzzy strict preference, indifference and incomparability relations, *Fuzzy Sets and Systems* 47, 313–318.

Perny P. (1990) Construction and exploitation of relationnal systems of fuzzy preferences, Proceedings of the 3rd IPMU conference, Paris, 431-434.

Perny P. (1992) Modélisation, agrégation et exploitation de préférences planes dans une problématique de rangement, Thèse de doctorat, Université Paris-Dauphine.

Perny P. (1994) How to deal with partial information in outranking methods ? An approach based on possibility theory, XIVth EURO conference, 3–6 July, Jerusalem, Israel.

Perny P. (1995) Defining fuzzy covering relations for decision aid, in: B. Bouchon, R. Yager, L. Zadeh (Eds.), *Fuzzy Logic and soft Computing, Advances in Fuzzy Systems - Applications and Theory*,**4**, 109–218.

Perny P. (1997) Multicriteria Filtering Methods based on Agreement and Discordance Principles, *Annals of Operations Research*, to appear.

Perny P. and Roy B. (1992), The use of fuzzy Outranking Relations in Preference modeling, *Fuzzy Sets and Systems* **49**, 33–53

Pirlot M. and Vincke Ph. (1997), *Semiorders: Properties, Representations, Applications*, Kluwer Academic Publishers, 1997.

Roberts F.S. (1971) Homogeneous families of semiorders and the theory of probabilistic consistency, *Journal of Mathematical Psychology*, 8, 248-263.

Roberts F.S. (1979) Measurement theory, *Encyclopedia of mathematics and its application*, vol. 7, Addison-Wesley.

Roubens M. and Vincke Ph. (1985) *Preference modeling*, (Springer-Verlag, Berlin).

Roubens M. and Vincke Ph. (1988) Fuzzy possibility graphs and their application to ranking fuzzy numbers, in *Non-Conventional Preference Relations in Decision Making*, J. Kacprzyk and M. Roubens (Eds.), 119-128.

Roy B. (1978) ELECTRE III : Un algorithme de classement fondé sur une représentation floue des préférences en présence de critères multiples, *Cahiers du Centre d'Etudes de Recherche Opérationnelle*, 20, 3–24.

Roy B. (1996) *Multicriteria Methodology for Decision Aiding*, Kluwer Academic Publishers, Dordrecht.

Roy B. and Bouyssou D. (1993) *Aide Multicritère à la Decision* : Méthodes et Cas. (Economica, Paris).

Roy B. and Vincke Ph. (1984) Relational systems of preference with one or more pseudo-criteria : some new concepts and results, *Management Science*, 30, 1323–1335.

Roy B. and Vincke Ph. (1987) Pseudo-orders : definition, properties and numerical representation, *Mathematical Social Sciences*, 14, 263–274.

Sen A.K. (1986), Social Choice Theory in : K. J. Arrow, M. D. Intriligator (Eds.), *Handbook of Mathematical Economics vol. III*, (Elsevier Sciences Publishers B. V., North-Holland), Chap. 22, 1073–1181.

Tversky A. and Simonson I. (1993) Context-dependent preferences, *Management Science* 39, 47–59.

Valverde L. (1985) On the structure of F-indistinguishability operators, *Fuzzy Sets and Systems* 17, 313–328.

Van de Walle B., (1996) *The existence and characterization of fuzzy preference structures*, Ph.D. thesis, University of Gent (in Dutch).

Van de Walle B. , De Baets B. and Kerre E., (1997) A plea for the use of Łukasiewicz triplets in fuzzy preference structures. Part 1: General argumentation, *Fuzzy Sets and Systems*, in press.

Vincke Ph. (1980) Vrais, quasi, pseudo et précritères dans un ensemble fini : propriétés et algorithmes, *Cahiers du Lamsade*, 27, Université de Paris-Dauphine.

Vincke, Ph. (1992), *Multicriteria Decision-aid*, J.Wiley and sons.

2 FUZZY AGGREGATION OF NUMERICAL PREFERENCES

Michel Grabisch
Sergei A. Orlovski
Ronald R. Yager

Abstract: The problem of aggregating numerical values is addressed in this chapter. The first part deals with the aggregation of criteria into a single one. Properties which are suitable for this case are presented, together with the most common aggregation operators. A special section is devoted to ordered weighted averaging (OWA) operators, and fuzzy integrals. Then, relation between properties, links between operators, characterization of some operators are presented. An important section is devoted to the behavioral analysis of OWA operators and fuzzy integrals, linking values of parameters with the attitude of the decision maker. The last section of this first part is concerned with the problem of identification of operators in a practical problem, a key issue in every application. The second part is devoted to special aspects in aggregation of preferences, in a multiattribute context. The Pareto principle is explained, and then the agreement-discordance principle, which is the basis of ELECTRE III and IV, is addressed.

2.1 THE AGGREGATION PROBLEM

The main object of chapter 2 is concerned with aggregation. It comes naturally after chapter 1, dealing with the modelling of preference, and before chapter 3, of which concern is the choice problem, i.e. how to exploit aggregated preferences.

Aggregation refers to the process of combining values (numerical or non numerical) into a single one, so that the final result of aggregation takes into account in a given fashion all the individual aggregated values. In order to clarify this definition, we elaborate on the words *values* and *fashion*.

In decision making, values to be aggregated are typically preference or satisfaction degrees. A preference degree tells to what extent an alternative A is preferred to an alternative B (see chapter 1), and thus is a *relative* appraisal. By contrast, a satisfaction degree expresses to what extent a given alternative is satisfactory with respect to a given criterion, or a kind of distance to a prototype which may represent the *ideal* alternative for the decision maker. It is an *absolute* appraisal. In a more general way and depending on the application, values to be aggregated can be confidence degrees in the fact that a given alternative is true, expert's opinions, similarity degrees, etc.

Values can belong to either a single numerical or non numerical scale, but the existence of a weak order relation on the set of all possible values is the minimal requirement which has to be satisfied in order to perform aggregation. Thus, *nominal* scales, i.e. enumerated values without any structure, such as {*cabbage, turnip, horseradish, cucumber*} are not suitable. A typical example of non numerical scale is a *linguistic scale* often used in fuzzy set theory, such as {*small, medium, large*}, or {*unsuitable, acceptable, average, good, excellent*}. Numerical scales can be of ordinal or cardinal type. On an ordinal scale, numbers have no other meaning than defining an order relation on the scale, and distances or differences between values cannot be defined. This is the case when people are asked to evaluate a product on an integer scale from 1 to 5, where for example 1 corresponds to "very bad" and 5 to "excellent". Here, we have no reason to say that the distance between 1 ("very bad") and 2 ("bad") is the same than the distance between 2 and 3 ("medium"). On the other hand, we are allowed to define operations other than comparison on a cardinal scale. Here again, the nature of the operation depends on the kind of scale: interval scales, where the position of the zero is a matter of convention (e.g. the temperature expressed on the Celsius scale) allow to perform only linear operations, so that for example multiplication is not allowed, while ratio scales, where a true zero exists (e.g. the Kelvin scale for temperature) allow to perform ratios and multiplication. We don't further detail these measurement theoretical aspects, and the interested reader is asked to refer to the monographs of Krantz *et al.* (Krantz et al., 1971) and Roberts (Roberts, 1979) for a complete treatment of the question.

Another distinction among values is related with precision. Precise values are simply elements of the scale, while imprecise values are subsets (fuzzy or not) of the scale, which means that the true (but unknown) value lies somewhere in this subset. Imprecise values can be considered on any type of scales, either numerical or non numerical. Section 2.2.9 will briefly address this issue.

Once values are defined we can aggregate them and obtain a new value defined on the same scale, but this can be done in many different ways according to what is expected from the aggregation operation, what is the nature of the values to be aggregated, and what kind of scale has been used. We elaborate on these three points.

Aggregation operators can be roughly divided into three classes, each possessing very distinct behavior or *semantic*. *Conjunctive* operators combine values as if they were related by a logical "and" operator. That is, the result of combination is high if and only if all the values are high. Triangular norms are the suitable operators for doing conjunctive aggregation. On the other hand, *disjunctive* operators combine values as an "or" operator, so that the result of combination is high if some (at least one) values are high. The most common disjunctive operators are triangular conorms. Between conjunctive and disjunctive operators, there is room for a third category, namely *averaging* operators. They are located between *minimum* and *maximum*, which are the bounds of the t-norm and t-conorm families. Averaging operators have the property to be *compensative*, that is, low values can be compensated by high values, so that the result of combination will be medium. There are of course other families of operators which do not fit into one these categories.

Properties of the aggregation operator can be dictated by the nature of the values to be aggregated. For example, in some multicriteria evaluation methods, the aim is to assess a global absolute score to an object (or alternative) given a set of partial scores with respect to different criteria. Clearly, it would be unnatural to give as global score a value which is lower than the lowest partial score, or greater than the highest score, so that only averaging operators are allowed. Now, if preference degrees coming from transitive (in some sense) relations are combined, it may be requested that the result of combination remains transitive. Another example concerns the aggregation of opinions in voting procedures. If, as usual, the voters are anonymous, the aggregation operator must be commutative.

The third point concerns the kind of scale which is used. As explained above, depending on the kind of scale, allowed operations on values are restricted, and the aggregation operator should belong to them. For example, aggregation on ordinal scales should be limited to operations involving comparisons only, such as medians, and order statistic, while linear operations, as well as *minimum* and *maximum* are allowed on interval scales. In practice however, when values are numerical, people often overlook the problem, and implicitly suppose that the underlying scale is absolute, without restriction on operators. But this is controversial since experiments have been done for measuring membership degrees of fuzzy sets, which tend to prove that the underlying scale is far from being absolute, but is rather an interval scale (Norwich and Turksen, 1984). A related topic concerns what is called the *meaningfulness* of operators. An operator is said to be meaningful if the ranking of alternatives induced by the aggregation does not depend on scale transformation. This means that, e.g. for multicriteria evaluation, using scores defined on a [0,100] scale or on [-2,3] scale has no influence on the ranking of alternatives.

This chapter will detail all the matters mentionned above, especially in section 2.2. The chapter is roughly divided into two parts. The first part deals with the problem of aggregation . For simplicity and in order to be self-contained, the presentation is done in the case of aggregation of satisfaction degrees, but the material is applicable to preferences as well. The second part deals with aggregation of preferences, and thus necessitates the reading of chapter 1. However, in this part, only aspects which are particular to the aggregation of preferences (and not satisfaction degrees) are mentionned. Also, the reader is encouraged to consult chapter 4 on group decision making for a different aspect of aggregation.

2.2 CONSTRUCTION OF ONE COMPREHENSIVE CRITERION

2.2.1 Introduction

This section will address in detail the problem of constructing a single comprehensive criterion from several criteria, and thus is centered on the problem of aggregation . The word *comprehensive* refers to the fact that the criterion resulting from combination is supposed to be representative of all the original criteria, and reflects the decision maker's attitude. As said above, although the presentation is done in the case of satisfaction degrees (absolute scores), the following aggregation methods apply as well to preferences.

We introduce some notations and conventions. We suppose we have a multicriteria decision making problem defined by a set A of alternatives denoted by lowercase letters $a, b, \ldots$, and a set of n criteria $h_1, \ldots h_n$. A criterion h_i is a mapping from the set of alternatives A to a measurement scale L. The value $h_i(a)$ is called the *partial score with respect to criterion* i. It is assumed that the measurement scale is common for all the criteria, and we will consider, otherwise indicated, that the scale is a bounded cardinal one, say the interval $[0,1]$, without further precision on its nature (interval scale, ratio scale, etc.). The choice of the interval $[0,1]$ is not restrictive if we consider that scores are defined up to a positive linear tranformation, i.e. $\phi(x) = \alpha x + \beta$, with $\alpha > 0$, as it is the case for example in multiattribute utility theory. For the sake of convenience, we will denote by $a_i = h_i(a)$ the partial score of a with respect to criterion i, so that an alternative a can be represented by a n-dimensional vector $[a_1 \cdots a_n]$ in $[0,1]^n$. We do not address here the way of constructing a criterion h_i, and we suppose that the scores are given beforehand.

An aggregation operator $\mathcal{H}$ is a mapping from L^n to L, which computes from a vector of partial scores a *global score* of an alternative a, which is $\mathcal{H}(a_1 \ldots, a_n)$. If necessary, we use the notation $\mathcal{H}^{(n)}$ to stress the dependency on the number of terms in the aggregation.

$\mathfrak{S}(n)$ denotes the set of all permutations on a n elements index set, and a particular permutation is denoted by σ. The particular permutation which arranges all the elements by increasing values is denoted simply by $(\cdot)$. Thus, if $a_1, \ldots, a_n$ are real numbers, then $a_{(1)}, \ldots, a_{(n)}$ denotes the sequence arranged in increasing order.

$\wedge, \vee$ denote respectively the minimum and maximum operations.

We assume in the following exposition that scores are precise values. The case of fuzzy scores will be addressed in section 2.2.9.

For more details or other points of view on aggregation operators, the reader is referred to (Dubois and Prade, 1984; Dubois and Prade, 1985; Fodor and Roubens, 1994; Grabisch et al., 1995; Mizumoto,1989a; Mizumoto, 1989b; Yager, 1991), which represent a selection of good surveys of the topic.

2.2.2 Properties for Aggregation

We list here the main properties which can be requested for aggregation. We divide them in mathematical and "behavioral" properties, the latter being general requirements which are useful in a practical problems. In the former, we distinguish between commonly used properties which are well known, and more elaborated properties whose definition could seem to be somewhat strange, but which appear naturally when trying to characterize operators. This section is mainly based on (Fodor and Roubens, 1994, Grabisch, 1995a, Grabisch et al., 1995).

elementary mathematical properties

- if 0 and 1 are the extremal values, then

  $$\mathcal{H}(0,\ldots,0) = 0, \quad \mathcal{H}(1,\ldots,1) = 1.$$

  A stronger requirement is idempotence (I) (or unanimity, agreement):

  $$\mathcal{H}(a,a,\ldots,a) = a, \forall a.$$

- continuity (C): $\mathcal{H}$ is a continuous function of $a_1,\ldots,a_n$.
- monotonicity (M) (non decreasingness) with respect to each argument:

  $$a_i' > a_i \Rightarrow \mathcal{H}(a_1,\ldots,a_i',\ldots,a_n) \geq \mathcal{H}(a_1,\ldots,a_i,\ldots,a_n).$$

  This means that increasing a partial score cannot decrease the result. The monotonicity is strict (SM) if strict inequality holds.

- neutrality (or commutativity, anonymity) (N)

  $$\mathcal{H}(a_1,\ldots,a_n) = \mathcal{H}(a_{\sigma(1)},\ldots,a_{\sigma(n)}), \quad \forall \sigma \in \mathfrak{S}(n).$$

  This is required when combining criteria of equal importance or anonymous expert's opinions.

- compensativeness (Co)

  $$\bigwedge_{i=1}^{n} a_i \leq \mathcal{H}(a_1,\ldots,a_n) \leq \bigvee_{i=1}^{n} a_i.$$

  This is a natural property when one tries to evaluate globaly an alternative on the basis of a set of partial scores of evaluation.

- associativity (A). Suppose an operator $\mathcal{H}$ is defined for two arguments. Associativity permits to extend it for more arguments unambigously.

  $$\mathcal{H}^{(3)}(a_1,a_2,a_3) = \mathcal{H}^{(2)}(\mathcal{H}^{(2)}(a_1,a_2),a_3) = \mathcal{H}^{(2)}(a_1,\mathcal{H}^{(2)}(a_2,a_3)).$$

- stability for the same ordinal transformation (SO). Suppose ϕ is a continuous, strictly increasing mapping from $\mathbb{R}$ to $\mathbb{R}$. Then

$$\mathcal{H}(\phi(a_1), \ldots, \phi(a_n)) = \phi(\mathcal{H}(a_1, \ldots, a_n)).$$

A particular case which is useful in practice is the case of positive linear transformation. The property (SPL) writes:

$$\mathcal{H}(ra_1 + t, \ldots, ra_n + t) = r\mathcal{H}(a_1, \ldots, a_n) + t,$$

with $r > 0$, $t \in \mathbb{R}$. These properties are closely related to the choice of a measurement scale.

- ϕ-comparison meaningfulness (ϕ-C). ϕ is as above a continuous, strictly increasing mapping from $\mathbb{R}$ to $\mathbb{R}$. Then

$$\mathcal{H}(a_1, \ldots, a_n) < \mathcal{H}(b_1, \ldots, b_n) \Rightarrow \\ \mathcal{H}(\phi(a_1), \ldots, \phi(a_n)) < \mathcal{H}(\phi(b_1), \ldots, \phi(b_n)).$$

This is a weaker requirement than SO, but which is sufficient in decision making, as far as only the ranking of alternatives has importance. ϕ^{-1} being also continuous and strictly increasing, this property is equivalent to

$$\mathcal{H}(a_1, \ldots, a_n) = \mathcal{H}(b_1, \ldots, b_n) \Leftrightarrow \\ \mathcal{H}(\phi(a_1), \ldots, \phi(a_n)) = \mathcal{H}(\phi(b_1), \ldots, \phi(b_n)).$$

As above, it is interesting to consider the particular case when ϕ is positive linear (PL-C).

more elaborated mathematical properties

- decomposability (D) (Kolmogoroff (Kolmogoroff, 1930))

$$\mathcal{H}^{(n)}(a_1, \ldots, a_k, a_{k+1}, \ldots, a_n) = \mathcal{H}^{(n)}(\underbrace{a, \ldots, a}_{k \text{ times}}, a_{k+1}, \ldots, a_n),$$

where $a = \mathcal{H}^{(k)}(a_1, \ldots, a_k)$.

- linkage (L) (Fodor and Roubens (Fodor et al., 1995))

$$\mathcal{H}^{(n+1)}(\mathcal{H}^{(n)}(a_1, \ldots, a_n), \ldots, \mathcal{H}^{(n)}(a_{n+1}, \ldots, a_{2n})) = \\ \mathcal{H}^{(n)}(\mathcal{H}^{(n+1)}(a_1, \ldots, a_{n+1}), \ldots, \mathcal{H}^{(n+1)}(a_n, \ldots, a_{2n})).$$

- stability under positive linear transformation with same unit, comonotonic zeroes (SPLUC) (Grabisch (Grabisch, 1995; Grabisch et al., 1995)):

$$\mathcal{H}(ra_{\sigma(1)} + t_{\sigma(1)}, \ldots, ra_{\sigma(n)} + t_{\sigma(n)}) = r\mathcal{H}(a_{\sigma(1)}, \ldots, a_{\sigma(n)}) + T(t_{\sigma(1)}, \ldots, t_{\sigma(n)}),$$
$$\forall a_1 \leq \cdots \leq a_n, \quad \forall r > 0, \quad \forall t_1 \leq \cdots \leq t_n \in \mathbb{R}, \quad \forall \sigma \in \mathfrak{S}(n).$$

Remark that SPLUC implies SPL.

The usefulness of these definitions will later become apparent.

behavioral properties

- possibility of expressing weights of importance on criteria if this is necessary.
- possibility of expressing the behavior of the decision maker. More specifically, we speak here of the tendency of the decision maker, i.e. if he is for example conjunctive or disjunctive oriented, or if he considers some criteria are absolutely necessary to be satisfied (*veto* criteria), or some criteria are sufficient to be satisfied (*pass* criteria). In fact, two decision makers with same partial scores $a_1, \ldots, a_n$, same weights on criteria, could still have different behaviors. We can give as example two typical behaviors: tolerant and intolerant. Tolerant decision makers can accept that only *some* criteria (at least one) are met (this corresponds to a disjunctive behavior, whose extreme exemple is *max*). On the other hand, intolerant decision makers demand that *all* criteria have to be equally met (conjunctive behavior, whose extreme example is *min*).
- possibility of expressing an interaction between criteria. For example, possible interactions between criteria are *redundancy* (two criteria are redundant if they express more or less the same thing) and *support* or *reinforcement* (two criteria with little importance when taken separately, become very important when considered jointly). This point will be detailed further in section 2.2.7.
- possibility of an easy semantical interpretation, i.e. being able to relate the values of parameters defining $\mathcal{H}$ to the behavior implied by $\mathcal{H}$ in a decision making point of view. This point is crucial and forbids the use of a "black box" methodology, e.g. neural nets.

2.2.3 Common Aggregation Operators

We give here a non exhaustive list of well known aggregation operators. Special aggregation operators will be developed in subsequent sections.

Aggregation operators can be roughly classified into several categories: conjunctive, disjunctive, compensative, non-compensative, and weighted operators, according to the way criteria are aggregated.

2.2.3.1 conjunctive operators. Conjunctive operators perform an aggregation where criteria are connected by a logical "and". Thus the global score is high if and only if all the partial scores are high. Taking a two-places operator, this can be represented by the property $\mathcal{H}(1, a) = a$ for every $a \in [0, 1]$. If we impose also nondecreasingness, commutativity, and associativity (so that n places operators can be defined unambiguously from 2 places operators), then the class of operators satisfying these requirements is the *t-norm* family. These operators are often denoted by $*$. Main properties of t-norms are:

- $*(0, a) = 0, \forall a.$
- the minimum operator is a t-norm and $*(a, b) \leq a \wedge b$, so that minimum is the greatest t-norm.

T-norms generally do not satisfy properties which are often requested for multicriteria aggregation, as idempotence, compensativeness, scale invariance, etc.

We do not detail further these operators since they have been treated elsewhere in this handbook. The interested reader can also consult (Fodor and Roubens, 1994; Mizumoto, 1989a).

2.2.3.2 disjunctive operators. Disjunctive operators perform an aggregation where criteria are connected by a logical "or", and are in this sense dual of conjunctive operators. Here the global score will be low if and only if all partial scores have low values. This is expressed by the property $\mathcal{H}(0, a) = a$, $\forall a \in [0, 1]$. If we impose in addition commutativity, non decreasingness and associativity, then we obtain the family of *t-conorms* (or *s-norms*), often denoted by $\perp$. Main properties are:

- if $\perp(a, b)$ is a t-conorm, then $*(a, b) := 1 - \perp(1 - a, 1 - b)$ is a t-norm.
- $\perp(1, a) = 1, \quad \forall a \in [0, 1].$
- the maximum operator is a the smallest t-conorm.

As t-norms, t-conorms do not possess suitable properties for criteria aggregation. (see again (Fodor and Roubens, 1994, Mizumoto, 1989a) for further details on t-conorms).

2.2.3.3 compensative operators. Following the previous definition of compensativeness (Co), compensative operators are comprised between minimum and maximum, so clearly they are neither conjunctive nor disjunctive. In this kind of operators, a bad (resp. good) score on one criterion can be compensated by a good (resp. bad) one on another criterion. If we add the properties of monotonicity (non decreasingness) and idempotence, then we get the family of *averaging operators*, where usually border operators minimum and maximum are excluded.

Averaging operators are suitable for combining scores in multicriteria evaluation problems. We give some common examples below.

mean operators the most widely used mean operators are the arithmetic mean $1/n\sum_{i=1}^{n} a_i$, the geometric mean $\prod_{i=1}^{n} a_i^{1/n}$, the harmonic mean $(1/n\sum_{i=1}^{n} 1/a_i)^{-1}$, and the root power mean (Dyckhoff and Pedrycz, 1984) $(1/n\sum_{i=1}^{n} a_i^{\alpha})^{1/\alpha}$. In fact, all these are particular cases of the *quasi-arithmetic mean*, defined by:

$$M_f(a_1, \ldots, a_n) = f^{-1}\left(\frac{1}{n}\sum_{i=1}^{n} f(a_i)\right),$$

where f is any continuous strictly monotonic function. General properties are:

- every quasi-arithmetic mean is idempotent, continuous, strictly monotonic, neutral, compensative and decomposable.
- only the arithmetic mean satisfies the stability for linear transformation.

median and order statistic as mean operators are clearly cardinal in nature (i.e. only cardinal information can be aggregated), order statistics are suitable for ordinal information. The k *order statistic* OS_k is defined as follows.

$$\mathrm{OS}_k(a_1, \ldots, a_n) = a_{(k)},$$

following our notations defined in section 2.2.1. Particular cases are the minimum ($k = 1$) and the maximum ($k = n$) operators, and the *median* ($k = (n+1)/2$ if n is odd[1]), denoted $\mathrm{med}(a_1, \ldots, a_n)$.

General properties of order statistics are:

- every order statistic is idempotent, continuous, monotonic, neutral, compensative, stable for ordinal transformation, ϕ-comparison meaningful and satisfies ordered linkage.
- in addition, medians are associative.

Other examples of averaging operators will be given below (OWA, fuzzy integals).

2.2.3.4 non compensative operators. Conjunctive, disjunctive and compensative operators form a large disjoint covering of operators on $[0, 1]$, but there are some operators which do not belong to one of these categories. These are often more or less of the average type, but they may extend beyond the minimum and maximum operators.

symmetric sums They have been introduced by Silvert (Silvert, 1979). Symmetric sums are continuous, non decreasing with respect to each argument, commutative, verifying $\mathcal{H}(0, 0) = 0$, $\mathcal{H}(1, 1) = 1$, and have the characteristic property that a reversal of the scale has no effect on the evaluation, i.e. $\mathcal{H}(a_1, \ldots, a_n) = 1 - \mathcal{H}(1 - a_1, \ldots, 1 - a_n)$. Silvert has shown that any symmetric sum is of the form:

$$\mathcal{H}(a_1, a_2) = \left(1 + \frac{g(1 - a_1, 1 - a_2)}{g(a_1, a_2)}\right)^{-1},$$

where g is any increasing continuous function, with $g(0,0) = 0$.

compensatory operators Zimmermann and Zysno [Zimmermann and Zysno, 1984] have proposed to mix a t-norm and a t-conorm to a proportion $\gamma \in [0,1]$ in order to produce a kind of compensation between criteria:

$$\mathcal{H}(a_1, \ldots, a_n) = (\prod_{j=1}^{n} a_j)^{1-\gamma} (\bigoplus_{j=1}^{n} a_j)^{\gamma},$$

where $\bigoplus$ denotes the probabilistic sum, defined by $\bigoplus_{j=1}^{n} a_j := 1 - \prod_{j=1}^{n}(1 - a_j)$. These mixed operators have in fact limited properties: continuity, monotonicity, neutrality and associativity only among the above list of properties.

2.2.3.5 weighted operators. Most of applications in multicriteria decision require weights of importance on criteria, thus implying an extension of usual non-weighted operators, which can be done in several more or less arbitrary ways (exponentiation, etc). Usually, the introduction of weights on criteria causes the loss of neutrality, but this is not always the case (see OWA operators below). Minimum and maximum operators have been extended by Dubois and Prade (Dubois and Prade, 1986, Dubois et al., 1988), in a way which is consistent with possibility theory:

Definition 1 *Let $w_1, \ldots, w_n$ be a set of weights so that $\vee_{i=1}^{n} w_i = 1$. The weighted minimum and weighted maximum operators are defined by*

$$\mathrm{wmin}_{w_1, \ldots, w_n}(a_1, \ldots, a_n) = \bigwedge_{i=1}^{n} [(1 - w_i) \vee a_i],$$

$$\mathrm{wmax}_{w_1, \ldots, w_n}(a_1, \ldots, a_n) = \bigvee_{i=1}^{n} [w_i \wedge a_i].$$

The family of quasi-arithmetic means can be easily generalized by:

$$M^f_{w_1, \ldots, w_n}(a_1, \ldots, a_n) = f^{-1}\left(\sum_{i=1}^{n} w_i f(a_i)\right),$$

where weights are normalized so that $\sum_{i=1}^{n} w_i = 1$. The most widely used example is the *weighted arithmetic mean* or *weighted sum* ($f = Id$). Weighted quasi-arithmetic means are no more decomposable. General considerations on the use of weighted operators in decision making can be found in the works of Yager (Yager, 1978, Yager, 1981, Yager, 1987).

Other weighted operators are presented below (OWA, fuzzy integrals).

2.2.4 Ordered Weighted Averaging (OWA) Operators

In (Yager, 1988) Yager introduced a family of aggregation operators called the ordered weighted averaging (OWA) operators (Yager and Kacprzyk, 1997). Theseoperators are in the class of mean operators, they are idempotent, monotonic and commutative. These operators have a natural framework for the inclusion of the types of behavioral properties discussed in the earlier section. We shall see that the OWA aggregation operators allow us to easily adjust the degree of "anding" and "oring" implicit in an aggregation. Furthermore, their structure admits easy semantic interpretation using linguistic quantifiers (Zadeh, 1983). Using these operators we can allow a decision making to specify linguistically their agenda for aggregating a collection of criteria and then provide a formal implementation of this agenda.

Definition 2 *An* OWA operator *of dimension n is a mapping denoted* $OWA_W : \mathbb{R}^n \longrightarrow \mathbb{R}$, *that has an associated n-dim vector* $W = [w_1\ w_2\ \cdots\ w_n]^T$ *such that*

1. $w_i \in [0, 1]$
2. $\sum_i w_i = 1.$

where $OWA_W(a_1, \ldots, a_n) = \sum_j w_j b_j$, *with* b_j *the jth largest of the* a_i *(or, equivalently* $b_j \equiv a_{(n-i+1)}$ *with our notations).*

A fundamental aspect of this operation is the re-ordering step, in particular a score a_i is not associated with a particular weight w_i but rather a weight is associated with a particular ordered position of score. This ordering step introduces a nonlinearity into the aggregation process.

> EXAMPLE: assume $W = [0.4\ 0.3\ 0.2\ 0.1]^T$. Consider the aggregation $\text{OWA}_W(0.7, 1, 0.3, 0.6)$. In this case carrying out the reordering process we get that $b_1 = 1$, $b_2 = 0.7$, $b_3 = 0.6$ and $b_4 = 0.3$. Then performing the aggregation $\sum_j w_j b_j$ we get
>
> $$\text{OWA}_W(0.7, 1, 0.3, 0.6) = (0.4)(1)+(0.3)(0.7)+(0.2)(0.6)+(0.1)(0.3) = 0.76$$

OWA operators are idempotent, continuous, monotonic, neutral, compensative, and stable for positive linear transformations. They also satisfy ordered linkage.

Some important special cases of OWA aggregations can be pointed out:

1. for W^* defined such that $w_1 = 1$ and $w_i = 0$ for all other weights, OWA is equivalent to the maximum operator.

2. for W_* defined such that $w_n = 1$ and $w_i = 0$ for all other weights, OWA is equivalent to the minimum operator.

3. for W_A defined such that $w_i = 1/n$ for all i, the arithmetic mean is recovered.

4. for W_k defined such that $w_{n-k+1} = 1$ and $w_i = 0$ for all other weights, the k-order statistic is recovered. The median can also be recovered by defining W_{med} as follows:

- for n odd: $w_{\frac{n+1}{2}} = 1$, $w_i = 0$ otherwise,
- for n even: $w_{\frac{n}{2}} = w_{\frac{n}{2}+1} = 0.5$, $w_i = 0$ otherwise.

5. the so-called "olympic aggregation" is obtained by assigning $w_1 = w_n = 0$ and $w_i = \frac{1}{n-2}$ for all other weights. Here we are essentially disregarding the highest and lowest scores. A more general version of this can be had by assigning the value zero to the top and bottom m weights and uniformally dividing the total of one to the remaining weights.

In (Yager, 1993) Yager provides a comprehensive discussion of different classes of OWA operators.

2.2.5 Fuzzy Integrals

As the integral of a function in a sense represents its average value, an integral in the discrete case, with suitable normalization, is a particular case of averaging operator. On the other hand, the concept of integral has been revisited many times, and in particular in the fuzzy community. Extending the concept of Lebesgue integral, that is, an integral with respect to a measure, Sugeno (Sugeno, 1974) has proposed the concept of fuzzy measure and fuzzy integral in 1974. Many other communities have proposed similar ideas, in particular in the field of game theory, utility theory, robust Bayesian statistics, etc. Due to the context, we restrict ourself to the discrete case, and avoid unnecessary mathematical developments (see e.g. (Denneberg, 1994, Grabisch et al., 1995) for more details).

Let us denote by $X = \{x_1, \ldots, x_n\}$ the set of criteria, and by $\mathcal{P}(X)$ the power set of X, i.e. the set of all subsets of X.

Definition 3 *A* fuzzy measure *on X is a set function $g : \mathcal{P}(X) \longrightarrow [0, 1]$, satisfying the following axioms.*

(i) $g(\emptyset) = 0$, $g(X) = 1$.

(ii) $A \subset B$ *implies* $g(A) \leq g(B)$, *for* $A, B \in \mathcal{P}(X)$.

$g(A)$ can be viewed as the weight of importance of the set of criteria A. Thus, in addition to the usual weights on criteria taken separately, weights on any combination of criteria are also defined.

A fuzzy measure is said to be *additive* if $g(A \cup B) = g(A) + g(B)$ whenever $A \cap B = \emptyset$, *superadditive* (resp. *subadditive*) if $g(A \cup B) \geq g(A) + g(B)$ (resp. $g(A \cup B) \leq g(A) + g(B)$) whenever $A \cap B = \emptyset$. If a fuzzy measure is additive, then it suffices to define the n coefficients (weights) $g(\{x_1\}), \ldots, g(\{x_n\})$ to define entirely the measure. In general, one needs to define the $2^n - 2$ coefficients corresponding to the 2^n subsets of X, except $\emptyset$ and X itself.

Another particular case of fuzzy measure is the *0-1 fuzzy measure*, whose values are either 0 or 1.

We introduce now the concept of discrete fuzzy integrals, viewed as aggregation operators. For this reason, we will adopt a connective-like notation instead of the usual integral form, and the integrand will be a set of n values $a_1, \ldots, a_n$ in $[0, 1]$.

Definition 4 *Let g be a fuzzy measure on X. The* discrete Sugeno integral *of $a_1, \ldots, a_n$ with respect to g is defined by :*

$$\mathcal{S}_g(a_1, \ldots, a_n) := \bigvee_{i=1}^{n} (a_{(i)} \wedge g(A_{(i)})),$$

with $A_{(i)} := \{x_{(i)}, \ldots, x_{(n)}\}$.

Another definition was proposed later by Murofushi and Sugeno (Murofushi and Sugeno, 1991), using a concept introduced by Choquet in capacity theory.

Definition 5 *Let g be a fuzzy measure on X. The* discrete Choquet integral *of $a_1, \ldots, a_n$ with respect to g is defined by*

$$\mathcal{C}_g(a_1, \ldots, a_n) := \sum_{i=1}^{n} (a_{(i)} - a_{(i-1)}) g(A_{(i)}),$$

with the same notations as above, and $a_{(0)} = 0$.

Sugeno and Choquet integrals are essentially different in nature, since the former is based on non linear operators (*min* and *max*), and the latter on usual linear operators. Both compute a kind of distorded average of $a_1, \ldots, a_n$. More general definitions exist but will not be considered here (see (Grabisch, 1995a, Grabisch et al., 1995)).

Some properties of fuzzy integrals are the following.

- the Sugeno and Choquet integral are idempotent, continuous, monotonic, and compensative operators. They both satisfy the linkage property (L) for ordered n-uples and all permutations of ordered n-uples.

- the Choquet integral is stable under positive linear transformations. The Sugeno integral does not share this property, but satisfies a similar property with *min* and *max* replacing product and sum. In this sense, it can be said that the Choquet integal is suitable for cardinal aggregation (where numbers have a real meaning), while the Sugeno integral seems to be more suitable for ordinal aggregation (where only order makes sense).

- if g is a 0-1 fuzzy measure, then the Choquet and the Sugeno integral take the following form (Murofushi and Sugeno, 1993):

  $$\mathcal{C}_g(a_1, \ldots, a_n) = \mathcal{S}_g(a_1, \ldots, a_n) = \bigvee_{A : g(A)=1} \left(\bigwedge_{x_i \in A} a_i \right).$$

  The result can be put in another way (Grabisch, 1994):

  $$\mathcal{C}_g(a_1, \ldots, a_n) = \mathcal{S}_g(a_1, \ldots, a_n) = a_{(k_\sigma)},$$

  where k_σ depends on the particular ordering of the arguments $a_1, \ldots, a_n$, that is, on which permutation $\sigma \in \mathfrak{S}$ such that $a_{\sigma(1)} \leq a_{\sigma(2)} \leq \cdots \leq a_{\sigma(n)}$.

2.2.6 Relation Between Operators and Characterization of Operators

There exists many relations between the properties defined in section 2.2.2, as well as among the different operators. Moreover, some operators have been characterized. We summarize here the main results in this topic, and give reference for further reading.

2.2.6.1 Relations between properties. (see Fodor and Roubens (Fodor and Roubens, 1994)) The following can be shown.

- compensativeness implies idempotency.
- idempotency and monotonicity imply compensativeness.
- stability for the same positive linear transformations implies idempotency.
- associativity and neutrality imply linkage.
- decomposability, idempotency and neutrality imply linkage.
- stability for the same ordinal transformation implies ϕ-comparison meaningfulness.
- ϕ-comparison meaningfulness and idempotency imply stability for the same ordinal transformations.

2.2.6.2 Relation between operators. The connection between OWA operators and various operators (in particular k-order statistic) has been already pointed out in section 2.2.4. We turn now to fuzzy integrals, since they include as particular cases many of the above defined compensative and weighted operators, so that they help at having a general view of them. Main results are summarized below (see Grabisch *et al.* (Grabisch, 1995a, Grabisch et al., 1995) and Murofushi and Sugeno (Murofushi and Sugeno, 1993)).

- a Choquet integral with respect to an additive measure g coincides with a weighted arithmetic mean, whose weights w_i are $g(\{x_i\})$.
- any OWA operator with weights $w_1, \ldots, w_n$ is a Choquet integral, whose fuzzy measure g is defined by

$$g(A) = \sum_{j=1}^{i} w_j, \quad \forall A \text{ such that } |A| = i,$$

  where $|A|$ denotes the cardinal of A. Reciprocally, any commutative Choquet integral is such that $g(A)$ depends only on $|A|$, and coincides with an OWA operator, whose weights are $w_i = g(A_i) - g(A_{i-1}), i = 2, \ldots, n$, and $w_1 = 1 - \sum_{i=2}^{n} w_i$. A_i denotes any subset such that $|A_i| = i$.
- the Sugeno and the Choquet integral contains all order statistics, thus in particular, *min*, *max* and the median. The fuzzy measure g corresponding

to the k order statistic is defined by $g(A) = 0$ if $|A| \leq n - k$, $g(A) = 1$ otherwise.

- weighted minimum and weighted maximum are particular cases of Sugeno integral. The fuzzy measure corresponding to $\text{wmax}_{w_1,\ldots,w_n}$ is defined by

$$g(A) = \bigvee_{x_i \in A} g(\{x_i\}),$$

while for $\text{wmin}_{w_1,\ldots,w_n}$

$$g(A) = 1 - \bigvee_{x_i \notin A} g(\{x_i\}).$$

In the above,$g(\{x_i\}) = w_i$. It can be noticed that g is a possibility measure in the case of wmax, and a necessity measure for wmin.

- any arbitrary combination of minimum and maximum can be represented by a Choquet or a Sugeno integral with respect to a suitable 0-1 fuzzy measure. For example, $(a_1 \wedge a_3) \vee (a_2 \wedge a_4)$ is represented by a fuzzy measure g with $g(\{x_1, x_3\}) = g(\{x_2, x_4\}) = 1$, the other coefficients being 0 (this is obvious from the property of 0-1 fuzzy measures given in section 2.2.5).

Figure 1 gives a summary of all set relations between various aggregation operators and fuzzy integrals.

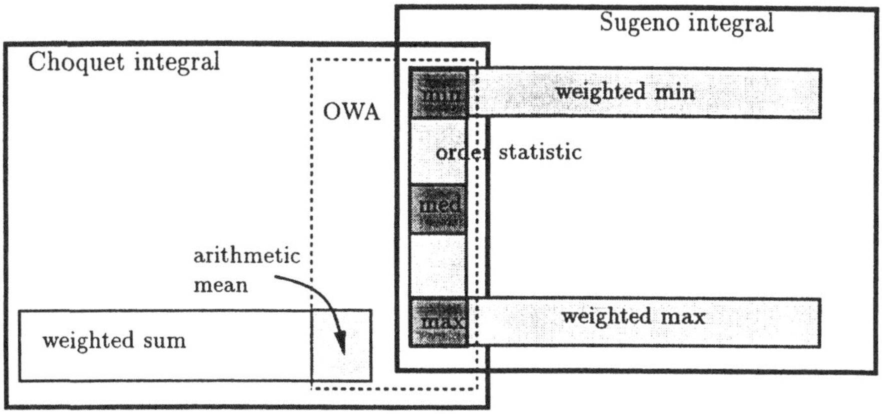

Figure 1 Set relations between various aggregation operators and fuzzy integrals

2.2.6.3 Characterization of aggregation operators. To characterize an operator is to give a minimal set of properties shared by this operator and only this one. This kind of result is very useful for a deep understanding of an operator, but in general they are hard to establish. Some of the most important results are given below.

characterization of k order statistic (Fodor and Roubens, 1995)

> *An operator is ϕ-comparison meaningful, neutral, compensative and continuous if and only if it is an order statistic.*

characterization of ϕ-comparison meaningful, neutral and compensative operators (Fodor and Roubens, 1995) This result is not properly a characterization since the reciprocal does not hold. Nevertheless, this property is worth mentioning.

> *A ϕ-comparison meaningful, neutral and compensative operator must be of the form*

$$\mathcal{H}(a_1, \ldots, a_n) \in \{a_1, \ldots, a_n\}, \quad \forall a_1, \ldots, a_n.$$

This property means that an operator which is insensitive to any scale change is forced to be ordinal in nature. An example of such an operator is given by fuzzy integral with respect to 0-1 fuzzy measures, as it was explained in section 2.2.5. However, these operators satisfy ϕ-comparison meaningfulness and compensativity, but not neutrality, unless they correspond to an order statistic.

characterization of quasi-arithmetic means This result is due to Kolmogoroff (Kolmogoroff, 1930).

> *An operator is idempotent, neutral, continuous, strictly monotonic and decomposable if and only if it is a quasi-arithmetic mean.*

characterization of associative averaging operators Let us consider the class of averaging operators, i.e. which are continuous, monotonic, idempotent, neutral, and different from minimum and maximum. The following has been shown by Dubois and Prade (Dubois and Prade, 1984).

> *An operator is an associative averaging operator if and only if it takes the form*

$$\mathcal{H}(a_1, \ldots, a_n) = med(\vee_{i=1}^n a_i, \alpha, \wedge_{i=1}^n a_i), \quad \alpha \in]0,1[.$$

characterization of OWA (Fodor et al., 1995)

> *The class of OWA operators corresponds to the operators which satisfy neutrality, monotonicity, idempotency and stability for positive linear transformation, same unit, independent zeroes and ordered values (i.e. SPLUC with $\sigma = Id$ only).*

characterization of Choquet integral (Grabisch et al., 1995)

> *The class of Choquet integrals corresponds to the operators which satisfy the properties of monotonicity, idempotence, and stability for positive linear transformation with same unit and comonotonic positive zeroes (SPLUC)*

characterization of Sugeno integral (Grabisch et al., 1995)

> *The class of Sugeno integrals corresponds to the operators which satisfy the properties of monotonicity, idempotence, and*

$$\mathcal{H}\left((r \wedge a_{\sigma(1)}) \vee t_{\sigma(1)}, \ldots, (r \wedge a_{\sigma(n)}) \vee t_{\sigma(n)}\right) =$$
$$\left(r \wedge \mathcal{H}(a_{\sigma(1)}, \ldots, a_{\sigma(n)})\right) \vee T(t_{\sigma(1)}, \ldots, t_{\sigma(n)}),$$
$$\forall 0 \leq a_1 \leq \cdots \leq a_n \leq 1, \forall r \in [0,1], \forall 0 \leq t_1 \leq \cdots \leq t_n \leq 1, \forall \sigma \in \mathfrak{S}.$$

2.2.7 Behavioral Analysis of Operators

Sofar we have focussed on mathematical properties of operators, and neglected what we called above the behavioral properties. This word is not to be taken in its full meaning, but refers more or less to properties related with the intuitive feeling of the way the operator does the combination. This includes different concepts, such as the degree of tolerance/intolerance, veto effects, importance of criteria, interaction between criteria, etc.

It is very difficult to relate these behavioral properties which are not precisely defined, to well cut mathematical properties. We know that k order statistics are more or less tolerant depending on the value of k, we know that a weighted sum is able to represent importance on criteria, but not interaction, but what about OWA, quasi-arithmetic means, weighted minimum, etc.? In a first section , we will define some quantities related to OWA operators, and then define more general concepts of importance of criteria and interaction between them in a second section. A third section is devoted to veto effects.

2.2.7.1 Measures of OWA operators. The OWA operator allows us, by appropriate choice of the weighting vector, to move continuously from the "and" (min) to "or" (max) type aggregation. In order to classify these OWA operators in regards to their location on this continuum a *measure of orness*, associated with any vector W, can be introduced. The orness measure is defined as

$$\text{orness}(W) = \frac{1}{n-1}\sum_{i=1}^{n}(n-i)w_i.$$

We note that for any W, orness(W) is always in the unit interval. Furthermore it is noted that the nearer W is to an "or", the closer its measure is to one, while the nearer it is to an "and", the closer its measure is to zero. Supporting this it can be easily shown that orness(W^*) $= 1$, orness(W_*) $= 0$, orness(W_A) $= 0.5$. More generally it is noted that an OWA operator with much of the non-zero weights at near the top will be an "orlike" operator, orness(W) > 0.5. At the other extreme when the weights are non-zero near the bottom the OWA operator will be "andlike", orness(W) < 0.5. Except for the extreme cases of orness(W) $= 1$ and 0 which are uniquely obtained by W^* and W_* respectively, numerous different vectors W can attain a given measure of orness.

To help further distinguish amongst the different OWA operators a measure, called the *dispersion* (or *entropy*) of an OWA vector W is defined as

$$\text{Disp}(W) = -\sum_i w_i \ln w_i.$$

The measure of dispersion can be used to indicate the degree to which we use all the information contained in the argument, $(a_1, \ldots, a_n)$ when calculating the aggregated value $F(a_1, \ldots, a_n)$. In particular the larger $\text{Disp}(W)$ the more of the information provided by the argument we use. Consider the median and the simple average, both of which have an orness value of 0.5. However we readily see that the median just uses the information contained in one argument value, whereas the simple average uses the information in all the arguments equally. This observation is supported by our measure of dispersion where $\text{Disp}(W_{\text{med}}) = 0$ and $\text{Disp}(W_A) = \ln 5$.

If F is an OWA aggregation with weights w_i the *dual* of F, denoted $\hat{F}$, is an OWA aggregation of the same dimension with weights $\hat{w}_i = w_{n-i+1}$. We can easily see that if F and $\hat{F}$ are duals then $\text{Disp}(\hat{F}) = \text{Disp}(F)$ and $\text{orness}(\hat{F}) = 1 - \text{orness}(F)$. Thus if F is orlike its dual is andlike.

2.2.7.2 Importance and interaction among criteria. It is possible to give a formal definition of importance of criteria and interaction between them within

the framework of fuzzy integrals, which fits well intuition. As fuzzy integrals include many of usual operators, this analysis sheds light on them on a behavioral point of view.

We recall that a fuzzy measure g defines weights on individual criteria by means of the coefficients $g(\{x_i\})$, but also on any group A of criteria by means of $g(A)$. This makes possible the representation of interaction between criteria. Intuitively, we could say the following about a pair of criteria x_i, x_j:

- x_i and x_j are considered to be without interaction (or to be *independent*) if the importance of the pair of criteria x_i, x_j is simply the sum of the individual importances. In terms of fuzzy measure, this means that $g(\{x_i, x_j\}) = g(\{x_i\}) + g(\{x_j\})$, that is, the fuzzy measure is additive with respect to criteria x_i and x_j.

- x_i and x_j exhibit a positive synergy between them (or are *complementary*) if the importance of x_i, x_j taken together is much greater than the importance of x_i, x_j alone. Consequently, we should have $g(\{x_i, x_j\}) > g(\{x_i\}) + g(\{x_j\})$, that is, a superadditive fuzzy measure for x_i and x_j.

- on the contrary, x_i and x_j are *redundant* or *substitutive*, or exhibit a negative synergy, if the importance of the pair is not very different from the importance of the individual criteria. Thus $g(\{x_i, x_j\}) < g(\{x_i\}) + g(\{x_j\})$ should hold (subadditive fuzzy measure).

Although this definition seems to be satisfactory, things get complicated since we must take into account all the 2^n coefficients of g to be meaningful. Indeed:

- the global importance of a criterion, say x_j is not solely determined by the value $g(\{x_j\})$, but also by all $g(A)$ such that $x_j \in A$.
- in the same spirit, the interaction between x_i and x_j is not only determined by the difference $g(\{x_i, x_j\}) - g(\{x_i\}) - g(\{x_j\})$ but also by all the coefficients $g(A)$ such that $\{x_i, x_j\} \subset A$.

Murofushi *et al.* (Murofushi, 1992, Murofushi and Soneda, 1993) (see also (Grabisch et al., 1995, Grabisch, 1996)) have proposed two kinds of indexes, based on game theory for the importance of criteria, and on multiattribute utility theory for the interaction between two criteria.

Definition 6 *Let g be a fuzzy measure on $X = \{x_1, \dots, x_n\}$. The* importance index *or* Shapley value *of criterion x_i with respect to g is defined by:*

$$\Lambda_i = \sum_{A \subset X \setminus \{x_i\}} \gamma_X(A)[g(A \cup \{x_i\}) - g(A)] \tag{1}$$

with $\gamma_X(A) = \frac{(|X|-|A|-1)! \cdot |A|!}{|X|!}$ where $|A|$ indicates the cardinal of A, and $0!=1$ as usual.

This definition has been proposed by Shapley in cooperative game theory (Shapley, 1953). These values have the property that $\sum_{i=1}^{n} \Lambda_i = 1$. It is convenient to scale these values by a factor n, so that an importance index greater than 1 indicates an attribute more important than the average.

Definition 7 *The* interaction index *between two criteria x_i and x_j with respect to a fuzzy measure g is defined by:*

$$I_{ij} = \sum_{A \subset X \setminus \{i,j\}} \xi_X(A) \cdot [g(A \cup \{x_i, x_j\}) - g(A \cup \{x_i\}) - g(A \cup \{x_j\}) + g(A)]$$

with $\xi_X(A) = \frac{(|X|-|A|-2)! \cdot |A|!}{(|X|-1)!}$.

A pair of criteria is said to be redundant or having a negative synergy (resp. complementary or having a positive synergy) if its interaction index is negative (resp. positive).

Shapley values and interaction indexes can be computed for various operators covered by fuzzy integrals, in particular the weighted mean, the weighted minimum and maximum and OWA. Table 1 summarizes the results for these operators, and particular cases of OWA which are interesting (see (Grabisch, 1997) for a deeper study). It can be noticed that the weighted minimum (resp. weighted maximum) exhibits synergy (resp. redundancy) between all pairs of criteria, but the interaction is zero for any OWA such that $w_1 = w_n$.

Generalizing these definitions, Grabisch has proposed an interaction representation of fuzzy measures, so that the interaction index is defined for any subset in X (see (Grabisch, 1996c, Grabisch, c, Grabisch, a)). It permits also to define the so-called k-additive fuzzy measures, which can range freely between additive measures ($k = 1$) and general fuzzy measures ($k = n$) (Grabisch, 1996c, Grabisch, c).

operator	Shapley value for x_i	interaction index for x_i, x_j
weighted sum	w_i	0
OWA	$1/n$	$\frac{w_n - w_1}{n-1}$
minimum	$1/n$	$1/(n-1)$
maximum	$1/n$	$-1/(n-1)$
weighted minimum	$\sum_{k=1}^{i} \frac{w_{(k)} - w_{(k-1)}}{n-k+1}$	$\sum_{k=1}^{\mathrm{argmin}(w_i,w_j)} \frac{w_{(k)} - w_{(k-1)}}{n-k}$
weighted maximum	$\sum_{k=1}^{i} \frac{w_{(k)} - w_{(k-1)}}{n-k+1}$	$\sum_{k=1}^{\mathrm{argmin}(w_i,w_j)} \frac{w_{(k-1)} - w_{(k)}}{n-k}$
k order statistic	$1/n$	0 if $k \neq 1, n$

Table 1 Shapley values and interaction indices for various operators

2.2.7.3 Veto and favor effects. Another interesting behavioral phenomenon in aggregation is the veto effect, and its counterpart, the favor effect. A criterion x_i is said to be a *veto* if its non satisfaction entails necessarily a low global score. In a symmetrical way, a criterion is said to be a *favor* if its satisfaction entails necessarily a high global score. This can be formalized as follows. Let us suppose that $\mathcal{H}$ is an aggregation operator. x_i is a veto for $\mathcal{H}$ if $\mathcal{H}$ can be decomposed as:

$$\mathcal{H}(a_1, \ldots, a_n) = a_i \wedge \mathcal{G}(a_1, \ldots, a_n).$$

Similarly, x_i is a favor for $\mathcal{H}$ if a suitable operator $\mathcal{G}$ exists, so that

$$\mathcal{H}(a_1, \ldots, a_n) = a_i \vee \mathcal{G}(a_1, \ldots, a_n).$$

Grabisch (Grabisch, 1996a, Grabisch, b) has shown that a veto effect on criterion i can be modelled by a Choquet integral[2] with respect to any fuzzy measure g satisfying $g(A) = 0$ whenever $i \notin A$. Similarly, a favor effect is obtained by any fuzzy measure g such that $g(A) = 1$ whenever $i \in A$. Let us denote by $g^{\wedge i}$ and $g^{\vee i}$ such fuzzy measures modelling veto and favor on criterion i respectively.

An interesting fact is that the interaction index defined above can reflect veto and favor effects. Denoting $I_{jk}^{\wedge i}$ the interaction index of $g^{\wedge i}$ for any pair of criteria j, k, it can be shown that $I_{ik}^{\wedge i} \geq 0$ for any $k \neq i$. In a similar way, the interaction index $I_{jk}^{\vee i}$ of $g^{\vee i}$ is such that $I_{ik}^{\vee i} \leq 0$ for any $k \neq i$. However, inspecting the sign of I_{ij} is not sufficient to detect veto and favor effects (see (Grabisch, a) for a complete study of this point).

The notion of veto and favor can be extended to groups of criteria, with the same results holding on the sign of the interaction indices for group of criteria.

2.2.8 Identification of Operators

Once a particular operator has been chosen for aggregation, relying properties required for a particular application, it remains to identi the parameters of the chosen operator, if any. It will be the case in particular for any weighted operator, such as weighted means, OWA and fuzzy integrals.

The problem of a significant identification of a weight vector in a multicriteria decision problem is not an easy task, even for the very simple weighted arithmetic mean. This elementary case has been studied at length, and many textbooks in multicriteria decision give methods for assessing weights in a weighted arithmetic means (see e.g. (Nijkamp et al., 1990)). One of the most popular method is the Analytic Hierarchy Process (AHP) of Saaty (Saaty, 1977, Saaty, 1980), which determines a relative importance of criteria by an eigen vector method. More recent approaches include ordinal regression methods, as proposed by Siskos (Siskos and Yannacopoulos, 1985, Zopounidis et al., 1995). A compilation of all these methods has been done by Mousseau (Mousseau, 1992). In this section we will present briefly some approaches related to OWA operators and fuzzy integrals.

2.2.8.1 Identification of OWA operators. As noted earlier the weights used in the OWA aggregation are used to reflect the decision makers desired agenda for aggregating the criteria. A number of approaches are available to accomplish this task. These approaches can be classified into three categories; those based on parametric considerations, those based on observations of passed performance and those based on semantic considerations.

Identification by parametric specification. The degree of orness, α, provides an important measure of the performance of an OWA operator. By allowing the selection of a particular value for α we are able to give the decision maker a useful handle for controlling the aggregation process as well as greatly reduce the possible implementations of the OWA operator. Furthermore by introducing other structural considerations we can often uniquely obtain a weighting vector.

In (O'Hagan, 1990), O'Hagan suggested an approach which makes use of the measures of orness and entropy introduced earlier. The first step is to choose a desired level of orness, α, to be associated with the aggregation. The second step is then to select from among the weighting vectors that attain this level of orness the one that has the maximum dispersion (entropy). As suggested by O'Hagan we can formulate this procedure in terms of the following mathematical programming problem:

$$
\begin{array}{ll}
\text{minimize} & \sum_{i=1}^{n} w_i \ln w_i \\
\text{subject to} & \left\{ \begin{array}{lll} \frac{1}{n-1}\sum_{i=1}^{n}(n-i)w_i & = & \alpha \\ \sum_{i=1}^{n} w_i & = & 1 \\ w_i \in [0,1], & & i = 1, \ldots, n \end{array} \right.
\end{array}
$$

We note that by just specifying one parameter, α, we uniquely get the weights. We can see that this approach is in the spirit of the maximum entropy techniques. This family of OWA operators are called the ME-OWA operators where the ME stands for Maximum Entropy. Other parametric approaches are discussed in (Yager and Filev, 1994, Yager et al., 1994).

Identification from observations. Another approach to obtain the OWA weights is to learn the weights from observations on the decision makers performance. In the following we shall describe an algorithm (Yager and Filev, 1994, Filev and Yager, 1994) which can be used to learn the weights associated with an OWA operator from observations. Our data will be a collection of m observations. Each observation is comprised of an n-tuple of values $(a_{k,1}, a_{k,2}, \ldots, a_{k,n})$, called the arguments, and an associated single value called the aggregated value, which we shall denote as d_k. Furthermore we shall let $(b_{k,1}, b_{k,2}, \ldots, b_{k,n})$ be the arguments in the kth observation presented in descending order. Our goal is to find an OWA operator, a weighting vector W, that best models the process of aggregation used in that data set.

Because of technical considerations which are discussed in detail in (Filev and Yager, 1994) we express each of the OWA weights in the form $w_i = e^{\lambda_i}/\sum_j e^{\lambda_j}$, $i = 1, \ldots, n$ and a gradient type technique to learn the λ_i's. Using the gradient descent method we obtain the following rule for updating the parameters

$$\lambda_i(l+1) = \lambda_i(l) - \beta w_i(l)(b_{k,i} - \hat{d}_k)(\hat{d}_k - d_k)$$

In the above β denotes the learning rate $(0 \leq \beta \leq 1)$, $\lambda_i(l+1)$ is our new estimate for λ_i, $w_i(l) = e^{\lambda_i(l)}/\sum_j e^{\lambda_j(l)}$ is the estimate of w_i after the lth iteration and

$$\hat{d}_k = \sum_{i=1}^{n} b_{k_i} w_i(l).$$

The process of updating the λ_i continues until we get small changes of parameter estimates $\Delta_i = |\lambda_i(l+1) - \lambda_i(l)|$, $i = 1, \ldots, n$.

Identification by semantic considerations. Yager (Yager, 1983; Yager, 1984) showed that linguistic quantifiers, a concept introduced by Zadeh (Zadeh, 1983) in natural language, can be used to semantically convey aggregation policies. Examples of these words are *almost all, few, many, most, nearly half, at least twenty percent* as well as the extreme cases of *all* and *at least one*. Given a collection of criteria and a linguistic quantifier Q, such as *most*, one can consider an aggregation of these criteria based on an imperative such as *most of the criteria should be satisfied.* Yager (Yager, a) called these *quantifier guided aggregations*. The use of quantifier guided aggregations can be considered as manifestation of a semantically based aggregation consideration. In the following we shall briefly describe how we can implement these types of aggregations using OWA operators.

In (Zadeh, 1983), Zadeh suggested that the class of linguistic quantifiers which say something about the proportion can be represented as fuzzy subsets of the unit interval I. Thus if $\mathcal{Q}$ is a linguistic quantifier, such as *most*, then $\mathcal{Q}$ can be represented as a fuzzy subset Q of I where for each $r \in I$, $Q(r)$ indicates the degree to which the proportion r satisfies the concept denoted by Q. Thus if Q is *most* then if $Q(0.8) = 1$ we would say that 80% is completely compatible with the idea conveyed by the linguistic quantifier *most*, while if $Q(0.5) = 0.7$ we are indicating that the proportion 50% is only 0.7 compatible with the concept of *most*.

An important class of these proportional linguistic quantifiers are those called regular non-decreasing which have the following properties

1. $\mathcal{Q}(0) = 0$
2. $\mathcal{Q}(1) = 1$
3. if $r_1 > r_2$ then $\mathcal{Q}(r_1) > \mathcal{Q}(r_2)$.

We note that these quantifiers are characterized by the fact that the satisfaction to the quantifier increases as the proportion increases. These quantifiers exhibit a kind of monotonicity. Among this class of linguistic quantifiers are concepts like *at least* α, *most* and *all*. The monotonicity of these quantifiers are particularly characteristic of the type of terms used in specifying aggregation agendas, where the satisfaction of more criteria by a proposed solution then required by a decision maker shouldn't degrade the overall satisfiability of a solution.

With the aid of the OWA operators we are able to provide an aggregation guided by these types of quantifiers. The aggregation of $(a_1, a_2, \ldots, a_n)$ guided by the quantifier $\mathcal{Q}$, which we denote as $\mathcal{Q}(a_1, a_2, \ldots, a_n)$, can be obtained by the following process.

1. obtain the OWA weighting vector W associated with the fuzzy subset Q representing the quantifier $\mathcal{Q}$,

$$w_i = Q\left(\frac{i}{n}\right) - Q\left(\frac{i-1}{n}\right), \quad i = 1, \ldots, n,$$

2. calculate

$$\mathcal{Q}(a_1, a_2, \ldots, a_n) = F_W(a_1, a_2, \ldots, a_n),$$

where $F_W(a_1, a_2, \ldots, a_n)$ is the OWA aggregation of the arguments using the weights found in the first step.

It is noted that the regularity and the increasing nature of Q guarantees the satisfaction of the two required conditions on the OWA weights: $\sum_i w_i = 1$ and $w_i \in [0, 1]$. In (Yager and Filev, 1994) it is shown how the process can be extended to other classes of linguistic quantifiers, nonincreasing and unimodal.

2.2.8.2 Identification of fuzzy integral operators. The parameters involved in a fuzzy integral are the 2^n coefficients of the fuzzy measure. The problem of identification is thus far more complex than for OWA and ordinary weighted operators, due to the exponential growing of the number of parameters. We can distinguish essentially three approaches (see (Grabisch et al., 1995) for more details).

Identification based on the semantics. It consists of guessing the coefficients of g, on the basis of semantical considerations. This could be for example:

- *importance of criteria.* This can be properly done by use of the Shapley value, or by the value of $g(\{x_j\})$ alone.
- *interaction between criteria.* The interaction index (def. 7) is suitable for this. For $n \leq 3$, the sign of $g(\{x_i, x_j\}) - g(\{x_i\}) - g(\{x_j\})$ should be sufficient, although not equivalent to the interaction index.
- *veto and favor effects.* This can be modelled via the interaction index (see section 2.2.7.3).
- *symmetric criteria.* Two criteria x_i, x_j are symmetric if they can be exchanged without changing the aggregation mode. Then, $g(A \cup \{x_i\}) = g(A \cup \{x_j\}), \forall A \subset X - \{x_i, x_j\}$. This reduces the number of coefficients.

This approach is practicable only for low values of n, and above all, if one has at his disposal an expert or decision maker who is able to tell the relative importance of criteria, and the kind of interaction between them, if any. This could be the case, in application of design of new products, where the marketing "defines" what should be the ideal product, in terms of aggregation of criteria.

Identification based on learning data. Considering the fuzzy integral model as a system, one can identify its parameters by minimizing an error criterion, provided learning data are available. Suppose that (z_k, y_k), $k = 1, \ldots, l$ are learning data where $z_k = [z_{k,1} \quad \cdots \quad z_{k,n}]^t$ is a n dimensional input vector, containing the partial scores of object k with respect to criteria 1 to n, and y_k is the global score of object k. Then, one can try to identify the best fuzzy measure g so that the squared error criterion is minimized.

$$E^2 = \sum_{k=1}^{l} (\mathcal{C}_g(z_{k,1}, \ldots, z_{k,n}) - y_k)^2. \tag{2}$$

It can be shown (Grabisch et al., 1995) that (2) can be put under a quadratic program form, that is

minimize $\frac{1}{2}\mathbf{u}^t\mathbf{D}\mathbf{u} + \mathbf{c}^t\mathbf{u}$
under the constraint $\mathbf{A}\mathbf{u} + \mathbf{b} \geq \mathbf{0}$

where $\mathbf{u}$ is a $(2^n - 2)$ dimensional vector containing all the coefficients of the fuzzy measure g (except $g(\emptyset)$ and $g(X)$ which are fixed), $\mathbf{D}$ is a $(2^n - 2)$ dimensional square matrix, $\mathbf{c}$ a $(2^n - 2)$ dimensional vector, $\mathbf{A}$ a $n(2^{n-1} - 1) \times (2^n - 2)$ matrix, and $\mathbf{b}$ a $n(2^{n-1} - 1)$ dimensional vector.

If there are too few learning data (there must be at least $n!/[(n/2)!]^2$ data: see (Grabisch et al., 1995) for details), matrices may be ill-conditioned. Moreover, the constraint matrix **A** is a sparse matrix, and become sparser as n grows, causing bad behavior of the algorithm. For all these reasons, including memory problems, time of convergence, the solution given by a quadratic program is not always reliable in practical situations.

Some authors have proposed alternatives to quadratic programming under the form of "heuristic" algorithms taking advantage of the peculiar structure of fuzzy measures, but often suboptimal. One of the most powerful in terms of performance, time and memory has been proposed by Grabisch (Grabisch, 1995b). The basic idea is that, in the absence of any information, the most reasonable way of aggregation is the arithmetic mean (provided the problem is cardinal), i.e. a Choquet integral with respect to an additive equidistributed fuzzy measure. Any input of information tends to move away the fuzzy measure from this equilibrium point. This means that, in case of few data, coefficients of the fuzzy measure which are not concerned with the data are kept as near as possible to the equilibrium point. This solves the problem of having too few data.

Combining semantics and learning data. Obviously, the combination of semantical considerations, which are able to reduce the complexity and provide guidelines, with learning data should lead to more efficient algorithms. An attempt in this direction is given by Yoneda *et al.* (Yoneda et al., 1993). The basic ideas are the following.

- objective: minimize the distance to the additive equidistributed fuzzy measure,

- constraints: the usual constraints implied by the monotonicity of fuzzy measures, and constraints coming from semantical considerations. These could be about relative importance of criteria, redundancy, support between criteria, etc.

The constraints being linear, this leads to a quadratic program as before.

2.2.9 Aggregation of Fuzzy Numbers

It may happen that scores cannot be given with precision, and in this case it is better to represent them by fuzzy numbers, taking into account the imprecision. Using the extension principle and suitable theorems related to it, it is not difficult to extend usual operators defined on $[0,1]^n$ to fuzzy numbers, since most operators are continuous and monotonic. The case of fuzzy integrals has been addressed in (Grabisch et al., 1995), and most of usual procedures of aggregation of fuzzy numbers can be found in the monograph of Cheng and Hwang (Chen and Hwang, 1992), including the pioneering works of Baas and Kwakernaak (Baas and Kwakernaak, 1977).

2.3 AGGREGATION OF FUZZY PREFERENCES IN PROBLEMS OF MULTIATTRIBUTE CHOICE

2.3.1 Introduction

Situations of choice with multiple preference relations in a set of alternatives may arise in two cases. In the first case, multiple preferences may reflect preferences of a single decision maker with respect to multiple attributes, or aspects, pertaining to every alternative. In the second case, there is a number of experts and each of the preference relations may reflect comparisons between alternatives from the viewpoint of the respective expert. In both cases, rational choices of alternatives are supposed to take into account all these preferences.

There are several approaches of a mathematical analysis of such situations. One of them is set in the context of multiperson decision making, and is called the theory of social choice and is concerned with the problem of aggregating the given preferences into a single preference relation. In effect, the principal goal of the theory of social choice is finding rules for amalgamating different preferences to ensure certain desirable properties of the resultant group preference. These rules are commonly pictured as criteria of "fairness" of choice. Problems of this class are discussed elsewhere in this book.

Another direction which is outlined here is more directly concerned with choices of alternatives from a given set based on given preference relations. Here, the choice problem appears to be intrinsically mixed with the aggregation problem. Although these two problems are in general considered apart, and the choice problem is the main subject of chapter 3 in this book, the ultimate goal of aggregation is nothing but choice of alternatives, and we do believe that a coherent approach must consider both aspects together. For this reason, we will develop here some material which could have some redundancy with other parts of this book.

The following paragraphs will illustrate two different approaches in this direction. The first one is logically similar to those used in multiobjective optimization problems and many concepts used in choice problems with a single preference relation are also relevant here. This leads to the Pareto principle of choice, which will be explained in section 2.3.2. The second approach, called the agreement-discordance or concordance-discordance principle, stems from the ELECTRE III and IV methods introduced by Roy (Roy, 1978, Roy and Bouyssou, 1994), and is firmly rooted in the context of multicriteria decision making. This approach will be presented in section 2.3.4.

2.3.2 Pareto Principle

In its purest form, a problem of choice with multiple preference relations is formulated as follows. Given a set of alternatives X and an n-tuple of preference relations $\mathcal{R} = \{R_1, ..., R_N\}$ on X^2 make rational choices of alternatives from X. Elements of $\mathcal{R}$ are regarded as individual preferences and $\mathcal{R}$ is called a profile of individual preferences on the set X. In applied problems individual preferences R_i may reflect preferences between alternatives with respect to a number of aspects or attributes. The problem lies in making a rational choice taking account of all these aspects or attributes.

In some problems additional information may be provided describing relative importances or significances of the attributes. Section 2.3.3 will describe an approach where importance between individual attributes is modelled by a fuzzy relation. An alternative method is to model importance through weights or fuzzy measures (see sections 2.2.3.5, 2.2.4, and 2.2.5).

Given such information, the common approach to solve the problem lies in constructing an aggregated preference relation R and considering as rational choices non-dominated alternatives with respect to this relation. Depending on the application, various rationality principles may be used for selecting alternatives and the aggregated preference relation should be such that corresponding non-dominated alternatives for the aggregated preference relation satisfy the selected rationality principle. Therefore, a rationality principle can be understood both as conditions that rational choices should satisfy and as conditions imposed on the corresponding aggregated relation.

One of the undisputable rationality principles in choice problems is the Pareto principle which in general terms may be stated as follows (case of crisp preference relations):

(i) if alternatives x and y are equivalent or incomparable with respect to each of the individual preferences, they must be such with respect to the aggregated relation.

(ii) if alternative x is better than alternative y with respect to at least one individual preference and is at least as good as y with respect to the other individual preferences, then x should be better than y with respect to the aggregated preference.

In set-theoretic terms the Pareto principle can be written as the following condition on the aggregated preference relation R:

$$\bigcap_i R_i \subseteq R \subseteq \bigcup_i R_i$$

Let us first illustrate the use of this principle in case of crisp individual preferences.

Suppose $R = \bigcap_{i=1,\ldots,N} R_i$ and let us determine non-dominated alternatives for such an aggregated preference relation. First, we construct the corresponding strict preference relation as $R^s = R \setminus R^{-1}$, or

$$R^s = \bigcap_{i=1,\ldots,N} R_i \setminus \bigcap_{i=1,\ldots,N} R_i^{-1}$$

A non-dominated alternative x with respect to R is such that

$$(y,\ x) \notin R^s, \text{ for any } y \in X$$

which means either that $(y,\ x) \notin R_i$ for at least one i, or that $(y,\ x) \in R_i$ and $(x, y) \in R_i$ for any i, i.e. y and x are equivalent to each other for any i. In other words, there is no alternative in X that would be better than x with respect to one or more individual preferences and not worse than x with respect to all the other individual preferences, i.e. x is a Pareto rational choice.

If we take $R = \bigcup_i R_i$ the corresponding strict preference relation has the form:

$$R^s = \bigcup_{i=1,\dots,N} R_i \setminus \bigcup_{i=1,\dots,N} R_i^{-1}$$

A non-dominated alternative x with respect to R is such that either $(y, x) \notin R_i$ for all i, or $(y, x) \in R_i$ for some i and $(x, y) \in R_i$ for some i. Again, we have that there is no alternative in X that would be better than x with respect to one or more individual preferences and not worse than x with respect to all the other individual preferences.

Clearly, in case of fuzzy relations, the result of applying the Pareto principle to processing individual preferences depends on the interpretation of the above set-theoretic operations of union, intersection, and set difference. Feasible classes of such interpretations are discussed elsewhere in this book. A thorough analysis of the two extreme cases of the above aggregation condition is presented in (Fodor and Ovchinnikov, 1995, Zhukovin, 1987). Other types of aggregation are also considered in (Aczel and Alsina, 1987, Dubois and Prade, 1991, Ovchinnikov, 1990, Trillas and Alsina, 1992).

Let us illustrate the use of the Pareto aggregation in the form of intersection of individual preferences for determining rational choices. We have a set X of alternatives and N fuzzy preference relations characterized by membership functions $m_i : X \times X \to [0, 1]$. We shall assume only that $m(x, x) = 1$ for any $x \in X$, i.e. that all the individual fuzzy preferences are reflexive.

The membership function of the aggregated fuzzy preference has the form:

$$m(x, y) = \min_{i=1,\dots,N} m_i(x, y)$$

Interpreting the respective fuzzy strict preference in the form

$$R^s = R \bigcap X \times X \setminus R^{-1}$$

we arrive at the following form of the membership function of the respective fuzzy strict preference:

$$m^s(x, y) = \min\{\min_{i=1,\dots,N} m_i(x, y), \max_{i=1,\dots,N} [1 - m_i(y, x)]\}$$

The fuzzy subset of non-dominated alternatives in this case is described by the membership function of the form:

$$m^{nd}(x) = 1 - \max_{y \in X} \min\{\min_{1,\dots,N} m_i(y, x), \max_{i=1,\dots,N} [1 - m_i(x, y)]\}$$

Suppose an alternative x has the degree of non-dominance α, i.e. $m^{nd}(x) = \alpha$. Then, from the above equation, we have that for any $y \in X$:

$$\min\{\min_{i=1,\dots,N} m_i(y, x), \max_{i=1,\dots,N} [1 - m_i(x, y)]\} \leq 1 - \alpha$$

which means that for any $y \in X$ either $m_i(y, x) \leq 1 - \alpha$ for some i, or $m_i(x, y) \geq \alpha$ for any i.

These properties of non-dominated alternatives represent a realization of the Pareto principle in case of fuzzy individual preferences using the type of aggregation and interpretations of the related set-theoretic operations as above. Other types of aggregation satisfying the Pareto principle and other interpretations of the set-theoretic operations involved will result in non-dominated alternatives having generally different properties. Logically, all types of non-dominated alternatives are acceptable as rational choices and the selection among them, i.e. the selection of the type of aggregation and types of the set-theoretic operations, should be based on specifics of the concrete applied problem.

2.3.3 Relation of Relative Importance

Let X be a given set of alternatives and P be a given set of attributes. Each alternative $x \in X$ is characterized, to a lesser or higher degree, by each of the attributes from the set P and alternatives can be compared with each other in each of the attributes. Such pairwise comparisons vary for different attributes and result in different fuzzy preference relations in the set X. Suppose for any attribute $p \in P$ a fuzzy preference relation $\mathcal{R}(p)$ in the set X is given and described by the membership function $m_{\mathcal{R}} : X \times X \times P \to [0, 1]$. For any two alternatives $x, y \in X$ and any attribute $p \in P$ the value $m_{\mathcal{R}}(x, y, p)$ of this function is understood as the degree to which the alternative x is considered to be "*not inferior to*" the alternative y with respect to the attribute p (if P is a group of experts then $\mathcal{R}(p)$ is understood as the fuzzy preference relation between alternatives from the viewpoint of the expert p). Therefore, $\{\mathcal{R}(p), p \in P\}$ is a profile of individual fuzzy preferences in the set X.

Elements (attributes or experts) of the set P differ in their relative importance for characterizing the alternatives. Suppose these relative importances are given in the form of a fuzzy relation of importance $\mathcal{I}$ in the set P described by the membership function $m_{\mathcal{I}} : P \times P \to [0, 1]$. For any two attributes $p, q \in P$ the value $m_{\mathcal{I}}(p, q)$ of this function is understood as a degree to which the attribute p is considered to be "*not less important*" than the attribute q.

The problem lies in rationally choosing alternatives from the set X based on the above formulated information. Depending on a concrete applied case, various approaches may be used to analyze this problem and the effectiveness of different approaches may be different in different cases. In what follows, we shall illustrate an approach described in (Orlovski, 1994).

Denote by $m^{nd}_{\mathcal{R}}(x, p)$ the membership function of the fuzzy set of non-dominated alternatives in the set $(X, \mathcal{R}(p)), p \in P$. In case the choice of alternatives were made based only on one attribute p, the rational choice would naturally be any alternative providing higher possible value of the respective function $m^{nd}_{\mathcal{R}}(x, p)$. But here we are to take into account multiple attributes differing in their relative importances.

As can easily be understood, for any fixed alternative $x^0 \in X$ the function $m^{nd}_{\mathcal{R}}(x^0, p)$, as a function of p, describes the fuzzy set of attributes in the set

P with respect to which the alternative x^0 is a non-dominated alternative in the set X. Let us take two alternatives $x, y \in X$ and consider the membership functions $m^{nd}_{\mathcal{R}}(x, p)$ and $m^{nd}_{\mathcal{R}}(y, p)$ (as the functions of p) of the respective fuzzy sets in the set P. If the first of these fuzzy sets "*is not less important*" than the second one it is only natural to accept that the alternative x "*is not less preferable*" than the alternative y. Clearly, the above linguistic expressions in quotation marks require mathematical formalization.

In other words, we shall extend the given fuzzy relation of importance $\mathcal{I}$ in the set P onto the class of fuzzy sets in the set P and consider the resulting fuzzy relation as the fuzzy preference relation in the set of alternatives X. Let us denote such extended fuzzy relation in the set X by $\mathcal{Q}$. Using the extension principle in its most traditional form and based on the above reasoning, we obtain the following expression for the membership function of $\mathcal{Q}$ (any $x, y \in X$):

$$m_{\mathcal{Q}}(x, y) = \sup_{p,q \in P} \min\{m^{nd}_{\mathcal{R}}(x, p), m^{nd}_{\mathcal{R}}(y, q), m_{\mathcal{I}}(p, q)\}.$$

This fuzzy preference is the aggregation of the profile of fuzzy individual preferences $\mathcal{R}(p)$ into a single fuzzy relation taking into account the information on the relative importances of the attributes in the form of the fuzzy relation $\mathcal{I}$.

Having constructed the fuzzy relation $\mathcal{Q}$, we have reduced the original choice problem with multiple fuzzy preferences to the problem with a single fuzzy preference relation. To solve it, we can determine the corresponding fuzzy set of non-dominated alternatives in the set $(X, \mathcal{Q})$.

Let us apply this logical scheme to an illustrative example.

Example 1. An individual wishes to make a choice among four models of suits of clothes: A, B, C, and D. Not quite relying on his taste, he invites four advisors (experts) $E1$, $E2$, $E3$, and $E4$ and values their opinions somewhat differently. He describes his relative evaluations of opinions of the advisors by means of the following matrix of the fuzzy relation of importance "*not less important than*":

	$E1$	$E2$	$E3$	$E4$
$E1$	1	0.4	0.6	0
$E2$	1	1	0.8	1
$E3$	0.2	1	1	1
$E4$	0.8	0	1	1

The opinions of the advisors with regard to preferences between the models of suits are described by the following matrices of fuzzy preference relations "*not worse than*":

Advisor $E1$:

	A	B	C	D
A	1	0.4	0.6	0
B	1	1	0.8	1
C	0.2	1	1	1
D	0.8	0	1	1

Advisor $E2$:

	A	B	C	D
A	1	0.1	0.5	0.3
B	0.8	1	0.8	0.8
C	0.5	0.3	1	0
D	0.8	0	0	1

Advisor $E3$:

	A	B	C	D
A	1	0	0.8	0
B	0	1	0	0
C	0.1	0	1	0.4
D	1	1	1	1

Advisor E4:

	A	B	C	D
A	1	1	0.9	0
B	0	1	1	1
C	0.4	0	1	0
D	0	0	0	1

To determine rational choices for this problem, first, we obtain the membership functions of the fuzzy sets of non-dominated alternatives for each individual advisor:

	A	B	C	D
$m^{nd}(\cdot,\ E1) =$	1	0.2	0	0
$m^{nd}(\cdot,\ E2) =$	0.3	1	0.5	0.2
$m^{nd}(\cdot,\ E3) =$	0	0	0.3	1
$m^{nd}(\cdot,\ E4) =$	1	0	0	0

Next we construct the matrix of the membership function of the aggregated fuzzy relation Q, as defined above:

	A	B	C	D
A	1	0.4	0.4	1
B	1	1	0.5	0.8
C	0.5	0.5	0.5	0.5
D	1	1	0.5	1

and then - the corresponding membership function for the fuzzy set of non-dominated alternatives in the set $(X,\ Q)$:

$$m_Q^{nd}(\cdot) = \frac{\begin{array}{cccc} A & B & C & D \end{array}}{\begin{array}{cccc} 1 & 0.4 & 0.9 & 0.8 \end{array}}$$

This function gives the resultant non-dominance degrees for our alternatives. A more careful analysis, however, reveals that the maximum non-dominance degree of alternative C with respect to all the experts (the above functions $m^{nd}(\cdot,\ E1,\ \ldots)$ is equal to 0.5 and, therefore, its overall degree should not exceed this value. Considering this, we can take as the final result the following "corrected" membership function:

$$m_Q^{ndc}(\cdot) = \frac{\begin{array}{cccc} A & B & C & D \end{array}}{\begin{array}{cccc} 1 & 0.4 & 0.5 & 0.8 \end{array}}$$

The highest non-dominance degree in this problem has the model A, therefore its choice should be considered rational according to this approach.

2.3.4 The Agreement-Discordance Principle for Aggregation

The approach described here has its roots in voting procedures, and has been the main foundations of methods such as ELECTRE III and IV introduced by Roy (Roy, 1978, Roy and Bouyssou, 1994). The material of this section is mainly based on the works of Perny and Roy (see (Perny, 1992, Perny, 1996, Perny and Roy, 1992)), and borrows concepts defined in chapter 1, such as outranking relations, indifference and preference relations. These relations are denoted as S, I and P respectively, and will be not redefined here.

The agreement-discordance principle lies upon concepts from voting problem and social choice theory, identifying criteria with voters, essentially the concepts of (absolute or relative) majority and the right of veto.

We start with the agreement principle. In preference oriented decision problems, the *agreement* principle also called *concordance* principle refers to the concept of majority and consists in evaluating the number of voters in favour of candidate a when opposed to candidate b. Let us denote by H_j a valued binary relation, which could be either S, I or P, related to criterion (or voter) j, and $H_j(a,b)$ expresses the preference (or indifference, depending on the nature of H_j) of a over b with respect to criterion j. Then the *overall agreement relation* $C_H(a,b)$ is defined as:

$$C_H(a,b) = \Psi(H_1(a,b), \ldots, H_n(a,b))$$

where Ψ is a (possibly weighted) compensative aggregation operator. A typical example is the family of weighted quasi-arithmetic means (see section 2.2.3.3). Clearly, $C_H(a,b)$ represents a kind of count of the pros and cons for the proposition a *is preferred to* b, or a *is not worse than* b, etc. depending on the precise nature of H_j.

We turn now to the discordance principle, which in other words expresses the respect of minorities. The main idea is to take into account minorities which are strongly conflicting with the majority choice. In a preference oriented problem, this amounts to check whether some voter or criterion wants to make use of his right of veto against the overall statement aHb, reflecting an advantage in favour of a. This can be formalized as follows. Let H be an overall valued binary relation, either S, P or I. For all $j = 1, \ldots, n$, the *veto threshold* v_j^H is a real-valued function such that, for any alternative a, $v_j^H(a_j)$ is the maximal value of a score difference on criterion j that could be compatible with the proposition aHb, that is:

$$\exists j \in \{1, \ldots, n\}, b_j - a_j > v_j^H(a_j) \Rightarrow \neg aHb.$$

The right of veto will apply against a if the score difference $b_j - a_j$ is seen as a decisive weakness that cannot be compensated by other possible strong points. The next step is to define a discordance relation.

Definition 8 *For any relation H (either S, P or I), for any criterion j whose scale is X_j, the valued binary relation D_j^H is said to be a* one-dimensional discordance relation *associated with H if there is a function d^H from $X_j \times X_j$ to $[0,1]$, defined as $d_j^H(a_j, b_j) = D_j^H(a,b)$ for all a, b, such that:*

(i) *$\forall y \in X_j, d_j^H(x,y)$ is a non increasing function of x,*

(ii) *$\forall x \in X_j, d_j^H(x,y)$ is a non decreasing function of y,*

(iii) $\forall x, y \in X_j, \min(d_j^H(x,y), H_j(x,y)) = 0$,

(iv) $\forall x, y \in X_j, y - x > v_j^H(x) \Rightarrow d_j^H(x,y) = 1$.

In ELECTRE III, one dimensional discordance relations associated with outranking relations are defined using the following formula:

$$d_j^S(x,y) = \min(1, \max(0, \frac{y - x - p_j(x)}{v_j^S(x) - p_j(x)}$$

for $j = 1, \ldots, n$, where p_j are preference thresholds, defined in the same spirit as veto thresholds. Similarly as it was done with the agreement principle, we build an *overall discordant relation* $D_H(a,b)$ by aggregating in a proper way all the individual discordant relations. Specifically:

$$D_H(a,b) = \Phi(D_1^H(a,b), \ldots, D_n^H(a,b))$$

where Φ is a (possibly weighted) aggregation operator satisfying

(i) Φ is monotone non decreasing with respect to all arguments,

(ii) $\Phi(0, 0, \ldots, 0) = 0$,

(iii) $\exists j \in \{1, \ldots, n\}$ such that $x_j = 1 \Rightarrow \Phi(x_1, \ldots, x_n) = 1$.

It is worth noting that conditions imposed here for the aggregation operator are somewhat different from those requested for the aggregation of agreement relations. Actually, $D_H(a,b)$ expresses the degree to which there exists at least one important and discordant criterion, implying that Φ is disjunctive in nature, such as t-conorms for example. An interesting choice for Φ is to take the dual of the geometric mean, defined as follows:

$$\Phi(a_1, \ldots, a_n) = 1 - \prod_{j=1}^{n} (1 - a_j)^{w_j},$$

or the weighted maximum (see section 2.2.3.5).

We introduce now the *agreement-discordance principle*, based on the previous definitions.

> **Agreement-discordance principle:** a proposition of type aHb holds if and only if the coalition of attributes or criteria in agreement with this proposition is strong enough, and there is no significant coalition discordant with this proposition.

This principle can be translated in the following logical equation:

$$H(a,b) = C_H(a,b) \wedge \neg D_H(a,b)$$

where $\wedge$ denotes the conjunction "and", and $\neg$ the negation. Usually, the conjunction is modelled by a t-norm $*$ and the negation by a strict negation operator N, so that:

$$H(a,b) = *(C_H(a,b), N(D_H(a,b))). \quad (3)$$

The idea of the agreement-discordance principle can be found in ELECTRE III and IV methods as already mentioned, but also in PROMETHEE (Brans et al., 1984; Brans and Vincke, 1985). In ELECTRE III, equation (3) was used for the outranking relation S, with the product operator as a t-norm, and the usual $1-(\cdot)$ as negation operator.

Properties of H coming from equation (3) are:

- independance: for any H being of the S, I or P type, $H(a,b)$ depends only on the scores a_i, b_i, $i = 1, \ldots, n$.
- respect of dominance: $\forall i \in \{1, \ldots, n\}, a_i \geq b_i \Rightarrow S(a,b) = 1$ and $\forall i \in \{1, \ldots, n\}, a_i = b_i \Rightarrow I(a,b) = 1$.
- for any H being of the S, I or P type, $H(a,b) \leq C_H(a,b)$.
- respect of the right of veto: if there exists j such that $b_j - a_j > v_j^H$, then $H(a,b) = 0$.
- compensativeness:

$$\bigwedge_{i=1}^{n} H_i(a,b) \leq H(a,b) \leq \bigvee_{i=1}^{n} H_i(a,b).$$

The above properties show the adequacy of the principle of agreement-discordance in multicriteria decision problems. Its roots in social choice theory makes it easy to grasp and to apply in practical problems.

Notes

1. When n is even, the usual extension is $\frac{1}{2}(a_{n/2} + a_{n/2+1})$.
2. A Sugeno integral can model veto effect as well, but this is obvious since the weighted minimum, which can express veto effects, is encompassed by the Sugeno integral.

References

Aczel, J. and Alsina, C. (1987). Synthesizing judgements: a functional equation approach. *Math. Modeling.*

Baas, S. and Kwakernaak, H. (1977). Rating and ranking of multiple-aspect alternatives using fuzzy sets. *Automatica*, 13:47–58.

Brans, J., Mareschal, B., and Vincke, P. (1984). Promethee: a new family of outranking methods in multicriteria analysis. In Brans, J., editor, *Operational Research'84*. North Holland.

Brans, J. and Vincke, P. (1985). A preference ranking organization method. *Management Science*, 31:647–656.

Chen, S. and Hwang, C. (1992). *Fuzzy Multiple Attribute Decision Making*. Springer-Verlag.

Denneberg, D. (1994). *Non-Additive Measure and Integral*. Kluwer Academic.

Dubois, D. and Prade, H. (1984). Criteria aggregation and ranking of alternatives in the framework of fuzzy set theory. In Zimmermann, H.-J., Zadeh,

L., and Gaines, B., editors, *Fuzzy Sets and Decision Analysis*, volume 20 of *TIMS Studies in the Management Sciences*, pages 209–240.

Dubois, D. and Prade, H. (1985). A review of fuzzy set aggregation connectives. *Information Sciences*, 36:85–121.

Dubois, D. and Prade, H. (1986). Weighted minimum and maximum operations in fuzzy set theory. *Information Sciences*, 39:205–210.

Dubois, D. and Prade, H. (1991). Fuzzy sets in approximate reasoning, part 1: inference with possibility distributions. *Fuzzy Sets and Systems*, 40:143–202.

Dubois, D., Prade, H., and Testemale, C. (1988). Weighted fuzzy pattern matching. *Fuzzy Sets & Systems*, 28:313–331.

Dyckhoff, H. and Pedrycz, W. (1984). Generalized means as model for compensative connectives. *Fuzzy Sets & Systems*, 14:143–154.

Filev, D. and Yager, R. (1994). Learning OWA operator weights from data. In *Third IEEE International Conference on Fuzzy Systems*, pages 468–473, Orlando.

Fodor, J., Marichal, J., and Roubens, M. (1995). Characterization of the ordered weighted averaging operators. *IEEE Tr. on Fuzzy Systems*, 3(2):236–240.

Fodor, J. and Ovchinnikov, S. (1995). On aggregation of t-transitive fuzzy binary relations. *Fuzzy Sets and Systems*, 72:135–145.

Fodor, J. and Roubens, M. (1994). *Fuzzy Preference Modelling and Multi-Criteria Decision Aid.* Kluwer Academic Publisher.

Fodor, J. and Roubens, M. (1995). On meaningfulness of means. *Journal of Computational and Applied Mathematics*, 64:103–115.

Grabisch, M. (Grabisch, a) Alternative representations of discrete fuzzy measures for decision making. *Int. J. of Uncertainty, Fuzziness, and Knowledge Based Systems.* to appear.

Grabisch, M. (Grabisch, b) Fuzzy integral as a flexible and interpretable tool of aggregation. In Bouchon-Meunier, B., editor, *Aggregation of evidence under fuzziness.* Physica Verlag. to appear.

Grabisch, M. (Grabisch, c) k-order additive discrete fuzzy measures and their representation. *Fuzzy Sets and Systems.* to appear.

Grabisch, M. (1994). Fuzzy integrals as a generalized class of order filters. In *European Symposium on Satellite Remote Sensing*, Roma, Italy.

Grabisch, M. (1995a). Fuzzy integral in multicriteria decision making. *Fuzzy Sets & Systems*, 69:279–298.

Grabisch, M. (1995b). A new algorithm for identifying fuzzy measures and its application to pattern recognition. In *Int. Joint Conf. of the 4th IEEE Int. Conf. on Fuzzy Systems and the 2nd Int. Fuzzy Engineering Symposium*, pages 145–150, Yokohama, Japan.

Grabisch, M. (1996a). Alternative representations of discrete fuzzy measures for decision making. In *4th Int. Conf. on Soft Computing*, Iizuka, Japan.

Grabisch, M. (1996b). The application of fuzzy integrals in multicriteria decision making. *European J. of Operational Research*, 89:445–456.

Grabisch, M. (1996c). k-order additive fuzzy measures. In *6th Int. Conf. on Information Processing and Management of Uncertainty in Knowledge-Based Systems (IPMU)*, pages 1345–1350, Granada, Spain.

Grabisch, M. (1997). Alternative representations of OWA operators. In Yager, R. and Kacprzyk, J., editors, *The Ordered Weighted Averaging Operators: Theory, Methodology, and Practice*, pages 73–85. Kluwer Academic.

Grabisch, M., Nguyen, H., and Walker, E. (1995). *Fundamentals of Uncertainty Calculi, with Applications to Fuzzy Inference*. Kluwer Academic.

Kolmogoroff, A. (1930). Sur la notion de moyenne. *Atti delle Reale Accademia Nazionale dei Lincei Mem. Cl. Sci. Fis. Mat. Natur. Sez.*, 12:323–343.

Krantz, D., Luce, R., Suppes, P., and Tversky, A. (1971). *Foundations of measurement*, volume 1: Additive and Polynomial Representations. Academic Press.

Mizumoto, M. (1989a). Pictorial representations of fuzzy connectives, part I: Cases of t-norms, t-conorms and averaging operators. *Fuzzy Sets & Systems*, 31:217–242.

Mizumoto, M. (1989b). Pictorial representations of fuzzy connectives, part II: case of compensatory operators and self-dual operators. *Fuzzy Sets & Systems*, 32:45–79.

Mousseau, V. (1992). Analyse et classification de la littérature traitant de l'importance relative des critères en aide multicritère à la décision. *Recherche Opérationnelle/Operations Research*, 26(4):367–389.

Murofushi, T. (1992). A technique for reading fuzzy measures (I): the Shapley value with respect to a fuzzy measure. In *2nd Fuzzy Workshop*, pages 39–48, Nagaoka, Japan. In Japanese.

Murofushi, T. and Soneda, S. (1993). Techniques for reading fuzzy measures (III): interaction index. In *9th Fuzzy System Symposium*, pages 693–696, Sapporo, Japan. In Japanese.

Murofushi, T. and Sugeno, M. (1991). A theory of fuzzy measures. Representation, the Choquet integral and null sets. *J. Math. Anal. Appl.*, 159(2):532–549.

Murofushi, T. and Sugeno, M. (1993). Some quantities represented by the Choquet integral. *Fuzzy Sets & Systems*, 56:229–235.

Nijkamp, P., Rietveld, P., and Voogd, H. (1990). *Multicriteria Evaluation in Physical Planning*. North Holland.

Norwich, A. and Turksen, I. (1984). A model for the measurement of membership and the consequences of its empirical implementation. *Fuzzy Sets and Systems*, 12:1–25.

O'Hagan, M. (1990). Using maximum entropy-ordered weighted averaging to construct a fuzzy neuron. In *24th Annual IEEE Asilomar Conf. on Signals, Systems and Computers*, pages 618–623, Pacific Grove, CA.

Orlovski, S. (1994). *Calculus of decomposable properties, fuzzy sets, and decisions*. Allerton Press.

Ovchinnikov, S. (1990). Means and social welfare functions in fuzzy binary relation spaces. In Kacpzyk, J. and Fedrizzi, M., editors, *Multiperson Decision Making Using Fuzzy Sets and Possibility Theory*, pages 143–154. Kluwer, dordrecht edition.

Perny, P. (1992). *Modélisation, agrégation et exploitation des préférences floues dans une problématique de rangement*. PhD thesis, Univ. Paris-Dauphine.

Perny, P. (1996). Multicriteria filtering methods based on agreement and discordance principles. Technical report, LAFORIA, Université Paris VI, France.

Perny, P. and Roy, B. (1992). The use of fuzzy outranking relations in preference modelling. *Fuzzy Sets & Systems*, 49:33–53.

Roberts, F. (1979). *Measurement Theory*. Addison-Wesley.

Roy, B. (1978). Electre III : un algorithme de classement fondé sur une représentation floue des préférences en présence de critères multiples. *Cahiers du Centre d'Etude de Recherche Opérationnelle*, 20(1):32–43.

Roy, B. and Bouyssou, D. (1994). *Aide multicritère à la décision : Méthodes et Cas*. Economica.

Saaty, T. (1977). A scaling method for priorities in hierarchical structures. *J. Math. Psychology*, 15:234–281.

Saaty, T. (1980). *The Analytic Hierarchy Process*. McGraw Hill, New York.

Shapley, L. (1953). A value for n-person games. In Kuhn, H. and Tucker, A., editors, *Contributions to the Theory of Games, Vol. II*, number 28 in Annals of Mathematics Studies, pages 307–317. Princeton University Press.

Silvert, W. (1979). Symmetric summation: a class of operations on fuzzy sets. *IEEE Tr. on Systems, Man, and Cybernetics*, 9:657–659.

Siskos, J. and Yannacopoulos, D. (1985). An ordinal regression method for building additive value functions. *Investigação Operacional*, 5(1):39–53.

Sugeno, M. (1974). *Theory of fuzzy integrals and its applications*. PhD thesis, Tokyo Institute of Technology.

Trillas, E. and Alsina, C. (1992). Some remarks on approximate entailment. *Int. J. Approximate Reasoning*, 6:525–533.

Yager, R. (Yager, a) Quantifier guided aggregation using OWA operators. *International Journal of Intelligent Systems.*

Yager, R. (1978). Fuzzy decision making using unequal objectives. *Fuzzy Sets and Systems*, 1:87–95.

Yager, R. (1981). A new methodology for ordinal multiple aspect decisions based on fuzzy sets. *Decision Sciences*, 12:589–600.

Yager, R. (1983). Quantifiers in the formulation of multiple objective decision functions. *Information Sciences*, 31:107–139.

Yager, R. (1984). General multiple objective decision making and linguistically quantified statements. *Int. J. of Man-Machine Studies*, 21:389–400.

Yager, R. (1987). A note on weighted queries in information retrieval systems. *J. of the American Society of Information Sciences*, 38:23–24.

Yager, R. (1988). On ordered weighted averaging aggregation operators in multicriteria decision making. *IEEE Trans. Systems, Man & Cybern.*, 18:183–190.

Yager, R. (1991). Connectives and quantifiers in fuzzy sets. *Fuzzy Sets & Systems*, 40:39–75.

Yager, R. (1993). Families of OWA operators. *Fuzzy Sets and Systems*, 55:255–271.

Yager, R. and Filev, D. (1994). *Essentials of Fuzzy Modeling and Control*. John Wiley and Sons, New York.

Yager, R., Filev, D., and Sadeghi, T. (1994). Analysis of flexible structured fuzzy logic controllers. *IEEE Transactions on Systems, Man and Cybernetics*, 24:1035–1043.

Yoneda, M., Fukami, S., and Grabisch, M. (1993). Interactive determination of a utility function represented by a fuzzy integral. *Information Sciences*, 71:43–64.

Zadeh, L. (1983). A computational approach to fuzzy quantifiers in natural languages. *Computing and Mathematics with Applications*, 9:149–184.

Zhukovin, V. (1987). A fuzzy multicriteria decision-making model. In Kacprzyk, J. and Orlovski, S. A., editors, *Optimization Models Using Fuzzy Sets and Possibility Theory*, pages 203–215. D. Reidel Publ. Co., Dordrecht, Boston, Lancaster, Tokyo.

Zimmermann, H.-J. and Zysno, P. (1980). Latent connectives in human decision making. *Fuzzy Sets & Systems*, 4:37–51.

Zopounidis, C., Despotis, D., and Stavropoulou, E. (1995). Multiattribute evaluation of greek banking performance. *Applied Stochastic Models and Data Analysis*, 11:97–107.

Yager, R.R. and Kacprzyk, J. Eds. (1997) The Ordered Weighted Averaging Operators: Theory and Application Kluwer, Boston.

3 THE USE OF FUZZY PREFERENCE MODELS IN MULTIPLE CRITERIA CHOICE, RANKING AND SORTING

Janos Fodor
Sergei Orlovski
Patrice Perny
Marc Roubens

Abstract: This chapter presents several approaches to derive a recommendation from a fuzzy relational preference model. Depending on the specificities of the decision problem, the recommendation can be a selection of the best alternatives, a ranking of these alternatives or a classification. In many decision problems involving several criteria and/or several experts with possibly conflicting evaluations of the alternatives, deriving a clear prescription is not an easy task since the overall fuzzy preference model is usually neither transitive nor complete. In this chapter, we present methods that have been specially designed to face this difficulty. We first consider the problem of choosing from a fuzzy preference relation. The classical notions of non-dominated alternatives and core are introduced (§ 3.1). Then several choice functions based on the use of scores, covering relations and kernels respectively are presented in sections (§ 3.2-3.5). Finally, a brief survey of ranking and sorting procedures is proposed (§ 3.6).

3.1 CHOICE BASED ON A FUZZY BINARY RELATION

Information providing a basis for preferring one alternatives to others in a decision situation may be given in various ways. In some situations, preference information is represented by utility functions or criterion functions allowing the alternatives to be evaluated and compared (see chapter 1). But this form of representing information is not always suitable. A alternative comparison model is information represented in the form of one or a number of preference relations in the set of alternatives. These preference relations can be derived from criterion functions as shown in the first chapter. Alternatively, preferences between alternatives can be elucidated through consultations with decision makers or experts. Indeed, in some cases, experts have a sufficient knowledge, or information, about the situation under analysis to be able to compare alternatives, even if their arguments are not formalized (because of the prohibitive complexity of such formalization or by other reasons).

Generally speaking, to model preferences by a preference relation means to specify preferences for all possible pairs $(a, \; b)$ of alternatives by providing answers to questions like: "*is a not inferior to b*?". Intending to obtain an ordinary preference relation in order to apply the concepts and logic of the mathematical theory of choice, we can accept only categorical answers like "*yes*" or "*no*".

But in most applied problems preference relations are fuzzy and experts are reluctant to make categorical assertions regarding preferences between alternatives, *i.e.* quite positively conclude, for example, that an alternative a is not inferior to an alternative b. If, nevertheless, an expert is obliged to produce such positive conclusions about preferences, he/she will have to sacrifice sometimes a larger part of his expertise. The resultant mathematical model may be lacking much of its adequacy to the real situation (see chapter 1).

Thus, modeling preferences by fuzzy relations provides more flexible language and procedures for revealing and processing preferences. In the elucidation of a fuzzy preference relation experts have a wider spectrum of possibilities being able to express intermediate degrees of preferences between alternatives, as compared to the categorical "*yes*" or "*no*".

The field of the mathematical analysis of choice problems with ordinary preference relations is elaborated very extensively. The scientific literature abounds with numerous approaches to the formal description of such problems as well as with formalizations of solution concepts and methods (see e.g. Aizerman and Malishevski (1981), Fishburn (1970), von Neumann and Morgenstern (1947), Rapoport (1989)).

Fuzzy preference relations in choice problems have never been suffering from the lack of attention of researchers, and almost all concepts and methods developed for ordinary relations have, to various degrees of logical consistency, been "fuzzified", or adapted, to the language of fuzzy relations. Examples of such formulations can be seen in Bortolan and Degani (1985), Bezdek *et al.* (1978), Bondareva (1990a, 1990b), Bouchon (1987), Fodor and Roubens (1994), Kim (1983), Kuzmin and Ovchinnikov (1980a, 1980b), Kuzmin and Travkin (1987), Orlovski (1978), Ovchinnikov (1987, 1988), Perny and Roy (1992), Roubens and Vincke (1987), Saaty (1978, 1980).

In the following subsection we introduce the concept of non-dominated alternatives that forms a part of, virtually, any solution concept for problems of rational choice with preference relations. First, we give a logical formulation of this concept and also of a more general concept of the related choice rule called the core and then interpret this formulation using membership functions of respective fuzzy relations and fuzzy sets.

Non-dominated alternatives. Let R be a preference relation defined on a set A and $\alpha(R)$ be the corresponding strict preference relation. Logically $\alpha(R)$ has the following form:

$$\alpha(R) = R \setminus R^{-1},$$

where R^{-1} is the inverse relation.

If $(a, b) \in \alpha(R)$ ($a\,\alpha(R)\,b$ is true) the alternative a is said to *dominate* the alternative b (a is preferred to b). Given a set of alternatives A, an alternative $a \in A$ is called *non-dominated* in the set A if we have $(b, a) \notin \alpha(R)$ for any alternative $b \in A$. In other words, if a is a non-dominated alternative, there is no alternative in the set A that dominates a.

Therefore, information in the form of a preference relation R enables us to narrow the class of rational choices in the set A down to the subset of non-dominated alternatives:

$$A^{nd} = \{a \in A \mid (b, a) \notin R \setminus R^{-1}, \ \forall b \in A\}.$$

Clearly, the narrower the set A^{nd} the less the ambiguity in the rational choice of alternatives.

Choice functions. There exists an alternative general framework for the analysis of decision problems based on the concept of choice function (see Aizerman and Malishevski (1981)). Informally speaking, a choice function is a correspondence that associates a set of selected alternatives with an initial set of alternatives. Given a basic set of alternatives A, the choice function is formally defined as a function $C : 2^A \rightarrow 2^A$, and for any initial set of alternatives $X \subseteq A$ the corresponding set $C(X)$ is understood as the set of alternatives the choice of which is considered rational by the logic imbedded in this function.

For a given non-strict preference relation R in a set A the choice of non-dominated alternatives corresponds to the choice function of the form (any $X \subseteq A$):

$$C_0(X) = \{a \in X \mid b\,\bar{\alpha}(R)a, \ \forall b \in X\},$$

where $\bar{\alpha}(R) = (A \times A) \setminus \alpha(R)$ is the complement of the strict preference relation $\alpha(R)$ corresponding to R.

The choice rule associating the choice function C_0 with a given preference relation is often called the *core* and is represented in the following way. Let us introduce sets:

$$\alpha(R)_a = \{b \in X \mid a\,\alpha(R)\,b\}, \quad \alpha(R)(X) = \bigcup_{a \in X} \alpha(R)_a.$$

Clearly, $\alpha(R)(X)$ is the set of all alternatives of the set X that are dominated by another alternative from the set X. Therefore, the set $C_0(X)$, as the subset of all non-dominated alternatives of the set X, can be represented as:

$$C_0(X) = X \setminus \alpha(R)(X)$$

and is in this form a choice function corresponding to a given relation $\alpha(R)$.

Various types of choice functions are characterized by properties reflecting conditions that rational choices are supposed to satisfy. The most simple and natural of such conditions are (for any X, $Y \subseteq A$):

1-(Heritage):

$$X \subseteq Y \implies C(Y) \bigcap X \subseteq C(X).$$

(Any alternative chosen from Y and belonging to X must be among the choices from X.)

2-(Condorcet):

$$\forall a \in Y, \ \bigcap_{b \in Y} C(\{a, b\}) \subseteq C(Y).$$

(Any alternative chosen by pairwise comparisons with all other alternatives must be among the choices.)

3-(Stability):

$$X \subseteq C(Y) \subseteq Y \implies C(X) = X.$$

(No alternatives are excluded by choosing among chosen alternatives.)

4-(Independence on irrelevant alternatives):

$$C(Y) \subseteq X \subseteq Y \implies C(X) = C(Y).$$

(Adding non-chosen alternatives does not affect the result of choice.)

5-(Concordance):

$$C(X) \cap C(Y) \subseteq C(X \cup Y)$$

(Any alternative chosen both from X and from Y must be among the choices from $X \cup Y$)

In principle, the concept of choice function provides an alternative framework for the formalization of solutions to choice problems without modeling preferences in the form of relations. A correspondence between relations and choice functions is called a *choice rule*. Formally, given a basic set of alternatives A, a choice rule is defined as the correspondence $F : \Re \to \Im$, where $\Re$ is a set of preference relations in A and $\Im$ is a set of choice functions defined on 2^A.

For any strict preference relation $\alpha(R)$ the choice rule, called the core, gives the choice function of the form C_0. As it can easily be shown (see Aizerman and Malishevski (1981)), the choice function C_0 satisfies the above conditions 1-(Heritage) and 2-(Condorcet). An interesting and useful property of the core is that any choice function satisfying the above two conditions uniquely determines the corresponding strict preference relation (Aizerman and Malishevski (1981)). In other words, any such choice function can be used to reconstruct the relation $\alpha(R)$ that gives this choice function under the core. Less formally, this means that any choice function of the considered class contains full information about the underlying strict preferences. Later on in this section we shall discuss this in the context of fuzzy preferences.

Having logically introduced the necessary concepts we shall now give them more concrete forms based on concrete membership functions of fuzzy preference relations.

Fuzzy set of non-dominated alternatives. Let A be a precisely described set of feasible alternatives and R be a FPR with the membership function $R: A \times A \to [0, 1]$. We denote by $\alpha(R)$ the membership function of a fuzzy strict preference relation $\alpha(R)$ corresponding to the FPR R. By definition, for any pair of alternatives $a, b \in A$ the value $\alpha(R)(b, a)$ of this function is the degree to which a is (strictly) dominated by b. Consequently, the value $1 - \alpha(R)(b, a)$ is the degree to which the alternative a is not (strictly) dominated by the alternative b. The minimum of the last function over all alternatives $b \in A$ is understood as the degree to which a is not dominated by any alternative $b \in A$. Denoting the value obtained by $m_R^{nd}(a)$ we can write ($a \in A$):

$$m_R^{nd}(a) = \inf_{b \in A} [1 - \alpha(R)(b, a)],$$

or

$$m_R^{nd}(a) = 1 - \sup_{b \in A} \alpha(R)(b, a). \tag{1}$$

The above function describes a fuzzy set in A that we shall refer to as the fuzzy set of non–dominated alternatives in the set (A, R) and denote it by A_R^{nd}. As it is introduced earlier in this section, the function $sup_{b \in A} \alpha(R)(b, a)$ is the membership function of the projection $\alpha(R)_I$ of the fuzzy relation $\alpha(R)$ onto the set A. Based on this and in virtue of relation (1), the fuzzy set A_R^{nd} can be written in the form $A_R^{nd} = R_I^{s*}$ where R_I^{s*} is the fuzzy set in A complemental to $\alpha(R)_I$.

Let us summarize the above in the following definition adapted from Orlovski (1978).

Definition 1 *For a set of alternatives A, for any FPR R defined on A, the fuzzy subset A_R^{nd} of non-dominated alternatives is the fuzzy set of A characterized by the membership function m_R^{nd} defined by equation (1).*

Considering the value $m_R^{nd}(a)$ as the degree of "non-dominance" of the corresponding alternative, it is natural to consider as rational choices alternatives that give the highest possible values of this function, i.e. alternatives for which the value of $m_R^{nd}(a)$ is closer possible to the value:

$$\sup_{a \in A} m_R^{nd}(a) = 1 - \inf_{a \in A} \sup_{b \in A} \alpha(R)(b, a).$$

Alternatives that give exactly this value, or in other words, alternatives from the set:

$$A^{nd} = \{a \in A \mid m_R^{nd}(a) = \sup_{b \in A} m_R^{nd}(b)\},$$

may be called *maximum non–dominated alternatives* in the set (A, R).

The higher the degree of non–dominance of an alternative the more attractive it is as the potential choice. In this section we consider alternatives having the highest possible non–dominance degree equal to 1. Such alternatives are not dominated to any positive degree by any other alternatives, and for this reason we refer to them as unfuzzy non–dominated alternatives or UND-alternatives. Clearly, UND-alternatives exist not in every set (A, R), and we give a number of existence conditions later on.

Given a set A and a FPR R defined in it, the set of UND-alternatives is described as follows (see Orlovski (1978) and Banerjee (1993)):

$$A_R^{und} = \{a \in A \mid m_R^{nd}(a) = 1\}. \tag{2}$$

Based on the definition of the function $m_R^{nd}(a)$, we can describe this set also in the form:

$$A_R^{und} = \{a \in A \mid \alpha(R)(b, a) = 0, \forall b \in A\} \tag{3}$$

Existence theorems. UND-alternatives have rather practically attractive properties that make these alternatives desired choices in any practical problem. But depending on the structure of a FPR in a concrete problem, such alternatives may or may not exist.

The form of the set of UND–alternatives depends on the form of expression of the fuzzy strict preference relation in terms of the membership function of the original FPR. Here we formulate existence conditions for UND-alternatives based on the fuzzy strict preference obtained using the Lukasiewicz t-norm. Several types of sufficient conditions are presented. In some of them a given FPR is not assumed to be transitive; however, the membership function of that FPR and the set of alternatives must satisfy certain algebraic and/or topological requirements. Other types of conditions are formulated only for a finite set of alternatives with a transitive FPR defined in it. Let us recall the following results proved in Orlovski (1978).

Theorem 1 *If a set A is convex and bicompact, a function $R(a, b)$ is continuous on the cartesian product $A \times A$, quasiconcave in a for any fixed $b \in A$, and quasiconvex in b for any fixed $a \in A$, then the set A_R^{und} is not empty.*

Theorem 2 *In a finite set A with a transitive FPR R defined in this set, the set A_R^{und} is not empty.*

Notice that if the conditions on the set A and the FPR R required in Theorems 1 and 2 are not satisfied, there may be no UND-alternatives in the set (A, R). The following sections will present various methods designed to overcome this difficulty.

Fuzzy Choice Rules and Choice Functions. As we have outlined above, choice functions represent an alternative framework for modeling problems of choice. For a basic set of alternatives A, a choice function C gives for any initial set $X \subseteq A$ a subset $C(X) \subseteq X$ of choices that corresponds to a rationality principle determined by this function.

Descriptions of the same problem in terms of preference relations and in terms of choice functions are, of course, related to each other, and the correspondence between them is called the choice rule.

One of the most studied choice rules is the core that for any given preference relation prescribes the choice of non-dominated alternatives in any given initial subset of alternatives. Under certain natural conditions the core possesses nice properties which, to a certain extent, amount to an equivalence between descriptions of a given problem using relations and using choice functions.

Bondareva (1990a, 1990b) was the first to discover that these properties of the core are valid also in the case of appropriately defined fuzzy relations and fuzzy choice functions.

Here we shall outline some of the results obtained in this rather promising, but not yet fully explored, direction of the analysis.

Following Bondareva (1990a), we shall consider problems of choice with ordinary initial subsets of alternatives of a basic set A. In this case a fuzzy choice function is defined as a function $C : 2^A \to F^A$ where F^A denotes a class of fuzzy sets in the set A. Under this function the image $C(X)$ of any subset of alternatives $X \subseteq A$ is a fuzzy set prescribed by the function C as the solution to the problem of choice.

Given a basic set of alternatives and a class of fuzzy strict (irreflexive) preference relations $\Re$ in A, a choice rule is defined as a function $F : \Re \times 2^A \to F^A$. Given a fuzzy relation P and a subset $X \subseteq A$, the image $F(P, X)$ is a fuzzy set of choices prescribed by this rule for P and X. Clearly, for any fixed relation P the function $F(P, X)$, considered as a function of X, is a fuzzy choice function.

Let us consider the choice rule that for any given fuzzy relation $P = \alpha(R)$ and a set X gives a fuzzy set of non–dominated alternatives. Following Bondareva (1990a), we shall refer to this rule as the fuzzy core and denote it by F_C. Using the notations introduced earlier in this section, we have that for any $\alpha(R)$ and X the fuzzy core satisfies the following relationship:

$$F_C(\alpha(R), X) = X_R^{nd},$$

where X_R^{nd} is the fuzzy set of non-dominated alternatives corresponding to the strict preference relation $\alpha(R)$ in the set (X, R).

To demonstrate the main result from Bondareva (1990a) we need to extend the properties 1-(Heritage) and 2-(Condorcet) to the case of fuzzy choice functions. For doing that let us introduce the following notations: if C is a fuzzy choice function then by $\mu_{C(X)}(a)$ we shall understand the value of the membership function of the fuzzy set $C(X)$ for an alternative $a \in X$. Similarly, by $F_C(\alpha(R), X)(a)$ we denote the value of the membership function of the fuzzy set $F_C(\alpha(R), X)$ where F_C is the fuzzy core.

The fuzzy counterparts of the Heritage, Condorcet, Stability, Independence and Concordance conditions are the following (for similar conditions, see Ovchinnikov and Ozernoi (1988) and Roubens (1989)):

Let X and Y be two crips subsets of A,

1-(Heritage):

$$X \subseteq Y \implies \forall a \in X,\ \mu_{C(Y)}(a) \leq \mu_{C(X)}(a)$$

2-(Condorcet):

$$\forall a \in Y,\ \inf_{b \in Y} \mu_{C(\{a,b\})}(a) \leq \mu_{C(Y)}(a)$$

3-(Stability):

$$(\forall a \in X, \mu_{C(Y)}(a) = 1) \implies (\forall a \in X, \mu_{C(X)}(a) = 1)$$

4-(Independence on irrelevant alternatives):

$$[X \subseteq Y \text{ and } (\forall a \in Y \setminus X, (\mu_{C(Y)}(a) = 0)] \implies [\forall a \in X, (\mu_{C(X)}(a) = \mu_{C(Y)}(a)]$$

5-(Concordance):

$$\forall a \in A,\ \min(\mu_{C(X)}(a), \mu_{C(Y)}(a)) \leq \mu_{C(X \cup Y)}(a)$$

The first property of the fuzzy core is that for any fixed strict fuzzy preference relation P the corresponding choice function $C(X) = F_C(P, X)$, $X \subseteq A$ satisfies the above conditions 1 and 2.

The second even more important property of the fuzzy core established in Bondareva (1990a) is that for any fuzzy choice function C on 2^A satisfying the above conditions 1 and 2 there exists a fuzzy strict preference relation P in A such that $F_C(P,\ X) = C(X)$ for any $X \subseteq A$, *i.e.* the core applied to the fuzzy relation P gives the fuzzy choice function from which the relation P was defined.

In this section we emphasized the particular interest of selecting the set of undominated alternatives from a fuzzy preference relation. As we will see in the following, this concept introduced by Orlovski (1978) has been used in various other works to define more sophisticated choice procedures. Let us remark that this approach is based on the definition of a non-domination score denote $m_R^{nd}(a)$ for any $a \in A$. In that sense, the core can be seen as a scoring method. However, because of its own importance it has been presented in a separate section. Let us introduce now more general choice methods based on the definition of scores from fuzzy preference relations.

3.2 CHOICE METHODS BASED ON SCORING FUNCTIONS

There are many ways of defining a scoring function from a fuzzy preference relation R. A general definition may be the following (see Perny (1992a), Fodor and Roubens (1994)):

Definition 2 *Let $A = \{a_1, a_2, \ldots, a_m\}$ be a set of alternatives and R be a fuzzy preference relation on A. A function f is said to be a scoring function on A for the relation R if it is a real-valued function defined on $[0,1]^{2m}$, non-decreasing of its m first arguments, non-increasing of its m last arguments and such that:*

$$\forall i \in \{1, \ldots, m\}$$
$$s(a_i, A, R) = f(R(a_i, a_1), \ldots, R(a_i, a_m), R(a_1, a_i), \ldots, R(a_m, a_i)) \tag{4}$$

where $s(a_i, A, R)$ is the score of alternative a_i in A according to relation R.

Thus, a scoring function is a mathematical tool allowing the relative strength of alternatives to be measured within the set A. Function f evaluates the extent to which each alternative dominates all the other elements in A and/or is dominated by none element belonging to A. An alternative a is declared "at least as good as" an alternative b within the set A if and only if $s(a, A, R) \geq s(b, A, R)$. As an illustration, let us mention various score functions which have been studied in the literature.

- *The leaving flow:* $s_1(a, A, R) = \sum_{c \in A} R(a, c)$
  has been considered in Roy (1978), Brans and Vincke (1985), Bouyssou and Perny (1992).
- *The (complemented) entering flow:* $s_2(a, A, R) = \sum_{c \in A}(1 - R(c, a))$
  has been considered in Roy (1978), Brans and Vincke (1985), Bouyssou and Perny (1992).
- *The net flow:* $s_3(a, A, R) = 1/ \mid A \mid \sum_{c \in A}[R(a, c) - R(c, a)]$
  has been considered in Roy (1978), Brans and Vincke (1985), Bouyssou (1992a, 1992b).
- *The min leaving flow:* $s_4(a, A, R) = \min_{c \in A} R(a, c)$
  has been considered in Bouyssou (1991), Pirlot (1994).
- *The (complemented) max entering flow:* $s_5(a, A, R) = 1 - \max_{c \in A} R(c, a)$
  has been considered in Orlovski (1978), Bouyssou (1991), Pirlot (1994).

Note that several extensions of these scoring functions may be considered as well. Some of them are investigated in Barett *et al.* (1990), and also in Fodor and Roubens (1994) where two families of functions based on t-norms and t-conorms are defined.

Notice also that a fuzzy choice function C_s may be associated with any scoring function s using the following equation:

$$\forall X \subset A, \ \mu_{C_s(X)}(a) = s(a, X, R)$$

Any α-cut of the fuzzy set $C_s(X)$ defines an ordinary choice function C_s. Among the various choice functions that could be considered for a given score s, the function providing minimal non-trivial choices is of particular interest because of its highly selective nature. This function is defined by:

$$\forall X \subset A, \ MC_s(X) = \{a \in X / s(a, X, R) = \max_{x \in X} s(x, X, R)\}$$

A particular example of such a choice function is the selection of the *maximum non-dominated* set, as shown by the following equation:

$$\forall X \subset A; \; X^{nd} = MC_{m_R^{nd}}(X) \tag{5}$$

3.3 CHOICE FUNCTIONS BASED ON COVERING RELATIONS

Covering relations, first defined by Riguet (1951), have been introduced to define nontrivial choice functions from intransitive preference relations in social choice theory, see for example Luce (1958), Sen (1971), Miller (1980) and Bordes (1983). Considering a crisp preference relation R on the set of alternatives A and the associated directed graph (A, V_R), usually three types of covering relations are defined. Informally speaking, in *forward covering* relations only the forward arcs (leaving the vertices) of (A, V_R) are taken into account, while *backward covering* relations use backward arcs (entering the vertices) of (A, V_R). Finally, a *comprehensive covering* relation is defined on the basis of both the leaving and entering arcs. One can formalize these ideas as follows:

Definition 3 *Let R be a crisp binary relation on the set A of alternatives. Then the* forward covering relation *R^r associated with R (called also the* right trace *of R) is defined on A as follows $(a, b \in A)$:*

$$aR^r b \quad \Leftrightarrow \quad bRc \text{ implies } aRc \text{ for all } c \in A. \tag{6}$$

The backward covering relation *R^ℓ associated with R (called also the* left trace *of R) is defined on A by*

$$aR^\ell b \quad \Leftrightarrow \quad cRa \text{ implies } cRb \text{ for all } c \in A, \tag{7}$$

$a, b \in A$. Finally, the comprehensive covering relation *R^t associated with R (the* trace *of R in other words) is defined as the intersection of R^r and R^ℓ.*

Remark: Notice the different terminology for the above defined three relations. Riguet (1951) used first the term 'covering relation', while Luce (1956) employed the name 'trace' first (for other interesting facts about traces, see the paper by Doignon *et al.* (1986)). In any case, we use the term 'covering relation' in this chapter.

It can be proved that forward, backward and comprehensive covering relations associated with an arbitrary crisp binary relation R are reflexive and transitive (i.e., covering relations are *partial preorders*). These are attractive properties of covering relations and are useful in choice theory. In addition, the following results are easy to prove:

$$\begin{aligned} R \quad &\text{is reflexive} \quad \Leftrightarrow R^\ell \subseteq R \Leftrightarrow R^r \subseteq R \\ R \quad &\text{is transitive} \quad \Leftrightarrow R^\ell \supseteq R \Leftrightarrow R^r \supseteq R. \end{aligned}$$

On the basis of the previous classical definitions and results, we need to extend the logical operations AND and IMPLIES to the fuzzy case, in order to define fuzzy covering relations associated with a fuzzy binary relation in a proper way. At the same time, we will also extend the reflexivity and transitivity properties.

Fuzzy covering relations. The extension of Definition 3 to the fuzzy case was proposed first by Doignon *et al.* (1986). Although R was a fuzzy binary relation, they defined R^r, R^ℓ and R^t still as crisp relations, as follows:

$$\begin{aligned} aR^\ell b &\Leftrightarrow R(c,a) \leq R(c,b)\ \forall c \in A, \\ aR^r b &\Leftrightarrow R(a,c) \geq R(b,c)\ \forall c \in A, \\ aR^t b &\Leftrightarrow aR^\ell b \text{ and } aR^r b. \end{aligned}$$

The next step was made by Fodor (1992), who defined covering relations associated with a fuzzy binary relation as fuzzy binary relations. To recall that approach, we need a brief review of triangular norms and residuated implications used in the extension to model AND and IMPLIES.

A *triangular norm* T (t-norm for short) is a symmetric, associative and nondecreasing function from $[0,1]^2$ to $[0,1]$ satisfying the boundary condition $T(1,x) = x$ for all $x \in [0,1]$. If T is a t-norm then the *R-implication* associated with T is defined by

$$I_T(x,y) = \sup\{z \mid T(x,z) \leq y\},$$

where R refers to the concept of residuation. For more details on t-norms and implications, see Chapter 2 in Volume 1 of this Handbook, and further references there.

In what follows we consider only t-norms that satisfy the following condition for all $x, y, z \in [0,1]$:

$$T(x,z) \leq y \quad \text{if and only if} \quad I_T(x,y) \geq z. \tag{8}$$

Note that condition (8) is equivalent to the left-continuity of T.
Now we list some t-norms with the corresponding R-implications which are used as illustrative examples in the paper.

(T1) If $T(x,y) = \min(x,y)$ then $I_T(x,y) = \begin{cases} 1 & \text{if } x \leq y \\ y & \text{if } x > y \end{cases}$.

This is the greatest t-norm and the related implication is the weakest one that can be defined by t-norms.

(T2) If $T(x,y) = xy$ then $I_T(x,y) = \min(1, y/x)$. T-transitivity with respect to this t-norm was investigated by Ovchinnikov (1984).

(T3) If $T(x,y) = \max(x+y-1, 0)$ then $I_T(x,y) = \min(1-x+y, 1)$. This t-norm is called the Lukasiewicz t-norm. It has zero divisors (that is, there exist $x, y > 0$ such that $T(x,y) = 0$).

(T4) If $T(x,y) = \begin{cases} \min(x,y) & \text{if } x+y > 1 \\ 0 & \text{otherwise} \end{cases}$

then $I_T(x,y) = \begin{cases} 1 & \text{if } x \leq y \\ \max(1-x, y) & \text{otherwise} \end{cases}$

This t-norm was found by Perny (1992a) and used in Perny and Roy (1992) in order to get ordinal definitions of relational systems of fuzzy preferences. It has been rediscovered during a study of implications by Fodor (1995) where it is called the nilpotent minimum. Although this T is only left-continuous, its associated R-implication has several nice properties (for more details see Fodor (1995)).

(T5) If $T(x,y) = \begin{cases} \min(x,y) & \text{if } \max(x,y) = 1 \\ 0 & \text{otherwise} \end{cases}$ then $I_T(x,y) = \begin{cases} 1 & \text{if } x < 1 \\ y & \text{if } x = 1 \end{cases}$
This is the weakest t-norm. It fails to satisfy condition (8).

Suppose that A is a given set of alternatives and R is a valued binary relation on A. That is, R is a function from A^2 to $[0,1]$. Let T be a t-norm satisfying (8). If R_1 and R_2 are valued binary relations on A, then their T-composition is defined by

$$(R_1 \circ_T R_2)(a,b) = \sup_{c \in A} T(R_1(a,c), R_2(c,b)). \tag{9}$$

The following properties are easily justified for any valued binary relations R_1, R_2, R_3, R_4 on A :

$$R_1 \circ_T (R_2 \circ_T R_3) = (R_1 \circ_T R_2) \circ_T R_3,$$

$$R_1 \subseteq R_3,\ \ R_2 \subseteq R_4 \ \text{ implies } \ R_1 \circ_T R_2 \subseteq R_3 \circ_T R_4,$$

where $R_i \subseteq R_j$ if and only if $R_i(a,b) \leq R_j(a,b)$ for all $a, b \in A$.

Definition 4 *A valued binary relation R on A is said to be T-transitive on A if*

$$R \circ_T R \subseteq R. \tag{10}$$

T-transitive valued relations are of special interest in the field of multicriteria decision aid when outranking methods are considered (Vincke (1992), Roy and Bouyssou (1993), Fodor and Roubens (1994), Perny (1995).

Note that, by (9), R is T-transitive on A if and only if the following condition holds:

$$\forall a, b, c \in A, \quad T(R(a,c), R(c,b)) \leq R(a,b) \tag{11}$$

By condition (8), T-transitivity of R is also equivalent to either

$$\forall a, b, c \in A, \quad R(c,b) \leq I_T(R(a,c), R(a,b)), \tag{12}$$

$$\textit{or} \quad \forall a, b, c, \in A, \quad R(a,c) \leq I_T(R(c,b), R(a,b)), \tag{13}$$

Definition 5 *Let R be any fuzzy binary relation on A. The* backward covering relation *associated with R (called also the* left trace *of R) is defined by:*

$$R^{\ell}(a,b) = \inf_{c \in A} I_T(R(c,a), R(c,b)). \tag{14}$$

The *forward covering relation* associated with R (called also the *right trace* of R) is given by:

$$R^r(a,b) = \inf_{c \in A} I_T(R(b,c), R(a,c)). \tag{15}$$

The *comprehensive covering relation* associated with R is given by:

$$R^t(a,b) = T(R^\ell(a,b), R^r(a,b)). \tag{16}$$

Obviously, this definition provides fuzzy extensions of the corresponding crisp relations introduced in Definition 3. Notice however that, the domain A being discrete and finite, the universal quantifier may be represented by any t–norm T and a more general extension of crisp covering relations could be the following:

$$\begin{array}{l} R^\ell_T(a,b) = T_{c \in A} I_T(R(c,a), R(c,b)) \\ R^r_T(a,b) = T_{c \in A} I_T(R(b,c), R(a,c)) \\ R^t_T(a,b) = T(R^\ell(a,b), R^r(a,b)) \end{array} \tag{17}$$

However, the following result has been obtained by Perny (1995).

Theorem 3 *The min t-norm is the only continuous t-norm making relation R^t_T min–transitive whatever R is. The same result hold for relations R^ℓ_T and R^r_T.*

Keeping in mind that $R^\ell_{min} = R^\ell$ and $R^r_{min} = R^r$, theorem 3 shows that the generalized definition does not really offers new possibilities when min–transitive relations are required. It may be interesting for a weaker version of transitivity. Let us remark also that R^ℓ is the greatest valued binary relation U on A such that $R \circ_T U \subseteq R$, while R^r is the greatest relation V such that $V \circ_T R \subseteq R$.

Main properties of traces are summarized in the next statement. For more details and proofs see Fodor (1992), Fodor and Roubens (1994), Fodor and Roubens (1995).

Theorem 4 *Let R be any valued binary relation on A. Then the following statements hold:*

(a) $R \circ_T R^\ell = R^r \circ_T R = R$

(b) *R^r and R^ℓ are reflexive and T-transitive relations (T-preorders in other words)*

(c) *R is reflexive* $\Leftrightarrow R^\ell \subseteq R \Leftrightarrow R^r \subseteq R$

(d) *R is T-transitive* $\Leftrightarrow R \subseteq R^\ell \Leftrightarrow R \subseteq R^r$

(e) *R is reflexive and T-transitive if and only if* $R = R^\ell = R^r$.

(f) $R(a,b) = \inf_{c \in A} I_T(R^\ell(b,c), R(a,c)) = \inf_{c \in A} I_T(R^r(c,a), R(c,b))$.

We recall the general representation theorem of T-transitive fuzzy binary relations established by Fodor and Roubens (1995). It is partially based on the formula in Theorem 4 (f).

Theorem 5 *A valued binary relation R on A is T-transitive on A if and only if there exist two families $\{f_\gamma\}_{\gamma \in \Gamma}$, $\{g_\gamma\}_{\gamma \in \Gamma}$ of functions from A to $[0,1]$ such that $f_\gamma \geq g_\gamma$, $\forall \gamma \in \Gamma$ and*

$$R(a,b) = \inf_{\gamma \in \Gamma} I_T(f_\gamma(a), g_\gamma(b)),$$

for all $a, b \in A$.

Another form of the representation is given as follows.

Theorem 6 *A valued binary relation R on A is T-transitive on A if and only if there exist two families $\{h_\gamma\}_{\gamma \in \Gamma}$, $\{k_\gamma\}_{\gamma \in \Gamma}$ of functions from A to $[0,1]$ such that $h_\gamma \geq k_\gamma$, $\forall \gamma \in \Gamma$ and*

$$R(a,b) = \inf_{\gamma \in \Gamma} I_T(h_\gamma(b), k_\gamma(a)),$$

for all $a, b \in A$.

To illustrate these results, we use $T = \min$. In this case, a fuzzy binary relation R is (min) transitive if and only if there exist two families $\{f_\gamma\}_{\gamma \in \Gamma}$, $\{g_\gamma\}_{\gamma \in \Gamma}$ of functions from A to $[0,1]$ such that $f_\gamma \geq g_\gamma$, $\forall \gamma \in \Gamma$ and

$$R(a,b) = \begin{cases} \inf_{\gamma \in \Gamma} g_\gamma(b) & \text{if there is } \gamma \text{ such that } f_\gamma(a) > g_\gamma(b) \\ 1 & \text{otherwise} \end{cases}.$$

An immediate consequence of Theorem 6 is the result of Valverde (1985) concerning reflexive and T-transitive relations (T-preorders in other words).

Corollary 1 *A valued binary relation R on A is reflexive and T-transitive on A if and only if there exists a family $\{f_\gamma\}_{\gamma \in \Gamma}$ of functions from A to $[0,1]$ such that*

$$R(a,b) = \inf_{\gamma \in \Gamma} I_T(f_\gamma(b), f_\gamma(a)),$$

for all $a, b \in A$.

Using fuzzy covering relations R^ℓ and R^r of R, nontrivial T-transitive relations contained in R can be obtained as follows. Suppose that T is a left-continuous t-norm.

Theorem 7 *For any valued binary relation R on A, both relations*

$$R \wedge R^\ell \text{ and } R \wedge R^r$$

are T-transitive and contained in R.

Note that if R is reflexive then $R \subseteq R \circ_T R$, thus it is T-transitive if and only if $R \circ_T R = R$. If R is reflexive, fuzzy covering relations R^r and R^ℓ are thus contained in R. However these two partial T-preorders are not maximal in the sense that there might be some T-transitive relation Q different from R^r (or R^ℓ) such that R^r (or R^ℓ) $\subseteq Q \subseteq R$.

Fodor and Roubens (1995) suggested a procedure to obtain a maximal T-transitive relation $\check{R}$ contained in a given fuzzy binary relation R. The rather sophisticated construction can be found in the above paper, together with some other important results concerning T-transitive fuzzy binary relations.

Interest of covering relations. Now, the interest of covering relations in choice problems is obvious. Thanks to theorems 3 and 7, we know that choosing the min t–norm leads to *min*–transitive fuzzy covering relations. Therefore, uncovered elements do exist and the following choice functions are well defined.

$$\begin{array}{rcll} UC^\ell(X) & = & X^{und}_{R^\ell} & \text{(backward uncovered elements)} \\ UC^r(X) & = & X^{und}_{R^r} & \text{(forward uncovered elements)} \\ UC^t(X) & = & X^{und}_{R^t} & \text{(uncovered elements)} \end{array} \tag{18}$$

Let us remark that, for any pair of t-norms T and T' such that $\forall x, y \in [0,1]$, $T(x,y) \leq T'(x,y)$, we have $R^r_{T'} \subseteq R^r_T$ and $R^l_{T'} \subseteq R^l_T$ where R^r_T and R^l_T (resp. $R^r_{T'}$ and $R^l_{T'}$) are the right and left traces of relation R for T (resp. T'). Therefore, choosing a very small t-norm (*e.g.* the drastic product, $T = T_5$) leads to very rich covering relations but their transitivity property is so weak that they are useless. On the contrary, choosing $T(x,y) = \min(x,y)$, we get a strong transitivity property (*i.e. the min–transitivity*) but the associated covering relations usually are poor and not very discriminating. Nevertheless, choice functions UC^ℓ, UC^r and UC^t can be used as a useful pre-filter at an early stage of the selection process. Moreover, the choice of a smaller t-norm (*e.g.* $T = T_3, T = T_4, T = T_5$) may provide original extensions of the fuzzy covering concept. But these extensions should be used with the maximal non-dominated choice function (and not with the core) to cope with the intransitivity problem.

3.4 CHOICE FUNCTIONS BASED ON TRANSITIVE CLOSURES

Transitive closure operations are an alternative tool to derive nontrivial choices from intransitive preference relations. They have been used in the same way as covering relations in social choice theory, (see for example Fishburn (1977)). Their specific interest is discussed in this section.

Definition 6 *Let R be a binary relation defined on a set A. Let $R^i = R^{i-1} \circ R$ be the ith power of R with $(R \circ R)(a,b) = \sup_{c \in A} \min(R(a,c), R(c,b))$ and $R^1 = R$. Then the transitive closure of R is the relation $\hat{R}$ characterized by the following equality:*

$$\hat{R} = \bigcup_{i=1}^{\infty} R^i \tag{19}$$

with $(R \bigcup R')(a,b) = \max(R(a,b), R'(a,b))$.

Theorem 8 *For any reflexive relation R defined on A with $\mid A \mid = m$, $\hat{R} = R^m$.*

Because reflexivity loops are immaterial in a preference graph, preference relations can always be completed so as to be reflexive. Then, a polynomial algorithm ($O(m^3)$) for the calculus of the transitive closure of a fuzzy preference relation is available. This is the following extension of the classical Floyd-Warshall algorithm (Floyd (1962) and Warshall (1962)) proposed for solving shortest path problems:

MINCLOSURE(R)

for $i \leftarrow 1$ to m
    do for $j \leftarrow 1$ to m
        do if $i = j$ then $R_{ij}^{(0)} \leftarrow 1$
            else $R_{ij}^{(0)} \leftarrow R_{ij}$
for $k \leftarrow 1$ to m
    do for $i \leftarrow 1$ to m
        do for $j \leftarrow 1$ to m
            do $R_{ij}^{(k)} \leftarrow \max(R_{ij}^{(k-1)}, \min(R_{ik}^{(k-1)}, R_{kj}^{(k-1)}))$

At the end of this procedure, $R^{(m)}$ is the transitive closure of R.

By construction, the $\hat{R}$ relation is shown to be the smallest *min*-transitive fuzzy relation containing R. Because of its transitivity, $\hat{R}$ can be used to define the following choice function

$$\begin{aligned} \forall X \subset A,\ C_{\hat{R}}(X) &= X_{\hat{R}}^{und} &= \{x \in X \mid m_{\hat{R}}^{nd}(x) = 1\} \\ & &= \{x \in X \mid \forall y \in X, \alpha(\hat{R})(y,x) = 0\} \end{aligned} \tag{20}$$

Usually, $\alpha(R)(a,b)$ is defined as a non-decreasing function of $R(a,b)$, non-increasing of $R(b,a)$ and such that :

$$\alpha(R)(a,b) > 0 \quad \Leftrightarrow \quad R(a,b) > R(b,a) \tag{21}$$

Some examples of such $\alpha(R)$ are :

- $\alpha_1(R)(a,b) = \max\{R(a,b) - R(b,a), 0\}$ (see Orlovski (1978); Ovchinnikov and Roubens (1991)).
- $\alpha_2(R)(a,b) = R(a,b)$ if $R(a,b) > R(b,a)$; $\alpha_2(R)(a,b) = 0$ otherwise (see Ovchinnikov (1981)).

- $\alpha_3(R)(a,b) = \min\{R(a,b), 1-R(a,b)\}$ if $R(a,b) > R(b,a)$ and $\alpha_3(R)(a,b) = 0$ otherwise (see Perny and Roy (1992)).

An interesting reason for considering these examples is the following:

Theorem 9 *Let A be a set of alternatives and R a min-transitive fuzzy binary relation defined on R. If*

$$\alpha(R)(a,b) = f(R(a,b), 1 - R(b,a))$$

where f is a non-decreasing function: $[0,1]^2 \rightarrow [0,1]$ in both arguments, is asymetric, then $\alpha(R)$ is min-*transitive.*

From this result obtained by Ovchinnikov and Roubens (1991), we immediately see that $\alpha_j(R)$ is transitive, $j = 1,2,3$. Considering equations (20) and (21), we get:

$$\forall X \subset A,\ C_{\hat{R}}(X) = \{x \in X \mid \forall y \in X, \hat{R}(x,y) \geq \hat{R}(y,x)\} \tag{22}$$

Thanks to theorem 2 we know this choice function to be well defined since $\hat{R}$ is min-transitive (by definition). Notice that, in real problems, R is not necessarily min-transitive and $C_R(X) = X_R^{und}$ may be ill-defined. This shows the specific interest of the transitive closure operation on R. However, the min-transitivity of R is not a necessary condition for $C_R(X)$ to be non-empty (for more details see Orlovski (1978), Montero and Tejada (1988)). Actually, the problem of defining non-trivial selections is more related with the possible existence of oriented cycles in the graph of $\alpha(R)$ than in the graph of R. Thus, another sufficient (but not necessary) condition for $C_R(X)$ to be non-empty is the min-transitivity of $\alpha(R)$. When $\alpha \in \{\alpha_1, \alpha_2, \alpha_3\}$ the min-transitivity of $\alpha(R)$ is a weaker condition than the min-transitivity of R as shown by theorem 9. This is the reason why the following choice function (also based on the transitive closure idea) has been proposed by Perny (1992a).

$$\begin{array}{rcll} \forall X \subset A,\ C_R(X) & = & X_{\tilde{R}}^{und} & = \{x \in X \mid m_{\tilde{R}}^{nd}(x) = 1\} \quad with\ \tilde{R} = \widehat{\alpha(R)} \\ Hence: & & C_R(X) & = \{x \in X \mid \forall y \in X, \alpha(\widehat{\alpha(R)})(y,x) = 0\} \end{array} \tag{23}$$

This choice function is well defined because $\widehat{\alpha(R)}$ is min-transitive. The $\alpha(\widehat{\alpha(R)})$ relation make explicit the acyclic part of strict preferences, within relation R. Thus, when R is asymetric and min-transitive (a fuzzy partial order), we have $\alpha(\widehat{\alpha(R)}) = R$. In the general case, relation $\alpha(\widehat{\alpha(R)})$ is a fuzzy partial order that can also be used in ranking problems (see § 3.5.1). The main properties of procedures based on the use of relation $\alpha(\widehat{\alpha(R)})$, are studied in Perny (1992a).

Transitive closure operations and covering relations are two different ways of answering to the problem of the possible vacuity of the set of undominated alternatives. Both of them are used to build a transitive relation from the non-transitive relation R. Such techniques seem natural since transitive relations of type $\hat{R}$, $\alpha(\widehat{\alpha(R)})$, R^r, R^ℓ and R^t keep a clear relationship with the information contained in the initial relation R.

3.5 CHOICE FUNCTIONS BASED ON KERNELS

Considering kernels is a third approach to the (fuzzy) choice problem. The basic idea when considering kernels is that the quality or relevance of the selection is not only thought as the sum of the individual qualities of the selected elements, but also as a collective property to be verified by the entire selected set. For a given alternative, belonging to an undominated set or reaching the maximal score is an individual quality evaluated by comparing the selected element with all the other elements in A. When defining the kernel of a preference relation, the new idea is to specify a collective consistency condition requiring complementarity among the selected alternatives. This idea is well fitted to decision situations where the problem is not only to indentify the best elements but also to determine a complementary coalition of alternatives. The best coalition is not always the one gathering the best elements.

In order to introduce the notion of kernel, we first recall some basic rationality concepts defined in the crisp case and then, we extend them to the fuzzy situation.

Let us consider an arbitrary crisp preference relation R with no a priori restriction (as transitivity, acyclicity, completeness,...). Consider Y as a subset of A and let us define the complementary set of Y, $\bar{Y} = A \setminus Y$, the successors of Y as

$$Y \circ R = \{a \in A : \exists b \in Y \text{ such that } bRa\}$$

and the composition $R^2 = R \circ R$; aR^2b iff $\exists c \in A$ such that aRc and cRb:

$Y \subseteq A$ is rational according to the three following rules iff ($Y \neq \emptyset$).

1. Y is externally undominated (GOCHA rule, see Schwartz 1986).
2. Y is internally undominated (internal stability, see von Neumann and Morgenstern (1953), Berge (1958)).
3. Y is externally stable (von Neumann and Morgenstern (1953), Berge (1958)).

We say that rule Δ_i ($i = 1, 2, 3$) is satisfied for the crisp binary relation R over Y if $\Delta_i(R)(Y) = 1$ with

$$\begin{array}{llll} \Delta_1(R)(Y) = 1 & \text{iff} & \forall a \in Y, \forall b \in \bar{Y} : \text{NOT}(bRa), & (\bar{Y} \circ R \subseteq \bar{Y}) \\ \Delta_2(R)(Y) = 1 & \text{iff} & \forall a \in Y, \forall b \in Y : \text{NOT}(bRa), & (Y \circ R \subseteq \bar{Y}) \\ \Delta_3(R)(Y) = 1 & \text{iff} & \forall b \in \bar{Y}, \exists a \in Y : aRb, & (Y \circ R \supseteq \bar{Y}) \end{array}$$

If we define $\Delta_{ij}(R)(Y) = 1$ iff both rules Δ_i and Δ_j are satisfied, we obtain different rational choice functions (non empty subsets):

- the set of undominated alternatives ND (studied by Aizerman (1985)):

$$\Delta_{12}(R)(ND) = 1 \text{ and } ND \text{ is maximal.}$$

- the kernel K (studied by von Neumann and Morgenstern (1953) in the framework of game theory and by Berge (1958) :

$$\Delta_{23}(R)(K) = 1.$$

- the G-kernel or stable core GK studied by von Neumann and Morgenstern (1947):

$$\Delta_{23}(R)(GK) = 1 \text{ and } \Delta_1(R)(GK) = 1.$$

- the quasi-kernel QK (studied by Berge (1958)) :

$$\Delta_2(R)(QK) = 1 \text{ and } \Delta_3(R \cup R^2)(QK) = 1$$

- the semi-kernel SK (studied by Berge (1973)) :

$$\Delta_2(R)(SK) = 1 \text{ and } \overline{\overline{SK} \circ R} \circ R \subseteq \overline{SK}$$

  (which is equivalent to $SK \circ R^{-1} \subseteq SK \circ R$).

- the subsolution SS (studied by Roth (1976)) :

$$\Delta_2(R)(SS) = 1 \text{ and } \overline{\overline{SS} \circ R} \circ R = \overline{SS}$$

  The subsolution is a semi-kernel which benefits from the complementary property : $\overline{\overline{SS} \circ R} \circ R \supset \overline{SS}$.

- the uncovered set U (studied by Fishburn (1977) and Miller (1980)) :

$$\Delta_{12}(R^r)(U) = 1 \text{ and } U \text{ is maximal}$$

  where R^r corresponds to the foreward covering relation (see sections 1 and 3).

The main question is to determine adequate choice functions that satisfy some rational choice axioms (see Aizerman (1985), Arrow (1963), Fishburn (1970) and (1977), Sen (1970), Schwartz (1986)). It is known that the only choice function that satisfies Heritage and Concordance axioms (see Aizerman and Malishevski, 1981) is the core. However, as mentioned above, the core is very often empty.

Similar problems exist about the kernel choice function. Possible non-existence or non-unicity of the kernel is the main drawback for the internal and external stable choice function. However, these problem vanish when acyclicity is assumed, because in this case the kernel exists and is unique (von Neumann and Morgenstern (1953)).

The kernel as a choice function received a significant place in the classical multiple criteria decision making methods ELECTRE I and II (see Roy (1968), Roy and Bertier (1973)) and Roy proposed to reduce the elementary circuits of the outranking graph in order to obtain acyclicity. Roy's approach was extended in the fuzzy case with ELECTRE IS and the definition of fuzzy outranking relations. The idea was first to cut the fuzzy relation at an appropriate level and then to find the kernel of the crisp cut. The main drawback of this procedure is the arbitrary choice of the cutting level as an early stage of the exploitation process.

Another approach could be to scan all level cuts of the outranking relation, to determine their kernel, and to analyze the robustness of the results. This is however not an easy task since "best" potential alternatives may appear at a first level, vanish at a higher second level and reappear at a third level.

In our opinion, fuzzy extensions of a classical choice function based on a crisp-concept can be obtained in two ways when dealing with fuzzy data:

(i) determine crisp subsets Y which satisfy the basic rules $\Delta_1, \Delta_2, \Delta_3$ to a certain degree.

(ii) define fuzzy rational subsets : every alternative $a \in A$ belongs to the choice function Y to some extent. One has to determine the degree of membership of alternative a to Y : $\mu_Y(a) \in [0,1]$.

If we consider a strong negation function N, we define

$$\begin{aligned}
\Delta_1(R)(Y) &= \inf_{a\in Y} \inf_{b\in \bar{Y}} NR(b,a) \\
\Delta_2(R)(Y) &= \inf_{a\in Y} \inf_{b\in Y} NR(b,a) \\
\Delta_3(R)(Y) &= \inf_{b\in \bar{Y}} \sup_{a\in Y} R(a,b)
\end{aligned}$$

and we determine for each combination (i,j), a degree of credibility

$$\Delta_{ij}(R)(Y) = \min\{\Delta_i(R)(Y), \Delta_j(R)(Y)\}.$$

Let us consider specifically Δ_{23} which is linked to the concept of kernel. Suppose that we are seeking a kernel $K(\alpha)$ with a degree of credibility $\Delta_{23}(R)(K) \geq \alpha$. $K(\alpha)$ must satisfy the following inequalities $(Nx = 1 - x)$:

$$\begin{aligned}
\inf_{a\in K} \inf_{b\in K} (1 - R(a,b)) &\geq \alpha \\
\inf_{b\in \bar{K}} \sup_{a\in K} R(a,b) &\geq \alpha
\end{aligned}$$

If R_α et $R_{>1-\alpha}$ represent respectively the α-cut and the strict $(1-\alpha)$-cut of the valued relation R, the preceding system of inequalities can be rewritten as :

$$\forall a \in K, \forall b \in K \quad : \quad \mathrm{Not}(bR_{>1-\alpha}a)$$

$$\forall b \in \bar{K}, \exists a \in K \quad : \quad aR_\alpha b$$

and $\{K(\alpha)\} = \{$Internally stable set of $G(A, R_{>1-\alpha}) \cap$ Externally stable set of $G(A, R_\alpha)\}$
where $G(A, \Gamma)$ represents a graph with a set of nodes A and an application Γ.

If $\alpha > .5$, it can be easily seen (Kitainik (1993)) that

$$K(\alpha;\, \alpha > .5) \subseteq K(A, R_{>.5}).$$

The kernel $K(\alpha^*)$ with maximal degree of credibility α^* is thus obtained by selecting among the kernels of the crisp graph related to the median cut of the valued relation R (if they exist!) the one with maximum degree $\Delta_{23}(R)(K)$.

Let us finally consider fuzzy kernels Y.The crisp rule Δ_2 can be interpreted in the following way :

$$Y \circ R \subseteq \bar{Y} \quad (a \in Y \circ R \Rightarrow a \in \bar{Y},\ \forall a \in A)$$

which is equivalent to

$$Y \circ R^{-1} \subseteq \bar{Y} \quad (a \in Y \circ R^{-1} \Rightarrow a \in \bar{Y},\ \forall a \in A)$$

when fuziness is introduced, $a \in \bar{Y}$ becomes $\mu_{\bar{Y}}(a)$ and $a \in Y \circ R$ becomes $\mu_{Y \circ R}(a)$. The implication operator is translated into a residual implication I (for example, the Gödel implication) and Δ_2 gives :

$$\begin{aligned} & I(\mu_{Y \circ R}(a), \mu_{\bar{Y}}(a)) = 1, \quad \forall a \in A \\ or \quad & \mu_{(Y \circ R)}(a) \leq \mu_{\bar{Y}}(a), \quad \forall a \in A. \end{aligned}$$

Moreover, $a \in Y \circ R$ means that $\exists b \in A \setminus a$ such that $b \in Y$ and $R(b, a)$ which gives in the valued case :

$$\mu_{Y \circ R}(a) = \sup_{b \in A \setminus \{a\}} \min\{\mu_Y(b), R(b,a)\}.$$

The non-symmetrical behaviour of R gives

$$\Delta_2^R \quad : \quad \sup_{b \in A \setminus \{a\}} \min\{\mu_Y(b), R(b,a)\} \leq \mu_{\bar{Y}}(a), \quad \forall a \in A$$

$$\Delta_2^{R^{-1}} \quad : \quad \sup_{b \in A \setminus \{a\}} \min\{\{\mu_Y(b), R^{-1}(b,a)\} \leq \mu_{\bar{Y}}(a), \quad \forall a \in A$$

The crisp rule Δ_3 is equivalent to

$$Y \circ R \supseteq \bar{Y} \quad (a \in \bar{Y} \Rightarrow a \in Y \circ R,\ \forall a \in A).$$

It is translated in valued terms as follows :

$$\Delta_3^R : \sup_{b \in A \setminus \{a\}} \min\{\mu_Y(b), R(b,a)\} \geq \mu_{\bar{Y}}(a), \quad \forall a \in A.$$

Kitainik (1993) has deeply explored the properties of the system $(\Delta_2^R \wedge \Delta_3^R)$, i.e.

$$\sup_{b \in A \setminus \{a\}} \min\{\mu_Y(a), R(b,a)\} = \mu_{\bar{Y}}(a), \quad \forall a \in A$$

to determine the existence and the structure of fuzzy kernels.
Bisdorff and Roubens (1995) have proposed to find solutions of the system $(\Delta_2^R \wedge \Delta_2^{R^{-1}} \wedge \Delta_3^R)$:

$$\begin{aligned} \sup_{b \in A \setminus \{a\}} \min\{\mu_Y(b), R(b,a)\} &= \mu_{\bar{Y}}(a), \quad \forall a \in A \\ \sup_{b \in A \setminus \{a\}} \min\{\mu_Y(b), R(a,b)\} &\leq \mu_{\bar{Y}}(a), \quad \forall a \in A \end{aligned}$$

with the aid of a constraint logic programming system CHIP supporting a finite domain solver (see Aggoun and Beldiceanu (1991), Van Hentenryck (1989), Dincbas, Van Hentenryck and Simonis (1990)). In order to avoid a great number of solutions, they consider only the maximal fuzzy kernels. μ_K is maximal iff for any other solution $\mu_{K'}$, either $\mu_K(a) \geq \mu_{K'}(a) \geq .5$, either $\mu_K(a) \leq \mu_{K'}(a) \leq .5$, for all alternative a belonging to A.

Finally, the sharpest fuzzy kernel may be defined using the following sharpness index

$$\sigma^2(K) = \sum_{a \in A} \{\mu_K(a) - .5\}^2$$

to be maximized.

In order to compare procedures (i) and (ii), let us consider the set of undominated alternatives ND. In the crisp case, this concept corresponds to $\Delta_{12}(R)(ND) = 1$:

$$\forall a \neq b \in A, \;\; a \in ND \; \Rightarrow \; not(bRa) \tag{24}$$

If one considers procedure (i) in the fuzzy case, ND is such that $\Delta_{12}(R)(Y)$ is maximum.

$$\begin{aligned} \Delta_{12}(R)(Y) &= \min\{\inf_{a \in Y} \inf_{b \in \bar{Y}} NR(b,a), \inf_{a \in Y} \inf_{b \in Y} NR(b,a)\} \\ &= \inf_{a \in Y} \inf_{b \neq a} NR(b,a) \end{aligned}$$

If $x(a)$ corresponds to the score $\inf_{b \neq a} NR(b,a)$,

$$\Delta_{12}(R)(Y) = \inf_{a \in Y} x(a)$$

and ND is obtained as the alternatives of maximal scores:

$$ND = \{a \in Y : \max_{a \in Y} x(a)\}$$

Let us now turn to procedure (ii). Relation (24) can be translated as:

$$I(\mu_{ND}(a), NR(b,a)) = 1, \ \forall b \neq a \in A$$

I being any residual implication, one obtains:

$$\mu_{ND}(a) \leq NR(b,a), \ \forall b \neq a \in A$$

or

$$\mu_{ND}(a) \leq \inf_{b \neq a} NR(b,a) \ = \ x(a) \tag{25}$$

Maximal solutions corresponding to inequalities (25) are:

$$\mu_{ND}(a) = \begin{cases} x(a) & \text{if } x(a) > 0.5 \\ 0 & \text{otherwise} \end{cases}$$

Let us show now how these concepts introduced in the context of choice may be used for other decision problems.

3.6 RANKING AND SORTING FROM FUZZY RELATIONS

3.6.1 Ranking Methods

In some cases, the decision situation cannot be thought only as a problem of selection. For instance, in order to allocate a grant to research projects, determining the very best project is not sufficient. In such a situation, the decision maker must take into account a strong financial constraint limiting the total amount of the grant or the total number of selected projects. He cannot be sure that a choice function will integrate this constraint. But when a ranking of alternatives is available, it is easy to scan the ordered list from the top to the bottom till one reaches the constraint. In this example and in many others, a ranking of projects by decreasing order of interest is more useful.

The problem of ranking alternatives by decreasing order of preference is far from being obvious when several criteria have to be considered. The fuzzy preferences resulting from the aggregation of several possibly conflicting one-dimensional preference relations are generally not transitive (see Leclerc (1984), Perny (1990, 1992a-b) and chapter 4 of this volume). Therefore, any ranking method must implicitly or explicitly twist the natural structure of overall preferences. Although transitivity was not necessary to define a selection of the best alternatives, it becomes the main requirement in ranking problems. This section presents some techniques allowing a partial or total ranking of alternatives from a fuzzy preference relation.

Scoring techniques. Considering orders induced by scores is the easiest way of building a ranking of alternatives from a binary relation. Let s be a scoring function (see equation (4)), the induced complete preorder is the binary relation $\succsim_s$ defined from s as follows:

$$\forall(a,b) \in A \times A, \ (a \succeq_s b \ \Leftrightarrow \ s(a,A,R) \geq s(b,A,R)) \tag{26}$$

A particular instance of this general technique has been implemented in the Promethee II method (see Brans and Vincke (1985)) where s is the net flow (see section 3.2). An axiomatic characterization of this ranking method has been obtained by Bouyssou (1992a).

This technique allows also partial orders to be obtained, by intersecting different complete orders resulting from complementary scores. For instance, let us consider the leaving and entering flows s_1 and s_2 introduced in section 3.2. The associated preorders can be intersected so as to make explicit the well established part of the overall ranking. This solution has been adopted in the Promethee I method (see Brans and Vincke (1985)) and an axiomatic characterization of the procedure has been obtained by Bouyssou and Perny (1992).

The main advantage of scoring procedures is their simplicity. However, the real interest of a scoring function in pairwise comparisons methods may be questionned, especially when relation R results from a non-compensatory ordinal aggregation procedure. As potential drawbacks, let us remark that:

- the incomparabilities in the graph of relation R are not easily representable by scores,
- the scoring functions are often compensatory and their use is still to be justified when R results from a non-compensatory aggregation of criteria.

Iterated Choice Sequences. An interesting question is how the core choice function can be used in order to rank alternatives on the basis of a non-transitive relation R ? Keeping in mind that choice function $F_C(R,X) = X_R^{nd}$ is also a scoring function (see section 3.2) the first answer coming to mind is to rank alternatives according to their non-domination score $m_R^{nd}(a)$, $a \in A$. However, the idea is wrong because non-domination scores only allow maximal alternatives to be distinguished from the rest of the list. For example, the non-domination score is unable to discriminate between the second and the third ranks.

Nevertheless, an interesting ranking can be obtained by involving the core choice function within an iterated choice sequence. An iterated choice sequence is a very natural way of involving a choice function in a ranking process. The idea is to make use of a choice function first to separate heading alternatives from the rest of the list and then to iterate the process on the rest of the list. Formally, an iterated choice sequence (ICS) can be defined by the following algorithm (see Perny, 1992a):

$ICS(R,A)$

$k \leftarrow 0$

$A^{(k)} \leftarrow A$

while $A^{(k)} \neq \emptyset$ do

        begin

$C_{k+1} \leftarrow F(R, A^{(k)})$
$A^{(k+1)} \leftarrow A^{(k)} \setminus C_{k+1}$
$k \leftarrow k+1$
end

where subsets $C_k, k = 1, \ldots, K$ denote equivalence classes of the resulting preorder $\succeq_R$, defined as follows:

$$\forall (a,b) \in A \times A \mid (a \in C_i \textit{ and } b \in C_j),\ i \leq j \ \Leftrightarrow \ a \succeq_R b)$$

$F(R, A^{(k)})$ denotes the crisp subset of $A^{(k)}$ determined by the choice function on the basis of relation R. Relation $\succeq_R$ is a complete ranking of alternatives resulting from a descending sequence of selections from the best to the worst. It is also possible to consider the reverse of the $\succeq_{R^{-1}}$ preorder returned by $ICS(R^{-1}, A)$. This alternative ranking results from an ascending sequence of selections from the worst to the best. The two preorders (descending and ascending) may differ one from the other. Hence, a partial preorder obtained by intersection can be considered as the reliable part of the ranking. This procedure has been implemented in the final stage of the Electre III method (see Roy (1978)), with a choice function based on the net flow. It also appears in the ranking algorithm of MAPPAC (see Matarazzo (1986)).

The ICS algorithm is a very natural idea. An ascending ICS is something very close to a *choice by elimination*. It seems so natural that it has been adopted by many practitioners during the past decades. Unfortunately, such a mechanism often hide a latent defect as shown in (Perny (1992a)). The problem is that iterated choices are easily non-monotonic when applied to a non-transitive relation. Let us show this phenomenon on a small example derived from Perny(1992a).

Let us consider the following fuzzy relation R:

	a	b	c	d
a	–	α	0	0
b	0	–	α	0
c	0	0	–	β
d	0	0	0	–

where α and β are two real numbers of the unit interval such that $\alpha > \beta$.

Using iterations of the core choice function $C_R(X) = X^{nd}$, we get:

$$a \succ_R b \succ_R c \succ_R d$$

Therefore c is preferred to d. Let us suppose now that the position of c is improved in the preference relation R. For instance, let us set now $R(c,a) = (\alpha + \beta)/2$. Applying the same algorithm to the new matrix, we get :

$$d \succ_R a \succ_R b \succ_R c$$

We get a wrong rank reversal since now $d \succ_R c$ instead of keeping unchanged $c \succ_R d$. An impovement of the position of c in the input of the procedure leads to a deterioration of the position of c in the output. This violates the following monotony condition introduced in Perny(1992a) and Bouyssou and Perny(1992).

Definition 7 *A ranking method is said to be monotonic if, for all valued relation R on A and all $a, b \in A$:*

$$a \succeq_R b \Rightarrow a \succeq_{R'} b$$

where R' is identical to R except that $[R(a,c) < R'(a,c)$ or $R(c,a) > R'(c,a))$ for some $c \in A \setminus \{a\}]$ or $[R(b,d) > R'(b,d)$ or $R(d,b) < R'(d,b)$ for some $d \in A \setminus \{b\}]$.

The above example reveals and illustrates an important drawback of many algorithms based on iterated choice sequences. Hence, the use of many ranking procedures based on iterated choices seems questionable, from a theoretical viewpoint, especially because other ranking methods from preference relation are monotonic (*e.g.* direct ranking methods based on a scoring function).

In our opinion, the main interest of the iterated choice principle is mainly pragmatic and algorithms implementing iterated choice sequences should be used as descriptive computational models. They are well fitted to describe and simulate the natural mental decision process of a decision maker facing a large set of alternatives.

Transitive approximations. Searching for a transitive approximation of relation R is a third approach to the ranking problem. The idea is to determine a transitive relation closely related to the intitial information contained in R. This approach is perhaps the more consistent with the idea of pairwise comparisons. Several transitive approximations of R have been considered. Among others, let us mention:

- closest transitive relations obtained by optimisation techniques (Cook *et al.* (1978, 1986); Montero and Tejada (1986)),
- transitive relations obtained by closure operations (see § 3.4 and Perny (1992a)),
- covering relations R^r, R^ℓ and R^t (see Fodor and Roubens (1995), Perny (1995)),
- maximal transitive relations contained in R (see Fodor and Roubens (1995)).

3.6.2 Sorting Methods

A sorting method aims at distinguishing several homogeneous categories of alternatives within A. Usualy, to each category is associated a specific decision. One can distinguish two different families of methods: *clustering methods* and *assignment methods*.

Clustering methods. They are especially fitted to decision problems where categories are not known *a priori*. A clustering method must determine homogeneous groups within the set A to form a partition of A into well distinguished categories. Such methods make use of similarity relations reflecting relative proximity of alternatives. One example of such similarity relations is given by considering the symmetric part of a fuzzy covering relation as proposed in Perny (1992a, 1995) and Fodor and Roubens (1995). The resulting relation contains a nested family of equivalence relations and thus is well suited to perform a hierarchical classification of alternatives.

Another interesting approach is to use a choice function so as to generate ordered categories. This alternative approach is illustrated by the following procedure proposed by Fodor and Roubens (1994) from the definition of cores of a fuzzy relation.

In some problems, it is advisable of considering the set of maximum non-dominated alternatives and the set of maximum non-dominating alternatives. We obtain the following categories:

- $C_1 = A^{nd}(R)$ as defined in section 3.1,
- $C_3 = A^{nd}(R^{-1})$ which corresponds to the set of maximum non-dominating alternatives calculated with the inverse relation R^{-1},
- $C_2 = A \setminus (C_1 \cup C_3)$.

These categories can be ordered according to $C_1 \succ C_2 \succ C_3$, if $C_1 \cap C_3 = \emptyset$. This approach may be extended to any choice function C_R and the use of kernels was proposed in Bisdorff and Roubens (1996).

Assignment methods based on filtering. These methods are used for decision problems where the set of categories is given *a priori* and characterized by a set of fictitious alternatives serving as reference points, each representing a typical element of a category, or each representing a frontier between two categories. In such problems, the membership of an alternative in a category depends on its intrinsic value, unlike clustering problems where categories are defined using relative comparisons of alternatives.

Several assignment methods have been introduced, all based on a multiple-criteria filtering process making use of valued or non-valued preference relations. The case of ordered categories has been first considered in Roy and Moscarola (1973), Roy (1981), Yu (1992) and Massaglia and Ostanello (1991) and then extended for a fuzzy assignment to categories by Slowinski (1993), Slowinski and Stefanowski(1994) and Perny (1997).

The assignement of alternatives to a category C defined by a set of reference points $Y = \{y_1, \ldots, y_q\}$ is performed by a choice function selecting alternatives within an area implicitely delimited by the elements of Y. The basic idea for the selection is filtering alternatives by evaluating their position relatively to the set Y representing the category. More precisely, the membership degree $\mu_C(a)$ of any alternative a in the category C is defined as follows (for more details see Perny (1997)):

- *Filtering by indifference:* $\mu_C(a) = V(I(a, y_1), \ldots, I(a, y_q))$ where $I(a, y_k) \in [0,1], k = 1, \ldots, n$ is the indifference index evaluating how a is similar or indifferent to y_k (see chapter 1), and V is a co-norm. This procedure is useful both for ordered or non-ordered categories.
- *Filtering by preference:* $\mu_C(a) = T(P(a, Y_-), NP(a, Y_+))$ or $\mu_C(a) = T(P(Y_+, a), NP(Y_-, a))$ where Y_- and Y_+ are sets of reference points representing the lower and upper frontier of category C respectively; $P(a, Y)$ denotes the preference of a over the set Y, defined as a conorm on quantities $P(a, y), y \in Y$; $P(Y, a)$ denotes the preference of Y over a, defined as a conorm on quantities $P(y, a), y \in Y$; T is a t-norm and N is a strict negation. This procedure is only convenient for problems with *ordered categories.*

Notice that the above assignment rules allow the definition of complex categories characterized by several reference points. The filtering by indifference procedure can even be implemented when prototypes defining categories are not known. In such cases, the set of reference points consist in a list of alternatives whose correct assignment is known (learning points). For example, a *k-nearest neighbour* algorithm based on fuzzy indifference relations has been proposed in Henriet and Perny (1996), with a large number of options.

When the overall preference or indifference relation is fuzzy, these filtering procedures define partial membership of alternatives in categories. This is well fitted to real decision problems where categories usually cannot be defined with sharp boundaries.

References

Aggoun A. and Beldiceanu N. (1991) Overview of the CHIP compiler system, in *Proc. ICLP 91*, MIT Press.

Aizerman M.A. (1985) New problems in the general choice theory: review of a research trend, *Social Choice and Welfare* **2**, 235–282.

Aizerman M.A. and Malishevski A.V. (1981) General theory of best variants choice: Some Aspects, *IEEE Trans. Automat. Control*, . **26**, 1030–1040.

Arrow K.J. (1963) *Social Choice and Individual Values*, John Wiley and Sons, Inc., New-York.

Banerjee A.(1993) Rational choice under fuzzy preferences: The Orlovsky choice function, *Fuzzy Sets and Systems*, **54**, 295–300.

Barrett C.R., Pattanaik, P.K., Salles, M. (1990), On choosing rationally when Preferences are fuzzy, *Fuzzy Sets and Systems* **2**, 197–212.

Berge Cl. (1958) Théorie des graphes et ses applications, Dunod, Paris.

Bezdek J., Spillman B. and Spillman R. (1978) A fuzzy relation space for group decision theory, *Fuzzy Sets and Systems* **1**, 255–268.

Bisdorff R and Roubens M (1996) On defining kernels on $\mathcal{L}$-fuzzy binary relations, IPMU Conference, Granada, July.

Bondareva O.N. (1990a) *Revealed Fuzzy Preferences*, Universitat Bielefeld/IMW, Institute of Math. Economics. Working paper no. 183.

Bondareva O.N.(1990b) Revealed fuzzy preferences, J. Kacprzyk and M. Fedrizzi, Eds., *Multiperson Decision Making Using Fuzzy Sets and Possibility Theory*, Kluwer Acad. Publ., 71–79.

Bordes G. (1983) On the possibility of reasonable consistent majoritarian choice: some positive results, *Journal of Economic Theory* **31**, 122–132.

Bortolan G. and Degani R. (1985) A review of some methods for ranking fuzzy subsets, *Fuzzy Sets and Systems*, **15**, 1–19, 1985.

Bouchon B. (1987) Preferences deduced from fuzzy questions, J. Kacprzyk and S. A. Orlovski, Eds., *Optimization Models Using Fuzzy Sets and Possibility Theory*, D. Reidel Publ. Co., Dordrecht, Boston, Lancaster, Tokyo, 110–122.

Bouyssou D. (1991) A note on the min in favor ranking method for valued preference relations, in: M. Cerny, D. Gluckaufova and D. Loula (Eds.), *Proc. International Workshop on Multicriteria Decision Making - Methods - Algorithms - Applications*, 16–25, Czechoslovak.

Bouyssou D. (1992a) Ranking methods based on valued preference relations: a characterization of the net flow method, *European Journal of Operationnal Research* **60**, 61–68.

Bouyssou D. (1992b) A note on the sum of differences choice function for fuzzy preference relations, *Fuzzy sets and systems* **47**, 197–202.

Bouyssou D. and Perny P. (1992) Ranking methods for valued preference relations: a characterization of a method based on leaving and entering flows, *European Journal of Operationnal Research* **61**, 186–194.

Brans J.P. and Vincke Ph. (1985) A preference ranking organization method, *Management Science* **31**, 6, 647–656.

Cook W.D. and Seiford L.M. (1978) Priority Ranking and Consensus Formation, *Management Science* **24**, 16, 1721–1732.

Cook W.D., Kress M. and Seiford L.M. (1986) An axiomatic Approach to Distance on Partial Orderings, *RAIRO Recherche Opérationnelle* **20**, 2, 115–122.

Doignon J.P, B. Monjardet, M. Roubens and Ph. Vincke (1986) Biorders families, valued relations and preference modelling, *J. Math. Psych.* **30**, 435–480.

Dincbas M., Van Hentenrynck P. and Simonis H. (1990) Solving large combinatorial problems in logic programming, *Journal of Logic Programming* **8**, 1 & 2.

Fishburn P.C.(1970) *Utility Theory for Decision Making*, John Wiley & Sons, Inc., New York, London, Sydney, Toronto.

Fishburn P.C. (1977) Condorcet Social Choice functions, *SIAM Journal of applied Mathematics* **33**, 469–489.

Floyd R. (1962), Algoritm 97: Shortest Path, *Communications of the ACM* **5**, 6, 345.

Fodor J.C. (1992) Traces of fuzzy binary relations, *Fuzzy Sets and Systems* **50**, 331–341.

Fodor J.C. (1995) Contrapositive symmetry of fuzzy implications, *Fuzzy Sets and Systems* **69**, 141–156.

Fodor J.C. and Roubens M. (1994), *Fuzzy Preference Modelling and Multicriteria Decision Support*, Kluwer Academic Publishers, Dordrecht.

Fodor J.C. and Roubens M. (1995), Structure of transitive valued binary relations, *Math. Soc. Sci.* **30**, 71–94.

Henriet L. and Perny P. (1996), *Méthodes multicritères non-compensatoires pour la classification floue d'objets*, LFA'96, 4–5 December, 9–15.

Kim J.B. (1983) Fuzzy rational choice functions, *Fuzzy Sets and Systems*, **10**, 37–43.

Kitainik L (1993) *Fuzzy Decision Procedures with Binary Relations. Towards a Unified Theory*, Kluwer Academic Publishers, Dordrecht.

Kuzmin V.B. and Ovchinnikov S.V. (1980a) Group decisions I - in arbitrary spaces of fuzzy binary relations, *Fuzzy Sets and Systems*, **4**, 53–62.

Kuzmin V.B. and Ovchinnikov S.V. (1980b) Design of group decisions II - in spaces of partial ordered fuzzy relations, *Fuzzy Sets and Systems*, **4**, 153–165.

Kuzmin V.B. and Travkin S.I. (1987) Fuzzy choice, J. Kacprzyk and S. A. Orlovski, Eds., *Optimization Models Using Fuzzy Sets and Possibility Theory*, D. Reidel Publ. Co., Dordrecht, Boston, Lancaster, Tokyo, 99–109.

Leclerc B. (1984) Efficient and binary consensus functions on transitively valued relations, *Mathematical Social Sciences*, 8, 45–61.

Luce R.D. (1958) A probabilistic theory of utility, *Econometrica* **26**, 193–224.

Massaglia R. and Ostanello A. (1991) N-TOMIC : A decision support for multicriteria decision Problems, in P. Korhonen (ed.):*International Workshop on Multicriteria Decision Support - Lecture notes in Economics and Mathematical Systems* **356**, Springer-Verlag, Berlin, 167–174.

Matarazzo B. (1986) Multicriterion analysis by means of pairwise actions and criterion comparisons (MAPPAC), *Applied Mathematics and Computation* **18**, 2, 119–141.

Matarazzo B. (1988), Preference ranking global frequencies in multicriterion analysis (PRAGMA), *European Journal of Operational Research* **36**, 1, 36–50.

Miller N.R. (1980) A new solution set of tournaments and majority voting : further graph-theoretical approaches to the theory of voting, *American Journal of Political Science* **24**, 68–96.

von Neumann J. and Morgenstern O. (1947) *Theory of Games and Economic Behavior*, J. Wiley, New York.

Montero F.J. and Tejada J. (1986) Some problems on the definition of fuzzy preference relations, *Fuzzy Sets and Systems* **20**, 45–53.

Montero F.J. and Tejada J. (1988) A necessary and sufficient condition for the existence of Orlovski's choice set, *Fuzzy Sets and Systems* **26**, 121–126.

Orlovski S.A. (1978) Decision-making with a fuzzy preference relation, *Fuzzy Sets and Systems*, . **1**, 155–167.

Ovchinnikov S. (1981) Structure of fuzzy binary relations, *Fuzzy Sets and Systems* **1**, 169–195.

Ovchinnikov S. (1984) Representation of transitive fuzzy relations, in: H. J. Skala, S. Termini and E. Trillas, eds., *Aspects of Vagueness*, D. Reidel Publishing Company, Dordrecht, 105–118.

Ovchinnikov S. (1987) Preference and choice in a fuzzy environment, J. Kacprzyk and S. A. Orlovski, Eds., *Optimization Models Using Fuzzy Sets and Possibility Theory*, D. Reidel Publ. Co., Dordrecht, Boston, Lancaster, Tokyo, 91–98.

Ovchinnikov S. (1991) Social Choice and Łukasiewicz logic, *Fuzzy sets and systems* **43**, 319–326.

Ovchinnikov S.V. and Ozernoy V.(1988) Using fuzzy binary relations for identifying noninferior decision alternatives, *Fuzzy Sets and Systems*, **25**, 21–32.

Ovchinnikov S.V. and Roubens M. (1991), On strict preference relations, *Fuzzy Sets and Systems*, **43**, 319–326.

Perny P. (1990) Building and Exploitation of relational Systems of fuzzy preferences, in: *Extended Abstracts of the third IMPU conference*, June 2–6, Paris, France, 431–433.

Perny P. (1992a) Modélisation, Agrégation et exploitation de préférences floues dans une problématique de rangement: bases axiomatiques, procédures et logiciel, PhD thesis, Paris-Dauphine University and LAFORIA document 9.

Perny P. (1992b) Sur le non respect de l'axiome d'independance dans les méthodes de type Electre, in:*Développements récents en Recherche Opérationnelle: Etat de l'Art, Cahiers du CERO* **34**, 211–232.

Perny P. (1995) Defining fuzzy covering relations for decision aid, in: B. Bouchon, R. Yager, L. Zadeh (Eds.), *Fuzzy Logic and soft Computing, Advances in Fuzzy Systems – Applications and Theory*, **4**, 109–218.

Perny P. (1997) Multicriteria Filtering Methods based on Concordance and Non-Discordance principles, *Annals of OR*, Special issue on Preference Modelling, to appear.

Perny P. and.Roy B. (1992) The use of fuzzy Outranking Relations in Preference Modelling, *Fuzzy sets and Systems* **49**, 33–53.

Pirlot M. (1994) A characterization of min as a procedure for exploiting valued preference relations and related results, working paper (Faculté Polytechnique de Mons).

Rapoport A. (1989) *Decision Theory and Decision Behavior. Normative and Descriptive Approaches*, Kluwer Acad. Publ., Dordrecht, Boston, London.

Riguet (1951) Les relations de Ferrers, *Comptes Rendus de l'Académie des Sciences de Paris* **232**, 1729–1730.

Roth A.E. (1976) Subsolutions and the supercore of cooperative games, *Mathematical of Operations Research*, **1**.

Roubens M. (1989) Some properties of Choice functions based on binary relations, *European Journal of Operational Research*, **40**, 309-321.

Roubens M. and Vincke Ph. (1987) Fuzzy preferences in an optimization perspective, J. Kacprzyk and S. A. Orlovski, Eds., *Optimization Models Using Fuzzy Sets and Possibility Theory*, D. Reidel Publ. Co, Dordrecht, Boston, Lancaster, Tokyo, 77- 90.

Roubens M. (1996) *Choice procedures in fuzzy multicriteria decision analysis based on pairwise comparisons*, Fuzzy Sets and Systems, *92*, **2**, to appear.

Roy B. (1968) Classement et choix en présence de points de vue multiples (la méthode ELECTRE), *Cahiers du CERO* **8**, 57–75.

Roy B. (1978) ELECTRE III : Un algorithme de classement fondé sur une représentation floue des préférences en présence de critères multiples, *Cahiers du CERO* **20**, 1, 3-24.

Roy B. (1981) A multicriteria analysis for trichotomic segmentation problems, in P. Nijkamp and J. Spronk (eds.), *Multiple Criteria Analysis: Operational methods*, Gower Press, 245–257.

Roy B. and Bertier P. (1973) La méthode ELECTRE II - Une application au media planning, *Operational Research'72, M. Ross (ED.)*, North-Holland Publishing Compagny, 9, 291–302.

Roy B. and Moscarola J. (1977) Procédure automatique d'examen des dossiers fondée sur une segmentation trichotomique en présence de critères multiples, *RAIRO Recherche Opérationnelle* **11**, 145–173.

Roy B. and Bouyssou D. (1994) *Aide Multicritère à la décision: Méthodes et Cas*, Economica, Paris.

Saaty T.L.(1978) Exploring the interface between hierarchies, multiple objectives and fuzzy sets, *Fuzzy Sets and Systems*, . **1**, 57–68.

Saaty T.L. (1980) *The Analytical Hierarchy Process*, McGraw Hill, New York.

Schwartz T. (1986) *The logic of collective choice*, Columbia University Press, New York.

Sen A.K. (1971) Choice functions and revealed preferences, *Review of Economic Studies* **38**, 307–313.

Slowinski R. (1993) Rough set learning of preferential attitude in multi-criteria decision making,. in: J.Komorowski & Z.Ras (eds.), *Methodologies for Intelligent Systems. Lecture Notes in Artificial Intelligence* **689**, Springer-Verlag, Berlin, 642–651.

Slowinski R. and Stefanowski J. (1994) Rough classification with valued closeness relation,. in: E.Diday, Y.Lechevallier, M.Schader, P.Bertrand, B.Burtschy (Eds.), *New Approaches in Classification and Data Analysis. Springer-Verlag*, Berlin, 482–489.

Valverde L. (1985) On the structure of F-indistinguishability operators, *Fuzzy Sets and Systems* **17**, 313–318.

Van Hentenryck P. (1989) *Constraint Satisfaction in Logic Programming*, MIT Press.

Vincke Ph. (1992) *Multicriteria Decision Aid*, Wiley.

Warshall S. (1962) A theorem on boolean Matrices, *Journal of the ACM*, **9**, 1, 11-12.

Yu W. (1992) *Aide Multicritére à la décision dans le cadre de la problématique du tri, Concepts méthodes et applications*, PhD thesis, Paris-Dauphine University.

4 GROUP DECISION MAKING UNDER FUZZINESS

Janusz Kacprzyk
Hannu Nurmi

Abstract: Group decision making, as meant in this paper, is the following choice problem which proceeds in a multiperson setting. There is a group of individuals (decisionmakers, experts, ...) who provide their testimonies concerning an issue in question. These testimonies are assumed here to be individual preference relations over some set of option (alternatives, variants, ...). The problem is to find a solution, i.e. an alternative or a set of alternatives, from among the feasible ones, which best reflects the preferences of the group of individuals as a whole. We will survey main developments in group decision making under fuzziness. First, we will briefly outline some basic inconsistencies and negative results of group decision making and social choice, and show how they can be alleviated by some plausible modifications of underlying assumptions, mainly by introducing fuzzy preference relations and, to a lesser extend, a fuzzy majority. Then, we will concentrate on how to derive solutions under individual fuzzy preference relations, and a fuzzy majority equated with a fuzzy linguistic quantifier (e.g., most, almost all, ...) and dealt with in terms of a fuzzy logic based calculus of linguistically quantified statements or via the ordered weighted averaging (OWA) operators. Finally, we will discuss a related issue of how to define a "soft" degree of consensus in the group under individual fuzzy preference relations and a fuzzy majority.

4.1 INTRODUCTION

In this section we will present first some general remarks on group decision making, and its related consensus reaching, and their modeling, with a rationale for the inclusion of some fuzzy elements into these models, mainly of fuzzy preference relations and fuzzy majorities.

Then, we will briefly sketch some formal tools that will be needed in our discussion, and which are somewhat specific, not generally dealt with in existing introductory texts. These are mainly fuzzy linguistic quantifiers and their related calculus of linguistically quantified propositions, and the ordered weighted averaging (or OWA, for short) operators, mainly from the point of view of their linguistic quantifier based aggregating properties which is crucial for the modeling of fuzzy majority.

4.1.1 Group Decision Making, Social Choice and Consensus Reaching – Basic Issues and the Role of Fuzziness

4.1.1.1 General remarks. The essence of decision making, which is one of the most crucial and omnipresent human activities, is basically to find a best alternative (option, variant, ...) from among some feasible (relevant, available, ...) ones. This general importance has naturally implied that decision making has become a subject of intensive research which has finally reached a "mature" stage of formal, mathematical models that try to formalize the human rational behavior. Initially, this rationality has been equated with the maximization of some utility (value) function, and this has lead to powerful formal results.

Unfortunately, more and more experiments have clearly indicated that the human behavior is rarely consistent with the maximization of a (expected) utility function, and some attempts to "soften" decision making models to make them more human-consistent have been made. The softening concerned with a plausible modification of assumptions on, say, the human preferences, axioms underlying the (expected) utility based approach, etc.; one can cite here, e.g., Aizerman (1985), many contributions in Kacprzyk and Fedrizzi (1990), Kacprzyk and Roubens (1988), Nurmi (1987), etc.

Potentials of fuzzy sets have been recognized and advocated in this respect quite early as well, and the reader may consult, e.g., Blin (1973) or Blin and Whinston (1974) for earlier works, and, e.g., Salles (1996) for later works.

However, decision making in real world rarely proceeds in such a simple setting as above, and normally one faces both multiple criteria and multiple decisionmakers. In this paper we will consider the latter situation. Among multiperson decision making settings, *group decision making* (which is meant here to be practically equivalent to social choice) is an important class. Its essence, for the purposes of our discussion, is basically as follows. There is a set of alternatives and a set of individuals who provide their testimonies concerning the alternatives. Usually, these testimonies are assumed to be *preferences* over the set of options, and this is also the case in this paper. The problem is to find a *solution*, i.e. an alternative (or a set of alternatives) which is best acceptable by the group of individuals as a whole. For a different point of departure, involving choice sets or utility functions, we may refer the interested reader to, e.g., Kim (1993), Salles (1996), Seo and Sakawa (1985) or Tanino (1990).

4.1.1.2 Group decision making – negative results. Unfortunately, the basic problem formulation of group decision making (social choice) sketched in Section 4.1.1.1 may only seem to be extremely simple, maybe even trivial, but it is not.

Since its very beginning group decision making has been plagued by negative results. Their essence is that no "rational" choice function satisfies all "natural", or plausible, requirements; so, each choice function has at least one serious drawback.

By far the best known negative result is the so-called Arrow's impossibility theorem (cf. Arrow, 1963) which may be briefly stated as follows. Suppose that there are $i = 1, 2, \ldots, m$ individuals whose individual connected and transitive relations over a set of n options $S = \{s_1, \ldots, s_n\}$ are R_i. We are looking for a social choice (welfare) function

$$f : \mathcal{R}_\infty \times \cdots \times \mathcal{R}_\backslash \longrightarrow \mathcal{R} \tag{1}$$

where $\mathcal{R}_)$ is a set of connected and transitive preference relations of individual i, and $\mathcal{R}$ is the set of connected and transitive social preference relations, both over S.

This choice function is required to satisfy the four basic and natural conditions which may be roughly stated as:

- f should have an unrestricted domain, i.e. it should be defined for all logically possible combinations of individual preferences;
- f should be independent of irrelevant options, i.e. the social preference relation with respect to any two options depends only on the individual preference relations concerning those two options, and not on those concerning other ones;
- the Pareto condition should be satisfied, i.e. if one option is preferred to another in all individual preference relations, then it should also be so in the social preference relation;
- the non-dictatorship condition should be satisfied, i.e. there should be no individual such that if one option is preferred to another in his or her individual preference relation, that it is also so in the social preference relation, for all options.

The Arrow's impossibility theorem states then that there is no social choice (welfare) function of the type (1) which satisfies all the above very natural requirements.

The Arrow's theorem has generated a rich literature (cf. Kelly, 1978), and implied two opposing attitudes. On the one hand, this result has been viewed as a final proof of the unachievability of democratic ideals. On the other hand, a somehow stringent and artificial character of its underlying requirements has often been pointed out. In any case, Arrow's results are discouraging to a considerable extent.

Another well-known negative result is due to Gibbard and Satterthwaite (cf. Gibbard, 1973) which concerns the so-called social decision functions, i.e.

$$f : \mathcal{R}_\infty \times \cdots \times \mathcal{R}_\backslash \longrightarrow 2^S \tag{2}$$

which map the individual preference relations into the set of options S.

The Gibbard and Satterthwaite's theorem states then that all (universal and non-trivial) social decision functions are either manipulable or dictatorial; that is, first of all, there is no election system that may encourage the voters to reveal their true preferences.

Though the above is a serious negative result, one should note, on the one hand, that all currently employed voting systems allow for ties, and hence are not realizations of social decision functions of type (2). On the other hand, the manipulability of a social decision function by a voter presupposes that he or she knows the entire preference profile, which is normally not the case in practice due to limited information availability.

Among other negative results, we may note, say, McKelvey's and Schofield's findings on the instability of solutions in spatial contexts, etc. (for more detail, see, e.g., Nurmi, 1982, 1983, 1987, 1988; Nurmi, Fedrizzi and Kacprzyk, 1990; Nurmi, Kacprzyk and Fedrizzi, 1996).

Basically, all these negative results might be summarized as follows: no matter which group choice procedure we employed, it would satisfy one set of plausible conditions but not another set of equally plausible ones. Unfortunately, this general property pertains to all possible choice procedures, so that attempts to develop new, more sophisticated choice procedures do not seem very promising in this respect. Much more promising seems to be to modify some basic assumptions underlying the group decision making process. This line of reasoning is also basically assumed here.

On the other hand, we will not deal in this paper with the alleviation of those shortcomings by employing probabilistic voting or some elements of Pawlak's (1982, 1991) rough sets, and we will refer the interested reader to, e.g., respectively, Intriligator (1973, 1982) for the former aspect, and Nurmi and Kacprzyk (1996), and Fedrizzi, Kacprzyk and Nurmi (1996) for the latter one.

4.1.1.3 Alleviating negative results in social choice by a fuzzification of preferences and majority. Since the process of decision making, notably of group type, is centered on the human beings, with their inherent subjectivity, imprecision and vagueness in the articulation of opinions, fuzzy sets have been used in this field for a long time.

A predominant research direction is here based on the introduction of an *individual* and *social fuzzy preference relation.*

Basically, suppose that we have a set of $n \geq 2$ options, $S = \{s_1, \ldots, s_n\}$, and a set of $m \geq 2$ individuals, $I = \{1, \ldots, m\}$. Then, an individual's $k \in I$ individual fuzzy preference relation in $S \times S$ assigns a value in the unit interval for the preference of one alternative over another.

Normally, there are also some conditions to be satisfied, as, e.g., reflexivity, connectivity, (max-min) transitivity, etc. One should however note that it is not clear which of these "natural" properties of preference relations should be assumed. We will briefly discuss this issue in Section 4.2, but the interested reader should consult, e.g., Salles (1996) or articles in Kacprzyk and Fedrizzi's (1990) or Kacprzyk and Roubens' (1988) volumes.

In this paper we assume that the individual and social fuzzy preference relations are defined in $S \times S$, i.e. assign to each pair of options a strength of preference of one over another as a value from $[0, 1]$. This will also be assumed in this paper. However, one should be aware that it may be viewed counter-intuitive, and a better solution would be to assume the values of the strength of preference belonging to some ordered set (exemplified by a set of linguistic values). This gives rise to some non-standard notions of soft preferences, orderings, etc. The best source for information on these and other related topics is Salles (1996).

The introduction of fuzzy preference relations may alleviate difficulties in social choice of the Arrow impossibility theorem type. For the traditional framework of numerically valued (with values in the unit interval) strengths of preferences, one may consult, e.g., Ovchinnikov (1990). On the other hand, for strengths of preferences with values in an ordered set, the interested reader is referred to, e.g., Barrett, Pattanaik and Salles (1992) or Salles (1996).

In this paper the fuzzy preferences will be employed only instrumentally, i.e. as a point of departure for procedures to find group decision making (social choice) solutions. Hence, we will not discuss them and their properties in more detail.

Another basic element underlying group decision making is the concept of a *majority* – notice that a solution is to be an option(or options) best acceptable by the group as a whole, that is by (at least!) *most* of its members since in practically no real nontrivial situation it would be accepted by all.

Some of the above mentioned problems, or negative result, with group decision making are closely related to a (too) strict representation of majority (e.g., at least a half, at least 2/3, ...). A natural line of reasoning is to somehow make that strict concept of majority closer to its real human perception by making it more vague. A good, often cited example in a biological context may be found in Loewer and Laddaga (1985):

> " ... It can correctly be said that there is a consensus among biologists that Darwinian natural selection is an important cause of evolution though there is currently no consensus concerning Gould's hypothesis of speciation. This means that there is a widespread agreement among biologists concerning the first matter but disagreement concerning the second ..."

and it is clear that a rigid majority as, e.g., more than 75% would evidently not reflect the essence of the above statement. However, it should be noted that there are naturally situations when a strict majority is necessary, for obvious reasons, as in all political elections.

A natural manifestations of such a "soft" majority are the so-called *linguistic quantifiers* as, e.g., most, almost all, much more than a half, etc. Such linguistic quantifiers can be, fortunately enough, dealt with by fuzzy-logic-based calculi of linguistically quantified statements as proposed by Zadeh (1983) and Yager (1983). Moreover, Yager's (1988) ordered weighted averaging (OWA) operators can be used for this purpose.

These calculi have been applied by the authors to introduce a fuzzy majority (represented by a fuzzy linguistic quantifier) into group decision making and consensus formation models (Fedrizzi and Kacprzyk, 1988; Kacprzyk, 1984, 1985b,c, 1986a, 1987a; Kacprzyk and Fedrizzi, 1986, 1988, 1989; Kacprzyk, Fedrizzi and Nurmi, 1990; Kacprzyk and Nurmi, 1988; Nurmi and Kacprzyk, 1990; Nurmi, Fedrizzi and Kacprzyk, 1990), and also in an implemented decision support system for consensus reaching (Fedrizzi, Kacprzyk and Zadrożny, 1988; Kacprzyk, Fedrizzi and Zadrożny, 1988).

In this paper we will present how fuzzy preference relations and fuzzy majorities can be employed for deriving solution of group decision making, and of degrees of consensus. Our discussion will be kept simple and constructive in the sense of discussing algorithms for determining solutions, and referring the interested reader to the source papers for more theoretical results.

We will basically follow a traditional, basic notation accepted in this volume, and will only mention more specific concepts and elements.

4.1.2 Fuzzy Linguistic Quantifiers and the Ordered Weighted Averaging (OWA) Operators

Our notation related to fuzzy sets will be standard, as adopted in most of the contributions in this handbook. A fuzzy set A in $X = \{x\}$, will be characterized by, and practically equated with its membership function $\mu_A : X \longrightarrow [0,1]$ such that $\mu_A(x) \in [0,1]$ is the grade of membership of $x \in X$ in A, from full membership to full nonmembership, through all intermediate values. For a finite $X = \{x_1, \ldots, x_n\}$ we write $A = \mu_A(x_1)/x_1 + \cdots + \mu_A(x_n)/x_n$ where "$\mu_A(x_i)/x_i$" is the pair "grade of membership – element" and "+" is meant in the set-theoretic sense. Moreover, we denote $a \wedge b = \min(a,b)$ and $a \vee b = \max(a,b)$. Other, more specific notation will be introduced when needed.

The only element of fuzzy logic, which is employed in this paper but which is probably less known, is the issue of *fuzzy linguistic quantifiers* and its related calculi of linguistically quantified statements. To make our next discussion readable and self-contained we will discuss this below.

A *linguistically quantified statement* may be exemplified by, say, "most experts are convinced" or "almost all good cars are expensive", and may be generally written as

$$Qy\text{'s are } F \tag{3}$$

where Q is a linguistic quantifier (e.g., most), $Y = \{y\}$ is a set of objects (e.g., experts), and F is a property (e.g., convinced).

We may assign to the particular y's (objects) a different importance (relevance, competence, ...), B, which may be added to (3) yielding a *linguistically quantified statement with importance qualification* generally written as

$$QBy\text{'s are } F \tag{4}$$

which may be exemplified by "most (Q) of the important (B) experts (y's) are convinced (F)".

From our point of view, the main problem is now to find the truth of such linguistically quantified statements, i.e. truth(Qy's are F) or truth(QBy's are F) knowing truth(y is F), for each $y \in Y$. Two basic fuzzy logic based calculi may be employed for this purpose: the ones due to Zadeh (1983) and to Yager (1983a, b). In the following we will present the essence of Zadeh's (1993) calculus since it is simpler and more transparent, hence better suited for the purposes of this paper. Our discussion will be kept as simple as possible, tailored to our particular needs. More information on fuzzy linguistic quantifiers, and their various representations and methods of handling, can be found in Part I, Chapter 2 in this volume.

4.1.2.1 A fuzzy-logic-based calculus of linguistically quantified statements. In Zadeh's (1983) method, a fuzzy linguistic quantifier Q is assumed to be a fuzzy set defined in $[0, 1]$. For instance, $Q =$ "most" may be given as

$$\mu_Q(x) = \begin{cases} 1 & \text{for } x \geq 0.8 \\ 2x - 0.6 & \text{for } 0.3 < x < 0.8 \\ 0 & \text{for } x \leq 0.3 \end{cases} \tag{5}$$

which may be meant as that if at least 80% of some elements satisfy a property, then *most* of them certainly (to degree 1) satisfy it, when less than 30% of them satisfy it, then *most* of them certainly do not satisfy it (satisfy to degree 0), and between 30% and 80% – the more of them satisfy it the higher the degree of satisfaction by *most* of the elements.

Clearly the above is an example of a *proportional* fuzzy linguistic quantifier (e.g., most, almost all, etc.), and we will deal with such quantifiers only since they are obviously more important for the modeling a fuzzy majority than the absolute quantifiers (e.g., about 5, much more than 10, etc.). The reasoning for the absolute quantifiers is however analogous.

Property F is defined as a fuzzy set in Y. For instance, if $Y = \{X, W, Z\}$ is the set of experts and F is a property "convinced", then F may be exemplified by $F =$ "convinced" $= 0.1/X + 0.6/W + 0.8/Z$ which means that expert X is convinced to degree 0.1, expert W to degree 0.6 and expert Z to degree 0.8. If now $Y = \{y_1, \ldots, y_p\}$, then it is assumed that truth(y_i is F) $= \mu_F(y_i)$, $i = 1, \ldots, p$.

The value of truth(Qy's are F) is determined in the following two steps (Zadeh, 1983):

$$r = \frac{1}{p} \sum_{i=1}^{p} \mu_F(y_i) \tag{6}$$

$$\text{truth}(Qy\text{'s are } F) = \mu_Q(r) \tag{7}$$

Basically, the expression (6) determines some mean proportion of elements satisfying the property under consideration, and (7) determines the degree to which this percentage satisfies the meaning of the fuzzy linguistic quantifier Q.

In the case of importance qualification, B is defined as a fuzzy set in Y, and $\mu_B(y_i) \in [0,1]$ is a degree of importance of y_i: from 1 for definitely important to 0 for definitely unimportant, through all intermediate values. For instance, $B =$ "important" $= 0.2/X + 0.5/W + 0.6/Z$ means that expert X is important (e.g., competent) to degree 0.2, expert W to degree 0.5, and expert Z to degree 0.6.

We rewrite first "QBy's are F" as "$Q(B$ and $F)y$'s are B" which leads to the following counterparts of (6) and (7):

$$r' = \frac{\sum_{i=1}^{p}[\mu_B(y_i) \wedge \mu_F(y_i)]}{\sum_{i=1}^{p} \mu_B(y_i)} \tag{8}$$

$$\text{truth}(QBY'\text{s are } F) = \mu_Q(r') \tag{9}$$

Example 1 Let Y = "experts" $= \{X, Y, Z\}$, F = "convinced" $= 0.1/X + 0.6/Y + 0.8/Z$, Q = "most" be given by (5), B = "important" $= 0.2/X + 0.5/Y + 0.6/Z$. Then: $r = 0.5$ and $r' = 0.92$, and truth("most experts are convinced")=0.4 and truth("most of the important experts are convinced")=1. □

The method presented is simple and efficient, and has proven to be useful in a multitude of cases, also in this paper.

4.1.2.2 Ordered weighted averaging (OWA) operators. Quite recently, Yager (1988) [see also Yager and Kacprzyk's (1997) book] has proposed a special class of aggregation operators, called the *ordered weighted averaging* (or OWA, for short) operators, which seem to provide an even better and more general aggregation in the sense of being able to simply and uniformly model a large class of fuzzy linguistic quantifiers.

An OWA operator of dimension p is a mapping $F : [0,1]^p \to [0,1]$ if associated with F is a weighting vector $W = [w_1, \ldots, w_p]^T$ such that: $w_i \in [0,1]$, $w_1 + \cdots + w_p = 1$, and

$$F(x_1, \ldots, x_p) = w_1 b_1 + \cdots + w_p b_p \tag{10}$$

where b_i is the i-th largest element among $\{x_1, \ldots, x_p\}$. B is called an ordered argument vector if each $b_i \in [0,1]$, and $j > i$ implies $b_i \geq b_j$, $i = 1, \ldots, p$.

Then

$$F(x_1, \ldots, x_p) = W^T B \tag{11}$$

Example 2 Let $W^T = [0.2, 0.3, 0.1, 0.4]$, and calculate $F(0.6, 1.0, 0.3, 0.5)$. Thus, $B^T = [1.0, 0.6, 0.5, 0.3]$, and $F(0.6, 1.0, 0.3, 0.5) = W^T B = 0.55$; and $F(0.0, 0.7, 0.1, 0.2) = 0.21$. □

For our purposes it is relevant how the OWA weights are found from the membership function of a fuzzy linguistic quantifier Q; an approach given in Yager (1988) may be used here:

$$w_k = \begin{cases} \mu_Q(k) - \mu_Q(k-1) & \text{for } k = 1, \ldots, p \\ \mu_Q(0) & \text{for } k = 0 \end{cases} \tag{12}$$

Some examples of the w_i's associated with the particular quantifiers are:

- If $w_p = 1$, and $w_i = 0$, for each $i \neq p$, then this corresponds to $Q =$ "all";
- If $w_i = 1$ for $i = 1$, and $w_i = 0$, for each $i \neq 1$, then this corresponds to $Q =$ "at least one",

and the intermediate cases as, e.g., a half, most, much more than 75%, a few, almost all, etc. may be obtained by a suitable choice of the w_i's between the above two extremes.

Thus, we will write

$$\text{truth}(Qy\text{'s are } F) = \text{OWA}_Q(\text{truth } y_i \text{ is } F) = W^T B \tag{13}$$

An important, yet difficult problem is the OWA operators with importance qualification, i.e. with importance coefficients associated with the particular data.

Suppose that we have a vector of data (pieces of evidence) $A = [a_1, \ldots, a_n]$, and a vector of importances $V = [v_1, \ldots, v_n]$ such that $v_i \in [0,1]$ is the importance of a_i, $i = 1, \ldots, n$, ($v_1 + \cdots + v_n \neq 1$, in general), and the OWA weights $W = [w_1, \ldots, w_n]^T$ corresponding to Q is determined via (12).

The case of an *ordered weighted averaging* operator with importance qualification, denoted OWA_I, is unfortunately not trivial. In a recent Yager's (1993) approach to be used here – which seems to be highly plausible, simple and efficient – the problem boils down to some redefinition of the OWA's weights w_i into $\overline{w}_i$. Then, (10) becomes

$$F_I(a_1, \ldots, a_n) = \overline{W}^T \cdot B = \sum_{j=1}^{n} \overline{w}_j b_j \tag{14}$$

We order first the pieces of evidence a_i, $i = 1, \ldots, n$, in descending order to obtain B such that b_j is the j-th largest element of $\{a_1, \ldots, a_n\}$. Next, we denote by u_j the importance of b_j, i.e. of the a_i which is the j-th largest; $i, j = 1, \ldots, n$. Finally, the new weights $\overline{W}$ are defined as

$$\overline{w}_j = \mu_Q\left(\frac{\sum_{k=1}^{j} u_k}{\sum_{k=1}^{n} u_k}\right) - \mu_Q\left(\frac{\sum_{k=1}^{j-1} u_k}{\sum_{k=1}^{n} u_k}\right) \tag{15}$$

Example 3 Suppose that $A = [a_1, a_2, a_3, a_4]$=$[0.7, 1, 0.5, 0.6]$, and U=$[u_1, u_2, u_3, u_4]$ =$[1, 0.6, 0.5, 0.9]$. Q="most" is given by (5).

Then, $B = [b_1, b_2, b_3, b_4] = [1, 0.7, 0.6, 0.5]$, and $\overline{W} = [0.04, 0.24, 0.41, 0.31]$, and $F_I(A) = \sum_{j=1}^{4} \overline{w}_j b_j = 0.067 \cdot 1 + 0.4 \cdot 0.7 + 0.333 \cdot 0.6 + 0.2 \cdot 0.5 = 0.6468$.□

For more information on the OWA operators we refer the reader to the recent Yager and Kacprzyk's (1997) book.

We have now the necessary formal means to proceed to our discussion of group decision making and consensus formation models under fuzzy preferences and a fuzzy majority.

Finally, let us mention that OWA-like aggregation operators may be defined in an ordinal setting, i.e. for non-numeric data (which are only ordered, and we will refer the interested reader to, e.g., Delgado, Verdegay and Vila (1993) or Herrera, Herrera-Viedma and Verdegay (1996).

4.2 GROUP DECISION MAKING UNDER FUZZY PREFERENCES AND A FUZZY MAJORITY

For our purposes, *group decision making* (equated here with social choice) proceeds in the following setting. We have a set of $n \geq 2$ options, $S = \{s_1, \ldots, s_n\}$, and a set of $m \geq 2$ individuals, $I = \{1, \ldots, m\}$. Each individual $k \in I$ provides his or her testimony as to the alternatives in S. These testimonies are assumed to be individual fuzzy preference relations defined over the set of alternatives S (i.e. in $S \times S$). Fuzzy preference relations are employed to reflect an omnipresent fact that the preferences may be not clear-cut so that conventional non-fuzzy preference relations may be not adequate (see, e.g., many articles in Kacprzyk and Roubens, 1988, Kacprzyk and Fedrizzi, 1990 or Kacprzyk, Nurmi and Fedrizzi, 1996).

An *individual fuzzy preference relation* of individual k, R_k, is given by its membership function $\mu_{R_k} : S \times S \longrightarrow [0, 1]$ such that

$$\mu_{R_k} = \begin{cases} 1 & \text{if } s_i \text{ is definitely preferred to } s_j \\ c \in (0.5, 1) & \text{if } s_i \text{ is slightly preferred to } s_j \\ 0.5 & \text{in the case of indifference} \\ d \in (0, 0.5) & \text{if } s_j \text{ is slightly preferred to } s_i \\ 0 & \text{if } s_j \text{ is definitely preferred to } s_i \end{cases} \qquad (16)$$

If card S is small enough (as assumed here), an individual fuzzy preference relation of individual k, R_k, may conveniently be represented by an $n \times m$ matrix $R_k = [r_{ij}^k]$, such that $r_{ij}^k = \mu_{R_k}(s_i, s_j)$; $i, j = 1, \ldots, n$; $k = 1, \ldots, m$. R_k is commonly assumed (also here) to be reciprocal in that $r_{ij}^k + r_{ji}^k = 1$; moreover, it is also normally assumed that $r_{ii}^k = 0$, for all i, j, k. Notice that we do not mention here other properties of (individual) fuzzy preference relations which are often discussed (cf. Salles, 1996) but which will not be relevant to our discussion.

The individual fuzzy preference relations, similarly as their nonfuzzy counterparts in traditional (non-fuzzy) group decision making, are a point of departure for most procedures for the derivation of solutions. As we have already mentioned, we will not deal with group decision making taking as a point of departure choice sets or utility functions (cf. Tanino, 1990).

More details on fuzzy preference relations and their use in decision making may be found in Part I, Chapters 2 and 3, of this volume.

Basically, two lines of reasoning may be followed here (cf. Kacprzyk, 1984–1986):

- a direct approach
$$\{R_1, \ldots, R_m\} \longrightarrow \text{solution} \tag{17}$$
that is, a solution is derived directly (without any intermediate steps) just from the set of individual fuzzy preference relations, and

- an indirect approach
$$\{R_1, \ldots, R_m\} \longrightarrow R \longrightarrow \text{solution} \tag{18}$$
that is, from the set of individual fuzzy preference relations we form first a social fuzzy preference relation, R (to be defined later), which is then used to find a solution;

and notice that the above schemes (17) and (18) correspond the the choice functions of the type (2).

A solution is here, unfortunately, not clearly understood – see, e.g., Nurmi (1981, 1982, 1983, 1987, 1988) for diverse solution concepts. More details related to the use of fuzzy preference relations as a point of departure in group decision making can also be found in, e.g., Nurmi (1981, 1982, 1988) and in many articles in Kacprzyk and Roubens (1988), Kacprzyk and Fedrizzi (1990), and Kacprzyk, Nurmi and Fedrizzi (1996).

In this paper we will only sketch the derivation of some more popular solution concepts, and this will show to the reader not only the essence of the particular solution concept but how a fuzzification may be performed so that the reader can eventually fuzzify other crisp solution concepts that may be found in the literature.

More specifically, we will show the derivation of some fuzzy cores and minimax sets for the direct approach, and some fuzzy consensus winners for the indirect approach. In addition to fuzzy preference relations, which are usually employed, we will also use a fuzzy majority represented by a linguistic quantifier as proposed by Kacprzyk (1984–1986a).

4.2.1 Direct Derivation of a Solution

We will first employ the direct approach (17), i.e.

$$\{R_1, \ldots, R_m\} \longrightarrow \text{solution}$$

to derive two popular solution concepts: fuzzy cores and minimax sets.

4.2.1.1 Fuzzy cores. The core is a very intuitively appealing and often used solution concept. Conventionally, the core is defined as a set of *undominated alternatives*, i.e. those not defeated in *pairwise comparisons* by a required majority (strict!) $r \leq m$, i.e.

$$C = \{s_j \in S : \neq \exists s_i \in S \text{ such that } r_{ij}^k > 0.5 \text{ for at least } r \text{ individuals}\} \quad (19)$$

The first attempt at a fuzzification of the core is due to Nurmi (1981) who has extended it to the *fuzzy α-core* defined as

$$C_\alpha = \{s_j \in S : \neq \exists s_i inS \text{ such that } r_{ij}^k > \alpha \geq 0.5 \text{ for at least } r \text{ individuals}\} \quad (20)$$

that is, as a set of alternatives not sufficiently (at least to degree α) defeated by the required (still strict!) majority $r \leq m$.

As we have already indicated, in many group decision making related situations is may be more adequate to assume that the required majority is imprecisely specified as, e.g., given by a fuzzy linguistic quantifier as, say, *most* defined by (3). This concept of a fuzzy majority has been proposed by Kacprzyk (1984–1986a), and it has turned out that it can be quite useful and adequate.

To employ a fuzzy majority to extend (fuzzify) the core, we start by denoting

$$h_{ij}^k = \begin{cases} 1 & \text{if } r_{ij}^k < 0.5 \\ 0 & \text{otherwise} \end{cases} \quad (21)$$

where here and later on in this section, if not otherwise specified, $i, j = 1, \ldots, n$ and $k = 1, \ldots, m$.

Thus, h_{ij}^k just reflects if alternative s_j defeats (in pairwise comparison) alternative s_i ($h_{ij}^k = 1$) or not ($h_{ij}^k = 0$).

Then, we calculate

$$h_j^k = \frac{1}{n-1} \sum_{i=1, i \neq j}^{n} h_{ij}^k \quad (22)$$

which is clearly the extent, from 0 to 1, to which individual k is not against alternative s_j, where 0 standing for definitely not against to 1 standing for definitely against, through all intermediate values.

Next, we calculate

$$h_j = \frac{1}{m} \sum_{k=1}^{m} h_j^k \quad (23)$$

which expresses to what extent, from 0 to 1 as in the case of (22), *all* the individuals are not against alternative s_j.

And, finally, we calculate

$$v_Q^j = \mu_Q(h_j) \quad (24)$$

is to what extent, from 0 to 1 as before, Q (say, most) individuals are not against alternative s_j.

The *fuzzy Q-core* is now defined (Kacprzyk, 1984–1986a, 1987a) as a fuzzy set

$$C_Q = v_Q^1/s_1 + \cdots + v_Q^n/s_n \tag{25}$$

i.e. as a fuzzy set of alternatives that are not defeated by Q (say, most) individuals.

Notice that in the above basic definition of a fuzzy Q-core we do not take into consideration to what degrees those defeats of one alternative by another are. They can be accounted for in a couple of plausible ways.

First and most straightforward is the introduction of a threshold into the degree of defeat in (21), for instance by denoting

$$h_{ij}^k(\alpha) = \begin{cases} 1 & \text{if } r_{ij}^k < \alpha \leq 0.5 \\ 0 & \text{otherwise} \end{cases} \tag{26}$$

where, again, $i, j = 1, \ldots, n$ and $k = 1, \ldots, m$.

Thus, $h_{ij}^k(\alpha)$ just reflects if alternative s_j sufficiently (i.e. at least to degree $1 - \alpha$) defeats (in pairwise comparison) alternative s_i or not.

Then, we calculate

$$h_j^k(\alpha) = \frac{1}{n-1} \sum_{i=1, i \neq j}^{n} h_{ij}^k(\alpha) \tag{27}$$

which is clearly the extent, from 0 to 1, to which individual k is not sufficiently (at least to degree $1 - \alpha$) against alternative s_j, where 0 standing for definitely not against to 1 standing for definitely against, through all intermediate values.

Next, we calculate

$$h_j(\alpha) = \frac{1}{m} \sum_{k=1}^{m} h_j^k(\alpha) \tag{28}$$

which expresses to what extent, from 0 to 1 as in the case of (22), *all* the individuals are not sufficiently (at least to degree $1 - \alpha$) against alternative s_j.

And, finally, we calculate

$$v_Q^j(\alpha) = \mu_Q[h_j(\alpha)] \tag{29}$$

is to what the extent, from 0 to 1 as before, Q (say, most) individuals are not sufficiently (at least to degree $1 - \alpha$) against alternative s_j.

The *fuzzy α/Q-core* is then defined (Kacprzyk, 1984–1986a, 1987a) as a fuzzy set

$$C_{\alpha/Q} = v_Q^1(\alpha)/s_1 + \cdots + v_Q^n(\alpha)/s_n \tag{30}$$

i.e. as a fuzzy set of alternatives that are not sufficiently (at least to degree $1 - \alpha$) defeated by Q (say, most) individuals.

We can also explicitly introduce the strength of defeat into (21). Namely, we can introduce a function exemplified by

$$\hat{h}_{ij}^k = \begin{cases} 2(0.5 - r_{ij}^k) & \text{if } r_{ij}^k < 0.5 \\ 0 & \text{otherwise} \end{cases} \tag{31}$$

where, again, $i, j = 1, \ldots, n$ and $k = 1, \ldots, m$.

Thus, $\hat{h}_{ij}^k$ just reflects how strongly (from 0 to 1) alternative s_j defeats (in pairwise comparison) alternative s_i.

Then, we calculate

$$\hat{h}_j^k = \frac{1}{n-1} \sum_{i=1, i \neq j}^{n} \hat{h}_{ij}^k \tag{32}$$

which is clearly how strongly, from 0 to 1, individual k is not against alternative s_j, where 0 standing for definitely not against to 1 standing for definitely against, through all intermediate values.

Next, we calculate

$$\hat{h}_j = \frac{1}{m} \sum_{k=1}^{m} \hat{h}_j^k \tag{33}$$

which expresses the strength, from 0 to 1, to which *all* the individuals are not against alternative s_j.

And, finally, we calculate

$$\hat{v}_Q^j = \mu_Q(\hat{h}_j) \tag{34}$$

is the strength, from 0 to 1, to which Q (say, most) individuals are not against alternative s_j.

The *fuzzy s/Q-core* is then defined (Kacprzyk, 1984–1986a, 1987a) as a fuzzy set

$$C_{s/Q} = \hat{v}_Q^1/s_1 + \cdots + \hat{v}_Q^n/s_n \tag{35}$$

i.e. as a fuzzy set of alternatives that are not strongly defeated by Q (say, most) individuals.

Example 4 Suppose that we have four individuals, $k = 1, 2, 3, 4$, whose individual fuzzy preference relations are:

$R_1 =$

	$j=1$	2	3	4
$i=1$	0	0.3	0.7	0.1
2	0.7	0	0.6	0.6
3	0.3	0.4	0	0.2
4	0.9	0.4	0.8	0

$R_2 =$

	$j=1$	2	3	4
$i=1$	0	0.4	0.6	0.2
2	0.6	0	0.7	0.4
3	0.4	0.3	0	0.1
4	0.8	0.6	0.9	0

$R_3 =$

	$j=1$	2	3	4
$i=1$	0	0.5	0.7	0.1
2	0.5	0	0.8	0.4
3	0.3	0.2	0	0.2
4	1	0.6	0.8	0

$R_4 =$

	$j=1$	2	3	4
$i=1$	0	0.4	0.7	0.8
2	0.6	0	0.4	0.3
3	0.3	0.6	0	0.1
4	0.7	0.7	0.9	0

Suppose now that the fuzzy linguistic quantifier is $Q =$ "most" defined by (5). Then, say:

$$C_{\text{"most"}} \cong 0.06/s_1 + 0.56/s_2 + 1/s_4$$
$$C_{0.3/\text{"most"}} \cong 0.56/s_4$$
$$C_{s/\text{"most"}} \cong 0.36/s_4$$

to be meant as follows: in case of $C_{\text{"most"}}$ alternative s_1 belongs to to the fuzzy Q-core to the extent 0.06. s_2 to the extent 0.56, and s_4 to the extent 1, and analogously for the $C_{0.3/\text{"most"}}$ and $C_{s/\text{"most"}}$. Notice that though the results obtained for the particular cores are different, for obvious reasons, s_4 is clearly the best choice which is evident if we examine the given individual fuzzy preference relations. □

Clearly, the fuzzy linguistic quantifier based aggregation of partial scores in the above definitions of the fuzzy Q-core, α/Q-core and s/Q-core, may be replaced by an ordered weighted averaging (OWA) operator based aggregation given by (12) and (13). This was proposed by Fedrizzi and Kacprzyk (1993), and Kacprzyk and Fedrizzi (1995a, b).

First, to derive the fuzzy Q-core, we start with h^k_{ij} (21) which reflects if alternative s_j defeats (in pairwise comparison) alternative s_i ($h^k_{ij} = 1$) or not ($h^k_{ij} = 0$).

Then, we calculate h^k_j (22) which is the extent, from 0 to 1, to which individual k is not against alternative s_j.

Next, we calculate h_j (23) which expresses to what extent, from 0 to 1 as in the case of (22), *all* the individuals are not against alternative s_j.

And, finally, we calculate

$$\overline{v}^j_Q = \text{OWA}_Q(h_j) \tag{36}$$

is to what extent, from 0 to 1 as before, Q (say, most) individuals are not against alternative s_j, and $\text{OWA}_Q(.)$ is the OWA based aggregation given by (13) where the OWA weights corresponding to Q are given by (12).

The *fuzzy Q-core* is then defined (cf. Kacprzyk, 1984–1986a, 1987a) as a fuzzy set

$$C_Q = \overline{v}^1_Q/s_1 + \cdots + \overline{v}^n_Q/s_n \tag{37}$$

i.e. as a fuzzy set of alternatives that are not defeated by Q (say, most) individuals.

And, analogously, by introducing a threshold into the degree of defeat in (21), $\alpha \in [0, 0.5)$,we start with $h^k_{ij}(\alpha)$ (26) which reflects if alternative s_j sufficiently (i.e. at least to degree $1 - \alpha$) defeats (in pairwise comparison) alternative s_i or not.

Then, we calculate $h^k_j(\alpha)$ (27) which is the extent, from 0 to 1, to which individual k is not sufficiently (at least to degree $1 - \alpha$) against alternative s_j, where 0 standing for definitely not against to 1 standing for definitely against, through all intermediate values.

Next, we calculate $h_j(\alpha)$ (28) which expresses to what extent, from 0 to 1 as in the case of (22), *all* the individuals are not sufficiently (at least to degree $1 - \alpha$) against alternative s_j.

And, finally, we calculate

$$\overline{v}^j_Q(\alpha) = \text{OWA}_Q[h_j(\alpha)] \tag{38}$$

is to what the extent, from 0 to 1 as before, Q (say, most) individuals are not sufficiently (at least to degree $1-\alpha$) against alternative s_j, and $\text{OWA}_Q(.)$ is the

OWA based aggregation given by (13) where the OWA weights corresponding to Q are given by (12).

The *fuzzy α/Q-core* is then defined (cf. Kacprzyk, 1984–1986a, 1987a) as a fuzzy set

$$C_{\alpha/Q} = \overline{v}_Q^1(\alpha)/s_1 + \cdots + \overline{v}_Q^n(\alpha)/s_n \tag{39}$$

i.e. as a fuzzy set of alternatives that are not sufficiently (at least to degree $1-\alpha$) defeated by Q (say, most) individuals.

Finally, we can also explicitly introduce the strength of defeat by using $\hat{h}_{ij}^k$ (31) which reflects how strongly (from 0 to 1) alternative s_j defeats (in pairwise comparison) alternative s_i.

Then, we calculate $\hat{h}_j^k$ (32) which is how strongly, from 0 to 1, individual k is not against alternative s_j, where 0 standing for definitely not against to 1 standing for definitely against, through all intermediate values.

Next, we calculate $\hat{h}_j$ (33) which expresses the strength, from 0 to 1, to which *all* the individuals are not against alternative s_j.

And, finally, we calculate

$$\hat{w}_Q^j = \text{OWA}_Q(\hat{h}_j) \tag{40}$$

is the strength, from 0 to 1, to which Q (say, most) individuals are not against alternative s_j,and $\text{OWA}_Q(.)$ is the OWA based aggregation given by (13) where the OWA weights corresponding to Q are given by (12).

The *fuzzy s/Q-core* is then defined (cf. Kacprzyk, 1984–1986a, 1987a) as a fuzzy set

$$C_{s/Q} = \hat{w}_Q^1/s_1 + \cdots + \hat{w}_Q^n/s_n \tag{41}$$

i.e. as a fuzzy set of alternatives that are not strongly defeated by Q (say, most) individuals.

The results obtained by using the OWA operators are similar to those for the usual fuzzy linguistic quantifiers.

Finally, let us notice that the individuals and alternatives may be assigned variable importance (competence) and relevance, respectively, and then the OWA based aggregation with importance qualification given by (14) may be used. This will not change however the essence of the fuzzy cores defined above, and will not be discussed here for lack of space.

4.2.1.2 Minimax sets. Another intuitively justified solution concept may be the minimax (opposition) set which may be defined for our purposes as follows.

Let $w(s_i, s_j) \in \{1, 2, \ldots, m\}$ be the number of individuals who prefer alternative s_i to alternative s_j, i.e. for whom $r_{ij}^k < 0.5$.

If now

$$v(s_i) = \max_{j=1,\ldots,n} w(s_i, s_j) \tag{42}$$

and

$$v^* = \min_{i=1,\ldots,n} v(s_i) \tag{43}$$

then the *minimax set* is defined as

$$M(v^*) = \{s_i \in S : v(s_i) = v^*\} \tag{44}$$

i.e. as a (nonfuzzy) set of alternatives which in pairwise comparisons with any other alternative are defeated by no more than v^* individuals, hence by the least number of individuals.

Nurmi (1981) extends the minimax set, similarly in spirit to his extension of the core (20), to the *α-minimax set* as follows. Let $w_\alpha(s_i, s_j) \in \{1, 2, \ldots, m\}$ be the number of individuals who prefer alternative s_i to alternative s_j at least to degree $1 - \alpha$, i.e. for whom $r_{ij}^k < \alpha \leq 0.5$.

If now

$$v_\alpha(s_i) = \max_{j=1,\ldots,n} w_\alpha(s_i, s_j) \tag{45}$$

and

$$v_\alpha^* = \min_{i=1,\ldots,n} v_\alpha(s_i) \tag{46}$$

then the *α-minimax set* is defined as

$$M_\alpha(v_\alpha^*) = \{s_i \in S : v_\alpha(s_i) = v_\alpha^*\} \tag{47}$$

i.e. as a (nonfuzzy) set of alternatives which in pairwise comparisons with any other alternative are defeated (at least to degree $1 - \alpha$) by no more than v^* individuals, hence by the least number of individuals.

A fuzzy majority is introduced into the above definitions of minimax sets as follows (Kacprzyk, 1985c, 1986a).

We start with (21), i.e.

$$h_{ij}^k = \begin{cases} 1 & \text{if } r_{ij}^k < 0.5 \\ 0 & \text{otherwise} \end{cases} \tag{48}$$

and

$$h_i^k = \frac{1}{n-1} \sum_{j=1, j \neq i}^{n} h_{ij}^k \tag{49}$$

is the extent, between 0 and 1, to which individual k is against alternative s_i.

Then

$$h_i = \frac{1}{m} \sum_{k=1}^{m} h_i^k \tag{50}$$

is the extent, between 0 and 1, to which all the individuals are against alternative s_i.

Next

$$t_i^Q = \mu_Q(h_i) \tag{51}$$

is the extent, from 0 to 1, to which Q (say, most) individuals are against alternative s_i, and

$$t_Q^* = \min_{i=1,\ldots,n} t_i^Q \tag{52}$$

is the least defeat of any alternative by Q individuals.

Finally, the Q-minimax set is

$$M_Q(t^*_Q) = \{s_i \in S : t^Q_I = t^*_Q\} \tag{53}$$

And analogously as for the α/Q-core (30), we can explicitly introduce the degree of defeat $\alpha < 0.5$ into the definition of the Q-minimax set.

We start with (26), i.e.

$$h^k_{ij}(\alpha) = \begin{cases} 1 & \text{if } r^k_{ij} < \alpha \leq 0.5 \\ 0 & \text{otherwise} \end{cases} \tag{54}$$

and

$$h^k_i(\alpha) = \frac{1}{n-1} \sum_{j=1, j\neq i}^{n} h^k_{ij}(\alpha) \tag{55}$$

is the extent, between 0 and 1, to which individual k is sufficiently (at least to degree $1-\alpha$) against alternative s_i.

Then

$$h_i(\alpha) = \frac{1}{m} \sum_{k=1}^{m} h^k_i(\alpha) \tag{56}$$

is the extent, between 0 and 1, to which all the individuals are sufficiently (at least to degree $1-\alpha$) against alternative s_i.

Next

$$t^Q_i(\alpha) = \mu_Q[h_i(\alpha)] \tag{57}$$

is the extent, from 0 to 1, to which Q (say, most) individuals are sufficiently (at least to degree $1-\alpha$) against alternative s_i, and

$$t^*_Q(\alpha) = \min_{i=1,\ldots,n} t^Q_i(\alpha) \tag{58}$$

is the least sufficient (at least to degree $1-\alpha$) defeat of any alternative by Q individuals.

Finally, the α/Q-minimax set is

$$M_{\alpha/Q}[t^*_Q(\alpha)] = \{s_i \in S : t^Q_I(\alpha) = t^*_Q(\alpha)\} \tag{59}$$

We can also introduce explicitly the strength of defeat similarly as in the case of the fuzzy s/Q-core (35).

We start with (31), i.e.

$$\hat{h}^k_{ij} = \begin{cases} 2(0.5 - r^k_{ij}) & \text{if } r^k_{ij} < 0.5 \\ 0 & \text{otherwise} \end{cases} \tag{60}$$

and

$$\hat{h}^k_i = \frac{1}{n-1} \sum_{j=1, j\neq i}^{n} \hat{h}^k_{ij} \tag{61}$$

is the extent, between 0 and 1, to which individual k is strongly against alternative s_i.

Then

$$\hat{h}_i = \frac{1}{m} \sum_{k=1}^{m} \hat{h}_i^k \tag{62}$$

is the extent, between 0 and 1, to which all the individuals are strongly against alternative s_i.

Next

$$\hat{t}_i^Q = \mu_Q(\hat{h}_i) \tag{63}$$

is the extent, from 0 to 1, to which Q (say, most) individuals are strongly against alternative s_i, and

$$\hat{t}_Q^* = \min_{i=1,\ldots,n} \hat{t}_i^Q \tag{64}$$

is the least strong defeat of any alternative by Q individuals.

Finally, the s/Q-minimax set is

$$M_{s/Q}(\hat{t}_Q^*) = \{s_i \in S : \hat{t}_I^Q = \hat{t}_Q^*\} \tag{65}$$

Example 5 For the same four individual fuzzy preference relations $R_1, \ldots, R_4$ as in Example 4, we obtain for instance:

$$M_{\text{"most"}}(0) = \{s_4\}$$
$$M_{0.3/\text{"most"}}(0) = \{s_1, s_2, s_4\}$$
$$M_{s/\text{"most"}} = \{s_1, s_2, s_4\}$$

□

The OWA based aggregation can also be employed for the derivation of fuzzy minimax sets given above.

First, we start with h_{ij}^k (48) and h_i^k (49) is the extent, between 0 and 1, to which individual k is against alternative s_i.

Then, h_i (50) is the extent, between 0 and 1, to which all the individuals are against alternative s_i.

Next, we calculate

$$\bar{t}_i^Q = \text{OWA}_Q(h_i) \tag{66}$$

which is the extent, from 0 to 1, to which Q (say, most) individuals are against alternative s_i, with $\text{OWA}_Q(.)$ being the OWA based aggregation defined by (12) and (13), and

$$\bar{t}_Q^* = \min_{i=1,\ldots,n} \bar{t}_i^Q \tag{67}$$

is the least defeat of any alternative by Q individuals.

Finally, the Q-minimax set is

$$M_Q(\bar{t}_Q^*) = \{s_i \in S : \bar{t}_I^Q = \bar{t}_Q^*\} \tag{68}$$

And, analogously, we can explicitly introduce the degree of defeat $\alpha < 0.5$, and start with $h^k_{ij}(\alpha)$ (54), and $h^k_i(\alpha)$ (55) is the extent, between 0 and 1, to which individual k is sufficiently (at least to degree $1-\alpha$) against alternative s_i.

Then, $h_i(\alpha)$ (56) is the extent, between 0 and 1, to which all the individuals are sufficiently (at least to degree $1-\alpha$) against alternative s_i.

Next

$$\bar{t}^Q_i(\alpha) = \text{OWA}_Q[h_i(\alpha)] \tag{69}$$

is the extent, from 0 to 1, to which Q (say, most) individuals are sufficiently (at least to degree $1-\alpha$) against alternative s_i, and

$$\bar{t}^*_Q(\alpha) = \min_{i=1,\ldots,n} \bar{t}^Q_i(\alpha) \tag{70}$$

is the least sufficient (at least to degree $1-\alpha$) defeat of any alternative by Q individuals.

Finally, the α/Q-minimax set is

$$M_{\alpha/Q}[\bar{t}^*_Q(\alpha)] = \{s_i \in S : \bar{t}^Q_I(\alpha) = \bar{t}^*_Q(\alpha)\} \tag{71}$$

We can also introduce explicitly the strength of defeat, and start with $\hat{h}^k_{ij}$ (60), and $\hat{h}^k_i$ (61) is the extent, between 0 and 1, to which individual k is strongly against alternative s_i.

Then, $\hat{h}_i$ (62) is the extent, between 0 and 1, to which all the individuals are strongly against alternative s_i.

Next

$$\hat{u}^Q_i = \text{OWA}(\hat{h}_i) \tag{72}$$

is the extent, from 0 to 1, to which Q (say, most) individuals are strongly against alternative s_i, and

$$\hat{u}^*_Q = \min_{i=1,\ldots,n} \hat{u}^Q_i \tag{73}$$

is the least strong defeat of any alternative by Q individuals.

Finally, the s/Q-minimax set is

$$M_{s/Q}(\hat{u}^*_Q) = \{s_i \in S : \hat{u}^Q_I = \hat{u}^*_Q\} \tag{74}$$

And, again, the results obtained by using the OWA based aggregation are similar to those obtained by directly employing Zadeh's (1983) calculus of linguistically quantified statements.

4.2.2 Indirect Derivation of a Solution – the Consensus Winner

Now we follow the scheme (18), i.e.

$$\{R_1, \ldots, R_m\} \longrightarrow R \longrightarrow \text{solution}$$

i.e. from the individual fuzzy preference relations we determine first a social fuzzy preference relation, R, which is similar in spirit to its individual counterpart but concerns the whole group of individuals, and then find a solution from such a social fuzzy preference relation.

It is easy to notice that the above direct derivation scheme involves in fact two problems:

- how to find a social fuzzy preference relation from the individual fuzzy preference relations, i.e.

$$\{R_1, \ldots, R_m\} \longrightarrow R$$

- how to find a solution from the social fuzzy preference relation, i.e.

$$R \longrightarrow \text{solution}$$

In this paper we will not deal in more detail with the first step, i.e. $\{R_1, \ldots, R_m\} \longrightarrow R$, and assume a (most) straightforward alternative that the social fuzzy preference relation $R = [r_{ij}]$ is given by

$$r_{ij} = \begin{cases} \frac{1}{m} sum_{k=1}^{m} a_{ij}^k & \text{if } i \neq j \\ 0 & \text{otherwise} \end{cases} \tag{75}$$

where

$$a_{ij}^k = \begin{cases} 1 & \text{if } r_{ij}^k > 0.5 \\ 0 & \text{otherwise} \end{cases} \tag{76}$$

Notice that R obtained via (75) need not be reciprocal, i.e. $r_{ij} \neq 1 - r_{ji}$, but it can be shown that $r_{ij} \leq 1 - r_{ji}$, for each $i, j = 1, \ldots, n$. For other definitions of R, see, e.g., Blin and Whinston (1973) or Ovchinnikov (1990).

We will discuss now the second step, i.e. $R \longrightarrow$ solution, i.e. how to determine a solution from a social fuzzy preference relation.

A solution concept of much intuitive appeal is here the consensus winner (Nurmi, 1981) which will be extended under a social fuzzy preference relation and a fuzzy majority.

We start with

$$g_{ij} = \begin{cases} 1 & \text{if } r_{ij} > 0.5 \\ 0 & \text{otherwise} \end{cases} \tag{77}$$

which expresses whether alternative s_i defeats (in the whole group's opinion!) alternative s_j or not.

Next

$$g_i = \frac{1}{n-1} \sum_{j=1, j \neq i}^{n} g_{ij} \tag{78}$$

which is a mean degree to which alternative s_i is preferred, by the whole group, over all the other alternatives.

Then

$$z_Q^i = \mu_Q(g_i) \tag{79}$$

is the extent to which alternative s_i is preferred, by the whole group, over Q (e.g., most) other alternatives.

Finally, we define the *fuzzy Q-consensus winner* as

$$W_Q = z_Q^1/s_1 + \cdots + z_Q^n/s_n \tag{80}$$

i.e. as a fuzzy set of alternatives that are preferred, by the whole group, over Q other alternatives.

And analogously as in the case of the core, we can introduce a threshold $\alpha \geq 0.5$ into (77), i.e.

$$g_{ij}(\alpha) = \begin{cases} 1 & \text{if } r_{ij} > \alpha \geq 0.5 \\ 0 & \text{otherwise} \end{cases} \tag{81}$$

which expresses whether alternative s_i sufficiently (at least to degree α) defeats (in the whole group's opinion!) alternative s_j or not.

Next

$$g_i(\alpha) = \frac{1}{n-1} \sum_{j=1, j\neq i}^{n} g_{ij}(\alpha) \tag{82}$$

which is a mean degree to which alternative s_i is sufficiently (at least to degree α) preferred, by the whole group, over all the other alternatives.

Then

$$z_Q^i(\alpha) = \mu_Q[g_i(\alpha)] \tag{83}$$

is the extent to which alternative s_i is sufficiently (at least to degree α) preferred, by the whole group, over Q (e.g., most) other alternatives.

Finally, we define the *fuzzy α/Q-consensus winner* as

$$W_{\alpha/Q} = z_Q^1(\alpha)/s_1 + \cdots + z_Q^n(\alpha)/s_n \tag{84}$$

i.e. as a fuzzy set of alternatives that are sufficiently (at least to degree α) preferred, by the whole group, over Q other alternatives.

Furthermore, we can also explicitly introduce the strength of preference into (77) similarly as in (31), for instance by defining

$$\hat{g}_{ij} = \begin{cases} 2(r_{ij} - 0.5) & \text{if } r_{ij} > 0.5 \\ 0 & \text{otherwise} \end{cases} \tag{85}$$

which expresses whether alternative s_i defeats (in the whole group's opinion!) alternative s_j or not, and if so (i.e. if $\hat{g}_{ij} > 0$), then $\hat{g}_{ij}$ gives the strength of this defeat.

Next

$$\hat{g}_i = \frac{1}{n-1} \sum_{j=1, j\neq i}^{n} \hat{g}_{ij} \tag{86}$$

which is a mean degree to which alternative s_i is strongly preferred, by the whole group, over all the other alternatives.

Then

$$\hat{z}_Q^i = \mu_Q(\hat{g}_i) \tag{87}$$

is the extent to which alternative s_i is strongly preferred, by the whole group, over Q (e.g., most) other alternatives.

Finally, we define the *fuzzy s/Q-consensus winner* as

$$W_{s/Q} = \hat{z}_Q^1/s_1 + \cdots + \hat{z}_Q^n/s_n \tag{88}$$

i.e. as a fuzzy set of alternatives that are strongly preferred, by the whole group, over Q other alternatives.

Example 6 For the same individual fuzzy preference relations as in Example 4, and using (75) and (76), we obtain the following social fuzzy preference relation

$$R = \begin{array}{r|cccc} & j=1 & 2 & 3 & 4 \\ \hline i=1 & 0 & 0 & 1 & 0.25 \\ 2 & 0.75 & 0 & 0.75 & 0.25 \\ 3 & 0 & 0.25 & 0 & 0 \\ 4 & 1 & 0.75 & 1 & 0 \end{array}$$

If now the fuzzy majority is given by Q = "most" defined by (5) and $\alpha = 0,8$, then we obtain

$$W_{\text{"most"}} = \tfrac{1}{15}/s_1 + \tfrac{11}{15}/s_2 + 1/s_4$$
$$W_{0.8/\text{"most"}} = \tfrac{1}{15}/s_1 + \tfrac{11}{15}/s_4$$
$$W_{s/\text{"most"}} = \tfrac{1}{15}/s_1 + \tfrac{1}{15}/s_2 + 1/s_4$$

which is to be read similarly as for the fuzzy cores in Example 4. Notice that here once again alternative s_4 is clearly the best choice which is obvious by examining the social fuzzy preference relation. □

The traditional aggregation based on Zadeh's (1983) calculus of linguistically quantified statements, which has been employed above in the derivation of the fuzzy consensus winners, may be replaced by its respective OWA based aggregation defined by (12) and (13), and denoted generically as $\text{OWA}_Q(.)$. This was proposed by Fedrizzi and Kacprzyk (1993), and Kacprzyk and Fedrizzi (1995a, b).

First, we start with g_{ij} (77) which expresses whether alternative s_i defeats (in the whole group's opinion!) alternative s_j or not.

Next, g_i (78) is a mean degree to which alternative s_i is preferred, by the whole group, over all the other alternatives.

Then,

$$\overline{z}_Q^i = \text{OWA}_Q(g_i) \tag{89}$$

is the extent to which alternative s_i is preferred, by the whole group, over Q (e.g., most) other alternatives.

Finally, we define the *fuzzy Q-consensus winner* as

$$\overline{W}_Q = \overline{z}_Q^1/s_1 + \cdots + \overline{z}_Q^n/s_n \tag{90}$$

i.e. as a fuzzy set of alternatives that are preferred, by the whole group, over Q other alternatives.

We can also introduce a threshold $\alpha \geq 0.5$ into (77) to obtain $g_{ij}(\alpha)$ (81) which expresses whether alternative s_i sufficiently (at least to degree α) defeats (in the whole group's opinion!) alternative s_j or not.

Next, $g_i(\alpha)$ (82) is a mean degree to which alternative s_i is sufficiently (at least to degree α) preferred, by the whole group, over all the other alternatives.

Then

$$\bar{z}_Q^i(\alpha) = \text{OWA}_Q[g_i(\alpha)] \tag{91}$$

is the extent to which alternative s_i is sufficiently (at least to degree α) preferred, by the whole group, over Q (e.g., most) other alternatives.

Finally, we define the *fuzzy α/Q-consensus winner* as

$$W_{\alpha/Q} = \bar{z}_Q^1(\alpha)/s_1 + \cdots + \bar{z}_Q^n(\alpha)/s_n \tag{92}$$

i.e. as a fuzzy set of alternatives that are sufficiently (at least to degree α) preferred, by the whole group, over Q other alternatives.

Furthermore, we can also explicitly introduce the strength of preference into (77) by defining $\hat{g}_{ij}$ (85) which expresses whether alternative s_i defeats (in the whole group's opinion!) alternative s_j or not, and if so (i.e. if $\hat{g}_{ij} > 0$), then $\hat{g}_{ij}$ gives the strength of this defeat.

Next, $\hat{g}_i$ (86) is a mean degree to which alternative s_i is strongly preferred, by the whole group, over all the other alternatives alternatives.

Then

$$\hat{y}_Q^i = \text{OWA}_Q(\hat{g}_i) \tag{93}$$

is the extent to which alternative s_i is strongly preferred, by the whole group, over Q (e.g., most) other alternatives.

Finally, we define the *fuzzy s/Q-consensus winner* as

$$W_{s/Q} = \hat{y}_Q^1/s_1 + \cdots + \hat{y}_Q^n/s_n \tag{94}$$

i.e. as a fuzzy set of alternatives that are strongly preferred, by the whole group, over Q other alternatives.

Again, the results obtained by using the OWA based aggregation are similar to those via the traditional Zadeh's (1983) calculus of linguistically quantified statements.

This concludes our brief exposition of how to employ fuzzy linguistic quantifiers to model the fuzzy majority in group decision making. For readability and simplicity we have only shown the application of Zadeh's (1983) calculus of linguistically quantified propositions. The use of Yager's (1983) calculus is presented in the source papers by Kacprzyk (1985b, c).

We will not present some other solution concepts as, e.g., minimax consensus winners (cf. Nurmi, 1981, Kacprzyk, 1985c) or those based on fuzzy tournaments which have been proposed by Nurmi and Kacprzyk (1991).

4.3 DEGREES OF CONSENSUS UNDER FUZZY PREFERENCES AND A FUZZY MAJORITY

In this section fuzzy linguistic quantifiers as representations of a fuzzy majority will be employed to define a degree of consensus as proposed in Kacprzyk (1987), and then advanced in Kacprzyk and Fedrizzi (1986, 1988, 1989), and Fedrizzi and Kacprzyk (1988) [see also Kacprzyk, Fedrizzi and Nurmi (1990, 1992a, b)]. This degree is meant to overcome some "rigidness" of the conventional concept of consensus in which (full) consensus occurs only when "all the individuals agree as to all the issues". This may often be counterintuitive, and not consistent with a real human perception of the very essence of consensus (see, e.g., the citation from a biological context given in the beginning of the paper).

The new degree of consensus proposed can be therefore equal to 1, which stands for full consensus, when, say, "most of the individuals agree as to almost all (of the relevant) issues (alternatives, options)".

Our point of departure is again a set of individual fuzzy preference relations which are meant analogously as in Section 4.2 [see, e.g., (refeq8)].

The degree of consensus is now derived in three steps:

- first, for each pair of individuals we derive a degree of agreement as to their preferences between *all* the pairs of alternatives,
- second, we aggregate these degrees to obtain a degree of agreement of each pair of individuals as to their preferences between Q_1 (a linguistic quantifier as, e.g., "most", "almost all", "much more than 50%", ...) pairs of relevant alternatives, and
- third, we aggregate these degrees to obtain a degree of agreement of Q_2 (a linguistic quantifier similar to Q_1) pairs of important individuals as to their preferences between Q_1 pairs of relevant alternatives, and this is meant to be the *degree of consensus* sought.

Notice that we assume here, as opposed to Section 4.2, that both the individuals and alternatives are assigned different degrees of importance and relevance. However, this may be useful in the context of consensus reaching, and a basic case with the same importance and relevance for all the individuals and alternatives will just be a special case of the one adopted in this paper.

The above derivation process of a degree of consensus may be formalized by using Zadeh's (1983) calculus of linguistically quantified statements and Yager's (1988, 1994) OWA based aggregation outlined in Section 4.1.

We start with the degree of strict agreement between individuals k_1 and k_2 as to their preferences between alternatives s_i and s_j

$$v_{ij}(k_1, k_2) = \begin{cases} 1 & \text{if } r_{ij}^{k_1} = r_{ij}^{K_2} \\ 0 & \text{otherwise} \end{cases} \tag{95}$$

where here and later on in this section, if not otherwise specified, $k_1 = 1, \ldots, m-1$; $k_2 = k_1 + 1, \ldots, m$; $i = 1, \ldots, n-1$; $j = i+1, \ldots, n$.

The relevance of alternatives is assumed to be given as a fuzzy set defined in the set of alternatives S such that $\mu_B(s_i) \in [0,1]$ is a *degree of relevance* of option s_i, from 0 for fully irrelevant to 1 for fully relevant, through all intermediate values.

The relevance of a pair of alternatives, $(s_i, s_j) \in S \times S$, may be defined, say, as

$$b_{ij}^k = \frac{1}{2}[\mu_B(s_i) + \mu_B(s_j)] \tag{96}$$

which is clearly the most straightforward option; evidently, $b_{ij}^B = b_{ji}^k$, and b_{ii}^k do not matter; for each i, j, k.

And analogously, the *importance of individuals*, I, is defined as a fuzzy set in the set of individuals such that $\mu_I(k) \in [0,1]$ is a *degree of importance* of individual k, from 0 for fully unimportant to 1 for fully important, through all intermediate values.

Then, the importance of a pair of individuals, (k_1, k_2), b_{k_1,k_2}^I, may be defined in various ways, e.g., analogously as (96), i.e.

$$b_{k_1,k_2}^I = \frac{1}{2}[\mu_I(k_1) + \mu_I(k_2)] \tag{97}$$

The degree of agreement between individuals k_1 and k_2 as to their preferences between *all* the pairs of alternatives is [cf. (8)]

$$v_B(k_1, k_2) = \frac{\sum_{i=1}^{n-1}\sum_{j=i+1}^{n}[v_{ij}(k_1,k_2) \wedge b_{ij}^B]}{\sum_{i=1}^{n-1}\sum_{j=i+1}^{n} b_{ij}^B} \tag{98}$$

The degree of agreement between individuals k_1 and k_2 as to their preferences between Q_1 relevant pairs of alternatives is

$$v_{Q_1}^B(k_1, k_2) = \mu_{Q_1}[v_B(k_1, k_2)] \tag{99}$$

In turn, the degree of agreement of *all* the pairs of important individuals as to their preferences between Q_1 pairs of relevant alternatives is

$$v_{Q_1}^{I,B} = \frac{2}{m(m-1)} \frac{\sum_{k_1=1}^{m-1}\sum_{k_2=k_1+1}^{m}[v_{Q_1}^B(k_1,k_2) \wedge b_{k_1,k_2}^I]}{\sum_{k_1=1}^{m-1}\sum_{k_2=k_1+1}^{m} b_{k_1,k_2}^I} \tag{100}$$

and, finally, the degree of agreement of Q_2 pairs of important individuals as to their preferences between Q_1 pairs of relevant alternatives, called the *degree of $Q1/Q2/I/B$-consensus*, is

$$con(Q_1, Q_2, I, B) = \mu_{Q_2}(v_{Q_1}^{I,B}) \tag{101}$$

Since the strict agreement (95) may be viewed too rigid, we can use the degree of sufficient agreement (at least to degree $\alpha \in (0,1]$ of individuals k_1 and k_2 as to their preferences between options s_i and s_j, defined by

$$v_{ij}^{\alpha}(k_1,k_2) = \begin{cases} 1 & \text{if } \mid r_{ij}^{k_1} - r_{ij}^{k_2} \mid \leq 1-\alpha \leq 1 \\ 0 & \text{otherwise} \end{cases} \tag{102}$$

where, $k_1 = 1,\ldots,m-1$; $k_2 = k_1+1,\ldots,m$; $i = 1,\ldots,n-1$; $j = i+1,\ldots,n$.

The degree of sufficient (at least to degree α) agreement between individuals k_1 and k_2 as to their preferences between all the relevant pairs of alternatives is

$$v_B^{\alpha}(k_1,k_2) = \frac{\sum_{i=1}^{n-1}\sum_{j=i+1}^{n}[v_{ij}^{\alpha}(k_1,k_2) \wedge b_{ij}^{B}]}{\sum_{i=1}^{n-1}\sum_{j=i+1}^{n} b_{ij}^{B}} \tag{103}$$

The degree of agreement between sufficient (at least to degree α) agreement between the individuals k_1 and k_2 as to their preferences between Q_1 relevant pairs of alternatives is

$$v_{Q_1}^{B,\alpha}(k_1,k_2) = \mu_{Q_1}[v_B^{\alpha}(k_1,k_2)] \tag{104}$$

In turn, the degree of sufficient (at least to degree α) agreement of *all* the pairs of important individuals as to their preferences between Q_1 pairs of relevant alternatives is

$$v_{Q_1}^{I,B,\alpha} = \frac{2}{m(m-1)} \frac{\sum_{k_1=1}^{m-1}\sum_{k_2=k_1+1}^{m}[v_{Q_1}^{B\alpha}(k_1,k_2) \wedge b_{k_1,k_2}^{I}]}{\sum_{k_1=1}^{m-1}\sum_{k_2=k_1+1}^{m} b_{k_1,k_2}^{I}} \tag{105}$$

and, finally, the degree of sufficient (at least to degree α) agreement of Q_2 pairs of important individuals as to their preferences between Q_1 pairs of relevant alternatives, called the *degree of $\alpha/Q1/Q2/I/B$-consensus*, is

$$con^{\alpha}(Q_1,Q_2,I,B) = \mu_{Q_2}(v_{Q_1}^{I,B,\alpha}) \tag{106}$$

We can also explicitly introduce the strength of agreement into (95), and analogously define the degree of strong agreement of individuals k_1 and k_2 as to their preferences between options s_i and s_j, e.g., as

$$v_{ij}^{s}(k1,k2) = s(\mid r_{ij}^{k_1} - r_{ij}^{k_2} \mid) \tag{107}$$

where $s : [0,1] \longrightarrow [0,1]$ is some function representing the degree of strong agreements as, e.g.,

$$s(x) = \begin{cases} 1 & \text{for } x \leq 0.05 \\ -10x + 1.5 & \text{for } 0.05 < x < 0.15 \\ 0 & \text{for } x \geq 0.15 \end{cases} \tag{108}$$

such that $x' < x'' \Longrightarrow s(x') \geq s(x'')$, for each $x', x'' \in [0,1]$, and there is no such an $x \in [0,1]$ that $s(x) = 1$.

The degree of strong agreement between individuals k_1 and k_2 as to their preferences between *all* the pairs of alternatives is then

$$v_B^s(k_1,k_2) = \frac{\sum_{i=1}^{n-1}\sum_{j=i+1}^{n}[v_{ij}^s(k_1,k_2) \wedge b_{ij}^B]}{\sum_{i=1}^{n-1}\sum_{j=i+1}^{n} b_{ij}^B} \tag{109}$$

The degree of strong agreement between individuals k_1 and k_2 as to their preferences between Q_1 relevant pairs of alternatives is

$$v_{Q_1}^{B,s}(k_1,k_2) = \mu_{Q_1}[v_B^s(k_1,k_2)] \tag{110}$$

In turn, the degree of strong agreement of *all* the pairs of important individuals as to their preferences between Q_1 pairs of relevant alternatives is

$$v_{Q_1}^{I,B,s} = \frac{2}{m(m-1)} \frac{\sum_{k_1=1}^{m-1}\sum_{k_2=k_1+1}^{m}[v_{Q_1}^{B,s}(k_1,k_2) \wedge b_{k_1,k_2}^I]}{\sum_{k_1=1}^{m-1}\sum_{k_2=k_1+1}^{m} b_{k_1,k_2}^I} \tag{111}$$

and, finally, the degree of agreement of Q_2 pairs of important individuals as to their preferences between Q_1 pairs of relevant alternatives, called the *degree of s/Q1/Q2/I/B-consensus*, is

$$con^s(Q_1,Q_2,I,B) = \mu_{Q_2}(v_{Q_1}^{I,B,s}) \tag{112}$$

Example 7 Suppose that $n = m = 3$, $Q_1 = Q_2 =$ "most" are given by (5), $\alpha = 0.9$, $s(x)$ is defined by (108), and the individual preference relations are:

$R_1 = [r_{ij}^1] =$

	$j=1$	2	3
$i=1$	0	0.1	0.6
2	0.9	0	0.7
3	0.4	0.3	0

$R_2 = [r_{ij}^2] =$

	$j=1$	2	3
$i=1$	0	0.1	0.7
2	0.9	0	0.7
3	0.3	0.3	0

$R_3 = [r_{ij}^3] =$

	$j=1$	2	3
$i=1$	0	0.2	0.6
2	0.8	0	0.7
3	0.4	0.3	0

If we assume the relevance of the alternatives to be $b_i^B = 1/s_1 + 0.6/s_2 + 0.2/s_3$, the importance of the individuals to be $b_k^I = 0.8/1 + 1/2 + 0.4/3$, $\alpha = 0.9$ and $Q =$ "most" given by (5), then we obtain the following degrees of consensus:

$$con(\text{"most"}, \text{"most"}, I, B) \cong 0.35$$
$$con^{0.9}(\text{"most"}, \text{"most"}, I, B) \cong 0.06$$
$$con^{s}(\text{"most"}, \text{"most"}, I, B) \cong 0.06$$

□

And, similarly as for the group decision making solutions shown in Section 4.2, the aggregation via Zadeh's (1983) calculus of linguistically quantified propositions employed above may be replaced by the OWA based aggregation given by (12) and (13). The procedure is analogous as that presented in Section 4.2, and will not be repeated here.

For more information on these degrees of consensus, see Fedrizzi and Kacprzyk (1988), Kacprzyk (1987a), and Kacprzyk and Fedrizzi (1986, 1988, 1989). Moreover, the use of Yager's fuzzy logic based calculus of linguistically quantified propositions is given in Kacprzyk and Fedrizzi (1989), while the use of the OWA operators is given in Fedrizzi and Kacprzyk (1993), Fedrizzi, Kacprzyk and Nurmi (1993), Kacprzyk and Fedrizzi (1995a) or Kacprzyk and Fedrizzi (1995b).

4.4 CONCLUDING REMARKS

In this paper we have briefly presented the use of fuzzy preference relations and fuzzy majorities in the derivation of group decision making (social choice) solution concepts and degrees of consensus. First, we briefly discussed some more general issues related to the role fuzzy preference relations and a fuzzy majority may play as a tool to alleviate difficulties related to negative results in group decision making exemplified by Arrow's impossibility theorem. Though very important for a conceptual point of view, these analyses are of a lesser practical relevance to the user who wishes to employ those fuzzy tools to constructively solve the problems considered.

Therefore, emphasis has been on the use of fuzzy preference relations and fuzzy majorities to derive more realistic and human-consistent solutions of group decision making. Reference has been given to other approaches and works in this area, as well as to the authors' previous, more foundational works in which an analysis of basic issues underlying group decision making and consensus formation has been included.

It is hoped that this work will provide the interested reader with some tools to constructively solve group decision making and consensus formation problems when both preferences and majorities are imprecisely specified or perceived, and may be modeled by fuzzy relations and fuzzy sets.

References

Aizerman, M.A. (1985). New problems in the general choice theory, *Social Choice and Welfare* 2, 235–282.

Arrow, K.J. (1963). *Social Choice and Individual Values.* Second Edition. Wiley, New York.

Barrett, C.R. and Pattanaik, P.K. (1990). Aggregation of fuzzy preferences. In J. Kacprzyk and M. Fedrizzi, (Eds.): *Multiperson Decision Making Models using Fuzzy Sets and Possibility Theory*, Kluwer, Dordrecht, pp. 155–162.

Barrett, C.R., Pattanaik, P.K. and Salles, M. (1986). On the structure of fuzzy social welfare functions. *Fuzzy Sets and Systems*, 19, 1–10.

Barrett, C.R., Pattanaik, P.K. and Salles, M. (1990). On choosing rationally when preferences are fuzzy. *Fuzzy Sets and Systems*, 34, 197–212.

Barrett, C.R., Pattanaik, P.K. and Salles, M. (1992). Rationality and aggregation of preferences in an ordinally fuzzy framework. *Fuzzy Sets and Systems*, 49, 9–13.

Basu,K, Deb, R. and Pattanaik, P.K. (1992) Soft sets: An ordinal formulation of vagueness with some applications to the theory of choice. *Fuzzy Sets and Systems*, 45, 45–58.

Bezdek, J.C., Spillman, B. Spillman, R. (1978). A fuzzy relation space for group decision theory, *Fuzzy Sets and Systems*, 1, 255–268.

Bezdek, J.C., Spillman, B. and Spillman, R. (1979). Fuzzy relation space for group decision theory: An application, *Fuzzy Sets and Systems* 2, 5–14.

Blin, J.M. (1974). Fuzzy relations in group decision theory, *J. of Cybernetics*, 4, 17–22.

Blin, J.M. and Whinston, A.P. (1973). Fuzzy sets and social choice, *J. of Cybernetics*, 4, 17–22.

Carlsson, Ch. et al. (1992). Consensus in distributed soft environments, *Europ. J. of Operational Research*, 61, 165–185.

Delgado, M., Verdegay, J.L. and Vila, M.A. (1993). On aggregation operations of linguistic labels, *Int. J. of Intelligent Systems*, 8, 351–370.

Fedrizzi, M. and Kacprzyk, J. (1988). On measuring consensus in the setting of fuzzy preference relations. In J. Kacprzyk and M. Roubens (Eds.): *Non-Conventional Preference Relations in Decision Making*, pp. 129–141.

Fedrizzi, M. and Kacprzyk, J. (1993). Consensus degrees under fuzzy majorities and preferences using OWA (ordered weighted average) operators, *Proc. of Fifth IFSA World Congress '93* (Seoul, Korea, July 1993), Vol. I, pp. 624-626.

Fedrizzi, M., Kacprzyk, J. and Nurmi, H. (1993). Consensus degrees under fuzzy majorities and fuzzy preferences using OWA (ordered weighted average) operators, *Control and Cybernetics*, 22, 71–80.

Fedrizzi, M., Kacprzyk, J. and Nurmi, H. (1996). How different are social choice functions: a rough sets approach, *Quality and Quantity*, 30, 87–99.

Fedrizzi, M., Kacprzyk, J., Owsiński, J.W. and Zadrożny, S. (1994). Consensus reaching via a GDSS with fuzzy majority and clustering of preference profiles, *Annals of Operations Research*, 51, 127–139.

Fedrizzi, M., Kacprzyk, J. and Zadrożny, S. (1988). An interactive multi-user decision support system for consensus reaching processes using fuzzy logic with linguistic quantifiers, *Decision Support Systems*, 4, 313–327.

Fishburn, P.C. (1990). Multiperson decision making: a selective review. In J. Kacprzyk and M. Fedrizzi (Eds.): *Multiperson Decision Making Models using Fuzzy Sets and Possibility Theory*, Kluwer, Dordrecht, pp. 3–27.

Gibbard A. (1973). Manipulation of schemes that mixvoting with chance, *Econometrica*, 45, 665–681.

Herrera, F., Herrera-Viedma, E. and Verdegay, J.L. (1996). A model of consensus in group decision making under linguistic assessments, *Fuzzy Sets and Systems*, 78, 73–88.

Herrera, F. and Verdegay, J.L. (1995). On group decision making under linguistic preferences and fuzzy linguistic quantifiers. In B. Bouchon-Meunier, R.R. Yager and L.A. Zadeh (Eds.): *Fuzzy Logic and Soft Computing*, World Scientific, Singapore, pp. 173–180.

Intriligator, M.D. (1973). A probabilistic model of social choice, *Review of Economic Studies*, 40, 553–560.

Intriligator, M.D. (1982). Probabilistic models of choice, *Mathematical Social Sciences*, 2, 157–166.

Kacprzyk, J. (1984). Collective decision making with a fuzzy majority rule, *Proc. of WOGSC Congress*, AFCET, Paris, pp. 153–159.

Kacprzyk, J. (1985a). Zadeh's commonsense knowledge and its use in multicriteria, multistage and multiperson decision making. In M.M. Gupta *et al.* (Eds.): *Approximate Reasoning in Expert Systems*, North–Holland, Amsterdam, pp. 105–121.

Kacprzyk, J. (1985b). Some 'commonsense' solution concepts in group decision making via fuzzy linguistic quantifiers. In J. Kacprzyk and R.R. Yager (Eds.): *Management Decision Support Systems Using Fuzzy Sets and Possibility Theory*, Verlag TÜV Rheinland, Cologne, pp. 125–135.

Kacprzyk, J. (1985c). Group decision-making with a fuzzy majority via linguistic quantifiers. Part I: A consensory-like pooling; Part II: A competitive-like pooling, *Cybernetics and Systems: an Int. J.*, 16, 119–129 (Part I), 131–144 (Part II).

Kacprzyk, J. (1986a). Group decision making with a fuzzy linguistic majority, *Fuzzy Sets and Systems*, 18, 105–118.

Kacprzyk, J. (1986b). Towards an algorithmic/procedural 'human consistency' of decision support systems: a fuzzy logic approach. In W. Karwowski and A. Mital (Eds.): *Applications of Fuzzy Sets in Human Factors*, Elsevier, Amsterdam, pp. 101–116.

Kacprzyk, J. (1987a). On some fuzzy cores and 'soft' consensus measures in group decision making. In J.C. Bezdek (Ed.): *The Analysis of Fuzzy Information*, Vol. 2, CRC Press, Boca Raton, pp. 119–130.

Kacprzyk, J. (1987b). Towards 'human consistent' decision support systems through commonsense-knowledge-based decision making and control models: a fuzzy logic approach, *Computers and Artificial Intelligence*, 6, 97–122.

Kacprzyk, J. and Fedrizzi, M. (1986). 'Soft' consensus measures for monitoring real consensus reaching processes under fuzzy preferences, *Control and Cybernetics*, 15, 309–323.

Kacprzyk, J. and Fedrizzi, M. (1988). A 'soft' measure of consensus in the setting of partial (fuzzy) preferences, *Europ. J. of Operational Research*, 34, 315–325.

Kacprzyk, J. and Fedrizzi, M. (1989). A 'human-consistent' degree of consensus based on fuzzy logic with linguistic quantifiers, *Mathematical Social Sciences*, 18, 275–290.

Kacprzyk, J. and Fedrizzi, M., Eds. (1990). *Multiperson Decision Making Models Using Fuzzy Sets and Possibility Theory*, Kluwer, Dordrecht.

Kacprzyk, J. and Fedrizzi, M. (1995a). A fuzzy majority in group DM and consensus via the OWA operators with importance qualification, *Proc. of CIFT'95 - Current Issues in Fuzzy Technologies* (Trento, Italy), pp. 128–137.

Kacprzyk, J. and Fedrizzi, M. (1995b). Consensus degrees under fuzziness via ordered weighted average (OWA) operators. In Z. Bien and K.C. Min (Eds.): *Fuzzy Logic and its Applications in Engineering, Information Sciences and Intelligent Systems*, Kluwer, Dordrecht, pp. 447–454.

Kacprzyk, J., Fedrizzi, M. and Nurmi, H. (1990). Group decision making with fuzzy majorities represented by linguistic quantifiers. In J.L. Verdegay and M. Delgado (Eds.): *Approximate Reasoning Tools for Artificial Intelligence*, Verlag TÜV Rheinland, Cologne, pp. 126–145.

Kacprzyk, J., Fedrizzi, M. and Nurmi, H. (1992a). Fuzzy logic with linguistic quantifiers in group decision making and consensus formation. In R.R. Yager and L.A. Zadeh (Eds.): *An Introduction to Fuzzy Logic Applications in Intelligent Systems*, Kluwer, Dordrecht, 263–280.

Kacprzyk, J., Fedrizzi, M. and Nurmi, H. (1992b). Group decision making and consensus under fuzzy preferences and fuzzy majority, *Fuzzy Sets and Systems*, 49, 21–31.

Kacprzyk, J., Fedrizzi, M. and Nurmi, H. (1996). 'Soft' degrees of consensus under fuzzy preferences and fuzzy majorities. In J. Kacprzyk, H. Nurmi i M. Fedrizzi (Eds.): *Consensus under Fuzziness*, Kluwer, Boston, pp. 55–83.

Kacprzyk, J., Fedrizzi, M. and Nurmi, H. (1997). OWA operators in group decision making and consensus reaching under fuzzy preferences and fuzzy majority. In R.R. Yager and J. Kacprzyk (Eds.): *The Ordered Weighted Averaging Operators: Theory and Applications*, Kluwer, Boston, pp. 193–206.

Kacprzyk, J. and Nurmi, H. (1989). Linguistic quantifiers and fuzzy majorities for more realistic and human-consistent group decision making. In G. Evans, W. Karwowski and M. Wilhelm (Eds.): *Fuzzy Methodologies for Industrial and Systems Engineering*, Elsevier, Amsterdam, pp. 267-281.

Kacprzyk, J., Nurmi, H. and Fedrizzi, M., Eds. (1996). *Consensus under Fuzziness*, Kluwer, Boston.

Kacprzyk, J. and Roubens, M., Eds. (1988). *Non-Conventional Preference Relations in Decision Making*, Springer–Verlag, Heidelberg.

Kacprzyk, J. and Yager, R.R. (1984a). Linguistic quantifiers and belief qualification in fuzzy multicriteria and multistage decision making, *Control and Cybernetics*, 13, 155–173.

Kacprzyk, J. and Yager, R.R. (1984b). 'Softer' optimization and control models via fuzzy linguistic quantifiers, *Information Sciences*, 34, 157–178.

Kacprzyk, J., Zadrożny, S. and Fedrizzi, M. (1988). An interactive user-friendly decision support system for consensus reaching based on fuzzy logic with linguistic quantifiers. In M.M. Gupta and T. Yamakawa (Eds.): *Fuzzy Computing*, Elsevier, Amsterdam, pp. 307–322.

Kacprzyk, J., Zadrony, S. and Fedrizzi, M. (1997). An interactive GDSS for consensus reaching using fuzzy logic with linguistic quantifiers. In D. Dubois, H. Prade and R.R. Yager (Eds.): *Fuzzy Information Engineering-A Guided Tour of Applications*, Wiley, New York, pp. 567–574.

Kelly, J.S. (1978) *Social Choice Theory*, Springer–Verlag, Berlin.

Kim, J.B. (1983). Fuzzy rational choice functions, *Fuzzy Sets and Systems*, 10, 37–43.

Kuzmin, V.B. and Ovchinnikov, S.V. (1980a). Group decisions I: In arbitrary spaces of fuzzy binary relations, *Fuzzy Sets and Systems*, 4, 53–62.

Kuzmin, V.B. and Ovchinnikov, S.V. (1980b). Design of group decisions II: In spaces of partial order fuzzy relations, *Fuzzy Sets and Systems*, 4, 153–165.

Loewer, B. and Laddaga, R. (1985). Destroying the consensus, in Loewer B., Guest Ed., Special Issue on Consensus, *Synthese*, 62 (1), pp. 79–96.

Nurmi, H. (1981). Approaches to collective decision making with fuzzy preference relations, *Fuzzy Sets and Systems*, 6, 249–259.

Nurmi, H. (1982). Imprecise notions in individual and group decision theory: resolution of Allais paradox and related problems, *Stochastica*, VI, 283–303.

Nurmi, H. (1983). Voting procedures: a summary analysis, *British J. of Political Science*, 13, 181–208.

Nurmi, H. (1984). Probabilistic voting, *Political Methodology*, 10, 81–95.

Nurmi, H. (1987). *Comparing Voting Systems*, Reidel, Dordrecht.

Nurmi, H. (1988). Assumptions on individual preferences in the theory of voting procedures. In J. Kacprzyk and M. Roubens (Eds.): *Non-Conventional Preference Relations in Decision Making*, Springer–Verlag, Heidelberg, pp. 142–155.

Nurmi, H., Fedrizzi, M. and Kacprzyk, J. (1990). Vague notions in the theory of voting. In J. Kacprzyk and M. Fedrizzi (Eds.): *Multiperson Decision Making Models Using Fuzzy Sets and Possibility Theory*, Kluwer, Dordrecht, pp. 43–52.

Nurmi, H. and Kacprzyk, J. (1991). On fuzzy tournaments and their solution concepts in group decision making, *Europ. J. of Operational Research*, 51, 223–232.

Nurmi, H., Kacprzyk, J. and Fedrizzi, M. (1996). Probabilistic, fuzzy and rough concepts in social choice, *Europ. J. of Operational Research*, 95, 264–277.

Ovchinnikov, S.V. (1990). Means and social welfare functions in fuzzy binary relation spaces. In J. Kacprzyk and M. Fedrizzi (Eds.): *Multiperson Decision*

Making Models using Fuzzy Sets and Possibility Theory, Kluwer, Dordrecht, pp. 143–154.

Pawlak, Z. (1982). Rough sets. *Int. Journal of Information and Computer Sciences*, 11, 341–356.

Pawlak, Z. (1991). *Rough Sets: Theoretical Aspects of Reasoning about Data.* Kluwer, Dordrecht.

Roubens, M. and Vincke, Ph. (1985). *Preference Modelling*, Springer–Verlag, Berlin.

Salles, M. (1996). Fuzzy utility. In S. Barbera, P.J. Hammond and C. Seidl (Eds.): *Handbook of Utility Theory*, Kluwer, Boston (forthcoming).

Seo, F. and Sakawa, M. (1985). Fuzzy multiattribute utility analysis for collective choice, *IEEE Trans. on Systems, Man and Cybernetics*, SMC–15, 45–53.

Szmidt, E. and Kacprzyk, J. (1996). Intuitionistic fuzzy sets in group decision making, *Notes on Intuitionistic Fuzzy Sets*, 2, 15–32.

Tanino, T. (1984). Fuzzy preference orderings in group decision making, *Fuzzy Sets and Systems*, 12, 117–131.

Tanino, T. (1990). On group decision making under fuzzy preferences. In J. Kacprzyk and M. Fedrizzi (Eds.): *Multiperson Decision Making Models using Fuzzy Sets and Possibility Theory*, Kluwer, Dordrecht, pp. 172–185.

Yager, R.R. (1983). Quantifiers in the formulation of multiple objective decision functions, *Information Sciences*, 31, 107–139.

Yager, R.R (1988). On ordered weighted averaging aggregation operators in multicriteria decision making, *IEEE Trans. on Systems, Man and Cybernetics*, SMC-18, 183-190.

Yager, R.R. (1993). On the Issue of Importance Qualifications in Fuzzy Multi-Criteria Decision Making. Tech. Report MII-1323, Machine Intelligence Institute, Iona College, New Rochelle, NY.

Yager, R.R. and Kacprzyk, J. (Eds.) (1997). *The Ordered Weighted Averaging Operators: Theory and Applications*, Kluwer, Boston.

Zadeh, L.A. (1983). A computational approach to fuzzy quantifiers in natural languages, *Computers and Maths. with Appls.*, 9, 149–184.

5 ELEMENTS OF FUZZY GAME THEORY

Antoine Billot

Abstract: This chapter presents in a first section the intuitional relationships between games and fuzziness; then, in a second section, we propose some arguments to justify and illustrate the two main concepts of fuzzy information and fuzzy coalition, on which cooperative and noncooperative analyses are founded; next, in the two following sections, we develop the noncooperative framework essentially from Butnariu's contributions, gathering the literature and presenting the results, and finally the cooperative model essentially from Aubin's works, Hüsseinov's and Billot's.

5.1 INTRODUCTION: RATIONAL BEHAVIOR AND GAMES

From a historical point of view, game theory is more or less one of the most smart children of economics. Several of its basic concepts were already leading to the ideas of equilibria whose existence conditions will be finally analyzed by game theory. Cournot's and Stackelberg's duopolies are as many historical notions that yet define the main concepts game theory finally uses and exhibits, as shown by Friedman 1990; for instance, the functions of reaction first studied by Stackelberg correspond to the famous 'response-map' of the noncooperative games; Debreu 1959 will later prove that a Walrasian general equilibrium is a Nash one, that they are equivalent in terms of social efficiency when the agents' cooperation is forbidden; Edgeworth's box generates equilibria which deal with the saddle-points of the first theorem of the theory.

However, von Neumann & Morgenstern's basic work - first published in 1944 - could not be summed up as a systematic mathematization of already existing results whose principles should be exclusively based on economics. The genesis of standard game theory was drived according to a fruitful dialectic movement between economics and mathematics, as the academic membership of its main contributors shows: Borel, von Neumann were mathematicians while Morgenstern was economist. What happened to Boolean theory, happened also to fuzzy one.

Methodologically, the chronology of the introduction of fuzzy set theory within game theory is very similar to that of the usual one. For instance, while Ponsard's first works on fuzzy behaviors were published, Butnariu's fuzzy fixed points theorems also appeared, in a way that recalls Kakutani and Arrow, since, in both cases fuzzy or not, the fixed point results are central in general equilibrium analysis. In a very coarse presentation, fuzzy microeconomics could be seen as an attempt to a generalized theory that integrates the standard one while expanding its explanation power thanks to a broader tool. However, very quickly, fuzzy economics and fuzzy game theory could be viewed as two autonomous new approaches directly linked to the descriptive improvement of the normative–standard-hard-defined-behavior-theories that are based upon a crisp notion of the individual rationality.

Hence, as soon as it is basically natural to adopt the view point of bounded rationality requirement - and it is the case now, for the great majority of economists and/or game theorists -, the resort to fuzzy modelling seems to be one of the possible ways. Actually, this means that the main arguments - often implicit - which are the true basis of this research program devoted to the setting up of a soft analysis - game theory being just one of the aspects to measure the influence of Zadeh's contribution inside economics - come historically from authors who developed in their own papers the very set of critiques on which the modern economics, including game theory, are founded. In a provocative way, we can say that Shackle - because of his notion of potential surprise which is a fuzzy measure - or Keynes - because of his works devoted to nonprobabilistic uncertainty -, or finally Simon - because of his bounded rationality proposal - are closer to these ideas than Nash, Harsanyi or Selten (the three Nobel-prized game theorists), even if the instrumental framework is the latters'.

The two operational notions that we commonly use in game theory are the following: *information* and *cooperation*. The first one appears soon in Borel's seminal paper in 1906 and designs the main conceptual tool for the noncooperative side. The second one comes directly from the *intuitional exchange box* due to Edgeworth in 1881, this last model being considered as the very first contribution for the cooperative side. Apart from these two basic articles and their followers, Zadeh 1965, in freeding the set algebra from the Boolean dichotomic membership, allows first, to relax the néo-classical rationality that was criticized by Simon 1954, thanks to the fuzzy relations design and second, to introduce more convexity by means of the natural property of the fuzzy referential set $[0,1]$ that replaces the Boolean nonconvex one, $\{0,1\}$.

Hence, it was not a surprise if the fuzzy apply to economics (and also that of the nonadditive probability theory which is a close neighbor of the fuzzy set one) has focused first on the player's behavior, especially in revisiting the rationality criterion in order to keep it relevant with the fuzzy relations of preference. Actually, the problem was to integrate some nonstandard (nontransitive and/or weakly reflexive) binary relations which were totally excluded because of their impossibility to yield a continuous numerical representation of the associated preferences. To show the paradoxical consistency of such decisions with the existence of a balanced solution - for a pure game as well as for an exchange economy - in modifying the notion of transitivity and that of completeness for a weakorder, was the previous step of a broader analysis in which game theory found its natural place. Thanks to the mathematical generalization of Kakutani's fixed point theorem, Butnariu could go on the achievement of a fuzzy noncooperative game theory in focusing on the beliefs formation about the individual credibility of the other agents' strategies. The third step starts with Aubin who interprets the possible fuzziness of the coalitions' membership as a new expression of the Aumann's intuition for the 'agents *continuum*'. This last concept, in a modified expression (replication vs continuum), was also fruitfully used by Debreu & Scarf 1963 in order to relax the impossibility theorem establishing the total vacuity of the core when the individuals' preferences are not convex. Hence, Aubin provides a large number of results leading to the existence of a smaller set of balanced issues for a cooperative framework generally presented as a pure exchange economy *à la* Edgeworth, therefore solving the problem of their multiplicity. In the same kind of 'philosophy', but to obtain the contrary result, i.e. to produce a list of conditions that guarantee the core nonemptiness, Billot 1990, 1995b used the fuzzy set theory in order to weaken the convexity requirement by means on one hand of a soft definition of that property which integrates the standard one when the referential subset remains Boolean and on the other hand of an aggregation rule that is based upon a fuzzy modelling of behaviors.

This paper follows the natural lines of the introduction in presenting in a first section the intuitional relationships between games and fuzziness; then, in a second section, we propose some arguments to justify and illustrate the two main concepts of fuzzy information and fuzzy coalition, on which cooperative and noncooperative analyses are founded; next, in the two following sections, we develop the noncooperative framework essentially from Butnariu's contributions, gathering the literature and presenting the results, and finally the cooperative model essentially from Aubin's works, Hüsseinov's and Billot's.

5.2 GAMES AND FUZZINESS

5.2.1 General Statements about Boolean Game Theory

Standard game theory as well as usual applications in economics - such as imperfect competition analysis or general equilibrium theorems with or without production (case of a pure exchange economy) - is based upon a distinction between the *agents* (pure players, consumers or producers) and the *objects* (strategies, commodities, goods or labor) that defines the two fundamental sets of a game, this last being an economy E (or a pure interactive framework). Formally, we note S the finite set of agents i, $i = 1, ..., m$, and X the finite set of objects j, $j = 1, ..., l$. A *commodity bundle* is a nonnegative vector $x = (x_1, ..., x_l)$ of $X^l \subset R^l$, it can be also considered as an individual *strategy* x^i for any agent $i \in S$. A muple $(x^i)_{i=1}^m$ is called an *issue of the game* (or an *allocation* in an economy E). In assuming different agents to face different sets - all of them belonging to X -, we define each individual *set of strategies* X_i as a subset of R^l. Some rules of behavior are introduced in order to link both sets, such as *instrumental* individual rationality (i.e. the way to exhibit the utility functions from the preference relations $(\succeq_i)_{i=1}^m$ defined upon the $(X_i)_{i=1}^m$ or the payoffs functions and the choice criterion) and *cognitive* individual (or/and collective) rationality (i.e. the design of the agents' beliefs about the decision environment and the way they update them) - for further details on the two kinds of rationality, see e.g. Walliser 1989. Under some standard axioms, transitivity, completeness and continuity, the relations $(\succeq_i)_{i=1}^m$ are order-preserved and numerically translated by the *utility* or *payoffs functions* $(u_i)_{i=1}^m$.

From each agent's view point, a game can be first characterized by means of his information, namely his knowledge about the *game structure* that we can denote as a triplet $(S, (X_i)_{i=1}^m, (u_i)_{i=1}^m)$, this last gathering notions such as players' name, players' set of strategies, players' payoffs functions untill the agent's knowledge about the game structure itself .(This kind of analysis linked to the *common knowledge hypothesis* due to Aumann 1976 corresponds to another framework, based on knowledge hierarchies. A lot of relationships can be proved to exist between the epistemic structures and the possibility-necessity measures, see Billot & Walliser 1995.) The agent's information can vary from the completeness to the total ignorance, describing all situations between that justify the use of some uncertainty measures. From that definition of a general game structure, the modern literature expands the notion to different sorts of new games: dynamic, repeated, differential, with side-payments or zero-sum, evolutionary, with finite automota....

There is another criterion which crosses the whole theory which is linked to the occurrence of the agents cooperation: sometimes, the modeller introduces the possibility for the players to join some coalitions in order to defend more efficiently their own interests. Hence, in this type of *cooperative* game, the balanced issue of the game, namely the m-vector of strategies $\vec{x} = (x^i)_{i=1}^m$ that corresponds to the equilibrium, must be allowed by every feasible *coalition*, i.e. each subset of $S - \emptyset$, from the smallest - each agent $\{i\}$ himself - to the biggest - all the players, S. As soon as one subset of S, at least, blocked the proposed issue, this last is not acceptable; it does not to belong to the *core* of the game, i.e. the set of feasible issues that are not blocked. In cooperative game theory, we can define the associated structure as the following triplet: $(2^S - \emptyset, (X_i)_{i=1}^m, (u_i)_{i=1}^m)$; one needs a condition to establish the circumstances under which a coalition blocks a proposed issue. Intuitively, as long as there exists an agent $i \in S$ such that his individual interest, i.e. the level of u_i defined on the Cartesian product $\prod_{i=1}^m X_i$ for a given admissible issue $\vec{x}$, is *lower* than what he can obtain in restricting $\vec{x}$ to the only components $\vec{x}^C$ related to a strict subset C of S, (namely he substitutes his basic exchange domain $X_i \prod_{j=1}^{m-1} X_j$ to $X_i \prod_{j \in C} X_j$, where C is a feasible coalition, i.e. $C \in 2^S - \emptyset$ and $\sharp C < \sharp S$), the issue $\vec{x}$ is said to be *blocked*. On the contrary, if, for a given issue $\vec{x}$, such a player does not exist, then $\vec{x}$ belongs to the Boolean Core of the economy, denoted $C(E)$. The most usual interpretation of $C(E)$ is the following: all the balanced issues are based upon a sort of social unanimity; hence, they ensure the stability of the game in forbidding its bursting into several subgames whose subequilibria are inconsistent with a global solution of the game, i.e. do not correspond to an admissible issue of G. A balanced solution guarantees the existence of a global solution for which all the local interests, i.e. those of the feasible coalitions, are not sufficiently strong to incite the agents to play just inside the coalitions.

5.2.2 General Statements about Fuzzy Game Theory

The basic instrumental tool that is used in noncooperative game theory is the *fixed point theorem* (Brouwer 1912 for the functions, Kakutani 1941 for the correspondences as a generalized mapping). Intuitively, the existence of a fixed point ensures that the contract process (in terms of economic exchange or in terms of strategic interaction) is completed since, from that step on, the aggregation of the individual interests yields the same solution - i.e. the same issue - that therefore can be considered as fixed. Since in Boolean standard noncooperative results such as Nash 1951, the application of Kakutani's theorem is equivalent to the logical existence of an equilibrium, one can immediatly infer in case of fuzzy noncooperative game theory that this technical result must be modified in order to integrate some fuzziness. Hence, it is not a surprise if the very first results in that domain were devoted to a fuzzy generalization of Kakutani's theorem, proposed by Bose & Sahani 1987, Butnariu 1978, 1979, 1982, Chitra & Subrahmanyam 1987, Heilpern 1980 or Kaleva 1985.

On the other side, namely in cooperative game theory, the basic instrumental tool is the mathematical definition of the agents' powerset, based on $2^S - \emptyset$, in order to build up the *set of feasible coalitions*, i.e. a subcollection of subsets of S, that are allowed to block the possible issues. Therefore, it is not also a surprise if the first fuzzy models in cooperative game theory try either to change the notion of the agent's membership to the coalitions or to take into account some fuzziness within the economic and/or the strategic behaviors. The major contributions to that field are Aubin 1974, 1976, 1979, 1981, 1986, Butnariu 1980, 1985, 1987, Billot 1990, 1992, 1995a (Ch. 3, pp. 63-103) or Husseïnov 1994. Let us note that, in this literature, the objectives are definitely not the same according to the Aubin-Husseïnov's approach, Butnariu's one or Billot's: for the two formers, the problem is to decrease the core's cardinal by strongening the blocking capacity of the coalitions or to define some new power indexes or values in a Shapley's style, while for the latter, it is to prove the existence of some alternative equilibria that are generally forbidden when the standard core is empty and therefore, to increase the core's size. In exhibiting central tools for both analysis, fixed point theorem for the cooperative side and coalitions/behaviors for the noncooperative one, it is easy to progress throughout these two sorts of contributions so as to discover the interpretative counterpart of the instrumental resort to fuzziness. Basically, when a game is said to become *fuzzy,* whatever its level of application (in Ponsard 1986, 1987 as well as Badard 1984 for fixed points with fuzzy numbers), it is always possible to distinguish two different meanings - information vs behavior - corresponding to two different frameworks - noncooperative vs cooperative - and theoretical requirements - expansion vs restriction of the equilibria set.

When restricted to the pure noncooperative game theory, the introduction of fuzziness consists in assuming some conceptual changes in the *information* design, especially with the new notion of *information exchange* (Butnariu 1979, 1982) that models the structure of the game convergence to a weaker concept of Nash equilibrium. Hence, it is possible to make appear some particular situations where the set of possible Nash equilibria is quite different in the sense that the usual one is included in the support of the fuzzy one (see e.g. Billot 1986 or 1995a, Ch. 3, Th. 18). Let us note that some models integrate directly the weakening of the information design inside the behaviors (see e.g. Ponsard 1986). When restricting to pure cooperative game theory, the introduction of fuzziness consists in assuming some conceptual changes in the *behaviors design*, especially with the extension of the coalition membership notion in order to increase the blocking capacities or by means of a less rational preferences system which allows to relax some convexity conditions that justify the standard theorems of core emptiness. Hence, in using the concept of a *fuzzy game,* denoted, we mean in a very general way that something inside the general game framework is assumed to be fuzzy, either the agents' information (leading to set up some fuzzy subsets of strategies which gather the theoretical data on both structure and knowledge) or their behaviors (inside the coalitions allowing some relaxed constraint on the membership level or in the way they modelled their preferences).

5.3 FUZZY INFORMATION AND FUZZY COALITIONS

5.3.1 Fuzzy Information

In defining a game as *a set of rules allowing the formal information exchange*, Butnariu 1979 introduces the idea that looking for a balanced solution is equivalent to searching for the *best exchange rule of information*. It clearly means that the basic conflict which defines the game in expressing the simultaneous existence of several individual interests that are socially inconsistent (in the sense that the players' strategies are threatening one and all) can be solved by means of an efficient exchange of information and therefore, that a conflict modelled as a game is essentially *subjective,* depending on each player's information and each player's perception, i.e. his knowledge and/or beliefs. This way to consider a game, even when noncooperative, is going to justify an amazing notion (already considered in standard voting games by Moulin 1979) of *nonconstraining collaboration* which exactly leads to the fuzzy equilibrium definition in stating each possible solution satisfying this nonconstraining condition as an *equilibrium point of the fuzzy game.* The whole structure of a n-person noncooperative fuzzy game is related to this common but implicit interest of searching for the best exchange information in order to exhibit a balanced solution that corresponds to an optimal allocation of the information. From a standard framework, the very location of the fuzziness influence consists in the n individual possibility distributions that translates, for each player, the level of credibility the player awards to the strategic choice he faces. Hence, after setting up a set of *strategic conceptions* from a fuzzy information set and a possibility distribution, each player can defines a *play* as the best answer to the fuzziness of the game based on the imperfection of the information. Let us note that resorting to fuzzy set theory in such a context proceeds from the nature of the individual expectations - here, possibilistic - on the other player's behavior; the case of probabilistic beliefs corresponds to the Mertens & Zamir 1985 approach, where the basic assumption about the imperfection of the information is very close of Butnariu's, even if the associated structure is epistemic and directly expressed in terms of hierarchies - which is obviously not the case in the fuzzy noncooperative game theory we present.

One of the most important features in this analysis is the new part that is given to the information, even fuzzy, since it becomes the main concept and the more appropriated tool for getting the equilibrium. In other words, it is always possible to interpret a fuzzy noncooperative game as a game where the problem is less the apparent inconsistency of the individual interests rather than the means to communicate and to carry information in an optimal way. While there is no social requirement in Boolean games - giving rise to some notions as 'Smith's invisible hand' - which allows to interpret these games as neutral frameworks where conflicts are naturally solved thanks to a mysterious will (actually, individual néo-classical rationality and perfect information), a fuzzy game can be considered as the very first attempt to mix up a sort of bounded rationality inside game theory - at both instrumental (Billot 1986, Ponsard 1986) and cognitive (all Butnariu's contributions) level a long time before the first Abreu & Rubinstein 1988 proposals -, with a social criterion, namely the condition of nonconstraining collaboration.

5.3.2 Fuzzy Behavior

In cooperative game theory, Boolean as well as fuzzy, there are two distinct kinds of behavior. The first one - we call it the *pure individual behavior* - is dealing with the way the agents/the players design their own preferences, namely the choice of the instrumental rationality level within comparisons (two Aristotelician values of truth, $\sharp\{0,1\} = 2$, in the Boolean case, an infinite number, $\sharp\,[0,1] = \infty$, in the fuzzy one), within memory consistency (full crisp transitivity vs f-transitivity or $\vee$-$\wedge$ transitivity) and within discrimination (full completeness in Boolean case vs pure delete of it in fuzzy one, thanks to the natural completeness of the '$\geq$' binary relation) - for further details, see e.g. Billot 1987 (Ch. 2, pp. 43-65), 1995a (Ch. 1, pp. 5-33) or 1995b. The second size of behavior - called the *social individual behavior* - deals with the notion of coalition membership; in the Boolean framework, because of the crisp set algebra, it is totally impossible to introduce some shades into the way people enter the groups, syndicate, team, party... The law is that of the excluded-third: namely either 'you are *in* ', either 'you are *out*'; either the membership level is 1, either it is 0. All the situations in between are not described and not taken into account, by definition. Therefore, by allowing on one hand the players to behave in a less rational, a less precise fashion, and on the other hand, by allowing them to belong more or less to the coalitions, cooperative game theory is two times ready to become fuzzy. None of the different contributions in that field really gathers the two possibilities of introducing simultaneously some fuzziness inside. One of the possible explanations is based upon the great difference in targets for the two major approaches; while some authors try to increase the equilibria number, the others try to decrease it. Hence, because of the full incompatibility of both researches, the formers just decide to choose the more appropriate tool, i.e. for them to relax the convexity property in designing behaviors in a softer way, and the latters just decide to allow the coalitions to become fuzzy which immediatly strongens their blocking power and therefore restricts the core cardinal.

Basically, the work share is as follows: on one hand, those who wish to expand the core, if it is precisely nonempty (Billot); on the other hand, those who wish to restrict it, if nonempty, for converging to a sole solution in a spirit of the uniqueness theorems (Aubin and Hüsseinov, except when the last studies the case of individual nonconvex preferences). In other words, in the first case, the problem is to find some conditions under which, when the usual core of a cooperative game is empty, the fuzzy correspondent one is not and in the second case, the problem is to solve a game in such a way that, after proving the existence of several solutions, there are no indifferent issues between which it is impossible to choose (in order to delete Buridan's situations). Another presentation of these two polar analyses inside the cooperative game theory, can be proposed which is more technical: first, a fuzzy cooperative game can be a direct alternative formulation of the general 'conflict of interests' concept, even if considered within an 'exchange economy' model. Second, it can be a mathematical improvement of the classical framework in order to exhibit new results by means of the fuzzy set theory in order to weaken the axiomatic requirement and/or to give rise to a better interpretation of them (this corresponds to the Aubin's field). In the first case, the problem is more to integrate some real phenomena inside the theory thanks to a softer tool and, in the second, the problem is to relax the assumption of *finite* structure. In both approaches, the convexity property is central even if it is not assumed on the same part of the mathematical structure.

5.4 NONCOOPERATIVE FUZZY GAME THEORY

If we consider the normal form of a game, the decision power is completely and precisely shared between the agents. However, if we consider the characteristic form of a game, it implies a cooperative rationality that allows cooperative solutions. If the game theory, in its normative aspect, gives an answer to the question 'how must we play ?', it permits to explain the different kinds of possible behaviors, prudent, sophisticated, passive and threatening. A noncooperative game is then a mathematical object that models a conflict between different agents (players). For example, when competing on a market where substituable goods are supplied, firm managers are typically players.

The fuzzy literature devoted to noncooperative games is mostly composed of Butnariu's works 1979, 1982, 1985 on fixed points in a fuzzy universe, Bose & Sahani 1987, Heilpern 1981, Chitra & Subrahmanyam 1987, Kaleva 1985 on cardinal fuzzy games and finally Ponsard 1986 on Nash-equilibrium in a fuzzy spatial universe. Besides, we present some results issued from a continuous representation of the lexicographic preorder under a condition of fuzziness that leads to a new notion of prudent equilibirum. We must note that most of the theorems (except in Ponsard's articles) need a referential space with a metric topological structure. In a purely economic context, this implies a cardinal representation of preferences. The theory of fuzzy games (especially fuzzy noncooperative games) can be summed up in a set of specific fixed points theorems and in a few cardinalistic models. First, we present some generalizations of Kakutani's theorem and an application to Nash-equilibria and second, we introduce fuzzy behaviors, in order to generate ordinal strategic utility functions.

5.4.1 Fixed Points and Nash-Equilibrium

5.4.1.1 Fuzzy Correspondences and Fixed Points.

We are going to explain fixed point theorems when the considered universe is fuzzy. Usual fixed point theorems concern the invariance of an element when the set is transformed either by a function (Brouwer 1912) or a correspondence (Kakutani 1941). What do the usual results become when any element can partially belong to the initial set ? For answering, we introduce the definitions of a *fuzzy correspondence* and that of a *fixed point of a fuzzy correspondence*.

Definition 1 : *A* ***fuzzy correspondence*** *on the referential set X is a mapping $\mu^X(.)$ of X towards 2_f^X, where 2_f^X is the powerset of all fuzzy subsets of X.*

We note $\mu^A(.)$ the membership function of the fuzzy subset A_f of X. If $\mu_{[.]}^X(.)$ is a fuzzy correspondence on X, we also note X_f^X the fuzzy subset of X^2 which has a membership function $\mu_{[x]}(y) \equiv \mu_x(y)$, for any (x, y) of X^2. We note F_X^μ the set of fuzzy correspondences defined on X.

Definition 2 : *A* ***fixed point of a fuzzy correspondence*** *$\mu_{[.]}^X(.)$ defined on X is an element x^0 of X such that $\forall x \in X$: $\mu_x(x^0) \geq \mu_x(x)$.*

The problem of existence for a fixed point in the case of a fuzzy correspondence $\mu_{[.]}^{X}(.) \in F_X^{\mu}$ is not a simple generalization of the very classic problem of the existence of a fixed point (e.g. Kakutani) for a usual correspondence. Actually, if we define a classic correspondence (CC) $F(.)$ of X towards 2^X (where 2^X is the powerset of usual subsets of X), then the element x^0 of X is a fixed point for $F(.)$ (namely $x^0 \in F(x^0)$) if and only if x^0 is a fixed point of the fuzzy correspondence $\mu_{[.]}(.) \in F_X^{\mu}$ defined such that:

$$\forall x \in X, f_x(.) = \mu_{[x]}(.), \tag{1}$$

where $f_x(.)$ is the characteristic function of a usual subset $F(x)$ of X and $\mu_{x^0} \neq \emptyset$.

Definition 3 : *Let A_f be a fuzzy subset belonging to 2_f^X. We say that A_f is a **convex** fuzzy subset if and only if its membership function $\mu^A(.)$ is quasi-concave: $\forall t \in [0,1], \forall x, y \in A_f$, then $\mu^A\left[tx + (1-t))\,y\right] \geq \bigwedge\left[\mu^A(x), \mu^A(y)\right]$.*

Definition 4 : *The fuzzy correspondence $\mu_{[.]}^{X}(.)$ is **convex** if for the vector topological space X, locally convex and Hausdorff-separated [1] and C a subset of X, nonempty, compact and convex, the fuzzy subset $\mu_x \in 2_f^X$ is a convex fuzzy subset of X.*

Definition 5 : *The fuzzy correspondence $\mu_{[.]}^{X}(.)$ is **closed** if and only if the membership function $\mu_x(.)$ is upper semicontinuous.*

Remark 1 : *If $\mu_{[.]}^{X}(.)$ is a fuzzy correspondence on a nonempty, compact and convex subset C, we assume $\mu_{[.]}^{X}(.)$ to be defined on X with $\mu_x = \emptyset$ if $x \in C$.*

We should note also that a topological space is Hausdorff-separated if it verifies the separation properties (it is the *canonical* form of the definition establishing that for any two points in the considered space, there exists at least two disjoint open sets containing respectively both points. The metric spaces (topology associated to a distance) always verify this separation property. It is not systematically true for pure topological spaces. One of the important corollaries of this separation property is that any single element is closed, i.e. the intersection of all closed neighborhoods. The following theorem includes two classic results by Brouwer and Kakutani (Butnariu 1982).

Theorem 1 : *Let X be a real vector topological space, locally convex and Hausdorff-separated. If $C \in 2^X$ is a compact, convex and nonempty subset of X and $F(.)$ a usual correspondence of C towards 2^C, having the two following properties:*
(1) $\forall x \in X$, $F(x)$ is convex, compact and nonempty,
(2) the graph set $F = \cup_{x \in C}(\{x\} \times F(x))$ is a closed set in $X \times X$, then, $\exists x^0$; $F(x^0) \ni x^0$.

We can see that if $X = R^m$ and $F(x) = \{f(x)\}$ with $f(.)$ a continuous function of C into itself, we find Brouwer's theorem, which means that any continuous function on a subset of R^m passes through the diagonal. If $X = R^m$ and $F(.)$ is a usual upper semicontinuous correspondence, it is Kakutani's theorem (see Border 1985, Chap.6 & 15). The general theorem with the help of which we prove the existence of a fixed point in a fuzzy universe deals (Butnariu 1980, 1982, 1985, Bose & Sahani 1987 and Kaleva 1985) with the application of Theorem 1 to the usual correspondence $R(.)$ associated to the fuzzy correspondence $\mu^X_{[.]}(.)$. Each fuzzy correspondence $\mu^X_{[.]}(.)$ defined on X where $\mu_x = \emptyset$ if $x \in C$ (the compact, convex and nonempty subset of X) can be associated to a usual correspondence $R(.)$ of C towards 2^C such that: $\forall x \in C$,

$$R(x) = \{y \in C; \mu_x(y) = \bigvee_{z \in C} \mu_x(z)\}. \tag{2}$$

Theorem 2 : *Let X be a real vector topological space, locally convex and Hausdorff-separated; let C be a nonempty, compact and convex subset of X. If $\mu^X_{[.]}(.)$ is a convex fuzzy correspondence, closed in C, then $\mu^X_{[.]}(.)$ has a fixed point in C.*

We can notice that Theorem 2 is exclusively based upon a topological structure which is not necessarily a metric. Butnariu's generalized theorem 1982 (the proof we gave is adaptated from Butnariu's one and recovered by Bose & Sahani 1987) implies -when we shall apply these results to the cooperative games-cardinal payoffs and utility functions that do not respect the independence axiom. Nevertheless, this theorem is an exception with Kaleva 1985 and Heilpern 1981 which are also founded on topological structures, but f-topological in the meaning of Chang 1968, namely founded upon the two particular notions of f-continuity and f-topology. There also exists fixed points theorems for fuzzy numbers (Badard 1984) but these have nothing to do with our analysis and rely upon fuzzy arithmetic, based upon possibility distributions (see Billot 1987b, Chap.1, 1991, 1995). The extensions of Theorem 2, realized by Bose & Sahani 1987, correspond to more particular specifications, related to a particular choice of metric; Hausdorff is one of them as Heilpern's one in his 1981 paper, which is a mix of a metric -usual distance- and a membership function. It is the same for Chitra & Subrahmanyam 1987 or Liu 1985 and that is why we have concentrated our efforts on Butnariu's theorem which remains valid for any topological structure, metric or not.

5.4.1.2 Nash-Equilibria for Fuzzy Games.

Here, the target consists essentially in generalizing Nash's 1950 contribution for the characterization of the conditions under which a noncooperative game has a balanced issue, namely satisfying everyone, or at least, satisfying enough players so that they continue to choose the same strategy. Actually, the reason why economic models (in their normative and microanalytic development) often need fixed point theorems, derives from the fact that a fixed point of an application or a correspondence describes an invariance situation, i.e. a state that always remains even if the mapping (that application or the correspondence) is mathematically active. If the mapping contains a particular discourse -which is the case in economics- for example related to the agents' rational behavior (consumers optimizing their utility, producers maximizing their profit as in a Walrasian economy), this invariance becomes the keeping symbol of a particular chosen state; this means a sufficiently satisfying situation for the agents to decide them not to change it. In this case, the agents are players, i.e. decision-makers whose choices are interactive. They have a set of strategies -prices, quantities, goods, etc...- into which they choose. We say a game to be fuzzy when players have fuzzy preferences. We consider a game with n players. We say that the cardinal of the set of players is n (i.e. if S is a society of n players: $\sharp S = n$). This game is -as Butnariu 1979 said it- '*a set of rules allowing the formal information exchange*' between the participants, namely the elements of S. It becomes obvious that looking for '*the best exchange rule of information*' depends on the nature of information but also on the subjective perception of each player, perception of the exchange rules or objects of the exchange. It was implicit, since the beginning of this work that $\underline{M} = [0,1]$ (even if Theorem 2 is independent of $\underline{M}$). Following Butnariu 1978, we denote U the universe of informations or the set of the objects. In the same way, $\mu^A(x)$ denotes the membership level of object $x \in U$ to the fuzzy subset A_f of U. By definition, Kaufmann 1973, the product of two fuzzy subsets A_f and B_f is a fuzzy subset defined such that:

$$[\mu^A \times \mu^B](x) = \begin{cases} \mu^A(y) \times \mu^B(z) \text{ if } x = (y,z) \in U \\ 0 \text{ otherwise.} \end{cases} \tag{3}$$

Let us note that this definition of the product of two fuzzy subsets allows the multiplication (therefore the addition) of membership levels. This definition is necessary to define the noncooperative game model that we present (see Butnariu 1979, 1982).

1. Let $\mu_{[.]}(.)$ be a fuzzy relation between A_f and B_f and $\eta_{[.]}(.)$ be another fuzzy relation between B_f and C_f. Then $[\eta \circ \mu]_{[.]}(.)$ describes a fuzzy relation between A_f and C_f defined as follows:

$$[\eta \circ \mu]_{[x]}(t) = \begin{cases} \bigvee_y [\mu_{[x]}(y) \times \eta_{[y]}(t); y \in U] \text{ if } (x,t) \in U \\ 0 \text{ otherwise.} \end{cases} \tag{4}$$

2. If $X_f \in 2_f^A$ (where 2_f^A is the set of all the fuzzy subsets included in A_f), then the image of X_f by $\mu_{[.]}(.)$ in a fuzzy subset μ_{X_f} whose membership function corresponds to $\mu_X(.)$ is defined as follows:

$$\mu_X(y) = \bigvee [\mu^X(x) \times \mu_{[x]}(y); x \in U]. \tag{5}$$

Here, we propose an additive form of membership function. We must say that if this function derives -not from the agents' subjective evaluation, but- from an objective measure of the model, the reluctance that we have related about cardinalism disappears. The concept of a *n-person fuzzy cooperative game* can be mathematically described as a data set:

$$G = \{St_1, ..., St_i, ..., St_n; Y_1, ..., Y_i, ..., Y_n; \Pi_1, ..., \Pi_i, ..., \Pi_n\} \quad (6)$$

where N is the set of players ($\sharp N = n$; $i : 1, ..., n$), such that for any player $i \in N$, the four following definitions (we call them the $4D$) are satisfied.

1. $St_i = \{st_1^i, ..., st_{n(i)}^i\}$ ($n(i) \in N$) is the player i's set of pure strategies.

2. Y_i is a usual subset of $R^{n(i)}$. Its elements are called 'the player i's *strategic arrangements*'. If $w^i = \{w_1^i, ..., w_{n(i)}^i\}$, then w_j^i is called the *weight* assigned by player i to the pure strategy st_j^i. A n-dimension vector w, $w = \{w^1, ..., w^n\} \in Z = \prod_{i \in N} Y_i$ is called *strategic choice* in G.

3. $\Pi_i \in 2^Z$ and $\forall w \in Z$, $\pi_i(w)$ is the *possibility level*[2] of the strategic choice w evaluated by the player i (see Billot 1987b, Chap.1).

4. $Z_i = \prod_{j \in N-\{i\}} Y_j$, $W_i = 2^{Z_i} \times Y_i$, and $s_i = (A_f^i, w^i)$ is called the player i's *strategic conception* in G. Moreover, the following axiom becomes necessary:

Axiom 1 : *If* $A_f^i \in 2^{Z_i}$ *and* $A_f^i \neq \emptyset$, *then* $\exists s_i \in W_i; \pi_i(A_f^i)(w^i) \neq \emptyset$, *which implies that* $\pi_i(A_f^i) \neq \emptyset$.

The first definition means that each player i has a set of pure strategies. Its cardinal can vary from an agent to another. The second proposition allows player i to mix different pure strategies. This permits the game to become convex, by going from ST_i to Y_i convex. Each player i assigns a level of possibility deriving from his belief on the strategic choice to the defined *strategic choice*, namely the vector of pure strategic combinations. This means, that their is no rigid relation (of an *exclusive* type, as it happens with probabilistic beliefs) between the player i's believes about all the strategic choices belonging to Z. In other words, $\pi_i(w)$ just measures the membership level of possibility (*I believe that it is very possible that...*) of w and not the probability of appearance of w. This possibility is purely subjective and does not influence the possibility that i assigns to the other strategic choices, what we sum up by saying that there is no mass-unity constraint; in other words, the sum of possibilities is not necessarily equal to 1 even though π_i is defined onto $[0, 1]$. The fourth proposition explains the player i's conception to be well defined with the fuzzy subset A_f^i issued from his possibility distribution upon the strategies chosen by the other players, acknowledging the fact that i had chosen w^i as mixed strategy. This intuitively corresponds to the *best answer* notion, even though it means here more or less good answers. Let us recall that Axiom 1 means for any information set dealing with the other players' behavior, that i is always able to build up a strategy -in Butnariu's words 1978, 1979, a *regular strategy-*, namely a strategic conception that may be used as an answer to fuzziness (actually, not well-known) of the other players' behavior. Let us note that solving the problem of the best information exchange in G is equivalent to find the best play in G, since a play s in G is a mathematical representation of information exchanges inside the game. Moreover, Definition 7 below means that the strategic conception s_i is better than $\dot{s}_i$ if the first one is considered as more possible than the second.

Definition 6 : *Let G be a n-person fuzzy noncooperative game satisfying the 4D and Axiom 1. A* **play** *is a vector $s = (s_1, ..., s_n) \in W = \prod_{i \in N} W_i$.*

Definition 7 : *Let $s_i = (A^i_f, w^i)$ and $\dot{s}_i = (\dot{A}^i_f, \dot{w}^i)$ be two i's strategic conceptions. We say that s_i is a* **better strategic conception** *than $\dot{s}_i$ (we write $\mu_{[s_i]}(\dot{s}_i) > \mu_{[\dot{s}_i]}(s_i)$) if and only if $\pi_i(A^i_f)(w^i) > \pi_i(\dot{A}^i_f)(\dot{w}^i)$.*

Definition 8 : *Let $s = (s_1, ..., s_n)$ and $\dot{s} = (\dot{s}_1, ..., \dot{s}_n)$ be two plays of a game where $\forall i \in N$, $\dot{s}_i = (\dot{A}^i_f, \dot{w}^i)$. We say that s is* **socially preferred** *to $\dot{s}$ (i.e. $\mu_{[s]}(\dot{s}) > \mu_{[\dot{s}]}(s)$) if and only if $\forall i \in N, \pi_i(A^i_f)(w^i) > \pi_i(\dot{A}^i_f)(\dot{w}^i)$.*

What does it all mean? Let us consider that the players have received some information. Each player i belonging to N can build up a fuzzy subset $A^i_f \in 2^{Z_i}$ and choose $w^i \in Y_i$ to define his strategic conception $s_i = (A^i_f, w^i)$. Each player does the same thing at the same time; the game is therefore based on a play s composed of the n strategic conceptions s_i. But all of them are *more or less* possible because of the rules of the game G. We can naturally consider as rational, the players who prefer strategic conceptions that are the *most possible*, therefore, the best possible plays. It is what Butnariu calls a *regular play*. The regularity of the play derives from the transition of the strategic choice possibility towards the individual conception and finally towards the play. Let us now consider $\dot{A}^i_f$, $i \in N$, to be defined by two players from their exchanges of information; a kind of *nonconstraining cooperation* (see for this concept Moulin 1979) which then implies each player i to define the best relative strategic conception ($s_i = \dot{s}_i$). It then becomes easy to consider that rational players choose $\dot{s}_i$ such that $\dot{s}$ forms the best possible play, i.e. the *socially most preferable* play that can be reached by using the given information deriving from $\prod_{i \in N} \dot{A}^i_f$.

Definition 9 : *A* **possible solution** *(PS) of the game G is a play $\dot{s}$, $\dot{s} = (\dot{s}_1, ..., \dot{s}_n)$, where $\forall i \in N$, $\dot{s}_i = (\dot{A}^i_f, \dot{w}^i)$, such that for any other play s, $s = (s_1, ..., s_n)$, where $\forall i \in N$, $s_i = (\dot{A}^i_f, w^i)$, the play s cannot be socially preferable to $\dot{s}$, i.e.: $\forall w^i \in Y_i, \forall i \in N, \pi_i(\dot{A}^i_f)(\dot{w}^i) \geq \pi_i(\dot{A}^i_f)(w^i)$.*

The rules defining the fuzzy games may allow the players to collaborate (that is to exchange some information) without inducing a *formal* coalition. In other words, this game structure, based on the individual and social comparisons of strategic conceptions, allows a description of the agents' collaboration without modifying the noncooperative nature of the game. This *nonconstraining collaboration* -we must insist on this concept since it greatly differs from the cooperation one that is related to interest *groups*- must be considered as a pure information exchange which does not imply the *collaborating* players's decisions to be consistent with the ones that would derive from direct cooperation. Mathematically, a *nonconstraining* collaboration of the players in a noncooperative game consists in playing as described below: $\dot{s} = (\dot{s}_1, ..., \dot{s}_n)$ where $\forall i \in N, \dot{s}_i = (\dot{A}^i_f, \dot{w}^i)$ with

$$\dot{\mu}^{A_i} = (w^1, ..., w^{i-1}, w^{i+1}, ..., w^n) = \begin{cases} 1 \text{ if } \forall j \in N - \{i\}, w^j = \dot{w}^j, \\ 0 \text{ otherwise.} \end{cases} \quad (7)$$

Definition 10 : *An* **equilibrium point** *of the game G is a possible solution (PS) ,* $\dot{s} = (\dot{s}_1, ..., \dot{s}_n)$ *where* $\dot{s}_i = (\dot{A}^i_f, \dot{w}^i), \forall i \in N$, *if the play* $\dot{s}$ *satisfies the nonconstraining condition.*

Definition 11 : *The* ***individual fuzzy preference*** *of a player* $i \in N$ *is the fuzzy relation* $\Pi^i_{[.]}(.)$ *on* $2^{W \times W}$ *(with W corresponding to* $2^{Z \times Y}$*) defined as follows:*

$$\Pi^i_{[s]}(\bar{s}) = \vee\{\pi_i(A^i_f)(\overline{w^i}), \pi_i(A^i_f)(w^i)\} \times \prod_{j \in N - \{i\}} \pi_j(A^j_f)(w^j),$$

for two plays $s = (s_1, ..., s_n)$ *and* $\bar{s} = (\bar{s}_1, ..., \bar{s}_n)$ *where s is preferred i, i.e.* $\Pi^i_{[s]}(\bar{s}) > \Pi^i_{[\bar{s}]}(s)$.

Definition 12 : *The* ***social preference*** *is a fuzzy relation* $\Pi^N_{[.]}(.)$ *on* $2^{W \times W}$ *defined as follows:* $\forall (s, \bar{s}) \in W \times W$ *(where* $W = 2^Z \times Y$*),*

$$\Pi^N_{[s]}(\bar{s}) = \prod_{i \in N} \Pi^i_{[s]}(\bar{s}).$$

As in the traditional approach, the noncooperative fuzzy game equilibrium corresponds to the fixed point of a correspondence defined from the individual preferences. We now define on one hand the individual fuzzy preference (according to Butnariu 1979) and on the other, the social fuzzy preference. We can notice $\Pi^i_{[.]}(.)$ to be a cardinal measure of i's preference. This preference evaluates which strategic conception is most possible according to the other players' ones. In the same way, we can notice $\Pi^N_{[.]}(.)$ to be a cardinal measure of preference.

Theorem 3 : *Let* $\bar{s} = (\bar{s}_1, ..., \bar{s}_n)$ *be a play of G where* $\bar{s}_i = (\overline{A^i_f}, \overline{w^i}), \forall i \in N$. *If* $\overline{A^i_f} \neq \emptyset$, *for any* $i \in N$, *then both following propositions are equivalent:*
(1) $\bar{s}$ *is a PS of G,*
(2) $\bar{s}$ *is a fixed point in* $\Pi^N_{[.]}(.)$.

Remark 2 : *Theorem 3 is no more true when one of the players (noted i) is such that* $\pi_i(\overline{A^i_f})(\overline{w^i}) = 0$. *Thus, it is necessary to exclude this occurrence.*

Lemma 1 : *There exists no agent i,* $i \in N$, *such that if* $\bar{s}$ *is a fixed point of* $\Pi^N_{[.]}(.)$, *then for* $\bar{s}_i = (\overline{A^i_f}, \overline{w^i})$: $\pi_i(\overline{A^i_f})(\overline{w^i}) = 0$.

Before commenting this result, let us remember this model to be firmly inspired by Butnariu 1979, 1982 for the very notion of a *strategic conception* and basic assumptions. For the other concepts, we free ourselves from Butnariu's works because the individual relations of preference like the fundamental sets of the game G are different. Nevertheless, the originality of such an approach and especially the introduction of the *nonconstraining cooperation* into noncooperative games, (often noticed but never explicitly introduced into usual games) are contained as a whole in Butnariu's papers devoted to fuzzy noncooperative games. The next theorem classically associates an equilibrium point to a fixed point of a correspondence. Actually, Theorem 3 allows to characterize the possible solutions of the game G by the set of fixed points of the social relation of preference. Henceforth, the possible solutions are equilibria if they verify the nonconstraining collaboration property: $\overline{s}$ is a Nash-equilibrium of G if and only if it satisfies eq (7). Let us remark $\overline{W}$ the set of all plays $\overline{s}$ of G to satisfy this property (which cancels the fuzzy aspect of the information sets). Such a game -reducing W into $\overline{W}$- is called with *perfect information*. This is obvious because the information becomes Boolean (i.e. $\underline{M} = \{0, 1\}$ and $\forall i \neq j$, $\overline{A^i_f} = \overline{A^j_f}$). The fuzzy aspect of the game just dealing with the plausibility -the *possibility* as particular case (see Billot 1991)- of strategic conceptions belonging to $\overline{W}$ and the individual relations of preference, we can conceive a concept of *balanced* solution which corresponds to a Nash-equilibrium. A play of G with perfect information is then totally defined by the following vector of strategic choices, $\overline{w} = (\overline{w}^1, ..., \overline{w}^n) \in Z = \prod_{i \in N} Y_i$.

Definition 13 : *We call* ***reduced social preference****, the fuzzy relation* $\Psi^N_{[.]}(.) \in 2^{Z \times Z}$ *such that* $\Psi^N_{[w]}(\dot{w}) = \Pi^N_{[s]}(\dot{s})$, *where* s *and* $\dot{s}$ *are plays with complete information respectively determined from vectors* w *and* $\dot{w}$.

Theorem 4 : *Let* $\overline{w} = (\overline{w}^1, ..., \overline{w}^n)$ *be a vector of* Z *and* $\overline{s}$ *a play with perfect information determined by* $\overline{w}$. *It is equivalent to write:*
(1) $\overline{s}$ *is Nash-equilibrium of* G,
(2) $\overline{w}$ *is a fixed point of the reduced social preference* $\Psi^N_{[.]}(.)$ *of* G.

The existence of a Nash-equilibrium for a noncooperative fuzzy game depends on the characteristics of the fuzzy correspondence $\Psi^N_{[.]}(.)$. Under the same conditions as in Theorem 2, we obtain the following existence theorem:

Theorem 5 : *Let* Z *(*$Z = \prod_{i \in N} Y_i$*) be a real convex topological space, Hausdorff-separated, nonempty and compact. If* $\Psi^N_{[.]}(.)$ *is a convex fuzzy relation, closed in* Z, *then* G *has a Nash-equilibrium.*

We clearly see that the method used for analyzing equilibria for noncooperative fuzzy games is identical to the one of a usual game. It is based on the equivalence between a fixed point and a Nash-equilibrium, the correspondence being fuzzy or not. The application of Theorem 5 to fuzzy economic games (Ponsard 1986, 1987) really concludes by relevant results according to the concept of nonconstraining collaboration. Nevertheless, we can regret that Theorem 5 needs, in order to be proved, a modification of the assumptions of fuzzy games which were described above. In particular, the assumption of complete information suggests a question with no answer: does there exist an equilibrium for a fuzzy game if the set of strategic choices remains fuzzy? We know (Aubin 1986) the standard behavior of a rational player in a noncooperative game not to be as precise as what we call a *cautious behavior* (minimizing losses) or a *sophisticated one* (based on the successive elimination of dominated strategies). Actually, the rules which define the game allow the appearance of Nash-equilibrium to be plausible. But, we are not necessarily able to deduce what strategy is going to be chosen by each player. This means that when a game has many (nonequivalent) Nash-equilibria, a player can rationally choose an equilibrium strategy if and only if he knows which equilibrium the other players have chosen. If we try to create a *symbolic* scenario explaining a noncooperative game process, we must delimit two different steps. First, the players have to agree on the choice of an issue (a play would write Butnariu), this would constitute a formal agreement. Then, all effective communication between players would be cut off. Finally, they decide their strategic choice (or their strategic conception) without being committed by the promise they made to play this strategy. If there are many chances for the n-tuple of strategies to be a Nash-equilibrium, there are also many chances for this issue to be effectively realized. It is the same in a fuzzy game when plays issued from the conceptions only depend on strategic choices, which corresponds to the case of perfect information (i.e. $\overline{A^i_f} = \overline{A^j_f}$, $\forall i \neq j$). It is clear that these kinds of scenarii, explaining the process of a noncooperative game solution, are related with an evident step of *nonconstraining collaboration* (what Border 1985 calls *secret cooperation*). We can sum up the very nature of this equilibrium -more largely the noncooperative games- by pointing out that noncooperative game models have a descriptive size more important than normative one. The advantage of Butnariu's noncooperative fuzzy games is to precise the influence of the information structure on the existence of an equilibrium. Actually, only perfect information allows the existence of a fuzzy correspondence fixed point into the game, i.e. an equilibrium. Hence, the *nonconstraining collaboration* -which is necessary in a usual game without being directly modelled- corresponds, in a fuzzy game, to a modification of $\overline{A^i_f}$, fuzzy subset of 2^{Z_i}, by exchanging some information. The existence of a Nash-equilibrium entirely depends on these exchanges. If they are sufficient, the information becomes the same for everyone. Otherwise, only possible solutions exist. Hence, in a fuzzy noncooperative game, the step of *nonconstraining collaboration* is more than contingent to the equilibrium; it completely determines the plausibility of existence of a Nash-equilibrium.

5.4.2 Prudent Behavior and Equilibria

5.4.2.1 NonCooperative Fuzzy Ordinal Game.

We have seen, during Section 4.1, that we could define a fuzzy game from the notion of a strategic conception which permits to sum up the information and the agent's choices according to his preferences. One of the most well-known corollaries of Nash-equilibrium theorem corresponds to the minimax theorem (von Neumann & Morgenstern 1944, ed.1970) which models the prudent behavior of a rational player in a zero-sum game with two-players (i.e. the gains of one player are the losses of the other). Here, we are going to prove that any prudent player can define a functional utility that ordinally translates his preferences in regard to strategies. It becomes obvious that the *nonconstraining collaboration* is no more pertinent as an operational concept because each agent can maximize his strategic utility without knowing what the other player does. Nevertheless, this strategic utility function does not a priori exist. Actually, the preference weakorder which precisely corresponds to a prudent approach of the game is the lexicographic preorder which is not continuous and prevents any utility function to translate it in a satisfying way (even if there exist *weak* types of representations for the lexicographic preorder). Thus, it is necessary to add an assumption, that we call *Local-nonDiscrimination*, in order to get a satisfying representation of the fuzzy lexicographic preorder by to a continuous utility function.

Definition 14 : *We call **usual lexicographic preorder**, noted $\succ^L$, the ordinal product of a totally preordered set by itself.* $\forall (a,b),(a',b') \in X^2$:

$$(a,b) \succ^L (a',b') \Leftrightarrow \begin{cases} (1)\ a > a' \text{ or} \\ (2)\ a = a' \text{ and } b > b'. \end{cases}$$

We see this decomposition not to lead to a partition of X^2. So, if we consider the structure $(R^2, \succeq^L)$ and if we suppose that there exists a utility function $u(.)$, we can write for any x of R:

$$J_x = [\bigwedge_{y \in R} u[(x,y)], \bigvee_{y \in R} u[(x,y)]]. \tag{8}$$

Because of the $\succeq^L$ definition, J_x is not a hole.[3] Let us now consider $x' \in R - \{x\}$. We have: $J_x \cap J_x = \emptyset$. If $x' < x$, we must only show that: $\bigwedge_{y \in R} u[(x',y)] > \bigvee_{y \in R} u[(x,y)]$. As $(x',y) \succeq^L (x,y)$ because $x' > x$, this means $\forall y \in R, u[(x,y)] < u[(x',y)]$. Hence, we can write: $\bigvee_{y \in R} u[(x,y)] < \bigwedge_{y \in R} u[(x',y)]$, therefore: $J_x \cap J_{x'} = \emptyset$. This shows that there would be, following the definition of $u(.)$, a bijection between R (*convex*) and a *countable* set of two by two disjoint intervals in R that would not be holes, which is absurd. This conclusion is the reason why $u(.)$ cannot be a real-valued continuous utility function representing the $\succeq^L$ lexicographic preorder. To sum up, the usual lexicographic preorder is not a preorder satisfying the assumption of preference continuity which is the basis of the existence of a continuous utility function preserving the preorder.

Definition 15 : *We call* ***fuzzy lexicographic preorder*** *noted* $\mu^{\succsim}_{[.]}(.)$, *the ordinal product of a usual preordered set* X, *(reflexive (a.s.)*[4] *and* f*-transitive*[5]*) by itself.* $\forall(a,b),(a\prime,b\prime)\in X^2$:

$$\mu^{\succsim}_{[(a,b)]}(a',b') > \mu^{\succsim}_{[(a',b')]}(a,b) \Leftrightarrow \left\{ \begin{array}{l} (1)\ a > a' \ or \\ (2)\ a = a' \ and\ b > b' mbox. \end{array} \right.$$

Let X be any given set. A topology T is a family of open sets of X such that:

1. X and $\emptyset$ are open sets: $X \in T$, $\emptyset \in T$.
2. The union of any family of open sets of X is an open set of X.
3. The intersection of a finite family of open sets is an open set.

Let (X,T) be a topological space and $x \in X$. A part V of X is called a *neighborhood of* x *in* X if there exists an open set $O \in T$ such that $x \in O \subseteq V$. We note that an open neighborhood (resp. closed) is an open part (resp. closed) of X containing x. Let $\succsim$ be a usual preorder $(\succ, \sim)$ on X, let V_x be the set of all the neighborhoods of x and assume:

$$\forall x \in X, \forall V(x) \in V_x : \exists y \in X \cap V(x); y \sim x. \tag{9}$$

This assumption, in the usual case, means that for any element x of X, there exists another element y of X, always belonging to the intersection of X with any neighborhood of x (especially the smallest) such that x and y are indifferent. If this assumption is satisfied, this means that it holds for *all the neighborhoods of* x *except* x *itself.*[6] Therefore, y can be as *close* of x as possible according to the topology T. We notice this *infinite proximity* (which reminds the principle of local-nonsatiation) to be implicitly based on the existence of an element y which belongs to the smallest neighborhood of x. We remark also that this asssumption is satisfied for any x of X. This means, by transitivity of indifference (this is particularly clear if X has a minimum), that we are in a generalized indifference situation. It looks like an assumption of *short-sightedness* because the agents hardly discriminate between two proximate objects and this bad ability of discrimination conveys from element to element.

Axiom 2 : *Let* $\mu_{[.]}(.)$ *be any fuzzy preorder on* X *and* V_x *the set of all the neighborhoods of* x, *then* $\forall x \in X, \forall V(x) \in V_x$ *we assume that there exists* $y \in X \cap V(x), y \neq x$ *such that* $\bigwedge[\mu_{[x]}(.)] = \mu_{[.]}(x) = \bigwedge[\mu_{[y]}(.)] = \mu_{[.]}(y)$.

Each player is supposed to consider two very close strategies x and y as indifferent. The eventual difference of the associated payoffs is such that the agent is not able to discriminate them. Even though mathematically different, both strategies lead to the same *satisfaction*. For example, let us consider two strategies which would be different only at the smallest payoff level. A prisoner (the dilemma) will not discriminate a fifteen year penalty from a fifteen year + *1 second* penalty, even though the usual lexicographic preorder would make that difference. Each agent has fuzzy preferences corresponding to $\mu^{\succsim}_{[.]}(.)$ and considers two very close objects (x, y) as indifferent (in the sense of Axiom 2) but this indifference is not of a unique type. It can vary in intensity between 0 and 1, even though in the usual case, it is conveyed in level 1 since the usual indifference means '$\mu^{\sim}_{[x]}(y) = \mu^{\sim}_{[y]}(x) = 1$'. Actually, in the fuzzy case, the transmission of indifference which implies a generalized indifference leads to a *multivalued indifference* which is not necessarily related to a single level of utility. Let us show that the fuzzy preorder $\mu^{\succsim}_{[.]}(.)$ under Axiom 2 does not induce any more a contradiction between a countable set and a convex set (as noticed by Debreu 1959, chap.4, footnote 2). Thanks to $\mu^{\succsim}_{[.]}(.)$, we build up a fuzzy subset X_f, associated to the referential X such that:

$$X_f = \{x \in X; \mu^X(.) = \bigwedge_y [\mu^{\succsim}_{[x]}(y) = \mu^{\succsim}_{[y]}(x)]\}. \tag{10}$$

Generally, x is only indifferent to itself for $\mu^{\succsim}_{[.]}(.)$. From now on, under Axiom 2, there exists another element $y \neq x$ that is indifferent in terms of membership level, i.e. in terms of utility level. The membership level is the minimum of the indifference levels $\{\mu^{\succsim}_{[x]}(x), \mu^{\succsim}_{[x]}(y)\}$ where $\mu^{\succsim}_{[x]}(y)$ depends on x and y. This derives from the difference between local and global preferences as suggested in Billot 1991, 1995. The agents are supposed to be always irrational and this irrationality is dominated by the *anti*-projector $\bigwedge$. We also know the utility function to be directly assimilated to the membership function $\mu^X(.)$ of the fuzzy subset X_f of X. We are going to prove that Axiom 2, when applied to the fuzzy lexicographic preorder, allows to cancel the absurd *biunivoque correspondence* between a countable set of disjoint and nonempty intervals and a convex set.

Lemma 2 : *Let $(R^2, \mu^{\succsim}_{[.]}(.))$ be a structure. The set I_a being defined for $a = (x, y)$ as $I_a = [\bigwedge_{y \in R} \mu^X(x, y), \bigvee_{y \in R} \mu^X(x, y)]$, under Axiom 2, $\exists a' = (x', y)$ with $\mu^{\succsim}_{[a]}(a') = \mu^{\succsim}_{[a']}(a)$, such that $I_a \cap I_{a'} \neq \emptyset$.*

Lemma 2 shows that Axiom 2 thickens the indifference classes, even if the preorder is lexicographic. The condition of membership to the intersection of any neighborhood with X introduces a supplementary notion of proximity which allows to link the maximum and the minimum of any set. This condition means that each element x is indifferent to another element y, belonging to the intersection of X with any neighborhood of x. Hence, there exists y which is infinitely close of x, and the agent cannot discriminate them. This demonstration implies the following theorem:

Theorem 6 : *Under Axiom 2, the fuzzy lexicographic preorder can be represented by a continuous utility function, defined on a real connected referential.*

5.4.2.2 Prudent Equilibria.

From now on, we are in the general case of decentralized behaviors. In other words, we only specify a fuzzy game thanks to preferences. The term *decentralized* derives from the fact that each player must choose lonely his strategy (and not a strategic conception). All communications are forbidden, there is no initial issue nor history (or memory) of the game. We consider a noncooperative n-person game in a normal form. Each player i ($i : 1, ..., n$) can choose into a convex set of strategies, X_i - this set is a usual real set - and knows his payoff-function $P_i(.,.)$, defined on the Cartesian product of the n sets of strategies , $\prod_{i=1}^{n} X_i$. We consider that $P_i(.,.)$ is continuous on $\prod_{i=1}^{n} X_i = X$. The game in its normal form corresponds to: $(X_1, ..., X_i, ..., X_n; P_1, ..., P_n)$. We are in the case where each agent i ignores everything of the $(n-1)$ other players' P_j. This means that i considers that all the $(n-1)$-tuple of strategies could be used by the other players. The *prudent* agent's rationality consists in *forecasting the worst* and being cautious. Let us consider an application $t(.)$ that *arranges* the different n-tuple of strategies. In fact, for any agent i, the choice of the strategy $x_i \in X_i$ leads to a payoffs vector associated to this strategy and related to the different strategies chosen by the $(n-1)$ other players. In other words, x_i is identical to the vector $P_i(x_i, \check{x})$ where the $(n-1)$-tuple $\check{x} \in \prod_{i=1}^{n-1} X_i$. The application $t(.)$ lets the vector $\vec{x}_i$ components be *ordered* in the increasing way:

$$\begin{array}{rccl} t(.) : & X_i & \to & X_i \\ & \vec{x} & \to & t(\vec{x}_i) = \vec{x}_i^t, \end{array} \tag{11}$$

where $P_i^o(x_i, \check{x}) > P_i^q(x_i, \check{x}) \Leftrightarrow o > q$. From now on, any vector x_i is an *arranged* vector $\vec{x}_i^t$. We therefore decide to keep on noting it x_i. It is obvious that the agent's prudent behavior corresponds to the lexicographic fuzzy preorder $\mu_{[.]}^{\succsim}(.)$. An arranged strategy x_i is preferred by i to another strategy y_i if and only if:

$$\forall t < q, P_i^t(x_i, \check{x}) = P_i^t(y_i, \check{x}) \text{ and } P_i^q(x_i, \check{x}) = P_i^q(y_i, \check{x}). \tag{12}$$

Theorem 6 allows to represent $\mu_{[.]}^{\succsim}(.)$ with a continuous strategic utility function (*strategic* because it is defined on X_i). We note it $s(.)$. Each player's program consists in maximizing $s(.)$ on X_i. The only question that we can ask is: does the strategy that maximizes $s(.)$ really exist? In the case of Theorem 6, we need the referential to be connected. We must give a supplementary condition.

Axiom 3 : *The set X_i of individual strategies is a compact set (closed and bounded).*

Theorem 7 : *Under Axioms 2 and 3, for all i, there exists $x_i \in X_i$ such that $\{Arg \bigvee_{X_i} s(x_i)\} \neq \emptyset$.*

Instead of directly studying the conditions of a Nash-equilibrium, we work here on the most important corollary of Nash's Theorem, namely von Neumann's Theorem which gives the sufficient conditions for a particular kind of noncooperative games - 2-person zero-sum games - to admit at least a balanced solution. We know that a saddle point (maximin) - or game value - is nothing else than a Nash-equilibrium of a zero-sum game. However, Axiom 2 by thickening the indifference classes of the lexicographic fuzzy preorder, increases the set of feasible equilibra. Henceforth, the balanced issues deriving from a careful behavior prevent any *nonconstraining collaboration*: the agents decide in a decentralized way. Nevertheless, we are going to prove that this kind of rationality can generate Nash-equilibria when the agents' behavior is fuzzy. We call *2-person zero-sum game*, a game in a normal form such that:

$$(X_1, X_2; P_1, P_2 = -P_1). \tag{13}$$

The first player's gains are the second player's losses. The strategic utility functions obtained under Axiom 2 are ordinal even though the payoffs are cardinal.

Theorem 8 : *Under Axioms 2 and 3, in the game $(X_1, X_2; P, -P)$ where $P(.,.)$ is continuous, the set $(\{Arg \bigvee_{X_1} s_1(x_i)\}, \{Arg \bigvee_{X_2} s_2(x_i)\})$ contains the usual saddle point of the game.*

Because the usual saddle point belongs to the subset of the nondominated optimal points of X, it follows an increase of the set of game equilibria. For any decentralized game equilibrium (without *nonconstraining collaboration*) to correspond to a Nash-equilibrium, implies the game to be *inessential*. This is the case of the 2-person zero-sum games. It means that the issue is Pareto-optimal. We can now establish the following theorem:

Theorem 9 : *Under Axioms 2 and 3, in the game $(X_1, ..., X_n; P_1, ..., P_n)$ supposed to be inessential, where $P_i(.,.)$ is continuous, the set of $\{Arg \bigvee_{X_i} s_i(x)\}$ contains the usual set of Nash-equilibria.*

In such an inessential game, there are no other balanced solutions than the Pareto-optimal ones. The prudent behavior only leads to Pareto-optimal solutions. Actually, a player using another strategy than prudent may be threatened. To be optimal, strategies must ensure the optimal payoff. They correspond to prudent strategies in an inessential game. Because an inessential game means that the issue based on these optimal strategies is Pareto-optimal, hence, this means any Pareto-optimal issue to be prudent, i.e. to correspond to a Nash-equilibrium. It is clear that the goal of noncooperative fuzzy games *à la* Butnariu and the one we presented afterwards are very different. In the first case, there is a modelling of the nonconstraining collaboration, in the second case, the players' behaviors are completely decentralized. There is therefore an implicit measure of noncooperation. The results obtained from fuzzy games (of Butnariu's type) essentially consist in describing rules for the information exchange untill it converges towards a Nash-equilibrium. It is obvious that these fuzzy games when completely decentralized do not have a balanced solution but only possible solutions. Instead, in the decentralized and ordinal approach that we present, the *inessential* game condition is sufficient to make the issue balanced. However, the main contribution in this chapter is to show on one hand the nonconstraining collaboration to be fundamental in the convergence process towards balanced solutions. On the other hand, by continuity of the fuzzy lexicographic preorder and the satisfying representation of this preorder with an associated continuous function, we substitute a n-maximisation for any *decentralized* game (and any inessential game). Here, in a conflict situation, the player remains optimizer: he does not *'recontract'* according to the other players choices. However, for a particular class of games, the inessential one, the set of Nash-equilibria is wider because the agents do not make the difference between two very close issues. In other words, under Axiom 2, the cost of discrimination between two theoretical objects - whatever they are - is more important than the eventual differential of utility. If we observe the mathematical process which lets $\mu^{\succsim}_{[.]}(.)$ continuous , we see that only a fuzzy preorder could do it, because it allows a mulitivalued indifference. Now, the group is placed in a conflict situation. What does the core notion become if players preferences are fuzzy, is a fuzzy core empty or not?

5.5 COOPERATIVE FUZZY GAME THEORY

In cooperative game theory, as already said before, there are two arguments - whose nature is clearly interpretative, hence not technical - which both lead to resort to fuzzy set theory: first, the improvement of the pure individual choice behavior description and second, the improvement of the individual membership behavior description inside the possible coalitions of a game. Technically, it is possible to sum up these two ambitions by considering that the central *unique* property which symbolizes the standard state of the Art (or the state of the standard Art) in cooperative game analysis is the convexity one and that it is necessary to deal with it if hoping some results: convexity for preferences as well as convexity for sets. The Boolean main result is the following: if the preferences are not convex, *ceteris paribus*, then the core is empty, if the preferences are convex, *ceteris paribus*, then the core is not empty. Hence, it seems to exist a sort of dummy 'trade-off' between the convexity of behaviors and the nonemptiness of the core: if you wish one side, you need to prevent the other one. If assuming such a trade-off, it should be possible (actually, it is already the Aumann's 1964 idea of a continuum of agents or that of a replicated finite exchange economy proposed by Debreu & Scarf 1963) to transfer the convexity property from the players' behaviors towards the players' set, by definition or replication in order to preserve some unblocked issues in the associated core.

We can distinguish three research subprograms, hence three models, each of them using fuzzy set theory in a specific way depending on the target. First (Aubin), the problem is to produce some conditions for the convergence towards a unique cooperative issue while the individual preferences remain Booelan and convex by means of a fuzzy coalition - i.e. convex standard preferences & fuzzy coalitions. Second (Husseïnov), it is to study the relevant framework, with fuzzy coalitions and nonconvex preferences, for an existence theorem - i.e.

nonconvex standard preferences & fuzzy coalitions. Next (Billot), it is to exhibit the conditions under which, when the preferences are fuzzy but not convex, one can expand the core size, actually to show how a standard core vacuity can yield a fuzzy solution with Boolean coalitions - i.e. nonconvex fuzzy preferences & standard coalitions

5.5.1 Fuzzy Structures with Boolean Preferences

5.5.1.1 Aubin's Model.

Scarf 1967 famous result was establishing the conditions under which a co-operative exchange economy owes a nonempty core, i.e. the set of feasible allocations that balanced the economy. In this theorem, he pointed out the dependence of the core existence on the convexity property of the individual's preferences weakorder. Without it, the other conditions remaining, the core is definitely empty. Hence, many contributions tried after to relax one of the Scarf's conditions in interpreting them previously as a finite structure: actually, the players set was finite and moreover, the individuals behavior was precisely at the origin of the core nonemptiness. This way, one could interpret the pair (players set, players'behavior) as the precise location of the problem. Debreu & Scarf 1963 proposed then to introduce the convexity requirement on the players set rather than on the individuals' behavior. To transform the set nature, they suggested to consider a r-fold replica of a given economy. Assuming each agent to give rise to several *children*, these ones being involved into some replicated economy, they show the existence of at least one balanced allocation, i.e. the core nonemptiness. On the other hand, Aumann 1964 proposed a large nonatomic model of an economy that yields the same kind of solvation method for the core nonemptiness by means of the *continuum of agents* hypothesis which is based on the direct requirement of convexity of the players set. Despite conceptual differences, these two contributions are both dedicated to the study of the possibility of characterization of unblockable allocations, i.e. allocations from the core of an economy via dual parameters, i.e. prices. Apart from these propositions, another way to change the mathematical nature of the players set was to extend the notion of *coalition* whose influence on the feasible equilibria is definitely basic. Aubin 1979 and Hüsseinov 1994 introduced and improved the definition of a *fuzzy coalition* in integrating some obvious real phenomena in a consistent mathematical definition that moreover allows the core to become nonempty. From the convexity of the players'set to the convexity of the players'coalitions, the field was rich and finally, Aubin showed the great flexibility of fuzzy sets tools in gathering in an elegant way all these different works.

Let us consider a finite number of commodities $j = 1, ..., l$ in an economy. A commodity bundle is viewed as a nonnegative vector $x = (x_j)_{j=1}^{l}$ and a commodity space is considered as a X-nonnegative orthant of space R^l. There are $i = 1, ..., m$ agents in the economy. Each agent is characterized by a commodity bundle x^i and a binary preference relation $\succeq_i$ defined on $X_i \subseteq X$ since directly considered by the agent i under the standard assumption of selfishness. We denote S the finite set of agents i. Hence, the structure of th economy is first described by $(S, (X_i)_{i=1}^m, (\succeq_i)_{i=1}^m)$. (Most of the notations are taken from Billot 1995b.)

Definition 16 : *$\forall i \in S$, the binary preference relation $\succeq_i$is said to be* ***convex*** *if $\forall a \in X_i$, the set*

$$\{x \in X_i : x \succeq_i a\}$$

is convex.

Definition 17 : *$\forall i \in S$, the binary preference relation $\succeq_i$ is said to be* ***continuous*** *if $\forall a \in X_i$, the sets*

$$\{x \in X_i : x \succeq_i a\} \text{ and } \{x \in X_i : a \succeq_i x\}$$

are closed.

Definition 18 : *$\forall i \in S$, the binary preference relation $\succeq_i$ is said to be* ***monotoneit*** *if $\forall x, y \in X_i$, $x \geq y$ (i.e. $x - y \in X_i$, $x \neq 0$) $\Rightarrow x \succeq_i y$.*

Definition 19 : *$\forall a \in X_i$, a is said to be a* ***$\succeq_i$-maximal element*** *if a is a point of global satiation.*

Any preference relation $\succeq_i$ is said to be *globally satiable* if it has no global satiation point. (For further details, see e.g. Debreu 1959, Chaps. 1 & 2.) Let us denote by Π the simplex of prices vector:

$$\Pi = \{p = (p^1, ..., p^l) \in X; p^1 + ... + p^l = 1\}. \tag{14}$$

The cost of any commodity bundle $x = (x_1, ..., x_l)$ at the prices $p = (p^1, ..., p^l)$ is defined on the scalar product of p and x: $px = p^1x_1 + ... + p^lx_l$ where each p^j is a relative price in an economy without money, i.e. a quantities ratio. Finally, the general framework consists in a quadruplet corresponding to a finite pure exchange economy denoted $E = (S, (X_i)_{i=1}^m, (\varpi^i)_{i=1}^m, (\succeq_i)_{i=1}^m)$ where ϖ^i corresponds to the initial endowments of the agent i, with $\Omega = \sum_{i=1}^m \varpi^i$.

Definition 20 : *Any vector $x = (x^i)_{i=1}^m$ of $\prod_{i=1}^m X_i$ is said to be an* ***allocation*** *of the economy E.*

Definition 21 : *An allocation $x = (x^i)_{i=1}^m$ of $\prod_{i=1}^m X_i$ is said to be* ***admissible*** *iff the condition*

$$\sum_{i=1}^m x^i = \Omega$$

is satisfied.

Definition 22 : *A pair $(\bar{x}(.), p)$ is said to be an* ***equilibrium of the economy*** ***E*** *if and only if*

$$\sum_{j=1}^{l} p^j \bar{x}^i_j = \sum_{j=1}^{l} p^j \varpi^i_j, \tag{15}$$

$$\forall y \; : \; \sum_{j=1}^{l} p^j y^i_j \leq \sum_{j=1}^{l} p^j \varpi^i_j \Rightarrow \bar{x}^i \succeq_i y^i. \tag{16}$$

We denote $B(E)$ the set of all equilibria of E.

Definition 23 : *Any vector $\alpha = (\alpha_1, ..., \alpha_m)$ consisting in m nonnegative numbers such that $\forall i \in S, \alpha_i \in M = [0,1]$ and $\exists i \in S$ for which $\alpha_i \neq 0$, is said to be a* ***fuzzy coalition****.*

First, note that the membership M is not necessarily equal to $[0,1]$. If $M = \{0,1\}$ then a fuzzy coalition is identified with an ordinary Boolean one (i.e. a Boolean coalition is just a particular case of a fuzzy coalition). Second, by convenience, we shall just say '*allocation*' for 'admissible allocation'. Third, we can remark that the choice of M is just constrained by the continuity assumption; for instance M can be equal to R^+.

Definition 24 : *The set $A = \{i \in S : \alpha_i > 0\}$ is said to be the* ***exclusive support*** *of the fuzzy coalition α.*

Definition 25 : *A fuzzy coalition α* ***blocks the allocation*** *x* ***against the allocation*** *y iff the two following conditions are satisfied:*

$$\sum_{i \in A} \alpha_i y^i = \sum_{i \in A} \alpha_i \varpi^i, \tag{17}$$

$$y^i \succ_i x^i, \forall i \in A. \tag{18}$$

One of the natural interpretation of the numbers α_i is a membership level to the crisp coalition A. That is the intuitive reason why Aubin has chosen to call α a *fuzzy coalition.* For further details about the fuzzy concept of membership, see Billot 1988, 1992. In order to convexify the economy without using the hypothesis of agents'continuity, the tool of fuzzy coalitions allows to change the very notion of an economy's core without changing the 'finite size of the agents'set postulate' in introducing some new possibilities of blocking by expanding the number of feasible coalitions in a fuzzy way. Since a fuzzy coalition is a more general concept than a Boolean one, the possibility for the allocations to be blocked by a fuzzy coalition is greater than that to be blocked by a Boolean coalition. Hence, the fuzzy core $\Delta(E)$ is always included (even in a fuzzy way) in the Boolean one $C(E)$ (i.e. the standard Scarf's one): $\Delta(E) \subseteq C(E)$. (For proofs, see Ekeland 1979, Billot 1988.)

Definition 26 : *The **Fuzzy Core** $\Delta(E)$ **of the Economy** E is the set of all admissible allocations that cannot be blocked by any fuzzy coalition.*

Furthermore, Aubin 1979 has shown the converse inclusion to hold under preferences convexity, which is quite intuitive. This link between the fuzzy core and the Boolean one recalls another possible relation between the size of the economy and the nonemptiness of its core in a way *à la* Debreu-Scarf-Aumann. Actually, it is always possible to show the existence of a one-to-one correspondence between a fuzzy core of a finite economy and a Boolean core of a continuum economy. More precisely, this corrrespondence becomes a bijection as soon as the preference relations $\succeq_i$ satisfy some conditions of monotonicity and continuity. First, we can associate each finite economy E to a continuum one $E_\infty = \{\varpi(t), \succeq_t, t \in [0,1]\,; \varpi(t) = \varpi^i, \succeq_t = \succeq_i, \forall t \in \Delta_i\}$, where $\Delta_i = [\frac{i-1}{m}, \frac{i}{m})$ for $(i = 1, ..., m-1)$, $\Delta_m = (\frac{m-1}{m}, 1]$. For any coalition $A \subset [0,1]$, let us denote $\sharp A$ the number of elements of A, and by $|A|$ the Lebesgue measure of a measurable subset A_∞ in R^1.

Theorem 10 : *Let an exchange economy E formed of m agents such that their preference relations $(\succeq_i)_{i=1}^m$ are convex. Any E-allocation x belongs to the fuzzy core $\Delta(E)$ of the finite economy E if and only if the E_∞-allocation $\bar{x}(.)$ such that $\overline{x}(t) = x^i$, for $i \in \Delta_i$, $(i = 1, ..., m)$, belongs to the Boolean core $C(E_\infty)$ of the continuum economy E_∞.*

Theorem 10 links the Aubin's fuzzy core $\Delta(E)$ of a finite economy to a part of the Boolean core $C(E_\infty)$ of the associated continuum economy in such a way that it allows to interpret explicitely *fuzzy coalitions and fuzzy core of a finite economy as, respectively, Boolean coalitions and Boolean core of a continuum economy*. In other words, the fuzziness that is introduced inside a finite economy thanks to a softer membership can be interpreted as a sort of replication, not in a vertical way (the number of agents remains) but rather in an horizontal one, by means of a multisized individual description of the agents' commitment within the crisp coalitions (actually, they correspond to the exclusive support of fuzzy coalitions).

Theorem 11 : *Let an exchange economy E formed of m agents such that their preference relations $(\succeq_i)_{i=1}^m$ are convex and monotone. If a E_∞-allocation $x(.)$ belongs to the Boolean core $C(E_\infty)$ of the continuum economy E_∞, then the E-allocation x defined by $x^i = \frac{1}{|\Delta_i|}\int_{\Delta_i} x(t)dt, i = 1, ..., m$, belongs to the fuzzy core $\Delta(E)$ of the finite economy E.*

Theorem 11 is founded upon an intermediate lemma proving that for any point of the agents' set it is always possible to find another point such that the two associated allocations are indifferent. (For further details, see Hüsseinov

1994, Lemma 1.) According to Theorem 11, we can remark that if an economy E_∞ corresponds to the continuum economy based on the finite one E, this means that for any allocation $x(.)$ belonging to $C(E_\infty)$, there exits another allocation $y(.)$ also belonging to $C(E_\infty)$ that first can be considered as equivalent to $x(.)$ (in the sense where $\forall t \in [0,1]$, $x(t) \sim_t y(t)$) and second, which is an image of some E-allocation x belonging to $\Delta(E)$. For instance, since the preferences are assumed to be convex, this means each equivalence class to be singleton; hence, the correspondence which associates x to $\bar{x}(.)$ is a strict bijection between $\Delta(E)$ and $C(E_\infty)$. The second step was less technical: Aubin 1979 showed that every allocation of $\Delta(E)$ is an equilibrium. This allows to link the first Aubin's result with the Aumann's Theorem about the coincidence of the core and the set of equilibrium allocations:

Theorem 12 : *Let a finite exchange economy E formed of m agents such that their preference relations $(\succeq_i)_{i=1}^m$ are convex, continuous and monotone and their total initial endowment vector Ω is strictly positive. Then, an allocation x belongs to $\Delta(E)$ if and only if it is an equilibrium allocation: $\Delta(E) = B(E)$.*

Theorem 12 can be interpreted in the following way: as soon as the individual preference relations are supposed to be convex, the Aubin's fuzzy core is a sufficient concept to allow that any unblockable allocation corresponds to an equilibrium of the associated finite economy. (Generally, crisp coalitions are unable to guarantee such a thing.) It is very easy to show that the hypothesis of preferences convexity is necessary to establish the converse proposition of the last theorem from which the equivalence between a fuzzy core of a finite economy and the Boolean one of the associated continuum economy is acquired. Hence, this interpretation, very appealing, of a fuzzy core, is definitely linked to the assumption of convexity. It seems particularly intuitive, since the existence of a Boolean core of a finite economy is depending on that property: hence, it means - and that is the most important thing - that any finite economy whose core is empty cannot lead to a nonempty fuzzy core by simple fuzzyfication of the previous feasible coalitions. In that way, fuzzy cooperative game theory (*à la* Aubin) is not able to produce conditions that yield a nonvacuity theorem if the associated Boolean finite economy does not admit any equilibrium.

5.5.1.2 Hüsseinov's Model.

The main argument of the Hüsseinov's contribution to the noncooperative fuzzy game theory is based upon a technical remark: if it is possible to amplify a particular class of coalitions in expanding the definition of a fuzzy coalition, then one can prove that in case of individual nonconvex Boolean preferences, there exist some unblocked allocations which correspond to feasible equilibria of the exchange economy. Methodologically, this model is an extension of Aubin's. The most spectacular change consists in the definition of a fuzzy coalition:

Definition 27 : *Any matrix $a = (\alpha_k^i)$, $(k = 1, ..., r)$, $(i = 1, ..., m)$ of dimension $r \times m$, where r is an arbitrary positive integer, with nonnegative terms α_i^k, at* *least one which is positive, is called a* ***H-fuzzy coalition*** *(H for Hüsseinov). We say that a H-fuzzy coalition* $a = (\alpha_k^i)$ *blocks the allocation* x*, if there exists allocations* $y_{(k)} = \left(y_{(k)}^i\right)_{i=1}^m$ *such that:*

$$\forall i \in S, \alpha_i^k \neq 0, \; y_{(k)}^i \succ_i x^i, \tag{19}$$

$$\sum_{i=1}^m \sum_{k=1}^r \alpha_i^k y_{(k)}^i = \sum_{i=1}^m \sum_{r=1}^r \alpha_i^k \varpi^i. \tag{20}$$

Generally, the interpretation of Aubin's coalition is the following: any agent i belonging to a coalition α participates in it with the share $(\alpha_i / \max_i \alpha_i)$ of his total resources. The interpretation for a H-fuzzy coalition is quite different: such a coalition a can be viewed as a *finite bundle of coalitions* $k = 1, ..., r$ in which the agent i participates with shares

$$\frac{\alpha_i^k}{\max_i \sum_{k=1}^r \alpha_i^k}, (k = 1, ..., r) \tag{21}$$

of his total resources. The principle here is a sort of Aumann's generalization by replication of the coalitions rather than that of the agents. Of course, the deduced definition of an extended fuzzy core is just the extension to the H-fuzzy coalitions of the blocking capacity:

Definition 28 : *The* ***Extended Fuzzy Core*** $\Lambda(E)$ ***of the finite Economy E*** *is the set of all admissible coalitions which are unblockable by any H-fuzzy coalition.*

Trivially, any fuzzy coalition is a particular case of a H-fuzzy coalition, for $r = 1$. This means the extended fuzzy core $\Lambda(E)$ to be included in $\Delta(E)$ since the blocking capacity increasases with respect to the number of fuzzy coalitions. Hüsseinov 1994 proves the converse in establishing the following theorem:

Theorem 13 : *Let a finite exchange economy* E *formed of* m *agents such that their preference relations* $(\succeq_i)_{i=1}^m$ *are convex, then* $\Lambda(E) = \Delta(E)$*.*

A way to read Theorem 13 is to focus onto the convexity property in showing, for instance, that while the preferences are not convex, the two notions of fuzzy core do not coincide. This remark leads Hüsseinov to weaken the axiomatic requirement about the individual preferences in substituting convexity with standard global nonsatiation.

Theorem 14 : *Let a finite exchange economy* E *formed of* m *agents such that their preference relations* $(\succeq_i)_{i=1}^m$ *are continuous and globally nonsatiable, then an* E*-allocation* x *belongs to* $\Lambda(E)$ *if and only if the associated* E_∞*-allocation* $x(.)$ *belongs to* $C(E)$*.*

In order to follow exactly the same kind of progress Aubin did, the following theorem proves the equality between the extended fuzzy core and the set of equilibria.

Theorem 15 : *Let a finite exchange economy E formed of m agents such that their preference relations $(\succeq_i)_{i=1}^m$ are continuous, monotone and their initial endowment vector Ω is strictly positive. Then, the E-allocation x belongs to $\Lambda(E)$ if and only if it is an equilibrium allocation: $\Lambda(E) = B(E)$.*

5.5.2 Boolean Structures with Fuzzy Preferences

5.5.2.1 Fuzzy Preorders and f-Convexity.

Let us assume the set X_i to be convex and thus connected. Let us consider a fuzzy relation of preference $\mu^i_{[.]}(.)$ be defined upon the Zadeh set $[0,1]$. The agent arranges the different elements of X_i thanks to $\mu^i_{[.]}(.)$, a f-transitive and reflexive (f.s.) fuzzy relation.[7] Moreover, we assume that if $\forall(x,y) \in X_i^2$, $\mu^i_{[x]}(y) \geq \mu^i_{[x]}(y)$, then $\mu_i^X(x) \geq \mu_i^X(y)$, in order to generate a preordered fuzzy subset, noted X^i_f: the agent's local preference $(\mu^i_{[.]}(.))$ does not contradict his global preference $(\mu_i^X(.))$. There is no locally irrational agent ($\mu^i_{[.]}(.)$ is always f-transitive). (For all details, see Billot 1995b, Chaps 1 & 2.) The game is now considered as a fuzzy game or, more precisely as a fuzzy exchange economy $E_f = \left(2^S - \emptyset, (X_i)_{i=1}^m, (\mu_i^X)_{i=1}^m, (\varpi_i)_{i=1}^m\right)$. If $\mu^i_{[.]}(.)$ is f-transitive and reflexive (f.s.), then $\mu^i_{[.]}(.)$ is a fuzzy preorder (f.s.). We also know that if the referential set X_i is connected, $\mu_i^X(.)$ is a continuous function of utility on X_i under the assumption of fuzzy preference continuity (Billot 1995a, 1995b). So, we suppose this assumption of continuity to be satisfied. As soon as two objects x and y have a membership degree to X^i_f which is superior to 0, there exists a convex combination which also has a membership degree superior to 0. (This result is established in Billot 1990.) The usual convexity means that if x and y belong to the usual set X_i, and if X_i is connected, then all the convex combinations of x and y belong to the usual set X_i. Intuitively, the notion of *weak fuzzy convexity* means: *if two elements of the referential set belong to the exclusive support of the fuzzy subset, then there is a convex combination of these two elements which also belongs to the exclusive support of the fuzzy subset.* Hence, when the membership function is continuous, it is easy to show that the associated exclusive support can always be *weak convex.*

Definition 29 : *A fuzzy subset X^i_f of a convex referential set X_i is **weak f-convex** iff: $\forall(x,y) \in (X_i^s)^2, \exists\alpha \in]0,1[; (\alpha x + (1-\alpha)y) \in X_i^s$ where X_i^s is the exclusive support of X^i_f.*

Definition 30 : *A fuzzy subset X^i_f of a convex referential set X_i is **strong f-convex** iff: $\forall(x,y) \in (X_i^s)^2, \forall\alpha \in]0,1[; (\alpha x + (1-\alpha)y) \in X_i^s$.*

Another way to present the strong f-convexity is to consider the two following propositions as equivalent:

$$[\forall \alpha \in]0,1[, \mu_i^X(\alpha x + (1-\alpha)y) > 0] \Leftrightarrow [\mu_i^X(x) > 0 \text{ and } \mu_i^X(y) > 0]. \qquad (22)$$

Lemma 3 : *Let X_i be convex and X_f^i any fuzzy subset of X_i. If the membership function $\mu_i^X(.)$ is continuous, then X_f^i is weak f-convex.*

Lemma 4 : *Let X_i be convex and X_f^i any fuzzy subset of X_i. If $\mu_i^X(.)$ is continuous and X_i^s convex, then X_f^i is strong f-convex.*

This f-convexity is quite different from convexity that is generally used in fuzzy literature. Actually, the more general definition of fuzzy convexity simply comes out of properties of an epigraph (see Moulin & Fogelman-Soulie 1979). If we consider a function $f(.)$ defined from a space E, towards R, we call *epigraph* of $f(.)$, noted *epi*(f), the following set: $\{(x,\lambda) \in E \times R; f(x) \leq \lambda\}$. The epigraph of $f(.)$ is convex if the function $f(.)$ is also convex. On the other hand, the nonexclusive complement set $\overline{epi(f)}_\lambda$, defined as follows: $\overline{epi(f)}_\lambda = \{(x,\lambda) \in E \times R; f(x) \geq \lambda\}$, is convex if and only if for E convex: $f(\alpha x + (1-\alpha)y) \geq Min\{f(x), f(y)\}$. It means that $f(.)$ is quasi-concave. In other words, a fuzzy subset X_f^i is convex (for the usual literature) if the membership function $\mu_i^X(.)$ is quasi-concave. The membership function $\mu_i^X(.)$ derives directly from the fuzzy preorders. The fuzzy subset derives from the individual preferences and it is convex only if the membership function is quasi-concave. This implies the fuzzy subset X_f^i, preordered by $\mu_{[.]}^i(.)$, to be convex if the membership function $\mu_i^X(.)$ is quasi-concave, i.e. if the preorders are convex. That way, there is a relation between the convexity of the fundamental set (which concerns the referential) and the convexity of associated preorders: this relation is basic for nonemptiness of the core in the cooperative games approach, fuzzy or not. The goal is simple. We try to work with the only f-convexity, immediately acquired if $\mu_i^X(.)$ is continuous, i.e. if X_i is connected and individual preorders continuous. The peripheral core (P.C.), is not identical to the usual core or Aubin's fuzzy core. To define it, we need two references: a planner interested in nonemptiness of the peripheral core and the ϵ-cores of the non fuzzy literature (Weber 1979, Wooders 1983).

5.5.2.2 The Peripheral Core.

First, we must explain, using a more intuitive analysis, what is the *peripheral core*. If the usual core (we call it *intra-muros core* because always included in the peripheral one) is empty - namely the game is not 'Booleanly' balanced - what is *practically* happening? In our approach, if individual preorders are not convex, the intra-muros core is empty. Actually, we try to suggest some original results in terms of equilibrium when individual fuzzy preorders are no more *convex*, but without any requirement on the replication of the economy, neither by the agents nor by the coalitions. If the game is not balanced - i.e.

the preorders are not convex -, the intra-muros core is empty. In an exchange economy, the concept of approached core allows new conditions which guarantee its nonemptiness (see Wooders & Zame 1984). The analysis of the approached core for games with a players continuum and without a balanced structure was made by Wooders 1983 who uses an assumption of replication of the economy. Hüsseinov's works (with fuzzy coalitions) showed the evident relation between the *extension of the players set* - by assumption or replication for the agents or the coalitions -, and the *condition of a balanced game*. In other words, we can expect balanced results for a cooperative game - without convexity for preorders - if we extend the cardinal of society from finite to infinite. Nevertheless, in such an economy (or such a game), the core is ϵ-approached. In terms of utility, the agents accept to loose an ϵ of satisfaction and thus they do not block up allocations which balance the game; hence, we can deduce an approached-core, noted ϵ-core. What happens if one do not convexify an economy with a finite number of players ? The answer that we propose is based, on one hand, on the f-convexity of the fuzzy subset associated to the player's preorder and on the other hand, on a game planner. The *last peripheral core* (i.e. the one which corresponds to the smaller but nonempty set of balanced issues) belongs to the same family as Owen's Least-Core 1982, implied by regression upon the ϵ-cores in a replicated economy. Here, we must not assimilate a coalition that can share some economic variable (internal market as in the labour theory) and a coalition directly issued from individual choices. The target of the last is just to generate a collective decision. This conception brings us to throw out the assumption of fuzzy coalitions. Aubin's assumption implied a restriction of the intra-muros core under a necessary condition of convexity for individual preorders. Our goal is not to restrain the core but, on the contrary, to allow some results in situations where it is naturally empty.

Any coalition is *a priori* feasible. However, we can intuitively suppose that some of the 2^m coalitions are not really active. That is the reason why the set of active coalitions, noted AC, is a usual subset of 2^S, the powerset of S. One calls AC a *structure of coalitions* where $AC \subseteq 2^S$ (hence, $\emptyset \in 2^S$ but trivially $\emptyset \notin AC$). Moreover, in order to *perfectly* describe a member i of X family, professor of Y university, supporter of Z union, we define 3 coefficients α_X, α_Y and α_Z such that:

$$\alpha^i_X + \alpha^i_Y + \alpha^i_Z = 1, \tag{23}$$

and we assume the Assistant-Professors or full-Professors to be represented by university Y (in which there is two different ranks) in an identical way:

$$\forall C \in AC, \forall (i,j) \in C^2, i \neq j; \alpha^i_c = \alpha^j_c, \tag{24}$$

which formally corresponds to:

Axiom 4 : *For any agent i belonging to C, a family of coefficients (α^i_c) corresponds to each active coalition C, $C \in AC$. α^i_c describes the i's individual* *fraction that he estimates represented by C (while belonging to C in a usual degree of membership, i.e. 1).*

Axiom 5 : *Each active coalition C, $C \in AC$, represents any member in the same way.*

The satisfaction of Axioms 4 and 5 yields what we call a *balanced structure*. For any subset C from S - especially any element of AC - we note R^C, the set of vectors of components u_i, where u_i is the utility level of the agent i, (for a given allocation). From this definition, we deduce: $R^m = R^S$. Let the application $\pi(.)$ be an operator such that for any vector $u = (u_i)_{i=1}^m$ of R^S, $\pi(u)$ corresponds to the only components that concern the members of a coalition C; it generates a vector of R^c:

$$\begin{array}{llll}\pi(.): & R^S & \rightarrow & R^C \\ & (u_i)_{i\in S} & \rightarrow & \pi(u) = (u_i)_{i\in C}.\end{array} \tag{25}$$

We call C-imputation any vector of R^C. A m-person cooperative game corresponds to the data - for any nonempty coalition from S - of a nonempty subset of R^C, noted $V(C)$ where:

$$V(C) - R^C \subset V(C). \tag{26}$$

One considers an exchange economy. Let us consider agents (in a finite number $m = \sharp S$) in a game situation. They have to share their initial endowments, noted Ω, composed with l goods divided between m individuals (i.e. $\Omega \in R^l_+$). We assume the initial detention of endowments and selfishness. Each agent i, $i \in S$, has a fuzzy preorder of preference (f.s.), i.e. $\mu^i_{[.]}(.)$, is f-transitive and reflexive (f.s.). We suppose the fuzzy preorder to be continuous and each consumption set X_i, $X_i \subset R^l_+$, to be convex (therefore connected). It allows the existence of a continuous fuzzy utility function on X_i, defined on $[0,1]$. We note $fu_i(.)$ this fuzzy utility function. We say the coalition C (of AC) to totally block the allocation $x \in R^l_{+n}$, if there exists another feasible allocation $y \in R^l_{+n}$, such that:

$$\Omega = \sum_{i\in S} y_i, \tag{27}$$

$$\pi_C[fu(y)] \gg \pi_C[fu(x)]. \tag{28}$$

Definition 31 : *The set of* ***Internal Feasible Allocations*** *of a coalition C, denoted IFA(C) corresponds to:*

$$IFA(C) = \{y \in R^l_{+n}; \forall i \in C, y_i \in R^l_+ \text{ and} \sum_{i\in C} y_i = \sum_{i\in C} \varpi_i\}. \tag{29}$$

The set of **Fuzzy Utility levels** that C can simultaneously ensure to its members, denoted $FU(C)$, corresponds to:

$$FU(C) = \{[fu_i(y_i)]; y \in IFA(C)\}. \tag{30}$$

The set $IFA(S)$ is the set of all feasible allocations. By definition, we suppose that $IFA(S) \subseteq \prod_{i \in S} X_i$. Thus $IFA(S)$ is convex and connected. We assume it to be compact. Each agent, member of the society S, is going to generate a fuzzy subset $IFA^i_f(S)$ - that *we assume the exclusive-support to be convex* - implied by his fuzzy preorder (f.s.): $\forall (x,y) \in (IFA(S))^2$, if i considers the relative qualities of allocation x (restricted by selfishness to x_i) to be greater than the ones of y, we have: $\mu_i^{IFA(S)}(x_i) > \mu_i^{IFA(S)}(y_i)$, where $\mu_i^{IFA(S)}(.)$ is the membership function of the fuzzy subset $IFA^i_f(S)$ included in the referential set $IFA(S)$. Since $\forall i \in S$, $\mu^i_{[.]}(.)$ is a continuous preorder (f.s.) and $IFA(S)$ is connected, the membership function $\mu_i^{IFA(S)}(.)$ is continuous.

Lemma 5 : *$\forall i \in S$, $IFA^i_f(S)$ is strong f-convex.*

By defining a fuzzy utility function $\mu_i^{IFA(S)}(.)$ on $IFA(S)$, each agent restrains his satisfaction to only admissible allocations. The possibilities of blocking are just of one type. In a given coalition, if the allocation x is always less satisfying than another internal feasible imputation, x is totally blocked. An allocation x belongs to the intra-muros core if no coalition blocks it.

Definition 32 : *The* ***Intra-Muros Core*** *(IMC) corresponds to:*

$$IMC = \{x \in IFA(S); \forall C \in AC, \exists i \in C, \mu_i^{IFA(S)}(x_i) \geq \mu_i^{IFA(S)}(y_i)\}. \quad (31)$$

We know IMC to be a usual set, that is: $\forall x \in IMC; \mu^{IMC}(x) = 1$. But the intra-muros core only describes allocations which are totally unblocked. This corresponds to a unitary membership degree. Let us now consider allocations that are *a priori* blocked but with a *more or less* level of satisfaction. If a feasible allocation x' is blocked, it does not belong to the intra-muros core: i.e. $\mu^{IMC}(x') = 0$ if and only if

$$\exists C \in AC; \forall i \in C, \exists y \in IFA(C) : \mu_i^{IFA(S)}(x_i) < \mu_i^{IFA(S)}(y_i). \quad (32)$$

As soon as a member of each allocation prefers a admissible *external* allocation, this allocation belongs to IMC. However, by introducing ordinal interpersonal comparisons of utility, because of the game planner, there are some rejected allocations that imply a level of satisfaction which is very closed but inferior to the level of utility corresponding to the elements of IMC. To sum up this information, the best tool is the intersection and/or the union of the different fuzzy subsets $IFA^i_f(S)$. Actually, taking the intersection of the different fuzzy subsets $IFA^i_f(S)$ is equivalent to confer the *minimum level of maximum levels* of satisfaction to the rejected allocation. Of course, it just concerns members of coalitions that belong to the AC structure. We assume furthermore:

Axiom 6 : *We can make interpersonal ordinal comparisons of utility.*

Axiom 7 : $\forall i \in C, \forall C \in AC, fu_i(\varpi_i) > 0$.

We can now define the Peripheral Core as follows:

Definition 33 : *We call* ***Peripheral Core*** *the fuzzy subset of $IFA(S)$ such that:*

$$PC_f = \{x \in IFA(S); \mu^{PC}(x) = \wedge_{C \in AC}[\vee_{i \in C} \mu_i^{IFA(S)}(x_i)]\}.$$

Theorem 16 : *Whatever is the nature of the agents' fuzzy preferences (convex or not) under Axioms 4 to 7, the exchange economy is balanced if, $\forall i \in S$, X_i is convex, $\mu^i_{[.]}(.)$ continuous and $IFA(S)^s_i$, the support of $IFA^i_f(S)$, convex (i.e. $PC_f \neq \emptyset$).*

Remark 3 : *It is obvious that any element of $IFA^i_f(S)$ belongs to PC_f ($\forall i \in S$).*

We have showed that an exchange economy is always balanced even though the agents' fuzzy preferences are not convex. In order to do this, there must exist, at least, one allocation of $\prod_{i \in S} X_i$ such that the coalitions members have levels of fuzzy utility greater than 0. This constraint is explained by Axiom 7 which is not so restrictive. It means that the agents assume initial endowments to have a nonnull level of utility. (Another way is defining fuzzy subsets on a membership set $\overline{M} =]0, 1]$).

Theorem 17 : *Under Axioms 4 to 7, if $IFA(S)$ is convex, and $\forall i \in S$, X_i convex and $\mu^i_{[.]}(.)$ continuous, the peripheral core of the exchange economy is nonempty.*

However, the peripheral core contains any allocation of $IFA(S)$ that satisfies Axiom 7. But, we can compare allocations of the peripheral core by looking at function $\mu^{PC}(.)$. If IMC is empty, there exists '*approximately balanced*' allocations that can solve the exchange problem in a given economy. To find a f-balanced solution, it is just necessary to find an internal allocation which is considered by any member of each coalition as defining a nonnull level of utility. It seems natural to assimilate this allocation to individual endowments. If we have a nonempty fuzzy subset that contains all the f-balanced solutions, we must choose the f-balanced allocations that most approach IMC. Because of the definition of PC_f, it implies a membership degree deriving from the utility of the less satisfied agent of the most satisfied agents, for each allocation of PC_f. Because of Axiom 6, a game planner can compare elements that compose PC_f. The principal tool used by the planner is α-cutting. First, we must prove two results.

Lemma 6 : *If $PC_f \neq \emptyset$ and $IMC \neq \emptyset$, then: $IMC \subset PC^s$.*

Lemma 7 : *Let $x \in IMC \neq \emptyset$ be any allocation. Under PIS, $\forall x' \in PC'_f$ (where PC'_f is such that $PC'^s = PC^s - IMC$): $\mu^{PC}(x') < \mu^{PC}(x)$.*

In Lemma 6, we assume any internal allocation in any given coalition not to generate a *maximum of satisfaction which can greater* than the maximum of satisfaction generated by any allocation belonging to IMC. We call this assumption *Postulate of Internal Satisfaction*:

Axiom 8 : $\forall C \in AC, \forall y \in IFA(C), \forall x \in IMC$:

$$\bigvee_{i \in C} \mu_i^{IFA(S)}(y) \leq \bigvee_{i \in C} \mu_i^{IFA(S)}(x).$$

Theorems 16 and 17 lead to consider any solution of PC'_f as *approximately balanced*, i.e. the allocations of the peripheral core when IMC is empty. However, even if it is not empty, the membership function of PC_f can discriminate between all of the allocations of IMC (because of Axiom 6). So, we can exhibit balanced solutions which are *most preferable*. Lemma 7 means that the game planner can search for a social allocation which is more satisfying in the sense that social utility (i.e. $\mu^{PC}(.)$) increases when, on one side, we approach IMC (if it exists), and on the other side, when we *order* the allocations of PC_f. This social utility, $\mu^{PC}(.)$, must be a continuous function defined on $[0,1]$. It means that PC_f is weak f-convex. Let us recall what is an α-cut: $\forall A_f$ a fuzzy subset of X, we call, A^α, 'α-cut of A_f' in α level, $\alpha \in]0,1]$, the following usual set: $A^\alpha = \{x \in X; \mu^A(x) \geq \alpha\}$.

Definition 34 : *We call α-**Approached Core**, noted PC^α, every α-cut of the peripheral core PC_f.*

Remark 4 : *We note that the greater is α, the smaller is the cardinal of the obtained usual set: let $(\alpha, \alpha') \in]0,1]^2$ with $\alpha > \alpha'$, then $\sharp PC^\alpha < \sharp PC^{\alpha'}$.*

The game planner is going to search for the level $\breve{\alpha}$ such that $\forall \alpha > \breve{\alpha}$, $\sharp PC^\alpha = 0$ where $\sharp PC^{\breve{\alpha}} \neq 0$, i.e. the *last α-cut of the peripheral core* that is nonempty.

Definition 35 : *We call **Last Core**, noted LC, the $\breve{\alpha}$-approached core.*

Lemma 8 : *If $IMC \neq \emptyset$, under PIS, $LC \subseteq IMC$.*

Lemma 9 : *If $IMC = \emptyset$, then $LC = PC^{\breve{\alpha}}$ where $\breve{\alpha} = \vee_{x \in PC'_f} \mu^{PC}(x)$.*

Lemma 10 : *If $PC^s \neq \emptyset$, then $LC \neq \emptyset$.*

Theorem 18 : *The conditions under which the Last Core of an exchange economy is nonempty are:*
(1) the set of feasible allocations is compact and convex,
(2) the Peripheral Core is nonempty and weak f-convex,
(3) Axiom 8 is satisfied.

Lemma 8 means the Last Core to be included or equal to IMC when individual preorders are convex. Lemma 9 explains the fact that the upper degree of the membership function $\mu^{PC}(.)$ defines the Last Core if IMC is nonempty. Lemma 10 assures nonemptinessness for the Last Core as soon as PC^s is nonempty. The introduction of α-approached cores is not here related to the concept of a replicated economy and/or a continuum of agents. The set of agents is finite. That explains the necessity of Axiom 6 of interpersonal comparison of utility. We wanted to weaken the assumption of convexity and we have reached our goal thanks to the f-convexity of a fuzzy subset. By definition, the peripheral core contains any socially feasible allocation for which the less satisfied of the most satisfied agents of society has a positive fuzzy utility. However, in any given coalition which is nonblocking, the most satisfied agent must prefer it to any internal allocation. Then, the most satisfied agent of the blocking coalition must not prefer it to any other internal allocation. In other words, a socially feasible allocation belongs to IMC under the condition of nonexistence of one internal allocation in at least one coalition that is preferred to it by every member of this coalition. For the peripheral core, we find these two steps. Each coalition considers every social feasible allocation. Then we deduce m possible levels of membership deriving from the most satisfied agents. Finally, we attribute to the considered allocation the level of satisfaction deriving from the less satisfied agent of the *preselected* agents. These different notions deal with the concepts of union and intersection of the $IFA^i_f(S)$, thanks to the fuzzy subsets algebra (Zimmermann 1985). In other words, if we are under usual conditions (convex preorders of preference), the game planner generates a *perfectly balanced solution* which has a better (or equal) social level than allocations of IMC.

Notes

1. A topological space is *Hausdorff-separated* if for any $x \in X$, the intersection of all closed neighborhoods of X is the single element $\{x\}$. A topological space X is *locally convex* if $\forall x \in X$, any neighborhood of X is convex.
2. We must consider also $\pi_i(.)$ as the membership function of Π_i.
3. Any interval $[x, y]$ of $(X, \succeq^L)$ such that $y \succ^L x$ and $]x, y[= \emptyset$ is a *hole* for $\succeq^L$.
4. A fuzzy preorder $\mu_{[.]}(.)$ is *reflexive (a.s.)* if and only if $\forall x \in X, \mu_{[x]}(x) = 1$.
5. A fuzzy preorder $\mu_{[.]}(.)$ is f-transitive if and only if $\forall (x, y, z) \in X^3$: $\mu_{[x]}(y) \geq \mu_{[y]}(x)$ and $\mu_{[y]}(z) \geq \mu_{[z]}(y)$ implies $\mu_{[x]}(z) \geq \mu_{[z]}(x)$.
6. If X is included in R, the natural topology of R prevents $\{x\}$ to be an open set, therefore a neighborhood of itself.
7. Fuzzy reflexivity (f.s.) means: $\forall x \in X_i; \mu^i_{[x]}(x) \in [0, 1]$.

References

Abreu D. & A. Rubinstein 1988. 'The Structure of Nash Equilibrium in Repeated Games with Finite Automata', *Econometrica*, **56**, pp. 1259-1282.

Aubin J.P. 1974. 'Coeur et Valeur des Jeux Flous', *Compte-rendus de l'Académie des Sciences de Paris*, **279**, pp. 891-894.

Aubin J.P.1976. 'Fuzzy Core and Equilibrium in Games Defined in Strategic Form', in *Directions in Large Scale Systems*, Y.C. Ho & S.K. Miller Eds, Plenum Press: New York.

Aubin J.P.1979. *Mathematical Methods in Economics and Game Theory*. North-Holland: Amsterdam.

Aubin J.P.1981. 'Locally Lipschitz Cooperative Games', *Journal of Mathematical Economics*, **8**, pp. 241-262.

Aubin J.P.1986. *L'Analyse non Linéaire et Ses Motivations Economiques*. Masson: Paris.

Aumann R.J. 1964. 'Values of Market with a Continuum of Traders', *Econometrica*, **83**, pp. 611-646.

Aumann R.J. 1976. 'Agreeing to Disagree', *The Annals of Statistics*, **4**, pp. 1236-1239.

Badard R. 1984. 'Fixed Point Theorems for Fuzzy Numbers', *Fuzzy Sets and Systems*, **13**, pp. 291-302.

Billot A.1986. 'A Contribution to A Mathematical Theory of Fuzzy Games' in *Fuzzy Economics and Spatial Analysis*, C. Ponsard & B. Fustier Eds, Librairie de l'Université, Dijon, pp. 47-56.

Billot A.1987. *Préférence et Utilité Floues*. Presses Universitaires de France: Paris.

Billot A. 1988. *Préférence Imprécise et Equilibres Economiques: Une Analyse Axiomatique*. Chs 3 & 4, Ph.D. in Economics, Université de Bourgogne, France.

Billot A. 1990. 'Peripherial Core of an Exchange Economy Represented as a Fuzzy Game' in *Multiperson Decision-Making Using Fuzzy Sets ans Possibility Theory*, J. Kacprzyck & M. Fedrizzi Eds, Klüwer: Boston, pp. 311-335.

Billot A.1991. 'Aggregation of Preferences: The Fuzzy Case', *Theory and Decision*, **30**, pp. 51-93.

Billot A.1992. 'From Fuzzy Set Theory to NonAdditive Probabilities: How Have Economists Reacted?', *Fuzzy Sets and Systems*, **49**, pp. 75-90.

Billot A.1995a. 'Fuzzy Utility Function: A New Elementary Proof', *Fuzzy Sets and Systems*, **74**, pp. 271-276.

Billot A.1995b. *Economic Theory of Fuzzy Equilibria*. Springer-Verlag: Berlin, New York.

Billot A.1996. 'Fuzzy Decision Theory', to appear in *Decision Under Uncertainty, International School of Economic Research*, J.D. Hey, F.Hahn & L.Luini Eds, Oxford University Press: Oxford.

Billot A.& B. Walliser 1995. 'A Mixed Knowledge Hierarchy', CERAS Working-Paper 95-17, Ecole des Ponts et Chaussées, Paris, France.

Border K.C. 1985. *Fixed Point Theorems with Applications to Economics and Game Theory*. Cambridge University Press, Cambridge.

Bose R.K. & D. Sahani 1987. 'Fuzzy Maapings and Fixed Point Theorems', *Fuzzy Sets and Systems*, **21**, pp. 53-58.

Brouwer L.E.J. 1912. 'Uber Abbildung von Mannigfaltikeiten', *Mathematische Annalen*, **71**, pp. 97-115.

Butnariu D.1978. 'Fuzzy Games. A Description of the Concept', *Fuzzy Sets and Systems,* **1**, pp. 181-192.

Butnariu D.1979. 'Solution Concepts for n-Person Fuzzy Games' in *Advances in Fuzzy Set Theory and Applications*, M.M. Gupta, R.K. Ragade & R.R. Yager Eds, Klüwer: Boston.

Butnariu D.1980. 'Stability and Shapley-Value for n-Person Fuzzy Games', *Fuzzy Sets and Systems,* **7**, pp. 63-72.

Butnariu D.1982. 'Fixed Points for Fuzzy Mappings', *Fuzzy Sets and Systems,* **14**, pp. 191-207.

Butnariu D.1985. 'NonAtomic Fuzzy Measures and Games', *Fuzzy Sets and Systems,* **17**, pp. 39-52.

Butnariu D.1986. 'Fuzzy Measurability and Integrability', *Journal of Mathematical Analysis and Applications,* **117**, pp. 385-410.

Butnariu D. 1987. 'Values and Cores of Fuzzy Games with Infinitely Many Players', *International Journal of Game Theory,* **16**, pp. 43-68.

Chitra A. & P.V. Subrahmanyam 1987. 'Fuzzy Sets and Fixed Points', *Journal of Mathematical Analysis and Applications,* **124**, 584-590.

Chang C.L. 1968. 'Fuzzy Topological Spaces', *Journal of Mathematical Analysis and Applications,* **17**, pp. 182-190.

Debreu G.1959. *Theory of Value.* Wiley: New York.

Debreu G. & H. Scarf 1963. 'A Limit Theorem on the Core of an Economy', *International Economic Review,* **4**, pp. 235-246.

Ekeland I.1979. *Economie Mathématique.* Hermann, Paris.

Friedman J. 1977. *Oligopoly and the Theory of Games.* North-Holland: Amsterdam.

Heilpern S. 1981. 'Fuzzy Mappings and Fixed Point Theorems', *Journal of Mathematical Analysis and Applications,* **83**, pp. 566-569.

Husseïnov F. 1994. 'Interpretation of Aubin's Fuzzy Coalitions and Their Extension: Relaxation of Finite Exchange Economy', *Journal of Mathematical Economics,* **23**, pp. 499-516.

Kakutani S. 1941. 'A Generalization of Brouwer's Fixed Point Theorem', *Duke Mathematical Journal,* **8**, pp. 416-427.

Kaufmann A. 1973. *Introduction à la Théorie des Sous-Ensembles Flous,* Vols 1 & 4. Masson: Paris.

Kaleva O. 1984. 'A Note on Fixed Points for Fuzzy Mappings', *Fuzzy Sets and Systems,* **15**, pp. 99-100.

Liu Y.M. 1985. 'A Note on Compactness in Fuzzy Unit Interval', *Kexue Tongbao,* **25**, pp. 33-35.

Mertens J.F. & S. Zamir 1985. 'Formulation of Bayesian Analysis for Games with Incomplete Information', *International Journal of Game Theory,* **14**, pp. 1-29.

Moulin H.1979. 'Dominance-Solvable Voting Schemes', *Econometrica,* **47**, pp. 249-269.

Moulin H. & F. Fogelman-Soulie 1979. *La Convexité dans les Mathématiques de la Décision.* Hermann:Paris.

Ponsard C. 1980. 'Fuzzy Economics Space', First World Regional Science Congress, Harvard University, Cambridge, Massachusetts.

Ponsard C. 1986. 'Foundations of Soft Decision Theory', in *Management Decision Support Systems Using Fuzzy Sets and Possibility Theory*, J. Kacprzyk & R.R. Yager Eds, Verlag T.U.V, Rheinland, pp. 27-37.

Ponsard C. 1987. 'Fuzzy Mathematical Models', *Fuzzy Sets and Systems*, **10**, pp. 302-313.

Simon H. 1964. 'Rationality', in A Dictionary of the Social Sciences, J. Gould & W.L. Kolb Eds, Free Press, Glencoe, pp. 573-574.

von Neumann J. & O. Morgenstern 1944. *Theory of Games and Economic Behavior*; Princeton University Press: Princeton.

Walliser B. 1989. 'Instrumental Rationality and Cognitive Rationality', *Theory and Decision*, **27**, pp. 7-36.

Weber S. 1979. 'On ϵ-Cores of Balanced Games', *International Journal of Game Theory*, **8**, pp. 241-250.

Wooders M.H. 1983. 'The ϵ-Cores of a Large Replica Game', *Journal of Mathematical Economics*, ***11***, pp. 277-300.

Zimmermann H.J. 1985. *Fuzzy Set Theory, and its Applications.* Kluwer, Nijhoff Publishing, Boston.

II Mathematical Programming

6 FUZZY LINEAR PROGRAMMING WITH SINGLE OR MULTIPLE OBJECTIVE FUNCTIONS

Heinrich Rommelfanger
Roman Słowiński

Abstract: The chapter presents an overview of methods for solving fuzzy linear programming (FLP) problems. It starts with the simplest case of linear programming models with soft constraints, called flexible programming. Then, it goes through most essential questions connected with FLP problems: modeling of fuzzy data, extended addition for aggregating fuzzy objectives and left-hand sides of fuzzy constraints, inequality relations between fuzzy left-hand sides and fuzzy right-hand sides of the constraints, treatment of fuzzy objectives and computation of a comprise solution in the multi-objective case. Finally, a survey of applications of FLP and of related bibliography is given.

6.1 INTRODUCTION

Empirical surveys reveal that Linear Programming (LP) is one of the most frequently applied OR techniques in real-world problems, (see e.g. Kivijärvi, Korhonen and Wallenius 1986; Lilien 1987; Tingley 1987; Meyer zu Selhausen 1989). However, given the power of LP one could have expected even more applications. This might be due to the fact that LP requires much of well-defined and precise data which involve high information costs. In real-world applications, certainty, reliability and precision of data is often illusory. Furthermore, the optimal solution of a LP problem depends on a part of constraints only, and thus, much of the information collected have little impact on the solution. Being able to deal with vague and imprecise data may greatly contribute to the diffusion and application of LP. The stochastic approach has never proved to be very efficient in handling vague and imprecise data.

However, since the seminal paper on "Fuzzy Sets" by Lotfi A. Zadeh in 1965, there exists a convenient and powerful way of modeling vague data without having recourse to stochastic concepts. The subject of this chapter is to review how fuzzy data can be integrated into LP problems.

The use of fuzzy models does not only avoid unrealistic modeling but also offers a chance for reducing information costs (see Rommelfanger 1995a). Then, in the first step of the interactive solution process, the fuzzy system is modeled by using only the information which the decision maker (DM) can provide without any expensive acquisition. Knowing a first "compromise solution" the DM can perceive which further information would be required and is able to justify the decision by comparing carefully additional advantages and arising costs. In doing so, the compromise solutions are improved step by step. This procedure obviously offers the possibility to limit the acquisition and processing of information to the relevant components and, therefore, information costs can be considerably reduced.

A general model of a fuzzy linear programming problem (FLP problem) is presented by the following system:

$$\tilde{C}_1 x_1 \oplus \tilde{C}_2 x_2 \oplus \cdots \oplus \tilde{C}_n x_n \to \widetilde{\max}$$

subject to (1)

$$\tilde{A}_{i1} x_1 \oplus \tilde{A}_{i2} x_2 \oplus \cdots \oplus \tilde{A}_{in} x_n \mathrel{\tilde{\leq}} \tilde{B}_i \ , \quad i = 1, \ldots, m$$

$$x_1, x_2, \ldots, x_n \geq 0 \ ,$$

where $\tilde{C}_j$, $\tilde{A}_{ij}$ and $\tilde{B}_i$ are (flat) fuzzy numbers and $\tilde{\leq}$ is a fuzzy inequality relation that will be defined later.

As each real number a can be modeled as a fuzzy number

$$\tilde{A} = \{(x, f_A(x)) \mid x \in \mathbf{R}\} \quad \text{with} \quad f_A(x) = \begin{cases} 1 & \text{if} \quad x = a \\ 0 & \text{otherwise} \end{cases} ,$$

the general system (1) includes the following special cases:

(i) the objective function is crisp, i.e.

$$z(\mathbf{x}) = c_1 x_1 + c_2 x_2 + \cdots + c_n x_n \to \max \tag{2}$$

(ii) some or all constraints are crisp, i. e.

$$g_i(\mathbf{x}) = a_{i1} x_1 + a_{i2} x_2 + \cdots + a_{in} x_n \leq b_i \tag{3}$$

(iii) some or all constraints are soft (flexible), i.e.

$$g_i(\mathbf{x}) = a_{i1} x_1 + a_{i2} x_2 + \cdots + a_{in} x_n \mathrel{\tilde{\leq}} \tilde{B}_i \ . \tag{4}$$

These special cases may be combined. In fact, they correspond to different semantics of the fuzzy sets. When the objective function(s) and the left-hand sides of the constraints are crisp, then fuzzy sets express preferences concerning attainment of goal(s) and satisfaction of soft constraints. This case corresponds to, so called, flexible programming. When the coefficients in the objective function(s) and in the constraints are fuzzy numbers, then fuzzy sets represent incomplete or vague state of data under the form of possibility distributions.

The application of FLP problems offers the advantage that the DM can model his/her problem in accordance with his/her current state of information. At the same time he/she is no longer able to use directly the well known simplex algorithms for computing a solution of his/her problem. Therefore, various procedures for calculating a compromise solution of an FLP problem (1) have been developed. They mainly differ from the assumptions made in order to boil down the FLP to an auxiliary classical mathematical programming problem, mostly to classical LP or goal programming problems.

In this chapter we are presenting a survey of procedures for solving FLP problems. At first, we deal with the simplest case, and historically the oldest one, LP models with soft constraints. We then tackle the essential points connected with general forms of FLP problems:

- modeling of fuzzy data,
- extended addition for aggregating fuzzy objectives and left-hand sides of fuzzy constraints,
- inequality relations between fuzzy left-hand sides and fuzzy right-hand sides in the constraints,
- treatment of fuzzy objectives,
- computation of a compromise solution.

Finally, we give a brief survey of applications of fuzzy linear programs published in the literature.

6.2 LINEAR PROGRAMMING WITH SOFT CONSTRAINTS

If the DM cannot or does not wish to precise with certainty the right-hand sides of some constraints, while being able to specify all coefficients of the left-hand sides as crisp numbers, we get the simplest form of an FLP model called *flexible programming*. Such problems with soft constraints of the type

$$g_i(\mathbf{x}) = g_i(x_1, x_2, ..., x_n) = a_{i1}x_1 + a_{i2}x_2 + \cdots + a_{in}x_n \mathrel{\tilde{\leq}} \tilde{B}_i$$

were discussed for the first time by Tanaka, Okuda and Asai (1974) who described the imprecise right-hand side $\tilde{B}_i$ by a fuzzy set with a support $[b_i, b_i + d_i] \subseteq \mathbf{R}$, $d_i \geq 0$ and a monotone decreasing membership function $\mu_{B_i}(\mathbf{x})$.[1]

Their approach is based on the general proposal of Bellman and Zadeh (1970) that a fuzzy decision $\tilde{D}$ is defined as a fuzzy set resulting from the intersection of the fuzzy constraint $\tilde{C}$ and the fuzzy goal $\tilde{G}$. If the fuzzy constraint and the fuzzy goal are characterized by the membership functions $\mu_C(\mathbf{x})$ and $\mu_G(\mathbf{x})$, the fuzzy decision is characterized by the membership function $\mu_D(\mathbf{x}) = \min(\mu_C(\mathbf{x}), \mu_G(\mathbf{x}))$. An optimal decision is defined as an alternative $\mathbf{x}$ which maximizes $\mu_D(\mathbf{x})$. As in this approach fuzzy constraints are treated in the same manner as fuzzy goals it is called a *symmetric* one.

Specifying a membership level α, $\alpha \in [0,1]$, Tanaka, Okuda and Asai restrict the set of feasible solutions of each soft constraint to the crisp set

$$C_\alpha^i = \{\mathbf{x} \mid \mu_{C^i}(\mathbf{x}) \geq \alpha\},\ i=1,\ldots,m.$$

In the usual case, where the right hand side $\tilde{B}_i$ is a fuzzy number of the LR-type $(b_i, 0, \beta_i)_{LR}$, the crisp set C_α^i is characterized by

$$C_\alpha^i = \left\{\mathbf{x} \,\middle|\, a_{i1}x_1 + a_{i2}x_2 + \cdots + a_{in}x_n \leq b_i + \beta_i R^{-1}(\alpha)\right\}$$

Then, the set of feasible solutions of an FLP with soft constraints is $C_\alpha = \bigcap_{i=1}^{m} C_\alpha^i$.

According to Tanaka, Okuda and Asai, the given objective function

$$z(\mathbf{x}) = c_1x_1 + \cdots + c_nx_n$$

is replaced by the fuzzy goal $\mu_G(\mathbf{x}) = \dfrac{z(\mathbf{x})}{\max\limits_{x \in C_0}\{z(\mathbf{x})\}}$.

To get an optimal solution of this FLP, we have to determine the optimal pair $(\alpha^*, \mathbf{x}^*)$ such that

$$\min(\alpha^*, \mu(\mathbf{x}^*)) = \sup_{\alpha} \min\left(\alpha, \max_{x \in C_\alpha}\{\mu_G(\mathbf{x})\}\right). \tag{5}$$

If the optimal α^* was determined a priori, the problem (5) could be reduced to an auxiliary crisp mathematical programming problem where the objective was to find $\mathbf{x}^*$ that maximizes $\mu_G(\mathbf{x})$ on the set C_{α^*}. In the general case, Tanaka, Okuda and Asai propose an iterative algorithm where, beginning with any $\alpha_1 \in [0,1]$, the values α_k and $\max\limits_{\mathbf{x} \in C_{\alpha_k}} \mu_G(\mathbf{x})$ converge to the optimum step by step.

A more integrative approach was proposed by Zimmermann (1976). For getting a common ground for comparing the constraints with the goals, he proposes to evaluate the performance of a soft constraint

$$g_i = g_i(\mathbf{x}) \,\tilde{\leq}\, \tilde{B}_i$$

by a membership function

$$\mu_{B_i}(g_i) \begin{cases} 1 & \text{if } g_i < b_i \\ \breve{\mu}_{B_i}(g_i) & \text{if } b_i \leq g_i \leq b_i + d_i \\ 0 & \text{if } b_i + d_i < g_i \end{cases} \tag{6}$$

expressing the individual satisfaction of the DM in relation to $g_i = g_i(x_1,\ldots,x_n)$, i=1,…,m, where $\breve{\mu}_{B_i}(g_i)$ is a monotone decreasing function of g_i taking values from the interval [0,1].

Other authors, e.g. Buckley (1988, 1995) and Luhandjula (1987), do not use this utility oriented interpretation but interpret $\mu_{B_i}(g_i)$ as the possibility that the i-th constraint is satisfied. Therefore, we can find in the literature the division into fuzzy programming and possibilistic programming (see Lai and Hwang 1992), however, this distinction is not clearly defined and its interpretation is different from above. In section 6.3, we will return to this point because it is related with the modeling of fuzzy data.

The composition of the functions $\mu_{B_i}(g_i)$ and $g_i = g_i(\mathbf{x}) = a_{i1}x_1 + \cdots + a_{in}x_n$ to $\mu_i(\mathbf{x}) = \mu_{B_i} \times g_i(\mathbf{x}) = \mu_{B_i}(g_i(\mathbf{x}))$ directly assigns a measure of the satisfaction of the i-th constraint to the solution $\mathbf{x} = (x_1, \ldots, x_n)$.

According to Zimmermann (1976, 1978), the inequality relation "$\tilde{\leq}$" in soft constraints (4) may be interpreted as

$$g_i(\mathbf{x}) \;\tilde{\leq}\; \tilde{B}_i \quad \Leftrightarrow \quad \begin{cases} g_i(\mathbf{x}) \leq b_i + d_i \\ \mu_{B_i}(\mathbf{x}) \rightarrow \max \end{cases}$$

i.e. each soft constraint adds an additional objective to the decision problem, called fuzzy objective.

There exist various proposals for modeling the function $\breve{\mu}_{B_i}(g_i)$ on the interval $[b_i, b_i+d_i]$, e.g.

a. **linear shape** (e.g. in (Zimmermann 1976); (Sommer 1978); (Chanas 1983); (Slowinski 1986); (Werners 1987); (Buckley 1988); (Lai and Hwang 1992).
b. **concave shape**
   a) by exponential functions in (Zimmermann 1978); (Sakawa 1983),
   b) by piecewise linear functions, e.g. in (Hannan 1981); (Rommelfanger 1984); (Nakamura 1984); (Sakawa and Yano 1990); (Inuiguchi, Ichihashi and Kume 1989); (Czyzak and Slowinski 1991),
c. **s- shape**
   a) by piecewise linear functions, e.g. in (Hannan 1981); (Rommelfanger 1984); (Czyzak and Slowinski 1991),
   b) by hyperbolic functions in (Leberling 1981, 1983); (Sakawa and Yano 1990),
   c) by hyperbolic inverse functions in (Sakawa and Yano 1990),
   d) by logistic functions in (Werners 1987), (Zimmermann 1996),
   e) by cubical functions.

Therefore, a linear programming system of the type

$$z(\mathbf{x}) = c_1x_1 + \cdots + c_nx_n \rightarrow \max$$

subject to (7)

$$a_{i1}x_1 + \cdots + a_{in}x_n \;\tilde{\leq}\; \tilde{B}_i \qquad i = 1, \ldots, m_1$$
$$a_{i1}x_1 + \cdots + a_{in}x_n \leq b_i \qquad i = m_1+1, \ldots, m$$
$$x_1, \ldots, x_n \geq 0$$

with m_1 soft constraints and $(m - m_1)$ crisp constraints may be described more precisely by a multiobjective optimization problem of the type

$$\max_{\mathbf{x} \in X_U} \left(z(\mathbf{x}), \mu_1(\mathbf{x}), \ldots, \mu_{m_1}(\mathbf{x})\right) \tag{8}$$

where $$X_U = \left\{ \mathbf{x} \in \mathbf{R}_0^n \;\middle|\; \begin{array}{lll} a_{i1}x_1 + \cdots + a_{in}x_n \leq b_i + d_i & , \; i = 1,2,\ldots,m_1 & \text{and} \\ a_{i1}x_1 + \cdots + a_{in}x_n \leq b_i & , \; i = m_1+1,\ldots,m & \end{array} \right\}$$

For comparing the given objective function $z(\mathbf{x})$ with the fuzzy objective functions $\mu_i(\mathbf{x})$, it is usually proposed to substitute $z(\mathbf{x})$ by a function $\mu_Z(\mathbf{x})$, where the quantification of $\mu_Z(\mathbf{x})$ is obtained by specifying the membership function $\hat{\mu}_Z(z)$ of the fuzzy set of the satisfying values $\tilde{Z} = \{(z, \hat{\mu}_Z(z)) \mid z \in \mathbf{R}\}$, i.e. $\mu_Z(\mathbf{x}) = \hat{\mu}_Z(z(\mathbf{x}))$.

Basic data for modeling the membership function $\hat{\mu}_Z(z)$ may be

$$\bar{z} = \max_{\mathbf{x} \in X_U} z(\mathbf{x}) \quad \text{and} \quad \underline{z} = \max_{\mathbf{x} \in X_L} z(\mathbf{x})$$

where X_U has already been defined for (8), and $X_L = \{\mathbf{x} \in \mathbf{R}_0^n | a_{i1}x_1 + \cdots, \cdots + a_{in}x_n \leq b_i \ \forall i = 1,2,..,m\}$ but in the literature we can also find other proposals (see e.g. Chanas 1983).

Then, the basic form of $\hat{\mu}_Z(z)$ is given by

$$\hat{\mu}_Z(z) = \begin{cases} 0 & \text{if} \quad z < \underline{z}^S \\ \breve{\mu}_Z(z) & \text{if} \ \underline{z}^S \leq z \leq \bar{z}^S \\ 1 & \text{if} \ \bar{z}^S < z \end{cases}$$

where $\breve{\mu}_Z(z)$ is a monotone increasing function of z, which may be modeled in analogy to $\mu_{B_i}(g_i)$, according to the preferences of the DM, and $\underline{z}^S \geq \underline{z}$ and $\bar{z}^S \leq \bar{z}$ are subjectively chosen boundaries.

In practical applications, a DM is not interested in finding the complete solution of the multiobjective problem (8), i.e. the set of all Pareto optimal solutions, but he/she needs a procedure which guides him/her to a so-called "compromise solution". If the compromise solution is understood as a fuzzy decision in the sense of Bellman and Zadeh (1970), then the total satisfaction of the DM may be described by

$$\lambda(\mathbf{x}) = \min(\mu_Z(\mathbf{x}), \mu_1(\mathbf{x}), \ldots, \mu_{m_1}(\mathbf{x})). \tag{9}$$

Empirical researches reveal that the min-operator is often too pessimistic (see e.g. (Thole, Zimmermann and Zysno 1979). However, not only the simple mathematical handling supports the use of this operator, but also the fact that the subjectively specified membership values are in general defined on an ordinal scale[2] which means that only simple data comparisons are possible. Therefore, the use of the mean value, as proposed by Sommer (1978), or the $\widetilde{\text{and}}$-operator, as proposed by Werners (1987), raises serious measurement problems. Other forms of trade-offs between the degrees of feasibility of the fuzzy constraints and goal satisfaction are proposed by Bellman and Zadeh (1973), Kacprzyk (1983) and Yager (1979).

Following Negoita and Sularia (1976), the optimization problem

$$\max_{\mathbf{x} \in X_U} \min(\mu_Z(\mathbf{x}), \mu_1(\mathbf{x}), \ldots, \mu_{m_1}(\mathbf{x})) \tag{10}$$

is clearly equivalent to

$$\lambda \to \max$$

subject to (11)

$$\lambda \le \mu_Z(\mathbf{x})$$
$$\lambda \le \mu_i(\mathbf{x}) \;,\quad i = 1, ..., m_1$$
$$x \in X_U \;,\quad \lambda \le 1 .$$

Using linear membership functions (see e.g. Zimmermann 1976; Werners 1987) or piecewise linear, concave membership functions (see Rommelfanger 1984), the problem (11) matches a classical LP model and can easily be solved by well-known algorithms.

The structure of this solution process suggests that it may easily be extended to multiobjective LP systems with crisp or soft constraints (see e.g. Zimmermann 1978; Leberling 1981; Rommelfanger 1984, 1990; Werners 1987).

Some authors recommend to organize the solution process as an interactive process. Werners (1987), for example, intends to ensure by this procedure that the calculated solution is a Pareto-optimal solution of (8), because it is not necessarily true when using the min-operator as a scalarizing function. According to Sakawa (1983, 1993), the DM determines membership functions for fuzzy goals and then specifies reference levels for the membership functions which can be changed from one iteration to another. The interactive solution process MOLPAL of Rommelfanger (1984, 1990) is based on the idea that, at the beginning, the DM is just able to approximate roughly the membership functions, whereas later he/she can specify the functions in detail referring to additional information from the solution process.

In the literature there also exist nonsymmetric approaches, where at first the set of feasible solutions is determined and then the optimal solution is selected by using the fuzzy goal. Below, the nonsymmetric procedure of Chanas (1983) is sketched; a more general case is discussed in (Orlovski 1984, 1985).

According to the nonsymmetric approach of Chanas (1983), we first solve the following parametric programming (12) for all values of the parameter $\theta \in [0,1]$.

$$z(\mathbf{x}) = c_1 x_1 + \cdots + c_n x_n \to \max$$

subject to (12)

$$a_{i1}x_1 + \cdots + a_{in}x_n \le \mu_{B_i}^{-1}(\theta) \qquad i = 1, ..., m_1$$
$$a_{i1}x_1 + \cdots + a_{in}x_n \le b_i \qquad i = m_1 + 1, ..., m$$
$$x_1, ..., x_n \ge 0$$

For a linear membership function, we have $\mu_{B_i}^{-1}(\theta) = b_i + d_i(1-\theta)$.

The optimal solutions $\mathbf{x}^*(\theta)$ and $z^*(\theta) = z(\mathbf{x}^*(\theta))$ are then presented to the DM, who constructs with this information a membership function $\mu_Z(\theta)$ expressing his/her satisfaction with the values $z^*(\theta)$, $\theta \in [0,1]$.

Then, the final optimal solution $\mathbf{x}^*(\theta^*)$ and $z^*(\theta^*) = z(\mathbf{x}^*(\theta^*))$ will exist at

$$\min\left[\mu_Z(\theta^*), 1-\theta^*\right] = \max_{\theta \in [0,1]}\left\{\min\left[\mu_Z(\theta), 1-\theta\right]\right\}. \qquad (13)$$

6.3 MODELING OF FUZZY DATA

In the literature, the flexible right-hand sides $\tilde{B}_i$ are modeled by L-R-type fuzzy numbers, where the left spread is zero and b_i is the largest value which is accepted with certainty as the right-hand side of the constraint i, i.e. $\tilde{B}_i = (b_i; 0; \beta_i)_{RR}$. Obviously, the meaning of $\tilde{B}_i$ as the greatest value for the right-hand side does not permit to model $\tilde{B}_i$ as a trapezoidal fuzzy number or as a fuzzy number with a positive left spread, as can be found in early papers on the subject.

Suitable reference functions are the functions presented above in section 6.1, which are confirmed by utility theory. Yet the essential problems are neglected: satisfaction values can be measured only on an ordinal scale and comparisons of the membership values of different objectives or constraints can only be attained with great difficulties. Obviously, the same objections exist when interpreting the μ_{B_i} as possibility distributions.

Furthermore, the interpretation and evaluation of the compromise solution λ_{Max} calculated from (10) are to question. It is a pity that the problem of specifying the membership functions is often ignored in the literature because this point is an essential criterion for the approval and application of fuzzy models. In our opinion there exist only few cases when the DM is capable of describing the precise form of a fuzzy number $\tilde{B}_i = (b_i; 0; \beta_i)_{LR}$ as this is assumed e.g. by Zimmermann (1976, 1996) or Sakawa and Yano (1985). In practice we will find only more or less successful approximations of the "true" shapes of membership functions. In particular, it is very difficult to model realistically the part of a membership function belonging to small membership values.

A practical way of getting suitable membership functions is the following procedure which was proposed by Rommelfanger (1989).

At first the DM specifies some prominent membership values and associates them with special meanings. After that he specifies values which belong to these selected membership degrees. This procedure shall be explained for a right-hand side $\tilde{B}_i = \{(y, \mu_{B_i}(y)) \mid y \in \mathbf{R}\}$, which is interpreted as the maximal quantity of stock at the DM's disposal.

$\alpha = 1 \; : \mu_{B_i}(y) = 1$ means that y belongs with certainty to the set of available values,

$\alpha = \lambda_A : \mu_{B_i}(y) \geq \lambda_A$ means that the DM is willing to accept y as an available value for the time being. A value y with $\mu_{B_i}(y) \geq \lambda_A$ has a good chance of belonging to the set of available values. Corresponding values of y are relevant to the decision. Obviously, a value y with $\mu_{B_i}(y) = \lambda_A$ is a sort of aspiration level.

$\alpha = \varepsilon\ :\ \mu_{B_i}(y) < \varepsilon$ means that y has only a very little chance of belonging to the set of available values. The DM is willing to neglect the values y with $\mu_{B_i}(y) < \varepsilon$.

Subsequently, the DM has to fix values $\bar{b}_i^{\lambda_A}$ and $\bar{b}_i^{\varepsilon}$ such that $\mu_{B_i}(\bar{b}_i^{\lambda_A}) = \lambda_A$ and $\mu_{B_i}(\bar{b}_i^{\varepsilon}) = \varepsilon$. Then the polygon line from $(b_i, 1)$ over $(\bar{b}_i^{\lambda_A}, \lambda_A)$ to $(\bar{b}_i^{\varepsilon}, \varepsilon)$ is a suitable representation of μ_{B_i} on the interval $[b_i, \bar{b}_i^{\varepsilon}]$. For all $y \notin [b_i, \bar{b}_i^{\varepsilon}]$ we set $\mu_{B_i}(y) = 0$; see Figure 1.

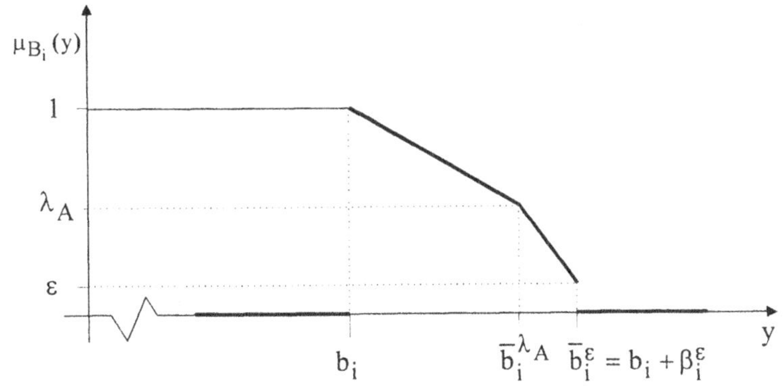

Figure 1 Membership function of $\tilde{B}_i$

Taking pattern from L-R-type fuzzy numbers, we symbolize a fuzzy number with this special membership function by $\tilde{B}_i = (b_i;\ 0;\ 0;\ \beta_i^{\lambda_A};\ \beta_i^{\varepsilon})^{\lambda_A, \varepsilon}$, where $\beta_i^{\lambda_A} = \bar{b}_i^{\lambda_A} - b_i$ and $\beta_i^{\varepsilon} = \bar{b}_i^{\varepsilon} - b_i$. If required, the DM can specify additional membership levels and additional points $(y,\ f_{B_i}(y))$ on the polygon line.

Fuzzy coefficients $\tilde{A}_{ij}$ or $\tilde{C}_{kj}$ do not usually include elements that may be realized with certainty. An appropriate point of reference for a fuzzy set $\tilde{A}_{ij}$ is the subset $[\underline{a}_{ij}, \bar{a}_{ij}] \in \mathbf{R}$ consisting of the real numbers with the highest chance of realization, i.e.

$$\mu_{A_{ij}}(y) \begin{cases} = 1 & \text{if } y \in [\underline{a}_{ij}, \bar{a}_{ij}] \\ < 1 & \text{else} \end{cases}$$

Accordingly, the DM should specify numbers $\underline{a}_{ij}^{\lambda_A}, \bar{a}_{ij}^{\lambda_A}, \underline{a}_{ij}^{\varepsilon}, \bar{a}_{ij}^{\varepsilon}$, so that

$$\mu_{A_{ij}}(y) \begin{cases} \geq \lambda_A & \text{if } y \in [\underline{a}_{ij}^{\lambda_A}, \bar{a}_{ij}^{\lambda_A}] \\ < \lambda_A & \text{else} \end{cases} \quad \text{and} \quad \mu_{A_{ij}}(y) \begin{cases} \geq \varepsilon & \text{if } y \in [\underline{a}_{ij}^{\varepsilon}, \bar{a}_{ij}^{\varepsilon}] \\ < \varepsilon & \text{else} \end{cases}.$$

The width of the intervals $[\underline{a}_{ij}^{\alpha}, \overline{a}_{ij}^{\alpha}]$, $\alpha = 1, \lambda_A, \varepsilon$ is inversely linked with the amount of information available to the DM.

Consequently, the polygon line from $(\underline{a}_{ij}^{\varepsilon}, \varepsilon)$ over $(\underline{a}_{ij}^{\lambda_A}, \lambda_A)$, $(\underline{a}_{ij}, 1)$, $(\overline{a}_{ij}, 1)$, $(\overline{a}_{ij}^{\lambda_A}, \lambda_A)$ to $(\overline{a}_{ij}^{\varepsilon}, \varepsilon)$ is a suitable representation of the membership function of $\tilde{A}_{ij}$ on the support of $[\underline{a}_{ij}^{\varepsilon}, \overline{a}_{ij}^{\varepsilon}]$; see Figure 2.

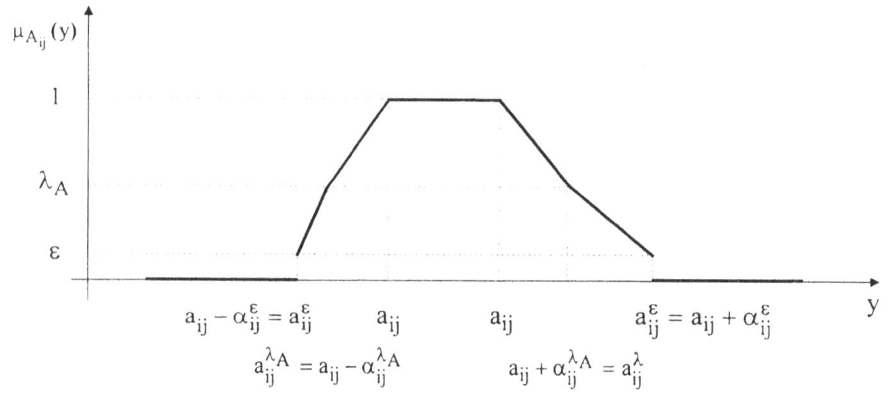

Figure 2 $\tilde{A}_{ij} = \left(\underline{a}_{ij};\overline{a}_{ij};\underline{\alpha}_{ij}^{\lambda_A};\underline{\alpha}_{ij}^{\varepsilon};\overline{\alpha}_{ij}^{\lambda_A};\overline{\alpha}_{ij}^{\varepsilon}\right)^{\lambda_A,\varepsilon}$

The special case in which $\underline{a}_{ij} = \overline{a}_{ij}$ is also imaginable, but in our opinion it is less realistic to assume that all coefficients $\tilde{A}_{ij}$ are unimodal fuzzy numbers. This assumption is characteristic to the approach of Kovacs (1994) based on centered fuzzy numbers (see section 6.7).

In comparison to the spreads of the right-hand sides $\tilde{B}_i$, the spreads $\underline{\alpha}_{ij}^{\varepsilon} = \underline{a}_{ij} - \underline{a}_{ij}^{\varepsilon}$ and $\overline{\alpha}_{ij}^{\varepsilon} = \overline{a}_{ij}^{\varepsilon} - \overline{a}_{ij}$ of the coefficients $\tilde{A}_{ij}$ are relatively small, so one often skips level λ_A and uses coefficients of the simple type $\tilde{A}_{ij} = (\underline{a}_{ij};\overline{a}_{ij};\underline{\alpha}_{ij}^{\varepsilon};\overline{\alpha}_{ij}^{\varepsilon})^{\varepsilon}$ or $\tilde{C}_j = (\underline{c}_j;\overline{c}_j;\underline{\gamma}_j^{\varepsilon};\overline{\gamma}_j^{\varepsilon})^{\varepsilon}$.

6.4 AGGREGATION OF THE LEFT-HAND SIDES OF FUZZY CONSTRAINTS

We are now passing to the case where the coefficients of the left-hand sides of the constraints are fuzzy intervals.
The left-hand side of a fuzzy constraint

$$\tilde{A}_{i1}x_1 \oplus \tilde{A}_{i2}x_2 \oplus \cdots \oplus \tilde{A}_{in}x_n \;\tilde{\leq}\; \tilde{B}_i \tag{14}$$

can be aggregated to a fuzzy set $\tilde{A}_i(\mathbf{x})$ by Zadeh's extension principle

$$f_{\tilde{A}*\tilde{B}}(z) = \sup_{z=x*y} T(f_A(x), f_B(y)), \quad z \in \mathbf{R} \tag{15}$$

where $*$ is a real operation $* : \mathbf{R} \times \mathbf{R} \rightarrow \mathbf{R}$ and $T: [0, 1] \times [0, 1] \rightarrow [0, 1]$ is any given t-norm.

In the literature, the Min t-norm is usually applied. Then, if all coefficients $\tilde{A}_{ij}$ of the i-th constraint are fuzzy intervals of the same L-R-type, the left-hand side can be consolidated to a fuzzy interval with the same reference functions. In particular, for coefficients of type $\tilde{A}_{ij} = (\underline{a}_{ij}; \bar{a}_{ij}; \underline{\alpha}^{\varepsilon}_{ij}; \bar{\alpha}^{\varepsilon}_{ij})^{\varepsilon}$, we get

$$\tilde{A}_i(\mathbf{x}) = \tilde{A}_{i1}x_1 \oplus \tilde{A}_{i2}x_2 \oplus \cdots \oplus \tilde{A}_{in}x_n = (\underline{a}_i(\mathbf{x}), \bar{a}_i(\mathbf{x}); \underline{\alpha}^{\varepsilon}_i(\mathbf{x}), \bar{\alpha}^{\varepsilon}_i(\mathbf{x}))^{\varepsilon} \quad \text{with}$$

$$\underline{a}_i(\mathbf{x}) = \sum_{j=1}^{n} \underline{a}_{ij}x_j, \quad \bar{a}_i(\mathbf{x}) = \sum_{j=1}^{n} \bar{a}_{ij}x_j, \quad \underline{\alpha}^{\varepsilon}_i(\mathbf{x}) = \sum_{j=1}^{n} \underline{\alpha}^{\varepsilon}_{ij}x_j, \quad \bar{\alpha}^{\varepsilon}_i(\mathbf{x}) = \sum_{j=1}^{n} \bar{\alpha}^{\varepsilon}_{ij}x_j.$$

Obviously, the spreads $\underline{\alpha}^{\varepsilon}_i(\mathbf{x})$ and $\bar{\alpha}^{\varepsilon}_i(\mathbf{x})$ extend, when the number and the values of the variables x_j increase. Thus, the left-hand side $\tilde{A}_i(\mathbf{x})$ gets fuzzier and fuzzier. We will come back to this problem in section 6.7.

6.5 INEQUALITY RELATIONS

A pivotal question while determining a solution of an FLP model is the interpretation of the inequality relation in fuzzy constraints $\tilde{A}_i(\mathbf{x}) \;\tilde{\leq}\; \tilde{B}_i$.

In the literature various concepts have been proposed for comparing fuzzy sets, see e.g. (Dubois and Prade 1983); (Bortolan and Degani 1985); (Rommelfanger 1986), but many of these techniques appeared to be of little interest for fuzzy mathematical programming. Special interpretations of the inequality relation in fuzzy constraints $\tilde{A}_i(\mathbf{x}) \;\tilde{\leq}\; \tilde{B}_i$ are suggested for instance by Negoita and Sularia (1976); Tanaka and Asai (1984); Ramik and Rimanek (1985); Slowinski (1986); Carlsson and Korhonen (1986); Luhandjula (1987); Rommelfanger (1988); Buckley (1988, 1989); Sakawa and Yano (1989); see the survey in (Rommelfanger 1989) and (Lai and Hwang 1992). Let us also mention a special comparison technique based on the area compensation determined by the membership functions of two fuzzy numbers being compared, characterized by Kolodziejczyk (1986), Chanas (1987), Roubens

(1990) and Fortemps and Roubens (1996). This technique has been used recently within multiobjective FLP by Fortemps and Teghem (1997).

In most of these approaches fuzzy constraints $\tilde{A}_i(\mathbf{x}) \tilde{\leq} \tilde{B}_i$ are replaced by one or two crisp linear constraints. To give an idea of these crisp surrogates, we present some of them in common terms, starting from the most pessimistic to a more and more optimistic attitude. For simplicity, we assume fuzzy interval of the type $\tilde{A}_i(\mathbf{x}) = (\underline{a}_i(\mathbf{x}); \bar{a}_i(\mathbf{x}); \underline{\alpha}_i(\mathbf{x}); \bar{\alpha}_i(\mathbf{x}))_{LR}$ and a fuzzy number $\tilde{B}_i = (b_i; 0; \beta_i)_{LR}$:

- $\bar{a}_i(\mathbf{x}) + \bar{\alpha}_i(\mathbf{x}) R^{-1}(\rho) \leq b_i \quad , \rho \in [0,1]$
  Tanaka and Asai (1984); Buckley (1988),
- $\left\{ \begin{array}{c} \bar{a}_i(\mathbf{x}) \leq b_i \\ \bar{a}_i(\mathbf{x}) + \bar{\alpha}_i(\mathbf{x}) R^{-1}(\varepsilon) \leq b_i + \beta_i R^{-1}(\varepsilon), \ \varepsilon \in [0,1[\end{array} \right\}$
  Ramik and Rimanek (1985); Rommelfanger (1989),
- $\bar{a}_i(\mathbf{x}) + \bar{\alpha}_i(\mathbf{x}) R^{-1}(\sigma) \leq b_i + \beta_i R^{-1}(\sigma), \ \sigma \in [0,1]$
  Carlsson and Korhonen (1986),
- $\left\{ \begin{array}{l} \underline{a}_i(\mathbf{x}) - b_i \leq \underline{\alpha}_i(\mathbf{x}) L^{-1}(\tau) + \beta_i R^{-1}(\tau), \tau \in [0,1] \text{ optimistic index} \\ \bar{a}_i(\mathbf{x}) + \bar{\alpha}_i(\mathbf{x}) R^{-1}(\eta) \leq b_i + \beta_i R^{-1}(\eta), \eta \in [0,1] \text{ pessimistic index} \end{array} \right\}$
  Slowinski (1986),
- $\left\{ \begin{array}{ll} \bar{a}_i(\mathbf{x}) \leq b_i + \delta\beta_i & (\delta + \varepsilon) \in [0,1], \ \delta, \varepsilon \geq 0 \\ \bar{a}_i(\mathbf{x}) + (1 - \varepsilon - \delta)\bar{\alpha}_i(\mathbf{x}) \leq b_i + (1-\varepsilon)\beta_i & \end{array} \right\}$
  Ramik and Rommelfanger (1993)[3],
- $\underline{a}_i(\mathbf{x}) - \underline{\alpha}_i(\mathbf{x}) L^{-1}(\alpha) \leq b_i + \beta_i R^{-1}(\alpha), \ \alpha \in [0,1]$
  Sakawa and Yano (1985, 1989); Luhandjula (1987).

In all these approaches, the parameters $\alpha, \delta, \varepsilon, \eta, \tau, \rho, \mu, \sigma$ can be used by the DM to control the degree of his/her satisfaction with fuzzy constraints in an interactive way. Obviously all these crisp inequality equations are linear ones.

Another surrogate has been proposed by Rommelfanger [1988]:

$$\tilde{A}_i(\mathbf{x}) \tilde{\leq}_R \tilde{B}_i \Leftrightarrow \left\{ \begin{array}{ll} \bar{a}_i(\mathbf{x}) + \bar{\alpha}_i^{\varepsilon}(\mathbf{x}) \leq b_i + \beta_i^{\varepsilon} & (16) \\ \mu_i(\mathbf{x}) = \mu_{B_i}(\bar{a}_i(\mathbf{x})) \rightarrow \max & (17) \end{array} \right.$$

It is composed of the "pessimistic index" (16), which is used by Slowinski (1986), and a new objective (17). The membership function μ_{B_i} is defined according to (5) and may be interpreted as a subjective evaluation of the needed quantity $\bar{a}_i(\mathbf{x}) = \sum_{j=1}^{n} \bar{a}_{ij} x_j$ with regard to the right-hand side $\tilde{B}_i$, see Figure 3.

The inequality relation "$\tilde{\leq}_R$" has the advantage that a possible surplus $\bar{a}_i(\mathbf{x})$ -b_i directly influences the decision process. Moreover, "$\tilde{\leq}_R$" coincides with the usual interpretation of the inequality relation in soft constraints (6); in the special case of deterministic inequalities it corresponds to the classical "$\leq$"-relation. Thus "$\tilde{\leq}_R$" is a general definition for inequality relations in optimization models.

When using the inequality relation "$\tilde{\leq}_R$", it is sufficient for all $\tilde{A}_{ij}$ to specify the values of $\bar{a}_{ij}$ and $\bar{a}^{\varepsilon}_{ij} = \bar{a}_{ij} + \bar{\alpha}^{\varepsilon}_{ij}$ only.

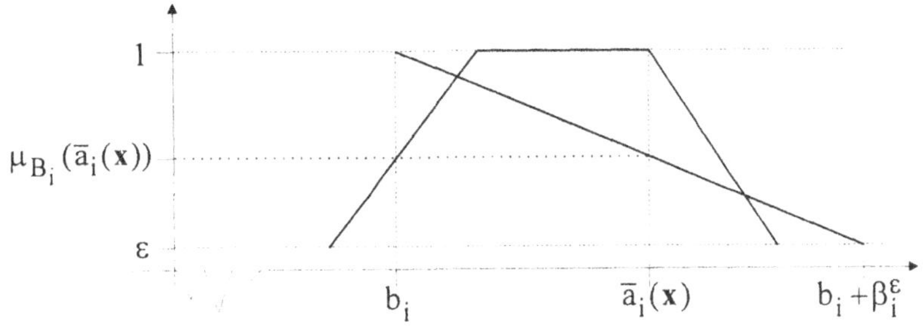

Figure 3 Inequality relation "$\tilde{\leq}_R$"

Obviously, the influence of the extended addition, based on the min t-norm, on the feasible solution of the optimization problem (1) depends on the interpretation of the inequality relation.

Using the pessimistic relation (16), as Slowinski (1986), Ramik and Rimanek (1985) (with $\varepsilon=0$) and Rommelfanger (1990) propose, the set of feasible solutions of the FLP problem (1) shrinks when the number and values of variables x_j increase, i.e. the pessimistic character of (4) is intensified by using the pessimistic min-norm.

On the other side, the optimistic character of other interpretations is intensified too. This is given for the α-possible feasibility of Luhandjula (1987) and the G-α-Pareto optimal solution of Sakawa and Yano (1985), but also the "optimistic index" of Slowinski (1986) loses its restrictive character when $\underline{\alpha}_i(\mathbf{x})$ increases.

6.6. MAXIMIZING FUZZY OBJECTIVES

It seems clear that a fuzzy objective function

$$\tilde{Z}(\mathbf{x}) = \tilde{C}_1 x_1 \oplus \cdots \oplus \tilde{C}_n x_n \rightarrow \widetilde{\max} \tag{18}$$

should be interpreted in multiobjective terms.

Even in the simple case in which the coefficients $\tilde{C}_j$ have the form

$\tilde{C}_j = (\underline{c}_j; \bar{c}_j; \underline{\gamma}^{\varepsilon}_j; \bar{\gamma}^{\varepsilon}_j)^{\varepsilon}$ and $\tilde{Z}(\mathbf{x})$ can be written as

$$\tilde{Z}(\mathbf{x}) = (\underline{c}(\mathbf{x}); \bar{c}(\mathbf{x}); \underline{\gamma}^{\varepsilon}(\mathbf{x}); \bar{\gamma}^{\varepsilon}(\mathbf{x}))^{\varepsilon} \tag{19}$$

with $\underline{c}(\mathbf{x}) = \sum_{j=1}^{n} \underline{c}_j x_j$, $\underline{\gamma}^{\varepsilon}(\mathbf{x}) = \sum_{j=1}^{n} \underline{\gamma}_j x_j$, $\bar{c}(\mathbf{x}) = \sum_{j=1}^{n} \bar{c}_j x_j$, $\bar{\gamma}^{\varepsilon}(\mathbf{x}) = \sum_{j=1}^{n} \bar{\gamma}_j x_j$,

the fuzzy objective function (18) implies that the four auxiliary objective functions

$$
\begin{array}{llll}
\underline{c}(\mathbf{x}) \to \max & & \underline{c}(\mathbf{x}) - \underline{\gamma}^{\varepsilon}(\mathbf{x}) \to \max \\
\bar{c}(\mathbf{x}) \to \max & & \bar{c}(\mathbf{x}) + \bar{\gamma}^{\varepsilon}(\mathbf{x}) \to \max
\end{array}
\tag{20}
$$

should be considered simultaneously on the set of feasible solutions X_U. In general, an "ideal solution" to this problem (i.e. a solution maximizing all auxiliary objectives at the same time) does not exist.

In the special case of triangular coefficients $\tilde{C}_j = (c_j; \underline{\gamma}_j; \bar{\gamma}_j)$ the set (20) boils down to the three objectives

$$z_1(\mathbf{x}) = c(\mathbf{x}) \to \max\ , \quad z_2(\mathbf{x}) = \underline{\gamma}(\mathbf{x}) \to \min\ , \quad z_3(\mathbf{x}) = \bar{\gamma}(\mathbf{x}) \to \max\ .$$

The first method for getting a "compromise solution" of (1) was proposed by Tanaka, Ichihashi and Asai (1984). They substitute the fuzzy objective by the crisp "compromise objective"

$$Z(\mathbf{x}) = \frac{1}{6}\sum_{j=1}^{n}(2\underline{c}_j + 2\bar{c}_j + \underline{\gamma}_j + \bar{\gamma}_j)x_j\ . \tag{21}$$

Carlsson and Korhonen (1986) proposed to substitute the maximized fuzzy objective function (18) by the crisp linear objective function

$$Z(\mathbf{x}) = (\bar{c}_1 + R_{C_1}^{-1}(\sigma)\bar{\gamma}_1)x_1 + \cdots + (\bar{c}_n + R_{C_n}^{-1}(\sigma)\bar{\gamma}_n)x_n \to \max \tag{22}$$

where σ is the same membership level chosen in the interpretation of the inequality relation by Carlsson and Korhonen[4], see section 6.5.

Moreover, a minimized fuzzy objective function

$$\tilde{Z}(\mathbf{x}) = \tilde{C}_1 x_1 \oplus \cdots \oplus \tilde{C}_n x_n \to \widetilde{\min}$$

should be substituted by

$$Z(\mathbf{x}) = (\underline{c}_1 - L_{C_1}^{-1}(\sigma)\underline{\gamma}_1)x_1 + \cdots + (\underline{c}_n - L_{C_n}^{-1}(\sigma)\underline{\gamma}_n)x_n \to \min\ . \tag{23}$$

Another procedure for computing a "compromise solution" of (1) is the "α-level related pair formation" which is based on few crisp objective functions analogous to (20), (Rommelfanger, Hanuscheck and Wolf 1989). It was developed for solving linear project investment problems with fuzzy data.

In this approach, it is assumed that the convex fuzzy coefficients $\tilde{C}_j$ of the linear fuzzy objective function

$$\tilde{Z}(\mathbf{x}) = \tilde{C}_1 x_1 + \cdots + \tilde{C}_n x_n \to \widetilde{\max} \tag{24}$$

can be completely described by few α-level sets

$$C_j^{\alpha} = \left[\underline{c}_j - \underline{\gamma}_j^{\alpha}; \bar{c}_j + \bar{\gamma}_j^{\alpha}\right] \qquad \text{for} \quad \alpha = \sigma_1, \ldots, \sigma_S \in [0,\ 1],\ S \in \mathbf{N}\ .$$

Rommelfanger, Hanuscheck and Wolf (1989) reduce the infinite set of objective functions in (24) to the two extreme cases of each α-level, i.e.

$$\left.\begin{array}{l} Z^{\sigma_1}_{\min}(\mathbf{x}) = (\underline{c}_1 - \underline{\gamma}_1^{\sigma_1})x_1 + \cdots + (\underline{c}_n - \underline{\gamma}_n^{\sigma_1})x_n \\ Z^{\sigma_1}_{\max}(\mathbf{x}) = (\overline{c}_1 + \overline{\gamma}_1^{\sigma_1})x_1 + \cdots + (\overline{c}_n + \overline{\gamma}_n^{\sigma_1})x_n \\ \vdots \\ Z^{\sigma_S}_{\min}(\mathbf{x}) = (\underline{c}_1 - \underline{\gamma}_1^{\sigma_S})x_1 + \cdots + (\underline{c}_n - \underline{\gamma}_n^{\sigma_S})x_n \\ Z^{\sigma_S}_{\max}(\mathbf{x}) = (\overline{c}_1 + \overline{\gamma}_1^{\sigma_S})x_1 + \cdots + (\overline{c}_n + \overline{\gamma}_n^{\sigma_S})x_n \end{array}\right| \rightarrow \max \qquad (25)$$

To get a compromise solution of this multicriteria problem on a crisp set X of feasible solutions, described by crisp linear constraints

$$g_i(\mathbf{x}) = a_{i1}x_1 + \cdots + a_{in}x_n \le b_i \qquad i = 1,\ldots,m$$
$$\mathbf{x} = (x_1,\ldots,x_n) \ge \mathbf{0}$$

the authors proposed to substitute, for each level α, the pair of extreme objective functions by two fuzzy objective functions with the membership functions

$$\mu^\alpha_k(\mathbf{x}) = \begin{cases} 1 & \text{if} \quad z^{\alpha*}_k < Z^\alpha_k(\mathbf{x}) \\ \dfrac{Z^\alpha_k(\mathbf{x}) - z^{\alpha L}_k}{z^{\alpha*}_k - z^{\alpha L}_k} & \text{if} \quad z^{\alpha L}_k \le Z^\alpha_k(\mathbf{x}) \le z^{\alpha*}_k \\ 0 & \text{if} \quad Z^\alpha_k(\mathbf{x}) < z^{\alpha L}_k \end{cases} \qquad (26)$$

where $k = \min$ or $\max$

$$z^{\alpha*}_{\min} = Z^\alpha_{\min}(\mathbf{x}^{\alpha*}_{\mathrm{Min}}) = \max_{\mathbf{x}\in X} Z^\alpha_{\min}(\mathbf{x})$$

$$z^{\alpha*}_{\max} = Z^\alpha_{\max}(\mathbf{x}^{\alpha*}_{\max}) = \max_{\mathbf{x}\in X} Z^\alpha_{\max}(\mathbf{x})$$

$$z^{\alpha L}_{\min} = Z^\alpha_{\min}(\mathbf{x}^{\alpha*}_{\max}) \quad \text{and} \quad z^{\alpha L}_{\max} = Z^\alpha_{\max}(\mathbf{x}^{\alpha*}_{\min})$$

Now, for getting a compromise solution of (1) the crisp mathematical program

$$\max_{\mathbf{x}\in X} \min\left(\mu^{\sigma_1}_{\min}(\mathbf{x}), \mu^{\sigma_1}_{\max}(\mathbf{x}), \ldots, \mu^{\sigma_S}_{\min}(\mathbf{x}), \mu^{\sigma_S}_{\max}(\mathbf{x})\right) \qquad (27)$$

has to be solved, which is equivalent to the crisp LP-problem (28).

$$\lambda \rightarrow \max$$

subject to (28)

$$\mu^{\sigma_s}_{\min}(\mathbf{x}) \ge \lambda, \qquad \mu^{\sigma_s}_{\max}(\mathbf{x}) \ge \lambda, \qquad s = 1,\ldots,S, \qquad \mathbf{x} \in X$$

In contrast to the approach of Rommelfanger et al., Delgado, Verdegay and Vila (1989) have restricted their solution method to one α-level, i.e. they set $S = 1$. On the other side, they do not reduce the objective function with the interval coefficients $\left[\left(\underline{c}_j - \underline{\gamma}_j^\alpha\right), \left(\overline{c}_j + \overline{\gamma}_j^\alpha\right)\right]$ to the two extreme objective functions

$$Z^\alpha_{\min}(\mathbf{x}) = (\underline{c}_1 - \underline{\gamma}_1^\alpha)x_1 + \cdots + (\underline{c}_n - \underline{\gamma}_n^\alpha)x_n$$
$$Z^\alpha_{\max}(\mathbf{x}) = (\overline{c}_1 + \overline{\gamma}_1^\alpha)x_1 + \cdots + (\overline{c}_n + \overline{\gamma}_n^\alpha)x_n$$

but to all crisp objective functions

$$Z^{\alpha}(\mathbf{x}) = c_1 x_1 + \cdots + c_n x_n \tag{29}$$

where $c_j \in \left[\left(\underline{c}_j - \underline{\gamma}_j^{\alpha}\right), \left(\bar{c}_j + \bar{\gamma}_j^{\alpha}\right)\right]$, $j = 1,\ldots, n$, and all combinations are allowed.

To get a compromise solution, these authors propose to work with a compromise objective function, which is a weighted sum of all (up to 2^n) crisp objective functions of type (29).

Chanas and Kuchta (1994) propose to calculate a fuzzy solution $\widetilde{X} = \left\{(\mathbf{x}, \mu_X(\mathbf{x})) \middle| \mathbf{x} \in \mathbf{R}^n_{0+}\right\}$ for linear programming problems with fuzzy coefficients $\widetilde{C}_j = (\underline{c}_j, \bar{c}_j, \underline{\gamma}_j, \bar{\gamma}_j)_{LL}$ in the objective function. They associate the problem

$$\widetilde{C}_1 x_1 \oplus \widetilde{C}_2 x_2 \oplus \cdots \oplus \widetilde{C}_n x_n \rightarrow \widetilde{\max}$$

subject to (30)

$$g_i(\mathbf{x}) = a_{i1} x_1 + \cdots + a_{in} x_n \leq b_i \qquad i = 1,\ldots, m$$
$$\mathbf{x} = (x_1,\ldots, x_n) \geq \mathbf{0}$$

with the following problem consisting of a family of interval linear programming problems:

$$C_1^{\alpha} x_1 + C_2^{\alpha} x_2 + \cdots + C_n^{\alpha} x_n \rightarrow \widetilde{\max}$$

subject to (31)

$$g_i(\mathbf{x}) = a_{i1} x_1 + \cdots + a_{in} x_n \leq b_i \qquad i = 1,\ldots, m$$
$$\mathbf{x} = (x_1,\ldots, x_n) \geq \mathbf{0}$$

where $C_j^{\alpha} = \left[\underline{c}_j - \underline{\gamma}_j^{\alpha}; \bar{c}_j + \bar{\gamma}_j^{\alpha}\right]$ is the α-level set of $\widetilde{C}_j$, $\alpha \in [0, 1]$, $j = 1,\ldots, n$.

For getting a solution of the interval linear program (31), Chanas and Kuchta (1994) use the concept of an $(t_0 - t_1)$ – optimal solution, $0 \leq t_0 < t_1 \leq 1$.

According to this concept, the problem (31) is transformed to the following parametric bicriteria linear programming problem, where $\theta = L^{-1}(\alpha)$:

$$\left.\begin{aligned} f_1(\mathbf{x}) &= \sum_{j=1}^{n}\left(\underline{c}_j + t_0(\bar{c}_j - \underline{c}_j) + (t_0(\underline{\gamma}_j + \bar{\gamma}_j) - \underline{\gamma}_j)\theta\right)x_j \\ f_2(\mathbf{x}) &= \sum_{j=1}^{n}\left(\underline{c}_j + t_1(\bar{c}_j - \underline{c}_j) + (t_1(\underline{\gamma}_j + \bar{\gamma}_j) - \underline{\gamma}_j)\theta\right)x_j \end{aligned}\right| \rightarrow \max$$

subject to (32)

$$g_i(\mathbf{x}) = a_{i1} x_1 + \cdots + a_{in} x_n \leq b_i \qquad i = 1,\ldots, m$$
$$\mathbf{x} = (x_1,\ldots, x_n) \geq \mathbf{0}$$

Now, a basic solution $\mathbf{x}^0$ of (32) is calculated, which is efficient for the set

$$<\theta_0, \theta_1> \cup <\theta_1, \theta_2> \cup \cdots \cup <\theta_{S-1}, \theta_S>,$$

where $<\theta_{s-1}, \theta_s>$ are subintervals of $[L^{-1}(1), L^{-1}(0)]$ which can be closed or open. Then, the membership degree of the basic solution $\mathbf{x}^0$ is

$$\mu_X(\mathbf{x}^0) = \sum_{s=1}^{S} \left(L(\theta_{s-1}) - L(\theta_s) \right).$$

Sakawa and Yano (1989) propose to calculate "G-α-Pareto-optimal solution" by restricting the coefficients $\widetilde{C}_j$ to α-level-sets $C_j^\alpha = [\underline{c}_j^\alpha, \overline{c}_j^\alpha]$.

This approach is formulated for the fuzzy linear programs with several objective functions. A general model of a fuzzy multiobjective linear programming problem (FMOLP problem) $(\widetilde{\mathbf{A}}, \widetilde{\mathbf{b}}, \widetilde{\mathbf{C}})$ is presented by the following system:

$$\left.\begin{array}{c} \widetilde{C}_{11}x_1 \oplus \widetilde{C}_{12}x_2 \oplus \cdots \oplus \widetilde{C}_{1n}x_n \\ \vdots \\ \widetilde{C}_{K1}x_1 \oplus \widetilde{C}_{K2}x_2 \oplus \cdots \oplus \widetilde{C}_{Kn}x_n \end{array}\right| \rightarrow \widetilde{\max}$$

subject to (33)

$$\widetilde{A}_{i1}x_1 \oplus \widetilde{A}_{i2}x_2 \oplus \cdots \oplus \widetilde{A}_{in}x_n \mathrel{\widetilde{\le}} \widetilde{B}_i , \qquad i = 1, \ldots, m$$

$$x_1, x_2, \ldots, x_n \ge 0 .$$

Sakawa and Yano (1989) reduce the FMOLP problem (33) to the α-MOLP problem

$$L_\alpha(\widetilde{\mathbf{A}}, \widetilde{\mathbf{b}}, \widetilde{\mathbf{C}}) = \left\{ (\mathbf{A}, \mathbf{b}, \mathbf{C}) \;\middle|\; \mu_{A_{ij}}(a_{ij}) \ge \alpha, \mu_{b_{ij}}(b_i) \ge \alpha, \mu_{C_{kj}}(c_{kj}) \ge \alpha \right\}, \tag{34}$$

which contains only the elements a_{ij}, b_i, c_{kj} of the m×n-matrix $\widetilde{\mathbf{A}} = \left(\widetilde{A}_{ij}\right)$, the vector $\widetilde{\mathbf{b}} = \left(\widetilde{B}_i\right)$ and the K×n-matrix $\widetilde{\mathbf{C}} = \left(\widetilde{C}_{ij}\right)$ with a membership degree greater than or equal to α, i.e., a_{ij}, b_i, c_{kj} are elements of the α-cuts $A_{ij}^\alpha = [\underline{a}_{ij}^\alpha, \overline{a}_{ij}^\alpha]$, $B_{ij}^\alpha = [\underline{b}_i^\alpha, \overline{b}_i^\alpha]$, $C_{ij}^\alpha = [\underline{c}_{kj}^\alpha, \overline{c}_{kj}^\alpha]$. $\mathbf{A} = \left(a_{ij}\right)$ is a crisp m×n-matrix, $\mathbf{b} = \left(b_i\right)$ a crisp m-vector, $\mathbf{C} = \left(c_{kj}\right)$ a crisp K×n-matrix, $\mathbf{a}_i = (a_{i1}, \ldots, a_{in})$ and $\mathbf{c}_k = (c_{k1}, \ldots, c_{kn})$ are crisp vectors.

An alternative $\mathbf{x}^* \in X(\mathbf{A}, \mathbf{b}) = \left\{ x \in \mathbf{R}_+^n \mid a_{i1}x_1 + \cdots + a_{in}x_n \le b_i \; , i = 1, \ldots, m \right\}$ is called an *α-Pareto-optimal solution* of $L_\alpha(\widetilde{\mathbf{A}}, \widetilde{\mathbf{b}}, \widetilde{\mathbf{C}})$ if no other alternative $\mathbf{x} \in X(\mathbf{A}, \mathbf{b})$, $(\mathbf{A}, \mathbf{b}, \mathbf{C}) \in L_\alpha(\widetilde{\mathbf{A}}, \widetilde{\mathbf{b}}, \widetilde{\mathbf{C}})$ exists such that

$$\mathbf{c}_k\mathbf{x} \ge \mathbf{c}_k\mathbf{x}^*, \; k = 1, \ldots, K \tag{35}$$

and for at least one k the inequality is fulfilled in the strong sense.

Sakawa and Yano generalize the α-MOLP problem to a G-α-MOLP *problem* by substituting each objective function by a monotone increasing membership function

$$\mu_{Z_k}(\mathbf{c}_k\mathbf{x}), \; k = 1, \ldots, K .$$

An alternative $\mathbf{x}^* \in X(\mathbf{A}, \mathbf{b})$ is called a G-α-*Pareto-optimal solution* of a G-α-MOLP *problem*, if no other alternative $\mathbf{x} \in X(\mathbf{A}, \mathbf{b})$, $(\mathbf{A}, \mathbf{b}, \mathbf{C}) \in L_\alpha(\widetilde{\mathbf{A}}, \widetilde{\mathbf{b}}, \widetilde{\mathbf{C}})$ exists such that

$$\mu_{Z_k}(\mathbf{c}_k\mathbf{x}) \geq \mu_{Z_k}(\mathbf{c}_k\mathbf{x}^*), k = 1,\ldots,K \tag{36}$$

and for at least one k the inequality is fulfilled in the strong sense.

In general, there exists an infinite number of G-α-Pareto-optimal solutions for each $\alpha < 1$. To determine a compromise solution, Sakawa and Yano propose to use an interactive process in which the DM has at each iteration step the occasion to specify a new value of α and the "reference membership values" $\overline{\mu}_{Z_k}$ for each objective $k = 1,\ldots, K$; see the detailed solution process in section 6.8.

A similar concept is the "β-possibility efficient solution" of Luhandjula (1987).

A vector $\mathbf{x}^0 \in X$ is called β-possible efficient for the mathematical program

$$\max_{\mathbf{x}\in X}\left(\tilde{Z}_1(\mathbf{x}),\cdots,\tilde{Z}_K(\mathbf{x})\right), \quad \text{where} \quad X \subset \mathbf{R}^n_+, \tag{37}$$

if there is no other vector $\mathbf{x} \in X$ and $l \in \{1,\ldots,K\}$ such that

$$\Pi\begin{pmatrix} \tilde{Z}_1(\mathbf{x}) \geq \tilde{Z}_1(\mathbf{x}^o),\ldots,\tilde{Z}_{l-1}(\mathbf{x}) \geq \tilde{Z}_{l-1}(\mathbf{x}^o),\tilde{Z}_l(\mathbf{x}) > \\ \tilde{Z}_l(\mathbf{x}^o),\tilde{Z}_{l+1}(\mathbf{x}) \geq \tilde{Z}_{l+1}(\mathbf{x}^o),\ldots,\tilde{Z}_K(\mathbf{x}) \geq \tilde{Z}_K(\mathbf{x}^o)\end{pmatrix} \geq \beta,$$

where Π denotes possibility.

Obviously, $\mathbf{x}^o$ is β-possible efficient for (37) if and only if $\mathbf{x}^o$ is an efficient solution of the mathematical program

$$\max_{x\in X}\left(\mathbf{C}_1^\alpha\mathbf{x},\cdots,\mathbf{C}_K^\alpha\mathbf{x}\right) \tag{38}$$

where $\mathbf{C}_k^\alpha = \left([\underline{c}_{k1}^\alpha,\overline{c}_{k1}^\alpha],\ldots,[\underline{c}_{kn}^\alpha,\overline{c}_{kn}^\alpha]\right)$ are vectors of the α-cuts of the coefficients $\tilde{C}_{kj}$.

For calculating an efficient solution of (38), Luhandjula proposes a procedure which is based on the lower and upper bounds $\underline{c}_{kj}^\alpha$ and $\overline{c}_{kj}^\alpha$ of the α-cuts of the coefficients $\tilde{C}_{kj}$.

As an ideal solution of (18) on a set of feasible solutions does not generally exist, Slowinski (1986) and Rommelfanger (1990) suggest to calculate a compromise solution, in an interactive way, a procedure which corresponds to the usual way of acting in practice.

In analogy to modeling a right-hand side $\tilde{B}_i$, a fuzzy aspiration level $\tilde{N}$ can be described as

$$\tilde{N} = (n;\ \nu^\varepsilon;0)^\varepsilon \ . \tag{39}$$

Then, the condition of satisfaction

$$\tilde{N} \mathrel{\tilde{\leq}} \tilde{Z}(\mathbf{x}) \tag{40}$$

is treated as an additional fuzzy constraint. In accordance to the chosen inequality-interpretation, (40) can be substituted by crisp inequalities or in case of "$\tilde{\leq}_R$", by a crisp linear inequality and a new objective function. By changing the aspiration levels, the set of feasible solutions is restricted step by step, (see Slowinski, 1986;

Rommelfanger 1990). Obviously, it is easy to extend this approach to multicriteria problems.

6.7 EXTENDED ADDITION, BASED ON YAGER'S T-NORM T_P

For getting a more realistic extended addition of the left-hand sides of fuzzy constraints and of fuzzy objectives, Rommelfanger and Keresztfalvi (1991) recommend the use of Yager's parametrized t-norm,

$$T_p(u,v) = \max\left\{0,\ 1-\left((1-u)^p + (1-v)^p\right)^{1/p}\right\} \quad u,v \in [0,1],\ p>0\ , \qquad (41)$$

which can be adapted to special situations. In case of n variables it has the form

$$T_p(t_1,\dots,t_n) = \max\left\{0,\ 1-\left(\sum_{i=1}^{n}(1-t_i)^p\right)^{1/p}\right\} \quad t_1,\dots,t_n \in [0,1]\ . \qquad (42)$$

In a special case when all coefficients $\tilde{A}_{ij}$ are trapezoidal fuzzy intervals of type $\tilde{A}_{ij} = (\underline{a}_{ij};\bar{a}_{ij};\underline{\alpha}_{ij}^{\varepsilon};\bar{\alpha}_{ij}^{\varepsilon})^{\varepsilon}$, the following theorem holds (see Rommelfanger and Keresztfalvi 1992).

Supposed the coefficients $\tilde{A}_{ij}$ of the left-hand side of the inequality constraints

$$\tilde{A}_{i1}x_1 \oplus \tilde{A}_{i2}x_2 \oplus \dots \oplus \tilde{A}_{in}x_n \mathrel{\tilde{\leq}} \tilde{B}_i \qquad i=1,\dots,m$$

are trapezoidal fuzzy intervals of type $\tilde{A}_{ij} = (\underline{a}_{ij};\bar{a}_{ij};\underline{\alpha}_{ij}^{\varepsilon};\bar{\alpha}_{ij}^{\varepsilon})^{\varepsilon}$.

If the addition is extended by Yager's t-norm T_P with $p \geq 1$, then

$$\tilde{A}_i(\mathbf{x}) = \tilde{A}_{i1}x_1 \oplus \tilde{A}_{i2}x_2 \oplus \cdots \oplus \tilde{A}_{in}x_n = \left(\underline{a}_i(\mathbf{x}),\bar{a}_i(\mathbf{x}),\underline{\alpha}_i^{\varepsilon}(\mathbf{x}),\bar{\alpha}_i^{\varepsilon}(\mathbf{x})\right)^{\varepsilon} \qquad (43)$$

is also a fuzzy interval with linear reference functions, such that

$$\underline{a}_i(\mathbf{x}) = \sum_{j=1}^{n}\underline{a}_{ij}x_j, \qquad \bar{a}_i(\mathbf{x}) = \sum_{j=1}^{n}\bar{a}_{ij}x_j$$

$$\underline{\alpha}_i^{\varepsilon}(\mathbf{x},p) = \left\|\left(\underline{\alpha}_{i1}^{\varepsilon}x_1,\dots,\underline{\alpha}_{in}^{\varepsilon}\right)\right\|_q = \left(\left(\underline{\alpha}_{i1}^{\varepsilon}x_1\right)^q + \cdots + \left(\underline{\alpha}_{in}^{\varepsilon}x_n\right)^q\right)^{1/q} \qquad (44)$$

$$\bar{\alpha}_i^{\varepsilon}(\mathbf{x},p) = \left\|\left(\bar{\alpha}_{i1}^{\varepsilon}x_1,\dots,\bar{\alpha}_{in}^{\varepsilon}x_n\right)\right\|_q = \left(\left(\bar{\alpha}_{i1}^{\varepsilon}x_1\right)^q + \cdots + \left(\bar{\alpha}_{in}^{\varepsilon}x_n\right)^q\right)^{1/q} \qquad (45)$$

where $q = \dfrac{p}{p-1} \geq 1$.

Looking for the consequences of using the extended addition based on t-norm T_p instead of the usually applied min-operator, we can state that the 1-level set $\left[\underline{a}_i(\mathbf{x}), \bar{a}_i(\mathbf{x})\right]$ does not change with the parameter p whereas the spreads $\underline{\alpha}_i^\varepsilon(\mathbf{x})$ and $\overline{\alpha}_i^\varepsilon(\mathbf{x})$ decrease if p decreases (q increases).

The extent of change will be evident by looking at the extreme cases.

- If p tends to infinity, then T_p tends to min T-norm and we come back to the usual extended addition;, see section 6.4. Thus, if $p \to \infty$ and $q=1$, then $\widetilde{A}_i(\mathbf{x})$ has the greatest spreads

$$\underline{\alpha}_i^\varepsilon(\mathbf{x},\infty) = \underline{\alpha}_{i1}x_1 + \cdots + \underline{\alpha}_{in}x_n \quad \text{and} \quad \overline{\alpha}_i^\varepsilon(\mathbf{x},\infty) = \overline{\alpha}_{i1}x_1 + \cdots + \overline{\alpha}_{in}x_n \ .$$

- If p=1 (and $q \to \infty$) then $T_p = T_L$ is the well known Lukasiewicz t-norm:

$$T_1\ (u, v) = T_L\ (u,v) = \max\ \{u, u+v-1\}.$$

In this case, $\widetilde{A}_i(\mathbf{x})$ has the smallest spreads

$$\underline{\alpha}_i^\varepsilon(\mathbf{x},1) = \max\{\underline{\alpha}_{i1}^\varepsilon x_1,\ldots,\underline{\alpha}_{in}^\varepsilon x_n\} \quad \text{and} \quad \overline{\alpha}_i^\varepsilon(\mathbf{x},1) = \max\{\overline{\alpha}_{i1}^\varepsilon x_n,\ldots,\overline{\alpha}_{in}^\varepsilon x_n\}.$$

- If $p \in]1,+\infty[$, the spreads $\underline{\alpha}_i^\varepsilon(\mathbf{x},p)$ and $\overline{\alpha}_i^\varepsilon(\mathbf{x},p)$ are strictly monotone increasing functions of p.

Therefore, if the set of feasible solutions of the inequality equation $\bar{a}_i(\mathbf{x}) + \overline{\alpha}_i(\mathbf{x},p) \le b_i + \beta_i^\varepsilon$ is denoted by $X_i(p)$, we have $X_i(p) \subset X_i(p')$ if $p > p'$, $p, p' \in]1, +\infty[$. Moreover, as $\|\ldots\|_q$ satisfies the triangle inequality, the sets $X_i(p)$ are convex sets for all $p \in [1,+\infty[$.

Using the Tp-norm based addition, the inequality relation "$\widetilde{\le}_R$" should be modified to

$$\widetilde{A}_i(\mathbf{x}) \ \widetilde{\le}_{KR}\ \widetilde{B}_i \iff \begin{cases} \bar{a}_i(\mathbf{x}) + \overline{\alpha}_i(\mathbf{x},p) \le b_i + \beta_i^\varepsilon \\ \mu_i(\mathbf{x}) = \mu_{B_i}(\bar{a}_i(\mathbf{x})) \to \max \end{cases} \tag{46}$$

Obviously, the inequality relation "$\widetilde{\le}_{KR}$" is identical with "$\widetilde{\le}_R$" for $p = \infty$, but in general the inequality interpretation varies with p.

Now, the DM has two ways for expressing his attitude towards risk:

- by specifying the values $\overline{\alpha}_{ij}^\varepsilon, \underline{\gamma}_{kj}^\varepsilon, \beta_i^\varepsilon, \nu_k^\varepsilon$.
- by choosing a value $p \in [1, +\infty[$ for each objective and for each constraint independently.

Another type of extended algebraic operations on fuzzy numbers was proposed by Kovacs (1994). For simplification, we only present the definition for the special case of so-called g-shape centered fuzzy numbers which are used by Kovacs in fuzzy linear programs. As g-shape centered fuzzy numbers are symmetric fuzzy numbers of L-R-type, i.e. of the type $\widetilde{A} = (a; \underline{\alpha}; \overline{\alpha})_{LL}$, they can be denoted by $\widetilde{A} = (a; \alpha)_L$.

The max- and min-extended algebraic operations are defined as

$$\widetilde{A} \circ^{(f)} \widetilde{B} = (a;\alpha)_L \circ^{(f)} (b;\beta)_L = (a \circ b; f(\alpha, b)), \tag{47}$$

where $f(\alpha, \beta)$ is either $\min(\alpha, \beta)$ or $\max(\alpha, \beta)$.

Obviously, using these max- and min-extended algebraic addition, the spreads $\min(\alpha, \beta)$ or $\max(\alpha, \beta)$ are smaller than the spreads $\alpha + \beta$, we have by using the usually min t-norm based extended addition.

6.8 SOLUTION PROCESS FOR GETTING A COMPROMISE SOLUTION

The discussion of the crucial points of fuzzy linear programming reveals that there exists an extensive offer of effective methods for substituting fuzzy linear programs by crisp optimization problems. For making the best choice, the DM has to consider the different assumptions of the existing procedures and to compare them with the actual decision problem. In any case, the solution should be determined step by step in an interactive process, in which additional information out of the decision process or from a wider domain knowledge should be used. In doing so, inadequate modeling of the real problem can be avoided and information costs can, in general, be decreased.

To give an idea of the different solution algorithms, three of the best-known procedures will be outlined below.

The procedure of Sakawa and Yano (1985, 1989)

The basic assumptions of this interactive solution method are:

- The fuzzy goals of the DM can be quantified by eliciting the corresponding membership functions in an interactive man-machine process.
- The preference function $\mu_D(\mu_{Z_1}(\mathbf{x}), \ldots, \mu_{Z_K}(\mathbf{x}))$, expressing the overall degree of satisfaction with the DM's multiple fuzzy objective functions $\mu_{Z_1}(\mathbf{x}), \ldots, \mu_{Z_K}(\mathbf{x})$, does exist, but it is only implicitly known to the DM. This means the "DM may not specify the entire function form of $\mu_D()$, but he/she can provide local information concerning his/her preferences. However, it is increasing and continuous." (Seo and Sakawa 1988, p.115).
- All objective functions $Z_k(\mathbf{x})$ and all left hand sides $g_i(\mathbf{x})$ of the constraints are continuously differentiable.

The compromise solution of the G-α-MOLP problem should be determined by solving the min-max problem

$$\min_{\mathbf{x} \in X(\mathbf{A},\mathbf{b})} \max_{k=1}^{K} \left(\overline{\mu}_{Z_k} - \mu_{Z_k}(\mathbf{c}_k' \mathbf{x})\right) \quad \text{for all } (\mathbf{A}, \mathbf{b}, \mathbf{C}) \in L_\alpha(\widetilde{\mathbf{A}}, \widetilde{\mathbf{b}}, \widetilde{\mathbf{C}}) \tag{48}$$

which is equivalent to

$$\begin{gathered} \nu \rightarrow \min \\ \text{subject to} \\ \overline{\mu}_{Z_k} - \mu_{Z_k}(\mathbf{c}_k \mathbf{x}) \leq \nu \\ \mathbf{A}\mathbf{x} \leq \mathbf{b} \\ \mathbf{x} \geq \mathbf{0} \end{gathered} \tag{49}$$

$$(\mathbf{A},\mathbf{b},\mathbf{C}) \in L_\alpha(\tilde{\mathbf{A}},\tilde{\mathbf{b}},\tilde{\mathbf{C}}).$$

Obviously, the proposal (48) is based on the minimum-regret-approach where the maximum operator is used for aggregation.

For getting a solution of the problem (49), Sakawa and Yano propose the following procedure:

As the membership functions μ_{Z_k} are, by definition, strictly monotone increasing functions, each constraint

$$\bar{\mu}_{Z_k} - \mu_{Z_k}(\mathbf{c}_k\mathbf{x}) \le \nu$$

is equivalent to

$$\mathbf{c}_k\mathbf{x} \le \mu_{Z_k}^{-1}(\bar{\mu}_{Z_k} - \nu). \tag{50}$$

Then, Sakawa and Yano define the following set valued functions

$$S_k(c_k) = \left\{(\mathbf{x},\nu) \middle| \mathbf{c}_k\mathbf{x} \le \mu_{Z_k}^{-1}(\bar{\mu}_{Z_k} - \nu)\right\} \text{ and } T_i(a_i,b_i) = \{\mathbf{x} | a_i\mathbf{x} \le b_i\},$$

which obviously have the following attributes:

(i) $\mathbf{c}_k \le \hat{\mathbf{c}}_k \Rightarrow S_k(\mathbf{c}_k) \supseteq S_k(\hat{\mathbf{c}}_k)$,

(ii) $\mathbf{a}_i \le \hat{\mathbf{a}}_i \Rightarrow T_i(\mathbf{a}_i, b_i) \supseteq T_i(\hat{\mathbf{a}}_i, b_i)$,

(iii) $b_i < \hat{b}_i \Rightarrow T_i(\mathbf{a}_i, b_i) \subseteq T_i(\mathbf{a}_i, \hat{b}_i)$.

Now, remembering that the α-level sets of $\tilde{\mathbf{a}}_i = (\tilde{A}_{i1},\ldots,\tilde{A}_{in})$, $\tilde{B}_i, \tilde{\mathbf{c}}_k = (\tilde{C}_{i1},\ldots\tilde{C}_{in})$ are crisp intervals denoted in section 6.6 by $\left[\underline{\mathbf{a}}_i^\alpha, \bar{\mathbf{a}}_i^\alpha\right]$, $\left[\underline{b}_i^\alpha, \bar{b}_i^\alpha\right]$, $\left[\underline{\mathbf{c}}_i^\alpha, \bar{\mathbf{c}}_i^\alpha\right]$, we can get an optimal solution by solving the optimization problem

$$\nu \to \min$$

subject to (51)

$$\bar{\mathbf{c}}_k^\alpha\mathbf{x} \le \mu_{Z_k}^{-1}(\bar{\mu}_{Z_k} - \nu), \qquad k = 1,\ldots, K$$

$$\underline{\mathbf{a}}_i^\alpha\mathbf{x} \le \bar{b}_i^\alpha, \qquad i = 1,\ldots,m$$

$$\mathbf{x} \ge \mathbf{0}$$

If ν is constant, the constraints of the problem (51) are linear. Therefore, an optimal objective value ν* can be calculated by an iterative procedure in which a value ν is given in the constraints and a standard LP can be used to check if ν may be a solution of the problem (51).

If a suitable ν* is found, a compromise solution **x*** of the G-α-MOLP problem can be calculated by solving the following crisp LP problem for any k of the K objectives:

$$\bar{\mathbf{c}}_k\mathbf{x} \to \max$$

subject to (52)

$$\bar{\mathbf{c}}_k^\alpha\mathbf{x} \le \mu_{Z_k}^{-1}(\bar{\mu}_{Z_k} - \nu^*), \qquad k = 1,\ldots, K$$

$$\underline{\mathbf{a}}_i^\alpha\mathbf{x} \le \bar{b}_i^\alpha, \qquad i = 1,\ldots,m$$

$$\mathbf{x} \ge \mathbf{0}$$

If the DM is not willing to accept the calculated compromise solution he/she can change the level α and/or the reference membership values. Sakawa and Yano assume that the DM is able to specify the membership functions of the coefficients of the objective functions so precisely that it is possible to calculate the trade-off rates between the objectives by calculating derivatives of the corresponding membership functions.

FLIP, the procedure of Slowinski (1986, 1990)

FLIP is an interactive procedure for solving fuzzy MOLP problems. It is assumed that for the particular objectives, the DM is in a position to define fuzzy aspiration levels, thought of as goals, denoted by $\tilde{N}_1,\ldots,\tilde{N}_K$. The main characteristics of FLIP are the following:

- All fuzzy coefficients of the MOLP problem, as well as the fuzzy goals, are given as L-R fuzzy numbers:

$$\tilde{C}_{kj}=\left(\underline{c}_{kj},\bar{c}_{kj},\underline{\gamma}_{kj},\bar{\gamma}_{kj}\right)_{LR},\quad \tilde{N}_k=\left(n_k,v_k,0\right)_{LR},\quad k=1,\ldots,K;\ j=1,\ldots,n,$$

$$\tilde{A}_{ij}=\left(\underline{a}_{ij},\bar{a}_{ij},\underline{\alpha}_{ij},\bar{\alpha}_{ij}\right)_{LR},\quad \tilde{B}_i=\left(b_i,0,\beta_i\right)_{LR},\quad i=1,\ldots,m;\ j=1,\ldots,n.$$

- In order to evaluate the degree of possibility for $\tilde{B}_i$ to be greater than or equal to $\tilde{A}_i(\mathbf{x})$ (see section 6.4), FLIP uses two indices, one called optimistic and another pessimistic (see section 6.5). In consequence, every fuzzy constraint is substituted by two crisp linear constraints, provided that reference functions L and R are strictly monotone decreasing.
- In order to maximize the consistency between

$$\tilde{C}_k(\mathbf{x})=(\underline{c}_k(\mathbf{x});\bar{c}(\mathbf{x});\underline{\gamma}_k(\mathbf{x});\bar{\gamma}_k(\mathbf{x}))_{LR}\quad\text{and}\quad \tilde{N}_k$$

  one has to maximize the ordinate $F_k(\mathbf{x})$ of the intersection point of the right slope of $\tilde{C}_k(\mathbf{x})$ with the left slope of $\tilde{N}_k$, i.e., assuming that the reference function R of fuzzy cost coefficients and the reference function L of fuzzy goals are the same, $F_k(\mathbf{x})=R\left(\dfrac{\bar{c}_k(\mathbf{x})-g_k}{\bar{\gamma}_k(\mathbf{x})-\varphi_k}\right)$.

  If reference function R is linear or piecewise linear, then $F_k(\mathbf{x})$ takes a linear fractional form: $F_k(\mathbf{x})=1-\dfrac{\bar{c}_k(\mathbf{x})-g_k}{\bar{\gamma}_k(\mathbf{x})-\varphi_k}$.

The resulting auxiliary deterministic multiobjective programming problem has the form:

$$\left(F_1(\mathbf{x}),\ldots,F_K(\mathbf{x})\right)\rightarrow\max$$

subject to (53)

$$\underline{a}_i(\mathbf{x})-b_i\le\underline{\alpha}_i(\mathbf{x})L^{-1}(\tau_i)+\beta_iR^{-1}(\tau_i),\quad \tau_i\in]0,1],\ i=1,\ldots,m$$

$$\bar{a}_i(\mathbf{x})+\bar{\alpha}_i(\mathbf{x})R^{-1}(\eta_i)\le b_i+\beta_iR^{-1}(\eta_i),\quad \eta_i\in[0,1],\ i=1,\ldots,m$$

$$\mathbf{x}\ge 0.$$

The constants τ_i and η_i are called "safety parameters" because they are responsible for the safety of the assertion that $\widetilde{B}_i$ is greater than or equal to $\widetilde{A}_i(\mathbf{x})$. These constants are used by the DM to control the risk of violation of the particular constraints.

If reference function R in $F_k(\mathbf{x})$ is linear or piecewise linear, this problem is a multiobjective linear fractional programming (MOLFP) one.

The MOLFP problem is solved using an interactive sampling procedure. In each calculation step of this procedure, a sample of non-dominated points (Pareto-optimal solutions) of the MOFLP problem is generated and then shown to the DM who is asked to select the one that fits best his/her preferences. If the selected point is not the final compromise, it becomes a central point of a non-dominated region which is sampled in the next calculation step. In this way, the sampled part of the non-dominated set is successively reduced (focusing phenomenon) until the most satisfactory efficient point (compromise solution) is reached. An important advantage of the method presented above is that the only optimization procedure to be used is a linear programming one. Moreover, it has a simple scheme and allows retractions to the points abandoned in previous iterations.

The interaction with the DM takes place at two levels: first when fixing the safety parameters and then in the course of the guided generation and evaluation of the non-dominated points of the auxiliary deterministic problem.

Let us point out that the fuzzy goals, used at the stage of defuzzification of fuzzy objectives, do not influence the set of non-dominated points of the auxiliary deterministic problem; they rather play the role of a visual reference than that of a preferential information influencing the set of generated proposals for the compromise solution.

An important feature of any software implementing a fuzzy multiobjective programming method is the presentation of candidate solutions in an interactive process. In the FLIP software, the Pareto-optimal solutions of the auxiliary deterministic problem are shown not only numerically, but also graphically, in terms of mutual positions of fuzzy intervals corresponding to original objectives and aspiration levels on the one hand, and to left- and right-hand sides of original constraints on the other hand. In this way, the DM gets quite a complete idea of the quality of each proposed solution.

The quality is evaluated taking into account the following characteristics:

- scores of fuzzy objectives in relation to the goals,
- dispersion of values of the fuzzy objectives due to uncertainty,
- safety of the solution or, using a complementary term, the risk of violation of the constraints.

So, the definition of the best compromise involves not only the scores on particular objectives but also the safety of the corresponding solution. It needs, indeed, a graphical display of objectives and constraints for any analyzed solution. The comparison of fuzzy left- and right-hand sides of the constraints, as well as evaluation of dispersion of the values of objectives, is practically infeasible on the basis of numerals only. The graphical presentation of proposed solutions is not only a "user's friendly" interface but the best way for a complete characterization of these solutions.

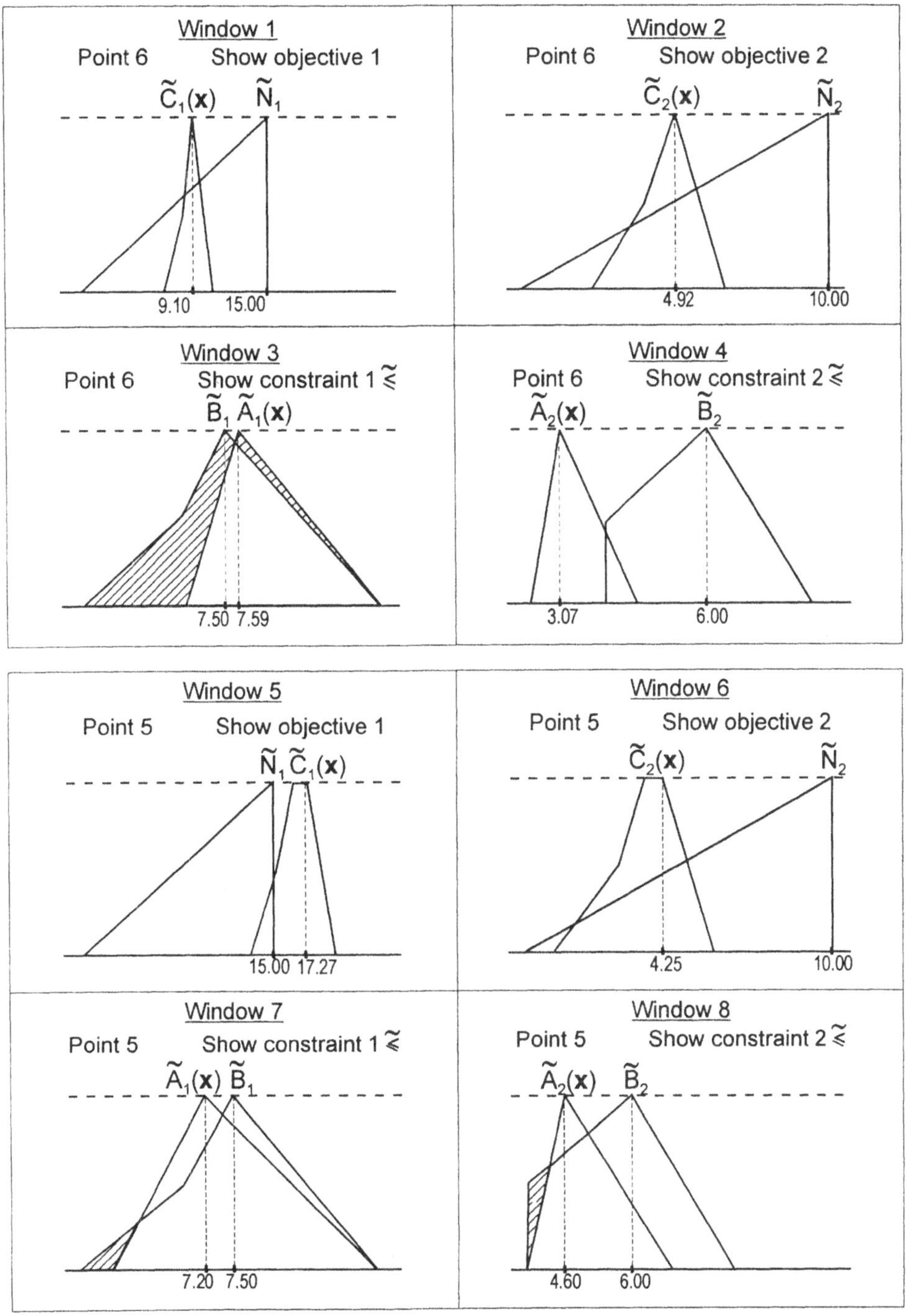

Figure 4 Exemplary screen displays of FLIP

In Figure 4, we show exemplary screen displays of FLIP for a didactic bi-objective fuzzy LP problem solved by Slowinski and Teghem (1990). The first four windows present one Pareto-optimal solution and the second four windows show another Pareto-optimal solution of the auxiliary deterministic problem, however, in terms of the original problem. Windows 1 and 2 show the objective functions for the first solution, while 5 and 6 view the objective functions of the second solution. Let us observe that, apart from dispersion of the scores on particular objectives, the pictures show the known Pareto dependency between two objectives. Windows 3, 4 and 7, 8 show two fuzzy constraints for both solutions, respectively. The hatched area corresponds to the risk of violation of the constraints.

There exists an implementation of FLIP in Visual Basic in the MS-Excel environment; it gives the possibility to the user to define the parameter p of the Yager's formula for the aggregation of fuzzy objectives and of fuzzy left-hand sides of fuzzy constraints (see section 6.7), as well as to define all safety parameters. The candidates for the best compromise solution are displayed both numerically and graphically.

FULPAL, the procedure of Rommelfanger (1990, 1995b)

The basic characteristics of FULPAL 2.0 are:

- FULPAL is based on the inequality relation "$\tilde{\leq}_{KR}$" defined by (46),
- The coefficients are modeled as $\tilde{A}_{ij} = (\underline{a}_{ij}; \bar{a}_{ij}; \underline{\alpha}_{ij}^{\varepsilon}; \bar{\alpha}_{ij}^{\varepsilon})^{\varepsilon}$ and
$$\tilde{C}_{kj} = (\underline{c}_{kj}; \bar{c}_{kj}; \underline{\gamma}_{kj}^{\varepsilon}; \bar{\gamma}_{kj}^{\varepsilon})^{\varepsilon},$$
- The right-hand sides of the constraints are modeled as $\tilde{B}_i = (b_i; 0, 0; \beta_i^{\lambda_A, \varepsilon}, \beta_i^{\varepsilon})^{\lambda_A, \varepsilon}$, where λ_A is the membership degree of the crisp aspiration level $b_i + \beta_i^{\lambda_A}$,
- FULPAL uses a satisfying approach, where the fuzzy aspiration levels $\tilde{N}_k = (n_k; v_k^{\lambda_A}, v_k^{\varepsilon}; 0, 0)^{\lambda_A, \varepsilon}$ for the objective functions are changed step by step by improving the crisp aspiration level $n_k - v_k^{\lambda_A}$,
- The total satisfaction of the DM with a solution **x** is expressed by the compromise objective function
$$\lambda(\mathbf{x}) = \min(\mu_{Z_1}(\underline{c}_1(\mathbf{x})), \ldots, \mu_{Z_K}(\underline{c}_K(\mathbf{x})), \mu_{B_1}(\bar{a}_1(\mathbf{x})), \ldots, \mu_{B_{m_1}}(\bar{a}_{m_1}(\mathbf{x}))),$$
- When cusing the values $n_k - v_k^{\varepsilon}$ and n_k of $\tilde{N}_k$ according to (39) and (40) for an objective k, the inequality relation "$\tilde{\leq}_R$" is used in $\tilde{Z}_k = \tilde{C}_{k1}x_1 \oplus \cdots \oplus \tilde{C}_{kn}x_n \geq \tilde{N}_k$.

Under these assumptions a compromise solution of a multiobjective fuzzy linear program of type

$$\left.\begin{array}{l}\tilde{Z}_1 = \tilde{C}_{11}x_1 \oplus \cdots \oplus \tilde{C}_{1n}x_n \\ \vdots \qquad \vdots \qquad\qquad \vdots \\ \tilde{Z}_K = \tilde{C}_{K1}x_1 \oplus \cdots \oplus \tilde{C}_{Kn}x_n\end{array}\right] \to \max$$

subject to (54)

$$\tilde{A}_{i1}x_1 \oplus \cdots \oplus \tilde{A}_{in}x_n \,\tilde{\le}\, \tilde{B}_i, \qquad i = 1,\ldots,m_1$$

$$\mathbf{x} = (x_1, x_2, \ldots, x_2) \in X = \left\{ \mathbf{x} \in \mathbf{R}_0^n \,\middle|\, a_{i1}x_1 + \cdots + a_{in}x_n \le b_i \quad \forall\, i = m_1 + 1, \ldots, m \right\}$$

can be calculated by solving the following crisp problem

$$\lambda \to \max$$

subject to (55)

$$\begin{array}{ll}\lambda \le \mu_{Z_k}(\underline{c}_k(\mathbf{x})) & k = 1,.., K \\ \lambda \le \mu_{B_i}(\bar{a}_i(\mathbf{x})) & i = 1,\ldots,m_1 \\ \underline{c}_k(\mathbf{x}) - \underline{\gamma}^{\varepsilon}(\mathbf{x}) \ge n_k - \nu_k^{\varepsilon} & k = 1,\ldots,K \\ \bar{a}_i(\mathbf{x}) + \bar{\alpha}_i^{\varepsilon}(\mathbf{x}, p) \le b_i + \beta_i^{\varepsilon} & i = 1,\ldots,m_1 \\ \mathbf{x} = (x_1, x_2, \ldots, x_2) \in X \;. & \end{array}$$

Following the algorithm FULPAL 2.0 we can assume that all the piecewise linear functions μ_{Z_k} and μ_{B_i} are concave membership functions. Then, if $p = 1$ or $p \to +\infty$, the problem (51) is a crisp linear program which can be solved by means of the well-known simplex algorithm. As far as the intermediate parameters $p \in]1, +\infty[$ are concerned, it is sufficient in practical applications to work with a linear approximation (see Rommelfanger 1995b). The advantage of this approximation procedure is that a compromise solution can be calculated by solving a crisp LP problem. To support and facilitate the application of FULPAL, a PC-software was written in Turbo Pascal in 1989 (see Rommelfanger 1991) which was improved in 1993. The software FULPAL 2.0, written by T. Keresztfalvi in C+ includes the simplex program LINX of C.I. Fabian (see Rommelfanger 1995b).

6.9 APPLICATIONS OF FUZZY LINEAR PROGRAMMING

Fuzzy Linear Programs were developed to tackle problems encountered in real-world applications. The following (non-exhaustive) list shows that the applications of FLP are numerous and diverse.

Agricultural Economics

Feed mix (Buckley 1988; Czyzak and Slowinski 1990)

Farm structure optimization problem (Czyzak 1990)

Regional Resource Allocation (Leung 1988; Mjelde 1986)

Water supply planning (Slowinski 1986, 1987)

Analysis of water use in agriculture (Owsinski, Zadrozny and Kacprzyk 1987)

Forest management (Pickens and Hof 1991)

Assignment Problems

Network location problem (Darzentas 1987)

Banking and Finance

Capital asset pricing model (Ostermark 1989)
Profit apportionment in concern (Ostermark 1988)
Bank hedging decision (Lai and Hwang 1992)
Project investment (Lai and Hwang 1992)

Environmental Management

Air pollution regulation problem (Sommer and Pollatschek 1978)
Water quality control (Esogbue and Guo 1991)

Manufacturing and Production

Aggregate production planning problem (Verdegay 1982; Lai and Hwang 1992)
Machine optimization problems (Trappey, Liu and Chang 1988)
Magnetic tape production (Wagenknecht and Hartmann 1987)
Optimal allocation of resources to production of metals (Ramík and Rimanek 1987)
Optimal system design (Zeleny 1986)
Crude oil manufacturing (Wagenknecht and Hartmann 1987)
Production-mix selection problem (Verdegay 1982)
Optimal scheduling (Baptistella and Ollero 1980)
Production scheduling (Carlsson and Korhonen 1986)
Quality control (chakraborty 1994)

Marketing

Media selection problem (Zimmermann 1997)

Personnel Management

Personnel Selection problem (Rubin and Narasimhan 1984)

Transportation

Transportation problem (Verdegay 1984)
Truck fleet (Zimmermann 1976)
Solid transportation problem (Bit 1993)
Transshipment problem (Perincherry and Kikucki 1990)

Acknowledgments

The authors wish to thank H. Tanaka and H.-J. Zimmermann for their helpful comments on the first draft of this chapter. The second author wishes also to acknowledge financial support from State Committee for Scientific Research, KBN research grant no. 8T11C 013 13 and from CRIT 2 - Esprit Project no. 20288.

Notes

1. Here we renounce to discuss the very simple approach where the fuzzy right hand side $\tilde{B}_i$ is substituted by a crisp boundary $b_i + \beta_i \cdot R_{\tilde{B}_i}^{-1}(\varepsilon)$ with $0 \leq \varepsilon \leq 1$ (see e.g. Verdegay 1982).
2. Obviously, it is possible to get membership functions on a ratio or even interval scale or to transform ordinal scaled membership values to a higher scale level by using the procedure for successive intervals, proposed by Diedrich, Messick and Tucker (1957). But in these cases extensive empirical studies and a lot of data are needed. Therefore, in real applications, a DM will abstain from carrying out this lavish procedure for all fuzzy data in a linear programming system.
3. This simple form of this inequality relation is stringent only in the case of the linear reference functions. In general, the inequality relation " $\tilde{\leq}^{R}_{\varepsilon,\delta}$ " consists of an infinite number of inequalities, but in practical applications it is sufficient to work with this approximation, see (Ramik and Rommelfanger 1996).
4. According to the approach of Carlsson and Korhonen an FLP problem (1) with fuzzy coefficients of the LR-type can be substituted by the crisp LP-problem

$$Z(\mathbf{x}) = (\bar{c}_1 + R_{\tilde{C}_1}^{-1}(\sigma)\bar{\gamma}_1)x_1 + \cdots + (\bar{c}_n + R_{\tilde{C}_n}^{-1}(\sigma)\bar{\gamma}_n)x_n \rightarrow \max$$

subject to

$$\bar{a}_i(\mathbf{x}) + \bar{\alpha}_i(\mathbf{x})R_{\tilde{A}_1}^{-1}(\sigma) \leq b_i + \beta_i R_{\tilde{A}_n}^{-1}(\sigma)$$

where $\sigma \in [0, 1]$ is a special membership level. By changing the parameter σ we obtain a set of solutions $(z^*(\sigma), x^*(\sigma) \mid \sigma)$ which should be presented to the DM. With this additional information the DM can select his/her preferred solution.

References

Bellman R.E. and Zadeh L.A. (1973). Decision-making in a fuzzy environment, *Management Sciences*, 17, 149-156.

Baptistella L.F.B. and Ollero A. (1980). Fuzzy methodologies for interactive multicriteria optimization, *IEEE Transactions on Systems, Man, and Cybernetics*, 10, 355-365.

Bit A.K. et al. (1993). Fuzzy programming approach to multiobjective solid transportation problem, *Fuzzy Sets and Systems*, 57, 183-194.

Bortolan G. and Degani R. (1985). Ranking fuzzy subsets, *Fuzzy Sets and Systems*, 15, 1-19.

Buckley J.J. (1988). Possibilistic linear programming with triangular fuzzy numbers, *Fuzzy Sets and Systems*, 26, 135-138.

Buckley J.J. (1989). Solving possibilistic linear programming problems, *Fuzzy Sets and Systems*, 31, 329-341.

Buckley J.J. (1995). Joint solution to fuzzy programming problems, *Fuzzy Sets and Systems*, 72, 215-220.

Carlsson C. and Korhonen P. (1986). A parametric approach to fuzzy linear programming, *Fuzzy Sets and Systems*, 20, 17-30.

Chanas S. (1983). The use of parametric programming in FLP, *Fuzzy Sets and Systems*, 11, 243-251.

Chanas S. (1987). Fuzzy optimization in networks, in Kacprzyk J., Orlowski S.A. (Eds.), *Optimization models using fuzzy sets and possibility theory*, Reidel, Dordrecht, 303-327.

Chanas S. (1989). Fuzzy programming in multiobjective linear programming - a parametric approach, *Fuzzy Sets and Systems*, 29, 303-313.

Chanas S. and Kuchta D. (1994). Linear programming problems with fuzzy coefficients in the objective function. in: Delgado M., Kacprzyk J. and Verdegay J.-L. and Vila M.A., *Fuzzy optimization*, Physica-Verlag, Heidelberg, 148-157.

Chakraborty T.K. (1994). A class of single sampling inspection plans based on possiblistic programming problem, *Fuzzy Sets and Systems*, 63, 35-43.

Campos L. and Verdegay J.L. (1989). Linear programming problems and ranking of fuzzy numbers, *Fuzzy Sets and Systems*, 32, 1-11.

Czyzak P.(1990). Application of the 'FLIP' method to farm structure optimization under uncertainty, in: Slowinski R. and Teghem J. (Eds.), *Stochastic versus fuzzy approaches to multiobjective mathematical programming under uncertainty*, Kluwer Academic Publishers, Dordrecht, 263-278.

Czyzak P. and Slowinski R. (1990). Solving multiobjective diet optimization problems under uncertainty, in: Korhonen P., Lewandowski A. and Wallenius J. (Eds.), Multiple criteria decision support, ser. LNEMS, vol. 356, Springer Verlag, Berlin, 272-281.

Czyzak P. and Slowinski R. (1991). 'FLIP'-Multiobjective fuzzy linear programming software with graphical facilities, in: Fedrizzi M., Kacprzyk J. and Roubens M. (Eds.), *Interactive fuzzy optimization*, ser. LNEMS, vol. 368, Springer Verlag, Berlin, 168-187.

Darzentas J. (1987). On fuzzy location models, in: Kacprzyk J. and Orlovski S.A. (Eds.), *Optimization models using fuzzy sets and possibility theory*, Reidel, Dordrecht, 328-341.

Delgado M., Kacprzyk J., Verdegay J.-L., Vila M.A. (Eds.) (1994). *Fuzzy optimization*, Physica-Verlag, Heidelberg.

Delgado M., Verdegay J.L. and Vila M.A. (1989). A general model for fuzzy linear programming, *Fuzzy Sets and Systems*, 29, 21-30.

Delgado M., Verdegay J.L. and Vila M.A. (1994). Fuzzy linear programming: From classical methods to new applications, in: Delgado M., Kacprzyk J., Verdegay J.-L. and Vila M.A. (Eds.), *Fuzzy optimization*, Physica-Verlag, Heidelberg, 111-134.

Diedrich G., Messick S.J. and Tucker L.R. (1957). General least squares solution for successive intervals, *Psychometrika* 22, 159-173.

Dubois D. and Prade H. (1980). *Fuzzy sets and systems. Theory and applications*, Academic Press, New York.

Dubois D. and Prade H. (1989). Ranking of fuzzy numbers in the setting of possibility theory, *Information Sciences*, 30, 183-224.

Esogbue A.O. and Guo K. (1991). Optimization of nonpoint source water pollution control planning using fuzzy mathematical programming, *Computer Management & Systems Science*, 67-70

Fortemps Ph., Roubens M. (1996). Ranking and defuzzification methods based on area compensation, *Fuzzy Sets and Systems*, 82, 319-330.

Fortemps Ph., Teghem J. (1997). Multi-objective fuzzy linear programming: the MOFAC method, *Cahier du LAMSADE*, Université de Paris Dauphine, Paris (to appear).

Fedrizii M., Kacprzyk J., Roubens M. (Eds.) (1991). *Interactive fuzzy optimization*, Springer Verlag, Berlin, Heidelberg.

Hannan E.L. (1981). Linear programming with multiple fuzzy goals, *Fuzzy Sets and Systems*, 6, 235-248.

Inuiguchi M., Ichihashi M. and Kume Y. (1990). A solution algorithm for fuzzy linear programming with piecewise linear membership functions, *Fuzzy Sets and Systems*, 34, 15-32.

Kacprzyk J. (1979). A branch and bound algorithm for the multistage control of a fuzzy system in a fuzzy environment, *Kybernetes*, 8, 139-147.

Kacprzyk J. (1983). *Multistage decision-making under fuzziness*, TÜV Rheinland, Köln.

Kacprzyk J. and Orlovski S.A. (Eds.) (1987). *Optimization models using fuzzy sets and possibility theory*, Reidel Dordrecht.

Kivijärvi H., Korhonen P. and Wallenius J. (1986). Operations research and its practice in Finland, *Interfaces*, 16, 53-59.

Kolodziejczyk W. (1986). Orlovsky's concept of decision-making with fuzzy preference relation – further results, *Fuzzy Sets and Systems*, 19, 11-20.

Kovacs M. (1994). Fuzzy linear programming with centered fuzzy numbers, in: Delgado M., Kacprzyk J., Verdegay J.-L. and Vila M.A. (Eds.), *Fuzzy optimization*, Physica-Verlag, Heidelberg, 135-147.

Lai Y.-J. and Hwang C.-L. (1992). *Fuzzy mathematical programming*. Methods and applications, Springer Verlag, Heidelberg.

Lai Y.-J. and Hwang C.-L. (1993). A new approach to some possibilistic linear programming problems, *Fuzzy Sets and Systems*, 49.

Lai Y.-J. and Hwang C.-L. (1994). *Fuzzy multiple objective decision making. Methods and applications*, Springer Verlag, Heidelberg.

Leberling H. (1981). On finding compromise solutions in multicriteria problems using the fuzzy min-operator, *Fuzzy Sets and Systems*, 6, 105-118.

Leung Y. (1988). Interregional equlibrium and fuzzy linear programming, *Environment and Planning*, A 20, 25-40 and 219-230.

Lilien G. (1987). MS/OR. A mid-life crises, *Interfaces*, 17, 53-59.

Luhandjula M.K. (1987). Multiple objective programming with possibilistic coefficients, *Fuzzy Sets and Systems*, 21, 135-146.

Meyer zu Selhausen H. (1989). Repositioning OR's products in the market, *Interfaces*, 19, 79-87.

Mjelde K.M. (1986). Fuzzy resource allocation, *Fuzzy Sets and Systems*, 19, 239-250.

Nakamura K. (1984). Some extensions of fuzzy linear programming, *Fuzzy Sets and Systems*, 14, 211-219.

Negoita C.V., Minoiu S. and Stan E. (1976). On considering imprecision in dynamic linear programming, *Economic Computation and Economic Cybernetics Studies and Research*, 3, 83-95.

Negoita C.V. and Ralescu, D. (1987). *Simulation, knowledge-based computing and fuzzy statistics*, Van Nostrand Reinhold, New York.

Negoita C.V. and Sularia M. (1976). On fuzzy mathematical programming and tolerances in planning, *Economic Computation and Economic Cybernetics Studies and Research*, 3, 3-15.

Orlovski S.A. (1984). Multiobjective programming problems with fuzzy parameters, *Control and Cybernetics*, 4, 175-184.

Orlovski S.A. (1985). Mathematical programming problems with fuzzy parameters, in: Kacprzyk J. and Yager R. R. (Eds.), *Management decision support systems using fuzzy sets and possibility theory*, TÜV Rheinland, Köln, 1985, 136-145.

Ostermark R. (1988). Profit apportionment in concerns with mutual ownership - an application of fuzzy inequalities, *Fuzzy Sets and Systems*, 26, 283-297.

Ostermark R. (1989). Fuzzy linear constraints in the capital asset pricing model, *Fuzzy Sets and Systems*, 30, 93-102.

Owsinski J.W., Zadrozny S. and Kacprzyk J. (1987). Analysis of water use and needs in agriculture through a fuzzy programming model, in: Kacprzyk J. and Orlovski S. A. (Eds.), *Optimization models using fuzzy sets and possibility theory*, Reidel, Dordrecht, 377-395.

Perincherry V., Kikuchi S. (1990). A fuzzy approach to the transshipment problem, *Uncertainty Modeling and Analysis*, 330-335.

Pickens J.B., Hof J.G. (1991). Fuzzy goal programming in forestry: an application with special solution problems, *FuzyySets and Systems*, 39, 239-246.

Ramik J. (1987). A unified approach to fuzzy optimization, *Reprints of the Second IFSA Congress in Tokyo,* Vol. 1, 128-130.

Ramik J. and Rimanek J. (1985). Inequality between fuzzy numbers and its use in fuzzy optimization, *Fuzzy Sets and Systems*, 16, 123-138.

Ramik J. and Rimanek J. (1987). Fuzzy parameters in optimal allocation of resources, in: Kacprzyk J. and Orlovski S.A. (Eds.), *Optimization models using fuzzy sets and possibility theory*, Reidel, Dordrecht, 1987, 359-374.

Ramik J. and Rommelfanger H. (1993). A single- and a multi-valued order on fuzzy numbers and its use in linear programming with fuzzy coefficients, *Fuzzy Sets and Systems*, 57, 203-208.

Ramik J. and Rommelfanger H. (1996). Fuzzy mathematical programming based on some new inequality relations, *Fuzzy Sets and Systems*, 81, 77-87.

Rommelfanger H. (1984). Concave membership functions and their application in fuzzy mathematical programming, in: *Proceedings of the Workshop on the Membership Function*, Ed. by EIASM Brussels, 1984, 88-101.

Rommelfanger H. (1989). Inequality relations in fuzzy constraints and its use in linear fuzzy optimization, in: Verdegay J.L. and Delgado M. (Eds.), *The interface between artificial intelligence and operational research in fuzzy environment*, Verlag TÜV Rheinland, Köln, 195-211.

Rommelfanger H. (1990). FULPAL - An interactive method for solving multiobjective fuzzy linear programming problems, in: Slowinski R. and Teghem J. (Eds.), *Stochastic versus fuzzy approaches to multiobjective mathematical programming under uncertainty*, Kluwer Academic Publishers, Dordrecht, 279-299.

Rommelfanger H. (1991). FULP - A PC-supported procedure for solving multicriteria linear programming problems with fuzzy data, in: Fedrizzi M., Kacprzyk J. and Rubens M. (Eds.), *Interactive fuzzy optimization*, Springer Verlag, Berlin, Heidelberg, 154-167.

Rommelfanger H. (1994a). *Fuzzy decision support systems*, Springer Verlag, Berlin, Heidelberg.

Rommelfanger H. (1994b). Some problems of fuzzy optimization with t-norm based addition, in: Delgado M., Kacprzyk J., Verdegay J.-L. and Vila M.A. (Eds.), *Fuzzy optimization*.. Physica-Verlag, Heidelberg, 158-168.

Rommelfanger H. (1995a) Reduction costs as a vital argument in favour of fuzzy models. In: *BUSEFAL*, 51, 30-38.

Rommelfanger H. (1995b). FULPAL 2.0 - An interactive algorithm for solving multicriteria fuzzy linear programs controlled by aspiration levels, in: Schweigert D. (Ed.), *Methods of multicriteria decision theory*, Pfalzakademie, Lamprecht, 21-34.

Rommelfanger H. (1996). Fuzzy linear programming and applications. *European Journal of Operational Research*, 92, 512-527.

Rommelfanger H., Hanuscheck R. and Wolf J. (1989). Linear programming with fuzzy objectives, *Fuzzy Sets and Systems*, 29, 31-48.

Rommelfanger H. and Keresztfalvi T. (1991). Multicriteria fuzzy optimization based on Yager's parametrized t-norm, *Foundations of Computing and Decision Sciences*, 16, 99-110.

Roubens M. (1990). Inequality constraints between fuzzy numbers and their use in mathematical programming, in: Slowinski R., Teghem J. (Eds.), *Stochastic versus fuzzy approaches to multiobjective mathematical programming under uncertainty*, Kluwer Academic Publishers, Dordrecht, 321-330.

Roubens M. and Teghem J. (1988). Comparison of methodologies for multicriteria feasibility constraint fuzzy and multiobjective stochastic linear programming, in: Kacprzyk J. and Fedrizzi M. (Eds.), *Combining fuzzy imprecision with probabilistic uncertainty in decision making*, Springer Verlag, Berlin, Heidelberg, 240-265.

Rubin P.A. and Narasimhan R. (1984). Fuzzy goal programming with nested priorities, *Fuzzy Sets and Systems*, 42, 115-129.

Sakawa M. (1983). Interactive computer programs for fuzzy linear programming with multiple objectives, *Int. J. Man-Machine Studies*, 18, 489-503.

Sakawa M. (1993). *Fuzzy sets and interactive multiobjective optimization*, Plenum Press, New York.

Sakawa M. and Yano H. (1985). Interactive decision making for multiobjective linear fractional programming problems with fuzzy parameters. *Cybernetics and Systems*, 16, 377-394.

Sakawa M. and Yano H. (1989). Interactive fuzzy satisficing method for multiobjective nonlinear programming problems with fuzzy parameters, *Fuzzy Sets and Systems*, 30, 221-238.

Sakawa M. and Yano H. (1990). Interactive decision making for multiobjective programming problems with fuzzy parameters, in: Slowinski R. and Teghem J. (Eds.), *Stochastic versus fuzzy approaches to multiobjective mathematical programming under uncertainty*, Kluwer Academic Publishers, Dordrecht, 191-228.

Seo F. and Sakawa M. (1988). *Multiple criteria decision analysis in regional planning*, Reidel Publishing Company, Dordrecht.

Slowinski R. (1986). A multicriteria fuzzy linear programming method for water supply system development planning, *Fuzzy Sets and Systems*, 19, 217-237.

Slowinski R. (1987). An interactive method for multiobjective linear programming method with fuzzy parameters and its application to water supply planning, in: Kacprzyk J. and Orlovski S.A. (Eds.), *Optimization models using fuzzy sets and possibility theory*, Reidel, Dordrecht, 396-414.

Slowinski R. (1990). 'FLIP': an interactive method for multiobjective linear programming with fuzzy coefficients, in: Slowinski R. and Teghem J. (Eds.), *Stochastic versus fuzzy approaches to multiobjective mathematical programming under uncertainty*, Kluwer Academic Publishers, Dordrecht, 249-262.

Slowinski R. (1997). Interactive fuzzy multiobjective programming, in: Climaco J. (Ed.), *Multicriteria analysis*, Springer-Verlag, Berlin, 202-212.

Slowinski R., Teghem J. (1988). Fuzzy versus stochastic approaches to multicriteria linear programming under uncertainty, *Naval Research Logistics*, 35, 673-695.

Slowinski R. and Teghem J. (1990). A comparison study of 'STRANGE' and 'FLIP', in: Slowinski R. and Teghem J. (Eds.), *Stochastic versus fuzzy approaches to multiobjective mathematical programming under uncertainty*, Kluwer Academic Publishers, Dordrecht, 365-393.

Slowinski R., Teghem J. (Eds.) (1990). *Stochastic versus fuzzy approaches to multiobjective mathematical programming under uncertainty*, Kluwer Academic Publishers, Dordrecht.

Sommer G. and Pollatschek, M.A. (1978). A fuzzy programming approach to an air pollution regulation problem, in: Trappl R., Klir G.J. and Pichler F.R. (Eds.), *Progress in cybernetics and systems research. General System Methodology, Mathematical System Theory, and Fuzzy Systems*, Biocybernetics and Theoretical Neurobiology, Vol III, Hemisphere Washington, 303-313.

Tanaka H. and Asai K. (1984). Fuzzy linear programming with fuzzy numbers, *Fuzzy Sets and Systems*, 13, 1-10.

Tanaka H., Ichihashi H. and Asai K. (1984). A formulation of linear programming problems based on comparison of fuzzy numbers, *Control and Cybernetics*, 13, 185-194.

Tanaka H., Okuda T., Asai K. (1974). On fuzzy-mathematical programming, *Journal of Cybernetics*, 3, 4, 37-46.

Thole U., Zimmermann H.-J. and Zysno P. (1979). On the suitability of minimum and product operators for the intersection of fuzzy sets, *Fuzzy Sets and Systems*, 2, 173-186.

Tingley G.A (1987). Can MS/OR sell itself well enough?, *Interfaces*, 17, 41-52.

Trappey J.F.C., Liu C.R. and Chang T.C. (1988). Fuzzy non-linear programming. Theory and application in manufacturing, *International Journal of Production Research*, 26, 957-985.

Verdegay J.L. (1982). Fuzzy mathematical programming, in: Gupta M.M. and Sanchez E. (Eds.), *Approximate reasoning in decision analysis*, North/Holland, Amsterdam, 231/236.

Verdegay J.L. (1984). Application of fuzzy optimization in operational research, *Control and Cybernetics*, 13, 229-239.

Wagenknecht M. and Hartmann K. (1987). Fuzzy evaluation of pareto points and its application to hydrocracking processes, in: Kacprzyk J. and Orlovski S.A. (Eds.), *Optimization models using fuzzy sets and possibility theory*, Reidel Dordrecht, 415-431.

Werners B. (1987). Interactive fuzzy programming system, *Fuzzy Sets and Systems*, 23, 131-147.

Yager R.R. (1979). Mathematical programming with fuzzy constraints and a preference on the objective, *Kybernetes*, 9, 109-114.

Yazenin A.V. (1987). Fuzzy and stochastic programming, *Fuzzy Sets and Systems*, 22, 171-180.

Zeleny M. (1986). Optimal system design with multiple criteria. De Novo Programming Approach, *Engineering Cost and Production Economics*, 10, 89-94.

Zimmermann H.-J. (1976). Description and optimization of fuzzy systems, *International Journal of General Systems*, 2, 209-216.

Zimmermann H.-J. (1978). Fuzzy programming and linear programming with several objective functions, *Fuzzy Sets and Systems*, 1, 45-55.

Zimmermann H.-J. (1993). Application of fuzzy set theory to mathematical programming, in: Dubois D., Prade H. and Yager R.R. (Eds.), *Readings in fuzzy sets for intelligent systems*, Morgan Kaufmann Publishers, Inc, 795-809.

Zimmermann H.-J. (1996). *Fuzzy sets, decision making and expert systems*, Kluwer Academic Publishers, Boston.

Zimmermann H.-J. and Zysno P. (1979). Latent connectives in human decision making, *Fuzzy Sets and Systems*, 3, 37-51.

Zimmermann H.-J. (1997). media selection and other applications of fuzzy linear programming, in: Dubois D., Prade H., Yager R. (Eds.), *Fuzzy information engineering. A guided tour of applications*. J. Wiley, New York, 595-617.

7 FUZZY NONLINEAR PROGRAMMING WITH SINGLE OR MULTIPLE OBJECTIVE FUNCTIONS

Masatoshi Sakawa

Abstract: In this chapter, being a nonlinear generalization of the previous one, nonlinear programming with single or multiple objective functions in a fuzzy environment is discussed. Firstly, nonlinear programming with a fuzzy goal and/or fuzzy constraints is briefly presented. Then multiobjective nonlinear programming and interactive multiobjective nonlinear programming, both incorporating fuzzy goals of the decision maker (DM), are explained in detail by putting special emphasis on the extended definition Pareto optimality. Finally, attention is focused on multiobjective nonlinear programming problems with fuzzy parameters (coefficients), which reflect the experts' ambiguous or fuzzy understanding of the nature of the parameters in the problem-formulation process. By extending the usual Pareto optimality concepts, interactive decision-making methods, both without and with the fuzzy goals of the DM, for deriving a satisficing solution from an extended Pareto optimal solution set are presented.

7.1 NONLINEAR PROGRAMMING WITH A FUZZY GOAL AND CONSTRAINTS

Consider the following nonlinear programming (NLP) problem represented in compact vector form:

$$\left.\begin{array}{ll} \text{minimize} & f(x) \\ \text{subject to} & g_i(x) \leq 0, \; i = 1, \ldots, m \end{array}\right\} \tag{1}$$

where $x = (x_1, \ldots, x_n)$ is an n-dimensional vector of decision variables, $f(x) = f(x_1, \ldots, x_n)$ and $g_i(x) = g_i(x_1, \ldots, x_n)$, $i = 1, \ldots, m$, are given real-valued nonlinear functions of n real variables $x_1, \ldots, x_n$ (Details of the theory and algorithms of nonlinear programming can be found in standard texts including Avriel (1976), Bazarra, Sherali and Shetty (1993) and Luenberger (1984)).

Similar to fuzzy linear programming as proposed by Zimmermann (1976), in contrast to the conventional nonlinear programming problem, it is possible to soften the rigid requirements of the decision maker (DM) to strictly minimize the objective function and strictly satisfy the constraints. In such a situation, the nonlinear programming problem may be softened into the following fuzzy version

$$\left.\begin{array}{ll} \widetilde{\text{minimize}} & f(x) \\ \text{subject to} & g_i(x) \precsim 0, \; i = 1, \ldots, m \end{array}\right\} \tag{2}$$

where the symbols "$\widetilde{\text{minimize}}$" and "$\precsim$" denote a relaxed or fuzzy version of those of "minimize" and "$\leq$" meaning that "the objective function should be minimized as much as possible" and "the constraints should be possibly well satisfied," respectively. Problem (2) is also called flexible programming problem (see chapter 6).

Such fuzzy requirements can be quantified by eliciting membership functions $\mu_i(g_i(x))$, $i = 0, \ldots, m$, from the DM for all the functions $f(x)$ and $g_i(x)$, $i = 1, \ldots, m$, where, for notational convenience, $\mu_0(f(x))$ is denoted by $\mu_0(g_0(x))$. By taking account of the rate of increased membership satisfaction, the DM must determine the subjective membership function $\mu_i(g_i(x))$ which is a strictly monotone decreasing function with respect to g_i in the following way:

$$\mu_i(g_i(x)) = \begin{cases} 1 & ; \; g_i(x) \leq g_i^1 \\ d_i(x) & ; \; g_i^1 \leq g_i(x) \leq g_i^0 \\ 0 & ; \; g_i(x) \geq g_i^0 . \end{cases} \tag{3}$$

As shown in Figure 1, g_i^1 or g_i^0 represents the value of g_i such that the grade of the membership function $\mu_i(g_i(x))$ is 1 or 0, and the grades of membership for the intermediate function values are expressed by a strictly monotone decreasing function $d_i(x)$ with respect to g_i.

Using these membership functions, if we follow the fuzzy decision of Bellman and Zadeh (1970), the problem of finding the maximizing decision is to choose x^* such that

$$\mu_D(x^*) = \max \min_{i=0,\ldots,m} \{\mu_i(g_i(x))\}. \tag{4}$$

By introducing an auxiliary variable λ, this problem can be transformed into the following equivalent conventional nonlinear programming problem:

$$\left.\begin{array}{ll} \text{maximize} & \lambda \\ \text{subject to} & \lambda \leq \mu_i(g_i(x)), \; i = 0, \ldots, m. \end{array}\right\} \tag{5}$$

Hence, it is evident that any existing algorithms of nonlinear programming can be applied to solve this problem and obtain the maximizing decision.

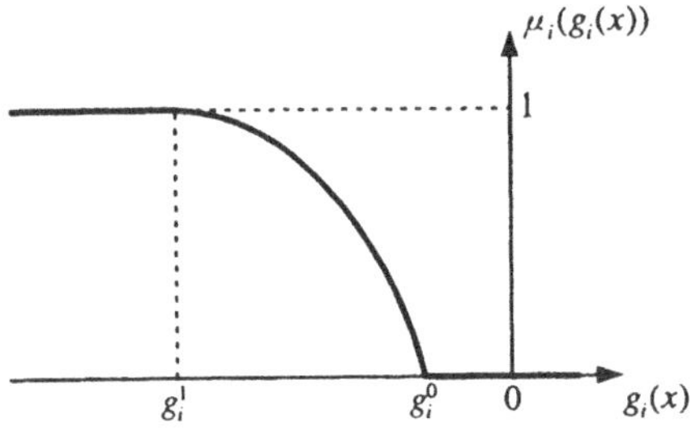

Figure 1 Strictly monotone decreasing membership function

7.2 MULTIOBJECTIVE NONLINEAR PROGRAMMING WITH FUZZY GOALS

There seems to be no fuzzy programming approach to cope with multiobjective nonlinear programming problems, except the theoretical papers of Feng (1983) and Buckley (1983), until Sakawa (1984a) first introduced multiobjective nonlinear programming with fuzzy goals as a natural extension of his linear programming with fuzzy goals (Sakawa 1983). In his method, assuming that the decision maker (DM) has a fuzzy goal for each of the objective functions in multiobjective nonlinear programming (MONLP) problems, the fuzzy goals of the DM are quantified using the five types of membership functions; linear, exponential, hyperbolic, hyperbolic inverse, and piecewise linear functions as proposed in the linear case. This case also fits to flaxible programming previously described. Then by selecting one of the three possible fuzzy decisions, the compromise or satisficing solution of the DM can be derived from among the Pareto optimal solution set. Furthermore, to cope with the fuzzy goals of the DM, such as "the objective function should be in the vicinity of some value," generalized multiobjective nonlinear programming (GMONLP) problems are introduced together with extended Pareto optimality. Although this method has been further extended to more powerful and attractive interactive programming versions, as will be discussed in the remainder of this section, it may be significant to overview it in the historic sense.

In general the multiobjective nonlinear programming (MONLP) problem is represented as the following vector-minimization problem:

$$\left.\begin{array}{ll} \text{minimize} & f(x) \triangleq (f_1(x), f_2(x), \ldots, f_k(x))^T \\ \text{subject to} & x \in X \triangleq \{x \in R^n \mid g_j(x) \leq 0,\ j = 1, \ldots, m\} \end{array}\right\} \tag{6}$$

where $f_1(x), \ldots, f_k(x)$ are k distinct objective functions of the decision vector x, $g_1(x), \ldots, g_m(x)$ are m inequality constraints and X is the feasible set of constrained decisions.

Here, for simplicity, assume that all $f_i(x)$, $i = 1, \ldots, k$, are convex and differentiable and the constrained set X is convex and compact.

Fundamental to the MONLP is the Pareto optimal concept, also known as a noninferior solution. Qualitatively, a Pareto optimal solution of the MONLP is one where any improvement of one objective function can be achieved only at the expense of another. Mathematically, a formal definition of a Pareto optimal solution to the MONLP is given as follows.

Definition 1 (Pareto optimal solution) x^* is said to be a Pareto optimal solution if and only if there does not exist another $x \in X$ such that $f_i(x) \leq f_i(x^*)$ for all i and $f_j(x) \neq f_j(x^*)$ for at least one j.

As can be seen from the definition, in general, a Pareto optimal solution consists of an infinite number of points. A Pareto optimal solution is sometimes called a noninferior solution since it is not inferior to other feasible solutions.

Assuming that the DM has fuzzy goals for each of the objective functions in the MONLP, similar to fuzzy multiobjective linear programming proposed by Zimmermann (1978), it is possible to soften the rigid requirements of the MONLP to strictly minimize the k objective functions under the given constraints. In such a situation, the MONLP may be softened into the following fuzzy version:

$$\left.\begin{array}{ll} \widetilde{\text{minimize}} & f(x) \triangleq (f_1(x), f_2(x), \ldots, f_k(x))^T \\ \text{subject to} & x \in X \triangleq \{x \in R^n \mid g_i(x) \leq 0,\ j = 1, \ldots, m\} \end{array}\right\} \tag{7}$$

where the symbol "$\widetilde{\text{minimize}}$" denotes a relaxed or fuzzy version of "minimize" with the interpretation that "the k objective functions should be minimized as much as possible" under the given constraints.

Such fuzzy requirements of the DM can be quantified by eliciting membership functions $\mu_i(f_i(x))$, $i = 1, \ldots, k$, from the DM for all the objective functions $f_i(x)$, $i = 1, \ldots, k$. In a minimization problem, a fuzzy goal stated by the DM may be to achieve "substantially less than or equal to some value p_i," and the DM is asked to determine the subjective membership function $\mu_i(f_i(x))$ which is a strictly monotone decreasing function with respect to f_i in the following way:

$$\mu_i(f_i(x)) = \begin{cases} 1 & ; \quad f_i(x) \leq f_i^1 \\ d_i(x) & ; \quad f_i^1 \leq f_i(x) \leq f_i^0 \\ 0 & ; \quad f_i(x) \geq f_i^0 \end{cases} \tag{8}$$

where, as in Figure 2, f_i^1 or f_i^0 represents the value of f_i such that the grade of the membership function $\mu_i(f_i(x))$ is 1 or 0 and the grades of membership for the intermediate function values are expressed by a strictly monotone decreasing function $d_i(x)$ with respect to f_i.

To elicit a membership function $\mu_i(f_i(x))$ from the DM for each of the objective functions $f_i(x)$ of the MONLP, Sakawa (1984a) suggested the following approach. First, calculate the individual minimum $f_i^{\min}$ and maximum $f_i^{\max}$ of each objective function $f_i(x)$ under the given constraints. Then by taking account of the calculated individual minimum and maximum of each objective function, together with the rate of increase of membership of satisfaction, the

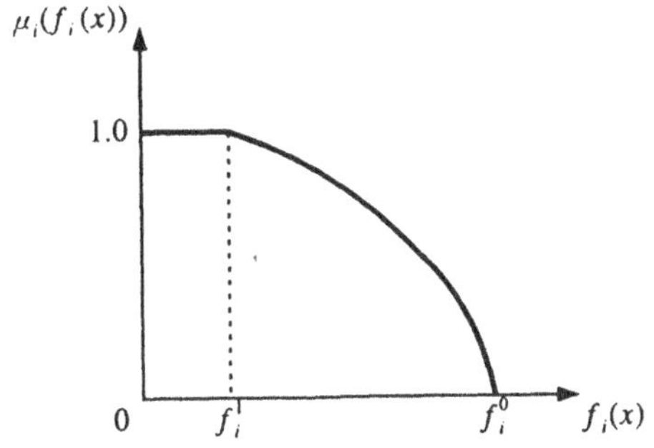

Figure 2 Membership function for minimization problem

DM is asked to select a membership function in a subjective manner from among the following five types of functions; linear, exponential, hyperbolic, hyperbolic inverse, and piecewise linear functions. The parameter values are determined through the interaction with the DM. Here, except for hyperbolic membership functions, it is assumed that $\mu_i(f_i(x)) = 0$ if $f_i(x) \geq f_i^0$ and $\mu_i(f_i(x)) = 1$ if $f_i(x) \leq f_i^1$.

(1) Linear membership function

$$\mu_i(f_i(x)) = [f_i(x) - f_i^0]/[f_i^1 - f_i^0]. \tag{9}$$

The linear membership function can be determined by asking the DM to specify the two points f_i^0 and f_i^1 within $f_i^{\max}$ and $f_i^{\min}$. Figure 3 illustrates the graph of the linear membership function.

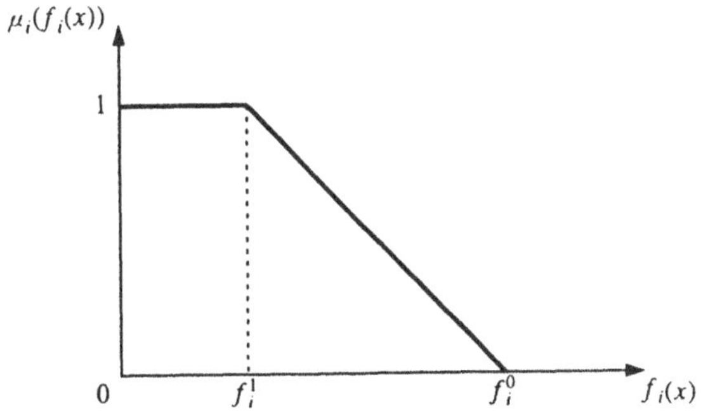

Figure 3 Linear membership function

(2) Exponential membership function

$$\mu_i(f_i(x)) = a_i[1 - \exp\{-\alpha_i(f_i(x) - f_i^0)/(f_i^1 - f_i^0)\}] \tag{10}$$

where $a_i > 1, \alpha_i > 0$ or $a_i < 0, \alpha_i < 0$, and α_i is a shape parameter. The exponential membership function can be determined by asking the DM to specify the three points f_i^0, $f_i^{0.5}$, and f_i^1 within $f_i^{\max}$ and $f_i^{\min}$, where f_i^a represents the value of $f_i(x)$ such that the degree of membership function $\mu_{f_i}(x)$ is a. Figure 4 illustrates the graph of the exponential membership function.

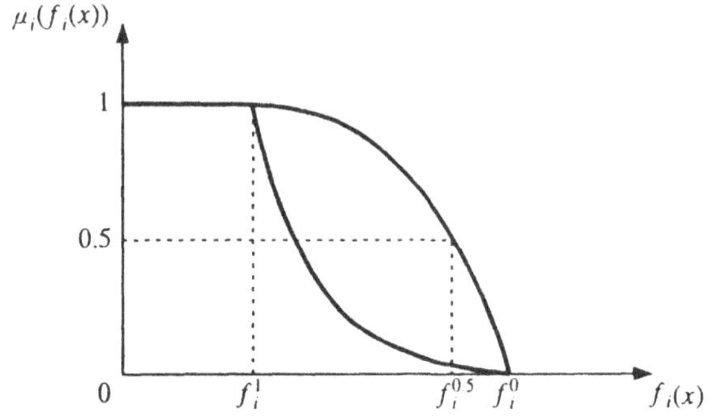

Figure 4 Exponential membership function

(3) Hyperbolic membership function

$$\mu_i(f_i(x)) = \frac{1}{2}\tanh((f_i(x) - b_i)\alpha_i) + \frac{1}{2} \tag{11}$$

where $\alpha_i < 0$ and α_i is a shape parameter and b_i is associated with the point of inflection. The hyperbolic membership function can be determined by asking the DM to specify the two points $f_i^{0.25}$ and $f_i^{0.5}$ within $f_i^{\max}$ and $f_i^{\min}$. Figure 5 illustrates the graph of the hyperbolic membership function.

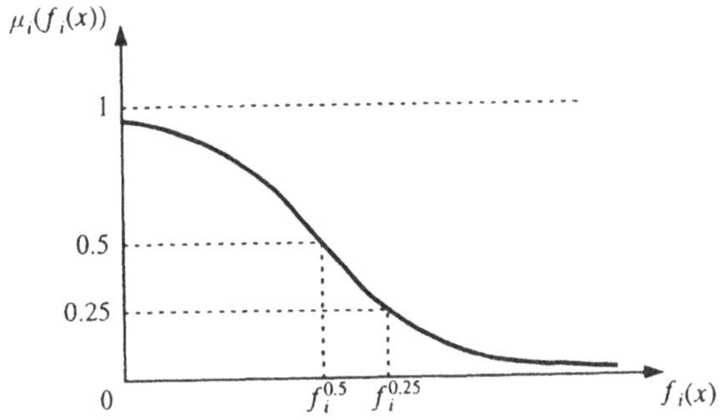

Figure 5 Hyperbolic membership function

(4) Hyperbolic inverse membership function

$$\mu_i(f_i(x)) = a_i \tanh^{-1}((f_i(x) - b_i)\alpha_i) + \frac{1}{2} \tag{12}$$

where $a_i > 0$ and $\alpha_i < 0$, and α_i is a shape parameter and b_i is associated with the point of inflection. The hyperbolic inverse membership function can be determined by asking the DM to specify the three points f_i^0, $f_i^{0.25}$, and $f_i^{0.5}$ within $f_i^{\max}$ and $f_i^{\min}$. Figure 6 illustrates the graph of the hyperbolic inverse membership function.

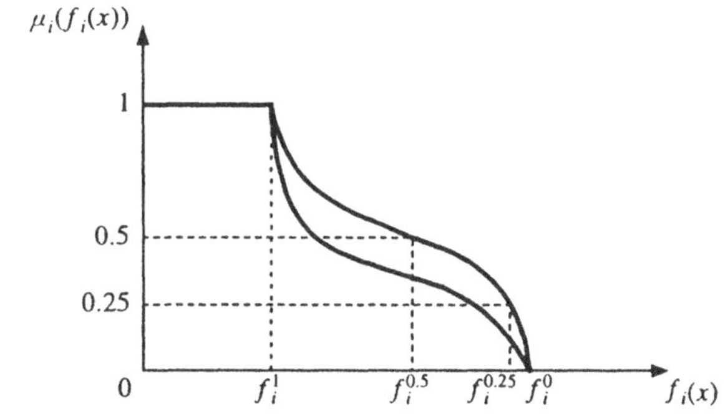

Figure 6 Hyperbolic inverse membership function

(5) Piecewise linear membership function

$$\mu_i(f_i(x)) = \sum_{j=1}^{N_i} \alpha_{ij}|f_i(x) - g_{ij}| + \beta_i f_i(x) + \gamma_i. \tag{13}$$

Here it is assumed that $\mu_i(f_i(x)) = t_{ir} f_i(x) + s_{ir}$ for each segment $g_{ir-1} \leq f_i(x) \leq g_{ir}$. The piecewise linear membership function can be determined by asking the DM to specify the degree of membership in each of several values of objective functions within $f_i^{\max}$ and $f_i^{\min}$. Figure 6 illustrates the graph of the piecewise linear membership function.

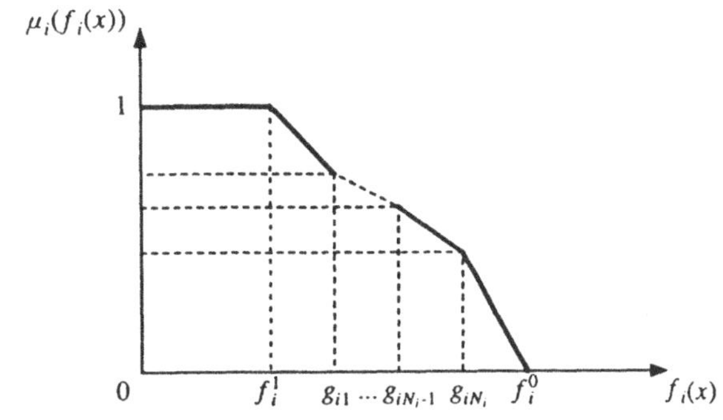

Figure 7 Piecewise linear membership function

Having determined the membership functions for each of the objective functions, Sakawa (1984a) proposed asking the DM to select one of the following three possible fuzzy decisions:

(1) a fuzzy decision (Bellman and Zadeh 1970) defined by

$$\min_{i=1,\ldots,k} (\mu_1(f_1(x)), \mu_2(f_2(x)), \ldots, \mu_k(f_k(x))), \tag{14}$$

(2) a convex-fuzzy decision (Bellman and Zadeh 1970) defined by

$$\sum_{i=1}^{k} \alpha_i \mu_i(f_i(x)), \qquad \sum_{i=1}^{k} \alpha_i = 1,\ \alpha_i \geq 0, \tag{15}$$

(3) a product-fuzzy decision (Zimmermann 1978) defined by

$$\prod_{i=1}^{k} \mu_i(f_i(x)). \tag{16}$$

In each case, following the maximizing decision which maximizes one of the three fuzzy decisions, the compromise or satisficing solution of the DM can be derived as follows.

If the DM selects the fuzzy decision or minimum operator, like Zimmermann's approach (1978), the resulting problem to be solved is equivalent to solving the following problem:

$$\left.\begin{array}{ll} \text{maximize} & \lambda \\ \text{subject to} & \lambda \leq \mu_i(f_i(x)),\ i = 1, \ldots, k \\ & x \in X. \end{array}\right\} \tag{17}$$

If the DM selects the convex-fuzzy decision, like Sommer's approach (1978), the problem to be solved is

$$\underset{x \in X}{\text{maximize}} \quad \sum_{i=1}^{k} \alpha_i \mu_i(f_i(x)). \tag{18}$$

If the DM selects the product-fuzzy decision or product operator, like Zimmermann's approach (1978), the problem to be solved becomes

$$\underset{x \in X}{\text{maximize}} \quad \prod_{i=1}^{k} \mu_i(f_i(x)). \tag{19}$$

Fundamental to the MONLP is the Pareto optimal concept and thus the DM must select a compromise solution from Pareto optimal solutions. The relationships between the optimal solutions of the above three types of problems and the Pareto optimal concept of the MONLP can be characterized by the following theorems.

Theorem 1 If x^* is a unique optimal solution to the problem (17), then x^* is a Pareto optimal solution to the MONLP.

Theorem 2 If x^* is an optimal solution to the problem (18) with $0 < \mu_i(f_i(x^*)) < 1$ holding for all i, then x^* is a Pareto optimal solution to the MONLP.

Theorem 3 If x^* is an optimal solution to the problem (19) with $0 < \mu_i(f_i(x^*)) < 1$ holding for all i, then x^* is a Pareto optimal solution to the MONLP.

The proofs of these theorems follow immediately from the definitions of (unique) optimality and Pareto optimality by contradictory arguments.

If x^* is an optimal solution to the problem (17), (18), or (19), and if none of the sufficiency conditions for Pareto optimality in Theorem 1, 2, or 3 is satisfied, then we can test the Pareto optimality for x^* by solving the following problem:

$$\left.\begin{array}{ll} \text{maximize} & \sum_{i=1}^{k} \varepsilon_i \\ \text{subject to} & f_i(x) + \varepsilon_i = f_i(x^*) \\ & x \in X,\ \varepsilon = (\varepsilon_1, \ldots, \varepsilon_k)^T \geq 0. \end{array}\right\} \tag{20}$$

Let $\bar{x}$ be an optimal solution to this problem. If all $\varepsilon_i = 0$, then x^* is a Pareto optimal solution. If at least one $\varepsilon_i > 0$, we adopt the solution $\bar{x}$ as the compromise or satisficing solution of the DM, because $\bar{x}$ not x^* is a Pareto optimal solution of the MONLP.

So far we have considered a minimization problem and consequently assumed that the DM has a fuzzy goal such as "$f_i(x)$ should be substantially less than or equal to p_i." In the fuzzy approaches, it is also possible to deal with the following general multiobjective nonlinear programming (GMONLP) problem first suggested by Sakawa (1984a):

$$\left.\begin{array}{lll} \text{fuzzy min} & f_i(x) & i \in I_1 \\ \text{fuzzy max} & f_i(x) & i \in I_2 \\ \text{fuzzy equal} & f_i(x) & i \in I_3 \\ \text{subject to} & x \in X & \end{array}\right\} \tag{21}$$

where $I_1 \cup I_2 \cup I_3 = \{1, 2, ..., k\}$, $I_i \cap I_j = \emptyset$, $i, j = 1, 2, 3$, $i \neq j$.

Here "fuzzy min $f_i(x)$" or "fuzzy max $f_i(x)$" represents the fuzzy goal of the DM such as "$f_i(x)$ should be substantially less than or equal to p_i or greater than or equal to q_i," and "fuzzy equal $f_i(x)$" represents the fuzzy goal of the DM such as "$f_i(x)$ should be in the vicinity of r_i."

Concerning the membership function for the fuzzy goal of the DM such as "$f_i(x)$ should be in the vicinity of r_i," it is obvious that a strictly monotone increasing function and a strictly monotone decreasing function corresponding to the left and right sides of r_i must be determined through interaction with the DM. In practice, the DM may select the left and right functions from among the five types of membership functions described previously (excluding hyperbolic functions). As an example, Figure 8 illustrates the graph of the possible shape of the fuzzy equal membership functions where the left function is linear and the right function is exponential.

When the fuzzy equal is included in the fuzzy goals of the DM, it is desirable that $f_i(x)$ should be as close as possible to r_i. As a result, the notion of Pareto optimal solutions defined in terms of objective functions is not applicable. With this in mind, Sakawa (1984a) first introduced the concept of fuzzy Pareto or M-

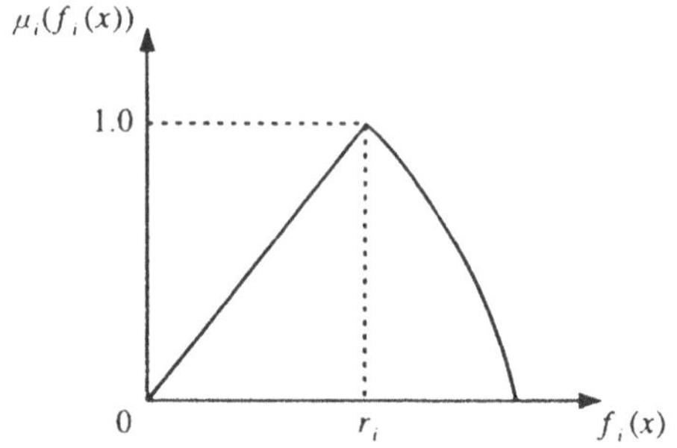

Figure 8 Fuzzy equal membership function

Pareto optimal solutions. These are defined in terms of membership functions instead of objective functions, where M refers to membership.

Definition 2 (M-Pareto optimal solution) $x^* \in X$ is said to be an M-Pareto optimal solution to the GMONLP if and only if there does not exist another $x \in X$ such that $\mu_i(f_i(x)) \geq \mu_i(f_i(x^*))$ for all i and $\mu_j(f_j(x)) \neq \mu_j(f_j(x^*))$ for at least one j.

After determining the membership functions for all types of fuzzy goals of the DM, the compromise or satisficing solution of the DM can be derived by selecting and solving one of the three types of problems (17), (18), or (19) according to the DM's decision.

Using the concept of M-Pareto optimality, the following fuzzy versions of Theorems 1, 2, and 3 can be obtained under slightly different conditions.

Theorem 4 If x^* is a unique optimal solution to the problem (17), then x^* is an M-Pareto optimal solution to the GMONLP.

Theorem 5 If x^* is an optimal solution to the problem (18), then x^* is an M-Pareto optimal solution to the GMONLP.

Theorem 6 If x^* is an optimal solution to the problem (19) with $\mu_i(f_i(x^*)) \neq 0$ holding for all i, then x^* is an M-Pareto optimal solution to the GMONLP.

Similar to the minimization case, a numerical test of M-Pareto optimality for x^* can be performed by solving the following problem:

$$\left.\begin{array}{ll} \text{maximize} & \sum_{i=1}^{k} \varepsilon_i \\ \text{subject to} & \mu_i(f_i(x)) + \varepsilon_i = \mu_i(f_i(x^*)),\ i = 1, \ldots, k \\ & x \in X,\ \varepsilon = (\varepsilon_1, \ldots, \varepsilon_k)^T \geq 0. \end{array}\right\} \quad (22)$$

Let $\bar{x}$ be an optimal solution to this problem. If all $\varepsilon_i = 0$, then x^* is an M-Pareto optimal solution. If at least one $\varepsilon_i > 0$, we adopt the solution $\bar{x}$ as the compromise or satisficing solution of the DM because M-Pareto optimality of $\bar{x}$ can be established. Further details involving a numerical example can be found in Sakawa (1984a).

7.3 INTERACTIVE FUZZY MULTIOBJECTIVE APPROACHES

In the multiobjective nonlinear programming with fuzzy goals discussed in the previous section, as well as the conventional fuzzy programming approaches proposed in the framework of Bellman and Zadeh (1970), it has been implicitly assumed that one of the possible three fuzzy decisions is the proper representation of the fuzzy preference of the decision maker (DM). Therefore, these approaches are preferable only when the DM feels that one of the three fuzzy decisions is appropriate when combining the fuzzy goals and/or constraints. However, such situations seem to happen infrequently, and consequently, it becomes evident that an interaction with the DM is necessary.

In this section, assuming that the DM has a fuzzy goal for each of the objective functions in multiobjective nonlinear programming problems, we present a few interactive fuzzy multiobjective nonlinear programming methods incorporating the desirable features of both goal programming and interactive programming approaches discussed thus far into the fuzzy approaches.

Consider the multiobjective nonlinear programming (MONLP) problem

$$\left.\begin{array}{ll}\text{minimize} & f(x) \triangleq (f_1(x), f_2(x), \ldots, f_k(x))^T \\ \text{subject to} & x \in X \triangleq \{x \in R^n \mid g_j(x) \leq 0,\ j = 1, \ldots, m\}\end{array}\right\} \quad (23)$$

and the generalized multiobjective nonlinear programming (GMONLP) problem

$$\left.\begin{array}{lll}\text{fuzzy min} & f_i(x) & i \in I_1 \\ \text{fuzzy max} & f_i(x) & i \in I_2 \\ \text{fuzzy equal} & f_i(x) & i \in I_3 \\ \text{subject to} & x \in X & \end{array}\right\} \quad (24)$$

where x is an n-dimensional vector of decision variables, $f_1(x), \ldots, f_k(x)$ are k distinct objective functions, $g_1(x), \ldots, g_m(x)$ are m inequality constraints, X is the feasible set of constrained decisions, and $I_1 \cup I_2 \cup I_3 = \{1, 2, \ldots, k\}$, $I_i \cap I_j = \emptyset$, $i, j = 1, 2, 3$, $i \neq j$.

As discussed before, fundamental to the MONLP is the concept of a Pareto optimal solution, which is also known as a noninferior solution. In practice, however, since only local solutions are guaranteed in solving a single objective optimization problem by any standard optimization technique, unless the problem is convex, instead of global Pareto optimal solutions, we deal with the local (M-) Pareto optimal solutions.

Definition 3 (local Pareto optimal solution) $x^* \in X$ is said to be a local Pareto optimal solution to the MONLP if and only if there exists a real number $\delta > 0$ such that x^* is Pareto optimal in $X \cap N(x^*, \delta)$, i.e., there does not exist another $x \in X \cap N(x^*, \delta)$ such that $f_i(x) \leq f_i(x^*)$ for all i and $f_j(x) \neq f_j(x^*)$ for at least one j, where $N(x^*, \delta)$ denotes the δ neighborhood of x^* defined by $\{x \in R^n \mid \|x - x^*\| < \delta\}$.

Definition 4 (local M-Pareto optimal solution) $x^* \in X$ is said to be a local M-Pareto optimal solution to the GMONLP if and only if there does not exist another $x \in X \cap N(x^*, \delta)$ such that $\mu_i(f_i(x)) \geq \mu_i(f_i(x^*))$ for all i

and $\mu_j(f_j(x)) \neq \mu_j(f_j(x^*))$ for at least one j, where $N(x^*, \delta)$ denotes the δ neighborhood of x^*.

Unfortunately, however, in general, (local) (M-) Pareto optimal solutions consist of an infinite number of points, and thus, the DM must select a (local) final solution from (local) (M-) Pareto optimal solutions as the satisficing solution.

Having elicited the membership functions $\mu_i(f_i(x))$ from the DM for each of the objective functions $f_i(x)$, $i = 1, \ldots, k$, the MONLP and/or the GMONLP can be converted into the fuzzy multiobjective optimization problem (FMOP) defined by

$$\underset{x \in X}{\text{minimize}} \quad (\mu_1(f_1(x)), \ldots, \mu_k(f_k(x))). \tag{25}$$

By introducing a general aggregation function

$$\mu_D(\mu(f(x))) = \mu_D(\mu_1(f_1(x)), \ldots, \mu_k(f_k(x))), \tag{26}$$

a general fuzzy multiobjective decision-making problem (FMDMP) can be defined by

$$\underset{x \in X}{\text{maximize}} \quad \mu_D(\mu(f(x))). \tag{27}$$

Observe that the value of $\mu_D(\mu(f(x)))$ can be interpreted as representing an overall degree of satisfaction with the DM's multiple fuzzy goals.

The fuzzy decision or the minimum operator of Bellman and Zadeh (1970)

$$\min_{i=1,\ldots,k} (\mu_1(f_1(x)), \ldots, \mu_k(f_k(x))) \tag{28}$$

can be viewed only as one special example of $\mu_D(\mu(f(x)))$.

In most of the conventional fuzzy approaches, it has been implicitly assumed that the minimum operator is the proper representation of decision makers' fuzzy preferences, and hence, the FMDMP has been interpreted as follows:

$$\underset{x \in X}{\text{maximize}} \quad \min_{i=1,\ldots,k} (\mu_1(f_1(x)), \ldots, \mu_k(f_k(x))) \tag{29}$$

or equivalently

$$\left.\begin{array}{ll} \text{maximize} & v \\ \text{subject to} & v \leq \mu_i(f_i(x)), \ i = 1, \ldots, k \\ & x \in X. \end{array}\right\} \tag{30}$$

There can be little doubt that the emphasis on this approach is only preferable when the DM feels that the minimum operator is the most appropriate situation. Needless to say, the DM does not always use the minimum operator in combinations of fuzzy goals and/or constraints. The most critical problem in the FMDMP is identifying an appropriate aggregation function representing the DM's fuzzy preferences. In circumstances where $\mu_D(\cdot)$ can be explicitly identifiable, then the FMDMP is reduced, as a result, to a standard mathematical programming problem. This, however, is a rare occurence and the alternative is an interaction with the DM to find a satisficing solution for the FMDMP.

Sakawa (1984b, 1986), Sakawa and Yano (1984, 1985a, 1985b, 1986a, 1986c) and Sakawa, Yumine, and Yano (1987) have suggested several interactive fuzzy multiobjective nonlinear programming methods by adopting (1) penalty scalarizing problems, (2) constraint problems, (3) goal programming problems, (4)minimax problems, and (5) augmented minimax problems as scalarizing methods for generating a candidate for the satisficing solution of the DM which is also (local) (M-) Pareto optimal. The DM then acts on this solution until satisfied with the current (local) (M-) Pareto optimal solution. Among them, in the following sections, we will discuss the interactive methods based on the last three scalarizing problems because of their desirable features.

7.4 INTERACTIVE FUZZY MULTIOBJECTIVE NONLINEAR GOAL PROGRAMMING

In interactive fuzzy multiobjective nonlinear goal programming proposed by Sakawa (1984b), after determining the membership functions $\mu_i(f_i(x))$ for each of the objective functions $f_i(x)$, $i = 1, \ldots, k$, the DM is asked to specify aspiration levels of achievement of goals in terms of membership values, called the goal membership levels, for all the membership functions. For the DM's goal membership levels $\bar{\mu}_i$, $i = 1, \ldots, k$, the corresponding (local) (M-) Pareto optimal solution is obtained by solving the following goal programming problem in the membership functions space:

$$\left.\begin{array}{ll} \text{maximize} & \sum_{i=1}^{k} d_i^- \\ \text{subject to} & \mu_i(f_i(x)) + d_i^- - d_i^+ = \bar{\mu}_i,\ i = 1, \ldots, k \\ & d_i^- \cdot d_i^+ = 0,\ i = 1, \ldots, k \\ & d_i^-,\ d_i^+ \geq 0,\ i = 1, \ldots, k \\ & x \in X \end{array}\right\} \quad (31)$$

where d_i^- and d_i^+ are, respectively, the negative and positive deviation variables, which provide us with a way to measure nonachievement of goal membership levels. This particular formulation may be called one-sided fuzzy nonlinear goal programming.

The relationships between the (local) optimal solutions of the fuzzy nonlinear goal programming problem (31) and the (local) Pareto optimal concept of the MONLP (23) can be characterized by the following theorems.

Theorem 7 If x^* is a (local) optimal solution to the fuzzy nonlinear goal programming problem with $0 < \mu_{f_i}(x^*) < 1$ and $d_i^+ = 0$ holding for all i, then x^* is a (local) Pareto optimal solution to the MONLP.

Theorem 8 If x^* is a (local) Pareto optimal solution to the MONLP with $0 < \mu_{f_i}(x^*) < 1$ holding for all i, then x^* is a (local) optimal solution to the fuzzy nonlinear goal programming with $d_i^+ = 0$ holding for all i.

The proofs of these theorems follow directly from the definitions of (local) optimality and (local) Pareto optimality by contradictory arguments.

If x^* is a (local) optimal solution to the fuzzy nonlinear goal programming problem, and if none of the sufficiency conditions for (local) Pareto optimality in Theorem 7 are satisfied (i.e., $^\exists i$, $\mu_{f_i}(x^*) = 0, 1$ or $d_i^+ \neq 0$), then, as discussed before, we can test the (local) Pareto optimality for x^* by solving the Pareto optimality test problem.

Using the concept of (local) M-Pareto optimality, the following (local) M-Pareto versions of Theorems 7 and 8 can be obtained immediately under slightly different conditions.

Theorem 9 If x^* is a (local) optimal solution to the fuzzy nonlinear goal programming problem with $d_i^+ = 0$ holding for all i, then x^* is a (local) M-Pareto optimal solution to the GMONLP.

Theorem 10 If x^* is a (local) M-Pareto optimal solution to the GMONLP, then x^* is a (local) optimal solution to the fuzzy nonlinear goal programming problem with $d_i^+ = 0$ holding for all i.

A numerical test of (local) M-Pareto optimality for x^*, similar to the minimization case, can be performed by solving the M-Pareto optimality test problem.

We can now present the interactive fuzzy nonlinear goal programming algorithm to derive the (local) satisficing solution for the DM from the (local) (M-) Pareto optimal solution set. The steps marked with an asterisk involve interaction with the DM.

Interactive fuzzy multiobjective nonlinear goal programming

Step 0: Calculate the individual minimum $f_i^{\min}$ and maximum $f_i^{\max}$ of each objective function $f_i(x)$ under the given constraints.

Step 1*: Elicit a membership function from the DM for each of the objective functions.

Step 2: Set all the initial goal membership levels to one.

Step 3: Solve the corresponding fuzzy nonlinear goal programming problem to obtain the (local) (M-) Pareto optimal solution and membership function values.

Step 4*: If the DM is satisfied with the current values of membership functions, stop. Then the current (local) (M-) Pareto optimal solution is the (local) satisficing solution of the DM. Otherwise, ask the DM to update the current goal membership levels to the new goal membership levels by considering the current values of the membership functions together with the current values of the negative deviation variables and return to Step 3.

Here it should be stressed for the DM that any improvement of one membership function can be achieved only at the expense of at least one of the other membership functions.

Further details, together with an application to water quality management, can be found in Sakawa (1984b).

7.5 INTERACTIVE FUZZY MULTIOBJECTIVE NONLINEAR PROGRAMMING

An interactive fuzzy multiobjective nonlinear programming method using minimax problems as a means of generating (local) (M-) Pareto optimal solutions was proposed by Sakawa, Yumine, and Yano (1987), realizing that the minimax problems are much more favorable than the constraint problems especially in the sense that each objective function is treated equally.

In their method, after determining the membership functions $\mu_i(f_i(x))$ for each of the objective functions $f_i(x)$, $i = 1, \ldots, k$, the DM is asked to specify the reference membership levels for all the membership functions. For the DM's reference membership levels $\bar{\mu}_i$, $i = 1, \ldots, k$, the corresponding (local) (M-) Pareto optimal solution which is, in the minimax sense, nearest to the requirement or better than that if the reference membership levels are attainable is obtained by solving the minimax problem

$$\underset{x \in X}{\text{minimize}} \quad \max_{i=1,\ldots,k} \quad \{\bar{\mu}_i - \mu_i(f_i(x))\} \tag{32}$$

or equivalently

$$\left.\begin{array}{ll} \text{minimize} & v \\ \text{subject to} & \bar{\mu}_i - \mu_i(f_i(x)) \leq v, \;\; i = 1, \ldots, k \\ & x \in X. \end{array}\right\} \tag{33}$$

The relationships between the (local) optimal solutions of the minimax problem and the (local) Pareto optimal concept of the MONLP can be characterized by the following theorems.

Theorem 11 If x^* is a unique (local) optimal solution to the minimax problem for some $\bar{\mu}_i$, $i = 1, \ldots, k$, then x^* is a (local) Pareto optimal solution to the MONLP.

Theorem 12 *If x^* is a (local) Pareto optimal solution to the MONLP with $0 < \mu_i(f_i(x^*)) < 1$ holding for all i, then there exists $\bar{\mu}_i$, $i = 1, ..., k$, such that x^* is a (local) optimal solution to the minimax problem.*

The proofs of these theorems follow directly from the definitions of (local) optimality and (local) Pareto optimality by making use of contradiction arguments.

If x^*, a (local) optimal solution to the minimax, is not unique, then we can test the (local) Pareto optimality for x^* by solving the Pareto optimality test problem.

Using the concept of (local) M-Pareto optimality, the following theorem, which is similar to Theorems 11 and 12, can immediately be obtained under slightly different conditions.

Theorem 13 x^* is a (local) M-Pareto optimal solution to the GMONLP if and only if there exists a $\bar{\mu}$ such that x^* is a unique (local) optimal solution to the minimax problem.

Similar to the minimization case, a numerical test of (local) M-Pareto optimality for x^* can be performed by solving the M-Pareto optimality test problem.

For the multiobjective nonlinear programming problem, to circumvent the necessity to perform the (local) (M-) Pareto optimality tests in the minimax problems, an interactive fuzzy multiobjective nonlinear programming method using augmented minimax problems has been proposed by Sakawa and Yano (1985b, 1986c). Therefore, in the following, we shall detail their method using augmented minimax problems instead of minimax problems. In their method, after determining the membership functions $\mu_i(f_i(x))$ for each of the objective functions $f_i(x)$, $i = 1, \ldots, k$, the DM is asked to specify the reference membership levels for all the membership functions. For the DM's reference membership levels $\bar{\mu}_i$, $i = 1, \ldots, k$, the corresponding (local) (M-) Pareto optimal solution, which is, in the minimax sense, nearest to the requirement or better than that if the reference membership values are attainable, is obtained by solving the augmented minimax problem

$$\underset{x \in X}{\text{minimize}} \left(\max_{i=1,\ldots,k} (\bar{\mu}_i - \mu_i(f_i(x))) + \rho \sum_{i=1}^{k} (\bar{\mu}_i - \mu_i(f_i(x))) \right), \qquad (34)$$

or equivalently

$$\left.\begin{array}{ll} \text{minimize} & v + \rho \sum_{i=1}^{k} (\bar{\mu}_i - \mu_i(f_i(x))) \\ \text{subject to} & \bar{\mu}_i - \mu_i(z_i(x)) \leq v, \ i = 1, \ldots, k \\ & x \in X. \end{array}\right\} \qquad (35)$$

The term augmented is adopted because the term $\rho \sum_{i=1}^{k} (\bar{\mu}_i - \mu_i(f_i(x)))$ is added to the standard minimax problem, where ρ is a sufficiently small positive scalar. Thus, the augmented minimax problem is a natural extension of the standard minimax problem and can be regarded as a modified fuzzy version of the augmented weighted Chebyshev norm problem of Steuer and Choo (1983) or Choo and Atkins (1983).

The relationships between the (local) optimal solutions of the augmented minimax problem and the (local) Pareto optimal concept of the MONLP can be characterized by the following theorems.

Theorem 14 If x^* is a (local) optimal solution to the augmented minimax problem for some $\bar{\mu}$, then x^* is a (local) Pareto optimal solution to the MONLP.

Theorem 15 If x^* is a (local) Pareto optimal solution to the MONLP with $0 < \mu_i(f_i(x^*)) < 1$ holding for all i, then there exists a $\bar{\mu}$ such that x^* is a (local) optimal solution to the augmented minimax problem.

Using the concept of (local) M-Pareto optimality, the following (local) M-Pareto version of Theorems 14 and 15 can also be obtained.

Theorem 16 x^* is a (local) M-Pareto optimal solution to the GMONLP if and only if there exists a $\bar{\mu}$ such that x^* is a (local) optimal solution to the augmented minimax problem.

An obvious advantage of the augmented minimax problem over the standard minimax problem is that, because of the presence of the augmented terms, Pareto optimality is guaranteed without the uniqueness assumption for the solution.

Now it is significant to compare the augmented minimax problem with the standard minimax problem, which is the special case of the augmented minimax problem when $\rho = 0$.

Added insight can be obtained by comparing the isoquant of the augmented minimax problem

$$\bar{\mu}_i - \mu_i(f_i(x)) + \rho \sum_{i=1}^{k} (\bar{\mu}_i - \mu_i(f_i(x))) = \text{constant}, \ i = 1, \ldots, k \tag{36}$$

with the isoquant of the minimax problem

$$\bar{\mu}_i - \mu_i(f_i(x)) = \text{constant}, \ i = 1, \ldots, k \tag{37}$$

in the membership function space as depicted in Figure 9.

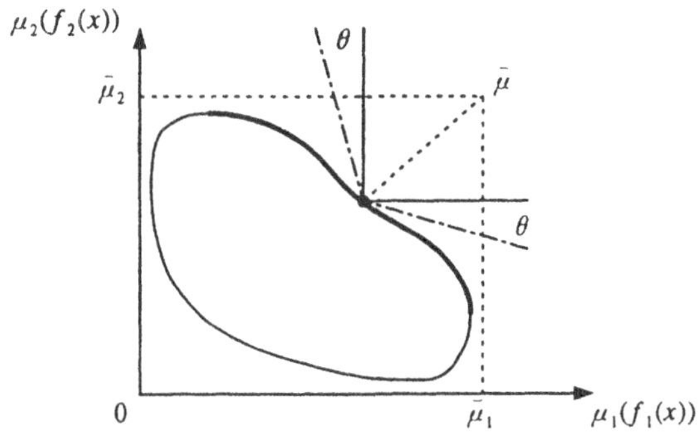

Figure 9 Isoquants of minimax and augmented minimax problems

Observing that, in Figure 9, the normal vectors of the isoquant of the augmented minimax problem and the minimax problem become $(-\rho, \ldots, -\rho, -1 - \rho, -\rho, \ldots, -\rho)$ and $(0, \ldots, 0, -1 - \rho, 0, \ldots, 0)$, respectively, it easily follows that the cosine of angle θ between these two normal vectors is given by

$$\cos\theta = (1 + \rho)/\sqrt{1 + 2\rho + k\rho^2}. \tag{38}$$

Hence, we have

$$\theta = \tan^{-1}(\sqrt{k-1}\ \rho/(1 + \rho)). \tag{39}$$

This relation shows that θ is monotonically increasing with respect to ρ. Thus, for a sufficiently small positive scalar, the augmented minimax problems overcome the possibility of generating weak Pareto optimal solutions as shown in Theorems 14, 15, and 16. Hence, augmented minimax problems are attractive for generating Pareto optimal solutions even if appropriate convexity assumptions are absent.

Naturally, ρ should be a sufficiently small, but computationally significant, positive scalar. For practical purposes, however, a computationally significant lower bound of ρ may be $\rho = 10^{-(a-b)}$ where a is the number of figures indicating the degree of precision of the computer and b is the number of figures in each membership value for which the DM can discriminate. In most cases, a computationally significant value of $\rho = 10^{-3} \sim 10^{-5}$ should suffice.

Now, given the (local) (M-) Pareto optimal solution for the reference membership levels specified by the DM by solving the corresponding augmented minimax problem, the DM must either be satisfied with the current (local) (M-) Pareto optimal solution or act on this solution by updating the reference membership levels. To help the DM express a degree of preference, trade-off information between a standing membership function $\mu_1(f_1(x))$ and each of the other membership functions is very useful. Such a trade-off between $\mu_1(f_1(x))$ and $\mu_i(f_i(x))$ for each $i = 2, \ldots, k$ is easily obtainable since it is related to the strict positive Lagrange multipliers of the augmented minimax problem.

Let the Lagrange multipliers associated with the constraints of the augmented minimax problem be denoted by λ_i, $i = 1, \ldots, k$. If all $\lambda_i > 0$ for each i, then, similar to the minimax problems, the following expression holds:

$$-\frac{\partial \mu_i(f_i(x))}{\partial \mu_1(f_1(x))} = \frac{\lambda_1 + \rho}{\lambda_i + \rho}, \quad i = 2, \ldots, k. \tag{40}$$

It should be pointed out here that, to obtain the trade-off rate information from this relation all the constraints of the augmented minimax problem must be active. Therefore, if there are inactive constraints, it is necessary to replace $\bar{\mu}_i$ for inactive constraints by $\mu_i(f_i(x^*))$ and solve the corresponding augmented minimax problem for obtaining the Lagrange multipliers.

We can now construct the interactive algorithm to derive the (local) satisficing solution for the DM from the (local) (M-) Pareto optimal solution set. The steps marked with an asterisk involve interaction with the DM.

Interactive fuzzy multiobjective nonlinear programming

Step 0: (Individual minimum and maximum) Calculate the individual minimum $f_i^{\min}$ and maximum $f_i^{\max}$ of each objective function $f_i(x)$ under the given constraints.

Step 1*: (Membership functions) Elicit a membership function $\mu_i(f_i(x))$ from the DM for each of the objective functions.

Step 2: (Initialization) Set the initial reference membership levels $\bar{\mu}_i^{(1)} = 1$, $i = 1, \ldots, k$, and set the iteration index $r = 1$.

Step 3: ((M-) Pareto optimal solution) Set $\bar{\mu}_i = \bar{\mu}_i^{(r)}$, solve the corresponding augmented minimax problem to obtain the (local) (M-) Pareto optimal solution $x^{(r)}$, $f(x^{(r)})$, and the membership function value $\mu(f(x^{(r)}))$ together with the trade-off rate information between the membership functions.

Step 4*: (Termination or new reference membership values) If the DM is satisfied with the current levels of $\mu_i(f_i(x^{(r)}))$, $i = 1, \ldots, k$, of the (local) (M-) Pareto optimal solution, stop. Then the current (local) (M-) Pareto optimal solution $f(x^{(r)}) = (f_1(x^{(r)}), \ldots, f_k(x^{(r)}))$ is the (local) satisficing solution for the DM. Otherwise, ask the DM to update the current reference membership levels $\bar{\mu}_1^{(r)}, \ldots, \bar{\mu}_k^{(r)}$ to the new reference membership levels $\bar{\mu}_1^{(r+1)}, \ldots, \bar{\mu}_k^{(r+1)}$ by considering the current levels of the membership functions together with the trade-off rates between the membership functions. Set $r = r + 1$ and return to Step 3.

Here it should be stressed to the DM that any improvement of one membership function can be achieved only at the expense of at least one of the other membership functions.

Finally, it should be noted that the DM often cannot specify the reference membership levels in an exact way. In such a situation, instead of trying to obtain point-valued estimates of the reference membership levels from the DM, it may be more appropriate to obtain fuzzy-valued assessments of the reference membership levels such as "it should be between $\underline{\mu}_i$ and $\bar{\mu}_i$ for $f_i(x)$." Details of appropriate modifications of the above algorithm using such fuzzy-valued assessments can be found in Sakawa (1986) and Sakawa and Yano (1985a).

7.6 MULTIOBJECTIVE NONLINEAR PROGRAMMING WITH FUZZY PARAMETERS

In the conventional model building approaches, when formulating a multiobjective nonlinear programming problem, the possible values of the parameters (coefficients) involved in the objective functions and constraints are required to be fixed at a value in an experimental and/or subjective manner through the experts' understanding of the nature of the parameters. Realizing that, in most real-world situations, the possible values of these parameters are often only imprecisely or ambiguously known to the experts and should be interpreted as fuzzy numbers, Orlovski (1984) first formulated a multiobjective nonlinear programming problem involving fuzzy parameters which can be viewed as the more realistic version of the conventional one. He presented two approaches to the formulated problems by making systematic use of the extension principle of Zadeh (1965) and demonstrated that there exist, in some sense, equivalent nonfuzzy formulations. Unfortunately, however, no interactive decision-making methods have been proposed.

Under these circumstances, to deal with multiobjective nonlinear programming problems with fuzzy parameters characterized by fuzzy numbers, Sakawa and Yano (1986b, 1987, 1989a, 1989b) recently presented several interactive decision-making methods for handling and tackling not only the experts' fuzzy understanding of the nature of parameters in the problem-formulation process but also the fuzzy goals of the DM by introducing the extended Pareto optimality concepts. These methods can be viewed as a natural generalization of the previous results for multiobjective nonlinear programming problems without fuzzy parameters by Sakawa, Yumine, and Yano (1987) and Sakawa and Yano (1985b, 1986c).

In general, the multiobjective nonlinear programming (MONLP) problem is represented as the following vector-minimization problem:

$$\left.\begin{array}{ll} \text{minimize} & f(x) \triangleq (f_1(x), f_2(x), \ldots, f_k(x))^T \\ \text{subject to} & x \in X \triangleq \{x \in R^n \mid g_j(x) \leq 0,\ j = 1, \ldots, m\} \end{array}\right\} \quad (41)$$

where x is an n-dimensional vector of decision variables, $f_1(x), \ldots, f_k(x)$ are k distinct objective functions of the decision vector $x, g_1(x), \ldots, g_m(x)$ are m inequality constraints, and X is the feasible set of constrained decisions.

However, as was pointed out by Orlovski (1984), it would certainly be more appropriate to consider that the possible values of the parameters in the description of the objective functions and the constraints of the MONLP, although in the conventional approaches they are fixed at some values, usually involve the ambiguity of the experts' understanding of the real system in the problem-formulation process. For this reason, in this section, we consider the following multiobjective nonlinear programming problem with fuzzy parameters (MONLP-FP):

$$\left.\begin{array}{ll} \text{minimize} & f(x, A) \triangleq (f_1(x, A_1), f_2(x, A_2), \ldots, f_k(x, A_k))^T \\ \text{subject to} & x \in X(B) \triangleq \{x \in R^n \mid g_j(x, B_j) \leq 0,\ j = 1, \ldots, m\} \end{array}\right\} \quad (42)$$

where $A_i = (A_{i1}, \ldots, A_{ip_i}), B_j = (B_{j1}, \ldots, B_{jq_j})$ represent, respectively, a vector of fuzzy parameters involved in the objective function $f_i(x, A_i)$ and the constraint function $g_j(x, B_j)$. These fuzzy parameters, which reflect the experts' ambiguous understanding of the nature of the parameters in the problem-formulation process, are assumed to be characterized as fuzzy numbers. In this section, we deal with a real fuzzy number M whose membership function $\mu_M(x)$ is defined as (Dubois and Prade 1978)

(1) A continuous mapping from R^1 to the closed interval $[0, 1]$
(2) $\mu_M(x) = 0$ for all $x \in (-\infty, c]$
(3) Strictly increasing and continuous on $[c, a]$
(4) $\mu_M(x) = 1$ for all $x \in [a, b]$
(5) Strictly decreasing and continuous on $[b, d]$
(6) $\mu_M(x) = 0$ for all $x \in [d, +\infty)$.

Now assume that A_{ir} and B_{js} in the MONLP-FP are fuzzy numbers whose membership functions are $\mu_{A_{ir}}(a_{ir})$ and $\mu_{B_{js}}(b_{js})$, respectively. For simplicity in the notation, define $a_i = (a_{i1}, \ldots, a_{ip_i})$, $b_j = (b_{j1}, \ldots, b_{jq_j})$, $a = (a_1, \ldots, a_k)$, $A = (A_1, \ldots, A_k)$, $b = (b_1, \ldots, b_m)$, $B = (B_1, \ldots, B_m)$.

Since the MONLP-FP involves fuzzy numbers both in the objective functions and the constraints, it is necessary to extend the notion of usual Pareto optimality in some sense. For that purpose, we first introduce the following α-level set to the fuzzy numbers A_{ir} and B_{js}.

Definition 5 **(α-level set)** The α-level set of the fuzzy numbers A_{ir}, $i = 1, \ldots, k$, $r = 1, \ldots, p_i$, and B_{js}, $j = 1, \ldots, m$, $s = 1, \ldots, q_j$, is defined as the ordinary set $(A, B)_\alpha$ for which the degree of their membership functions exceeds the level α:

$$(A,B)_\alpha = \{(a,b) \mid \mu_{A_{ir}}(a_{ir}) \geq \alpha,\ i=1,\ldots,k,\ r=1,\ldots,p_i; \\ \mu_{B_{js}}(b_{js}) \geq \alpha,\ j=1,\ldots,m,\ s=1,\ldots,q_j\}. \tag{43}$$

Now suppose that the decision maker (DM) considers that the degree of all of the membership functions of the fuzzy numbers involved in the MONLP-FP should be greater than or equal to a certain value of α. Then, for such a degree α, the MONLP-FP can be interpreted as the following nonfuzzy multiobjective nonlinear programming (MONLP-FP(a,b)) problem which depends on the coefficient values $(a,b) \in (A,B)_\alpha$:

$$\left.\begin{array}{ll} \text{minimize} & f(x,a) \triangleq (f_1(x,a_1), f_2(x,a_2),\ldots,f_k(x,a_k))^T \\ \text{subject to} & x \in X(b) \triangleq \{x \in R^n \mid g_j(x,b_j) \leq 0,\ j=1,\ldots,m\}. \end{array}\right\} \tag{44}$$

It is significant to note here that there exists an infinite number of such MONLP-FP(a,b) depending on the coefficient values $(a,b) \in (A,B)_\alpha$, and the values of (a,b) are arbitrary for any $(a,b) \in (A,B)_\alpha$ in the sense that the degree of all of the membership functions for the fuzzy numbers in the MONLP-FP exceeds the level α. However, from the viewpoint of the DM, if possible, it would be desirable to choose $(a,b) \in (A,B)_\alpha$ in the MONLP-FP(a,b) to minimize the objective functions under the constraints. For this reason, for a certain degree α, it seems to be quite natural for the DM to understand the MONLP-FP as the following nonfuzzy α-multiobjective nonlinear programming (α-MONLP) problem:

$$\left.\begin{array}{ll} \text{minimize} & f(x,a) \triangleq (f_1(x,a_1), f_2(x,a_2),\ldots,f_k(x,a_k))^T \\ \text{subject to} & x \in X(b) \triangleq \{x \in R^n \mid g_j(x,b_j) \leq 0,\ j=1,\ldots,m\} \\ & (a,b) \in (A,B)_\alpha. \end{array}\right\} \tag{45}$$

It must be observed here that, in the α-MONLP, the parameters (a,b) are treated as decision variables rather than constants.

Through the introduction of the α-MONLP, we can now define the concepts of α-Pareto optimality and local α-Pareto optimality as follows.

Definition 7 (α-Pareto optimal solution) $x^* \in X(b)$ is said to be an α-Pareto optimal solution to the α-MONLP if and only if there does not exist another $x \in X(b)$ and $(a,b) \in (A,B)_\alpha$ such that $f_i(x,a_i) \leq f_i(x^*,a_i^*)$, $i = 1,\ldots,k$, with strict inequality holding for at least one i, where the corresponding values of parameters a^* and b^* are called α-level optimal parameters.

Definition 8 (local α-Pareto optimal solution) $x^* \in X(b)$ is said to be a local α-Pareto optimal solution to the α-MONLP if and only if there does not exist another $x \in X(b) \cap N(x^*;r)$ and $(a,b) \in (A,B)_\alpha \cap N(a^*,b^*;r')$ such that $f_i(x,a_i) \leq f_i(x^*,a_i^*)$, $i=1,\ldots,k$, with strict inequality holding for at least one i, where the corresponding values of parameters a^* and b^* are called α-level local optimal parameters and $N(x^*;r)$ denotes the r neighborhood of x^* defined by $\{x \in R^n \mid \| x - x^* \| < r\}$.

Observe that (local) α-Pareto optimal solutions can be obtained through a direct application of the usual scalarizing methods for generating (local) Pareto optimal solutions just by regarding the decision variables in the α-MONLP as (x, a, b). However, as can be immediately understood from Definitions 7 and 8, in general, (local) α-Pareto optimal solutions to the α-MONLP consist of an infinite number of points, and the DM must select a (local) satisficing solution from an (local) α-Pareto optimal solution set based on a subjective value judgment. In the following, we first present interactive programming approaches to the MONLP-FP. We then present interactive fuzzy programming approaches to the MONLP-FP by incorporating the fuzzy goals of the DM for each of the objective functions of the α-MONLP.

7.7 INTERACTIVE NONLINEAR PROGRAMMING WITH FUZZY PARAMETERS

To generate a candidate for the (local) satisficing solution which is also (local) α-Pareto optimal, the DM is asked to specify the degree α of the α-level set and the reference levels of achievement of the objective functions, called reference levels. For the DM's degree α and reference levels $\bar{f}_i$, $i = 1, \ldots, k$, the corresponding (local) α-Pareto optimal solution, which is, in the minimax sense, nearest to the requirement or better than that if the reference levels are attainable, is obtained by solving the following minimax problem:

$$\left.\begin{array}{ll} \text{minimize} & \max\limits_{i=1,\ldots,k} (f_i(x, a_i) - \bar{f}_i) \\ \text{subject to} & (a, b) \in (A, B)_\alpha, \ x \in X(b) \end{array}\right\} \quad (46)$$

or equivalently

$$\left.\begin{array}{ll} \text{minimize} & v \\ \text{subject to} & f_i(x, a_i) - \bar{f}_i \leq v, \ i = 1, \ldots, k \\ & (a, b) \in (A, B)_\alpha, \ x \in X(b). \end{array}\right\} \quad (47)$$

The relationships between the (local) optimal solutions of the minimax problem and the (local) α-Pareto optimal concept of the α-MONLP can be characterized by the following theorems.

Theorem 17 If (x^*, v^*, a^*, b^*) is a unique (local) optimal solution to the minimax problem for some $\bar{f} = (\bar{f}_1, \ldots, \bar{f}_k)$, then x^* is a (local) α-Pareto optimal solution to the α-MONLP.

Theorem 18 If x^* is a (local) α-Pareto optimal solution and (a^*, b^*) is a (local) α-level optimal parameter to the α-MONLP, then there exists $\bar{f} = (\bar{f}_1, \ldots, \bar{f}_k)$ such that (x^*, v^*, a^*, b^*) is a unique (local) optimal solution to the minimax problem.

It should be noted here that, for generating (local) α-Pareto optimal solutions using the result in Theorem 17, uniqueness of solution must be verified. In general, however, it is not easy to check numerically whether a (local) optimal solution to the minimax problem is unique or not.

Consequently, to test the (local) α-Pareto optimality of a current (local) optimal solution (x^*, v^*, a^*, b^*), we formulate and solve the following nonlinear programming problem:

$$\left.\begin{array}{ll} \text{maximize} & \sum_{i=1}^{k} \varepsilon_i \\ \text{subject to} & f_i(x, a_i) + \varepsilon_i = f_i(x^*, a_i^*),\ \varepsilon_i \geq 0,\ i = 1, \ldots, k, \\ & x \in X(b),\ (a, b) \in (A, B)_\alpha. \end{array}\right\} \quad (48)$$

Let $(\bar{x}, \bar{a}, \bar{b})$, $\bar{\varepsilon}$ be a (local) optimal solution to this problem. If all $\bar{\varepsilon}_i = 0$, then x^* is a (local) α-Pareto optimal solution. If at least one $\bar{\varepsilon}_i > 0$, it can easily be shown that $\bar{x}$ is a (local) α-Pareto optimal solution.

However, in the framework of nonlinear programming, for circumventing the (local) α-Pareto optimality tests as in the minimax problems, it is recommended to use augmented minimax problems rather than minimax problems. To be more specific, for the DM's degree α and reference levels $\bar{f}_i$, $i = 1, \ldots, k$, the corresponding (local) α-Pareto optimal solution, which is, in the minimax sense, nearest to the requirement or better than that if the reference levels are attainable, is obtained by solving the following augmented minimax problem:

$$\min_{\substack{x \in X(b) \\ (a,b) \in (A,B)_\alpha}} \left\{ \max_{i=1,\ldots,k} (f_i(x, a_i) - \bar{f}_i) + \rho \sum_{i=1}^{k} (f_i(x, a_i) - \bar{f}_i) \right\} \quad (49)$$

or equivalently

$$\left.\begin{array}{ll} \text{minimize} & \left\{ v + \rho \sum_{i=1}^{k} (f_i(x, a_i) - \bar{f}_i) \right\} \\ \text{subject to} & f_i(x, a_i) - \bar{f}_i \leq v,\ i = 1, \ldots, k \\ & (a, b) \in (A, B)_\alpha,\ x \in X(b). \end{array}\right\} \quad (50)$$

The relationships between the (local) optimal solutions of the augmented minimax problem and the (local) α-Pareto optimal concept of the α-MONLP can be characterized by the following theorems.

Theorem 19 If (x^*, a^*, b^*) is a (local) optimal solution to the augmented minimax problem for some $\bar{f} = (\bar{f}_1, \ldots, \bar{f}_k)$, then x^* is a (local) α-Pareto optimal solution to the α-MONLP.

Theorem 20 If x^* is a (local) α-Pareto optimal solution and (a^*, b^*) is a (local) α-level optimal parameter to the α-MONLP, then there exists $\bar{f} = (\bar{f}_1, \ldots, \bar{f}_k)$ such that (x^*, a^*, b^*) is a (local) optimal solution to the augmented minimax problem.

It should be noted here that an obvious advantage of the augmented minimax problem over the usual minimax problem is that (local) α-Pareto optimality is guaranteed even if the uniqueness assumption for the solution is absent, because of the presence of the augmented term.

Now given the (local) α-Pareto optimal solution for the degree α and the reference levels specified by the DM by solving the corresponding augmented minimax problem, the DM must either be satisfied with the current (local) α-Pareto optimal solution or update the reference levels and/or the degree α. In order to help the DM express a degree of preference, trade-off information between a standing objective function and each of the other objective functions as well as between the degree α and the objective functions is very useful. Fortunately, such trade-off information is easily obtainable since it is closely related to the strict positive Lagrange multipliers of the augmented minimax problem.

To derive the trade-off information, we first define the Lagrangian function L for the augmented minimax problem as follows:

$$\begin{aligned} L(x, v, a, b, \lambda^f, \lambda^a, \lambda^b, \lambda^g, \bar{f}, \alpha) &= v + \rho \sum_{i=1}^{k} (f_i(x, a_i) - \bar{f}_i) \\ &+ \sum_{i=1}^{k} \lambda_i^f (f_i(x, a_i) - \bar{f}_i - v) + \sum_{j=1}^{m} \lambda_j^g g_j(x, b_j) \\ &+ \sum_{i=1}^{k} \sum_{r=1}^{p_i} \lambda_{ir}^a (\alpha - \mu_{A_{ir}}(a_{ir})) + \sum_{j=1}^{m} \sum_{s=1}^{q_j} \lambda_{js}^b (\alpha - \mu_{B_{js}}(b_{js})) \end{aligned} \tag{51}$$

Then the following sensitivity theorem, which is based on the implicit function theorem (Fiacco 1983), holds, where for notational convenience, the decision variables in the augmented minimax problem are denoted by $y = (x, v, a, b)$.

Theorem 21 Let $y^* = (x^*, v^*, a^*, b^*)$ be a unique local solution of the augmented minimax problem satisfying (1) y^* is a regular point of the constraints of the augmented minimax problem, (2) the second-order sufficiency conditions are satisfied at y^*, and (3) there are no degenerate constraints at y^*. Also, let $\lambda^* = (\lambda^{f^*}, \lambda^{a^*}, \lambda^{b^*}, \lambda^{g^*})$ denote the Lagrange multiplier vector corresponding to the constraints of the augmented minimax problem. Then there exist continuously differentiable vector valued functions $y(\cdot)$ and $\lambda(\cdot)$ defined on some neighborhood $N(\alpha^*)$ so that $y(\alpha^*) = y^*, \lambda(\alpha^*) = \lambda^*$, where $y(\alpha)$ is a unique local solution of the augmented minimax problem for any $\alpha \in N(\alpha^*)$ satisfying the three assumptions, and $\lambda(\alpha)$ is the Lagrange multiplier vector corresponding to the constraints of the augmented minimax problem. Furthermore, the following relation holds on some neighborhood $N(\alpha^*)$ of α^*:

$$\frac{\partial \left\{ v + \rho \sum_{i=1}^{k} (f_i(x, a_i) - \bar{f}_i) \right\}}{\partial \alpha} = \frac{\partial L}{\partial \alpha} = \sum_{i=1}^{k} \sum_{r=1}^{p_i} \lambda_{ir}^a + \sum_{j=1}^{m} \sum_{s=1}^{q_j} \lambda_{js}^b. \tag{52}$$

If all the constraints for v of the augmented minimax problem are active, namely, if $v(\alpha^*) = f_i(x(\alpha^*), a_i(\alpha^*)) - \bar{f}_i$, $i = 1, \ldots, k$, then the following theorem holds.

Theorem 22 Let all the assumptions in Theorem 21 be satisfied. Also assume that all the constraints for v of the augmented minimax problem are active.

Then it holds that

$$\left.\frac{\partial f_i(x,a_i)}{\partial \alpha}\right|_{\alpha=\alpha^*} = \frac{1}{1+\rho k}\left(\sum_{i=1}^{k}\sum_{r=1}^{p_i}\lambda_{ir}^{a^*} + \sum_{j=1}^{m}\sum_{s=1}^{q_j}\lambda_{js}^{b^*}\right). \tag{53}$$

Regarding a trade-off rate between $f_1(x)$ and $f_i(x)$ for each $i = 2, \ldots, k$, by extending the results in Haimes and Chankong (1979), it can be proved that the following theorem holds.

Theorem 23 Let all the assumptions in Theorem 21 be satisfied. Also assume that all the constraints for v of the augmented minimax problem are active. Then it holds that

$$-\left.\frac{\partial f_i(x,a_i)}{\partial f_1(x,a_1)}\right|_{\alpha=\alpha^*} = \frac{\lambda_1^{f^*}+\rho}{\lambda_i^{f^*}+\rho},\ i = 2, \ldots, k. \tag{54}$$

It should be noted here that, to obtain the trade-off rate information from these relations, all the constraints for v of the augmented minimax problem must be active. Therefore, if there are inactive constraints, it is necessary to replace $\bar{f}_i$ for inactive constraints by $f_i(x^*, a_i^*)$ and solve the corresponding augmented minimax problem for obtaining the Lagrange multipliers.

We can now construct the interactive algorithm to derive the (local) satisficing solution for the DM from the (local) α-Pareto optimal solution set. The steps marked with an asterisk involve interaction with the DM.

Interactive multiobjective nonlinear programming with fuzzy parameters

Step 0: (Individual minimum and maximum) Calculate the (local) individual minimum and maximum of each objective function under the given constraints for $\alpha = 0$ and $\alpha = 1$.

Step 1*: (Initialization) Ask the DM to select the initial value of α $(0 < \alpha < 1)$ and the initial reference levels $\bar{f}_i$, $i = 1, \ldots, k$.

Step 2: (α-Pareto optimal solution) For the degree α and the reference levels specified by the DM, solve the augmented minimax problem to obtain the α-Pareto optimal solution together with the trade-off rates between the objective functions and the degree α.

Step 3*: (Termination or updating) The DM is supplied with the corresponding (local) α-Pareto optimal solution and the trade-off rates between the objective functions and the degree α. If the DM is satisfied with the current objective function values of the (local) α-Pareto optimal solution, stop. Otherwise, the DM must update the reference levels and/or the degree α by considering the current values of the objective functions and α together with the trade-off rates between the objective functions and the degree α and return to step 2.

Here it should be stressed to the DM that (1) any improvement of one objective function can be achieved only at the expense of at least one of the other objective functions for some fixed degree α and (2) the greater value of the degree α gives the worse values of the objective functions for some fixed reference levels.

7.8 INTERACTIVE FUZZY NONLINEAR PROGRAMMING WITH FUZZY PARAMETERS

As discussed above, (local) α-Pareto optimal solutions consist of an infinite number of points, and the DM may be able to select a (local) satisficing or compromise solution from (local) α-Pareto optimal solutions based on subjective value judgments by making use of the interactive programming method presented in the previous section.

However, considering the imprecise nature of the DM's judgments, it is quite natural to assume that the DM may have imprecise or fuzzy goals for each of the objective functions in the α-MONLP. In a minimization problem, a fuzzy goal stated by the DM may be to achieve "substantially less than or equal to some value p_i." This type of statement can be quantified by eliciting a corresponding membership function.

In order to elicit a membership function $\mu_i(f_i(x, a_i))$ from the DM for each of the objective functions $f_i(x, a_i)$ in the α-MONLP, we first calculate the individual (local) minimum $f_i^{\min}$ and maximum $f_i^{\max}$ of each objective function $f_i(x, a_i)$ under the given constraints for $\alpha = 0$ and $\alpha = 1$. By taking into account the calculated individual (local) minimum and maximum of each objective function for $\alpha = 0$ and $\alpha = 1$ together with the rate of increase of membership satisfaction, the DM may be able to determine a subjective membership function $\mu_i(f_i(x, a_i))$ which is a strictly monotone decreasing function with respect to $f_i(x, a_i)$.

If we restricted ourselves to a minimization problem, it is quite natural to assume that the DM has a fuzzy goal such as "$f_i(x, a_i)$ should be substantially less than or equal to p_i." In the fuzzy approaches, as discussed in previous sections, we can further treat a more general case where the DM has two types of fuzzy goals, namely, fuzzy goals expressed in words such as "$c_i x$ should be in the vicinity of r_i" (called fuzzy equal) as well as "$c_i x$ should be substantially less than or equal to p_i or greater than or equal to q_i" (called fuzzy min or fuzzy max). Such a generalized α-MONLP (Gα-MONLP) problem can be expressed as

$$\left.\begin{array}{rll} \text{fuzzy min} & f_i(x, a_i) & i \in I_1 \\ \text{fuzzy max} & f_i(x, a_i) & i \in I_2 \\ \text{fuzzy equal} & f_i(x, a_i) & i \in I_3 \\ \text{subject to} & (a, b) \in (A, B)_\alpha, & x \in X(b) \end{array}\right\} \qquad (55)$$

where $I_1 \cup I_2 \cup I_3 = \{1, 2, ..., k\}$, $I_i \cap I_j = \emptyset$, $i, j = 1, 2, 3$, $i \neq j$.

When the DM's fuzzy goals include a fuzzy equal, then preferably the $f_i(x, a_i)$ is as close to r_i as possible. However, as a consequence of this, the (local) α-Pareto optimal solution, defined in terms of objective functions, is not applicable. It is because of this, we introduce a concept of (local) M-α-Pareto optimal solutions which is defined in terms of membership functions, where M refers to membership.

Definition 8 ((local) M-α-Pareto optimal solution) $x^* \in X(b^*)$ is said to be a (local) M-α-Pareto optimal solution to the Gα-MONLP if and only if there does not exist another $x \in X(b)\,(\cap N(x^*; r))$, $(a, b) \in (A, B)_\alpha$ $(\cap N(a^*, b^*; r'))$ such that $\mu_i(f_i(x, a_i)) \geq \mu_i(f_i(x^*, a_i^*)), i = 1, \ldots, k$, with strict inequality holding for at least one i, where the corresponding values of parameters a^* and b^* are called α-level (local) optimal parameters.

Having elicited the membership function $\mu_i(f_i(x, a_i)), i = 1, \ldots, k$, from the DM for each of the objective functions $f_i(x, a_i), i = 1, \ldots, k$, if we introduce a general aggregation function

$$\mu_D(\mu(f(x, a)), \alpha) = \mu_D(\mu_1(f_1(x, a_1)), \ldots, \mu_k(f_k(x, a_k)), \alpha) \tag{56}$$

a general fuzzy α-multiobjective decision-making problem (Fα-MODMP) can be defined by

$$\left.\begin{array}{ll} \text{maximize} & \mu_D(\mu_1(f_1(x, a_1)), \mu_2(f_2(x, a_2)), \ldots, \mu_k(f_k(x, a_k)), \alpha) \\ \text{subject to} & (x, a, b) \in P(\alpha), \ \alpha \in [0, 1] \end{array}\right\} \tag{57}$$

where $P(\alpha)$ is the set of M-α-Pareto optimal solutions and corresponding α-level optimal parameters to the Gα-MONLP.

Probably the most crucial problem in the Fα-MODMP is the identification of an appropriate aggregation function which well represents the human decision makers' fuzzy preferences. If $\mu_D(\cdot)$ can be explicitly identified, then the Fα-MODMP reduces to a standard mathematical programming problem. However, this seems to happen rarely, and as an alternative approach, interactive fuzzy programming methods under the following assumptions would be recommended.

(1) The fuzzy goals of the DM can be quantified by eliciting the corresponding membership functions through the interaction with the DM.

(2) $\mu_D(\cdot)$ exists and is known only implicitly to the DM, which means the DM cannot specify the entire form of $\mu_D(\cdot)$, but can provide local information concerning a preference. Moreover, it is strictly increasing and continuous with respect to $\mu_i(\cdot)$ and α.

(3) All $f_i(x, a_i)$, $i = 1, \ldots, k$, and all $g_j(x, b_j)$, $j = 1, \ldots, m$, are continuously differentiable in their respective domains.

Having determined the membership functions for each of the objective functions, to generate a candidate for the (local) satisficing solution which is also (local) M-α-Pareto optimal, the DM is asked to specify the degree α of the α-level set and the reference levels of achievement of the membership functions called the reference membership values. Observe that the idea of the reference membership values (e.g. Sakawa, Yumine, and Yano 1984a, b; Sakawa and Yano

1984a, 1985f) can be viewed as an obvious extension of the idea of the reference point of Wierzbicki (1980). For the DM's degree α and the reference membership values $\bar{\mu}_i$, $i = 1, \ldots, k$, the following minimax problem is solved to generate the (local) M-α-Pareto optimal solution, which is, in the minimax sense, nearest to the requirement or better than that if the reference membership values are attainable:

$$\min_{\substack{x \in X(b) \\ (a,b) \in (A,B)_\alpha}} \max_{i=1,\ldots,k} (\bar{\mu}_i - \mu_i(f_i(x, a_i))) \tag{58}$$

or equivalently

$$\left.\begin{array}{ll} \text{minimize} & v \\ \text{subject to} & \bar{\mu}_i - \mu_i(f_i(x, a_i)) \leq v, \; i = 1, \ldots, k \\ & (a, b) \in (A, B)_\alpha, \; x \in X(b). \end{array}\right\} \tag{59}$$

The relationships between the (local) optimal solutions of the minimax problem and the (local) M-α-Pareto optimal concept of the Gα-MONLP can be characterized by the following theorems.

Theorem 24 If (x^*, a^*, b^*) is a unique (local) optimal solution to the minimax problem for some $\bar{\mu}_i$, $i = 1, \ldots, k$, then x^* is a (local) M-α-Pareto optimal solution and a^*, b^* are α-level (local) optimal parameters to the Gα-MONLP.

Theorem 25 If x^* is a (local) M-α-Pareto optimal solution and a^*, b^* are α-level (local) optimal parameters to the Gα-MONLP, then there exists $\bar{\mu}_i$, $i = 1, \ldots, k$, such that (x^*, a^*, b^*) is a (local) optimal solution to the minimax problem.

Quite similar to the minimization case, a numerical test of (local) M-α-Pareto optimality for x^* can be performed by solving the following M-α-Pareto optimality test problem:

$$\left.\begin{array}{ll} \text{maximize} & \sum_{i=1}^{k} \varepsilon_i \\ \text{subject to} & \mu_i(f_i(x, a_i)) - \varepsilon_i = \mu_i(f_i(x^*, a_i)), \; \varepsilon_i \geq 0, \; i = 1, \ldots, k, \\ & x \in X(b), \; (a, b) \in (A, B)_\alpha. \end{array}\right\} \tag{60}$$

Let $(\bar{x}, \bar{a}, \bar{b})$, $\bar{\varepsilon}$ be a (local) optimal solution to this problem. If all $\bar{\varepsilon}_i = 0$, then x^* is a (local) M-α-Pareto optimal solution. If at least one $\bar{\varepsilon}_i > 0$, it can easily be shown that $\bar{x}$ is a (local) M-α-Pareto optimal solution.

To circumvent the necessity to perform the (local) M-α-Pareto optimality test in the minimax problems, for the nonlinear case, it is reasonable to use augmented minimax problems instead of minimax problems. For the DM's degree α and the reference membership values $\bar{\mu}_i$, $i = 1, \ldots, k$, the following augmented minimax problem is solved for generating the (local) M-α-Pareto optimal solution, which is, in the minimax sense, nearest to the requirement or better than that if the reference membership values are attainable:

$$\min_{\substack{x \in X(b) \\ (a,b) \in (A,B)_\alpha}} \left\{ \max_{i=1,\ldots,k} (\bar{\mu}_i - \mu_i(f_i(x, a_i))) + \rho \sum_{i=1}^{k} \Big(\bar{\mu}_i - \mu_i(f_i(x, a_i)) \Big) \right\} \tag{61}$$

or equivalently

$$\left.\begin{array}{ll} \text{minimize} & v + \rho \sum_{i=1}^{k} \left(\bar{\mu}_i - \mu_i(f_i(x, a_i))\right) \\ \text{subject to} & \bar{\mu}_i - \mu_i(f_i(x, a_i)) \leq v, \; i = 1, \ldots, k \\ & x \in X(b), \; (a, b) \in (A, B)_\alpha. \end{array}\right\} \quad (62)$$

The relationships between the (local) optimal solutions of the augmented minimax problem and the (local) M-α-Pareto optimal concept of the Gα-MONLP can be characterized by the following theorems.

Theorem 26 If (x^*, a^*, b^*) is a (local) M-α-Pareto optimal solution to the augmented minimax problem for some $\bar{\mu}_i$, $i = 1, \ldots, k$, then x^* is a (local) M-α-Pareto optimal solution and a^*, b^* are α-level (local) optimal parameters to the Gα-MONLP.

Theorem 27 If x^* is a (local) M-α-Pareto optimal solution and a^*, b^* are α-level (local) optimal parameters to the Gα-MONLP, then there exists $\bar{\mu}_i$, $i = 1, \ldots, k$, such that (x^*, a^*, b^*) is a (local) optimal solution to the augmented minimax problem for a sufficiently small positive ρ.

An obvious advantage of the augmented minimax problem over the usual minimax problem is that, because of the presence of the augmented term, (local) M-α-Pareto optimality is guaranteed without the uniqueness assumption for the solution.

Naturally, ρ should be a sufficiently small but computationally significant positive scalar. However, for practical purposes, a computationally significant lower value of ρ may be $\rho = 10^{(a-b)-c+1}$ where a and b are the figures of $\max_{i=1,\ldots,k}(\bar{\mu}_i - \mu_i(f_i(x, a_i)))$ and $\sum_{i=1}^{k}(\bar{\mu}_i - \mu_i(f_i(x, a_i)))$ in the first and second terms in the augmented minimax problem, respectively, and c is the precision figure of the computer. Usually, since the values of a and b are unknown in advance, if we roughly estimate $a = b$, then we have $\rho = 10^{-c+1}$. In most cases, a computationally significant value of $\rho = 10^{-3} \sim 10^{-5}$ should suffice.

Now, given the (local) M-α-Pareto optimal solution for the degree α and the reference membership values specified by the DM for solving the corresponding augmented minimax problem, the DM must either be satisfied with the current (local) M-α-Pareto optimal solution and α or update the reference membership values and/or the degree α. To help the DM express a degree of preference, trade-off information between a standing membership function and each of the other membership functions as well as between the degree α and the membership functions is very useful. Fortunately, such trade-off information is easily obtainable since it is closely related to the strict positive Lagrange multipliers of the augmented minimax problem.

To derive the trade-off information, we first define the Lagrangian function L for the augmented minimax problem as follows:

$$
\begin{aligned}
L = & v + \rho \sum_{i=1}^{k} \Big(\bar{\mu}_i - \mu_i(f_i(x, a_i))\Big) \\
& + \sum_{i=1}^{k} \lambda_i^{\mu}(\bar{\mu}_i - \mu_i(f_i(x, a_i)) - v) + \sum_{j=1}^{m} \lambda_j^{g} g_j(x, b_j) \\
& + \sum_{i=1}^{k} \sum_{r=1}^{p_i} \lambda_{ir}^{a}(\alpha - \mu_{A_{ir}}(a_{ir})) + \sum_{j=1}^{m} \sum_{s=1}^{q_j} \lambda_{js}^{b}(\alpha - \mu_{B_{js}}(b_{js})).
\end{aligned} \tag{63}
$$

Then, similar to Theorems 21, 22, and 23, the following sensitivity theorem holds.

Theorem 28 Let $y^* = (x^*, v^*, a^*, b^*)$ be a unique local solution of the augmented minimax problem satisfying (1) y^* is a regular point of the constraints of the augmented minimax problem, (2) the second-order sufficiency conditions are satisfied at y^*, and (3) there are no degenerate constraints at y^*. Also, let $\lambda^* = (\lambda^{\mu^*}, \lambda^{a^*}, \lambda^{b^*}, \lambda^{g^*})$ denote the Lagrange multiplier vector corresponding to the constraints of the augmented minimax problem. Then there exist continuously differentiable vector-valued functions $y(\cdot)$ and $\lambda(\cdot)$ defined on some neighborhood $N(\alpha^*)$ so that $y(\alpha^*) = y^*, \lambda(\alpha^*) = \lambda^*$, where $y(\alpha)$ is a unique local solution of the augmented minimax problem for any $\alpha \in N(\alpha^*)$ satisfying the three assumptions and $\lambda(\alpha)$ is the Lagrange multiplier vector corresponding to the constraints of the augmented minimax problem. Furthermore, the following relation holds on some neighborhood $N(\alpha^*)$ of α^*:

$$
\frac{\partial\left(v + \rho \sum_{i=1}^{k} \Big(\bar{\mu}_i - \mu_i(f_i(x, a_i))\Big)\right)}{\partial \alpha} = \frac{\partial L}{\partial \alpha} = \sum_{i=1}^{k} \sum_{r=1}^{p_i} \lambda_{ir}^{a} + \sum_{j=1}^{m} \sum_{s=1}^{q_j} \lambda_{js}^{b}. \tag{64}
$$

If all the constraints for v of the augmented minimax problem are active, namely, if $v(\alpha^*) = \bar{\mu}_i - \mu_i(f_i(x(\alpha^*), a_i(\alpha^*)))$, $i = 1, \ldots, k$, then the following theorem holds.

Theorem 29 Let all the assumptions in Theorem 28 be satisfied. Also assume that the constraints for v are active. Then it holds that

$$
-\left.\frac{\partial \mu_i(f_i(x, a_i))}{\partial \alpha}\right|_{\alpha=\alpha^*} = \frac{1}{1 + \rho k} \left(\sum_{i=1}^{k} \sum_{r=1}^{p_i} \lambda_{ir}^{a^*} + \sum_{j=1}^{m} \sum_{s=1}^{q_j} \lambda_{js}^{b^*} \right), \quad i = 1, \ldots, k. \tag{65}
$$

For trade-off rates between $\mu_1(f_1(x, a_1))$ and $\mu_i(f_i(x, a_i))$ for each $i = 2, \ldots, k$, by extending the results in Haimes and Chankong (1979), it can be proved that the following theorem holds.

Theorem 30 Let all the assumptions in Theorem 28 be satisfied. Also assume that the constraints for v are active. Then it holds that

$$-\left.\frac{\partial \mu_i(f_i(x, a_i))}{\partial \mu_1(f_1(x, a_1))}\right|_{y=y^*} = \frac{\lambda_1^{\mu^*} + \rho}{\lambda_i^{\mu^*} + \rho},\ i = 2, \ldots, k. \tag{66}$$

It should be noted here that to obtain the trade-off rate information from these relations, all the constraints for v of the augmented minimax problem must be active. Therefore, if there are inactive constraints, it is necessary to replace $\bar{\mu}_i$ for inactive constraints by $\mu_i(f_i(x^*, a_i^*)) + v^*$ and solve the corresponding augmented minimax problem for obtaining the Lagrange multipliers.

We can now construct the interactive algorithm to derive the (local) satisficing solution for the DM from the (local) M-α-Pareto optimal solution set. The steps marked with an asterisk involve interaction with the DM.

Interactive fuzzy multiobjective nonlinear programming with fuzzy parameters

Step 0: (Individual minimum and maximum) Calculate the (local) individual minimum $f_i^{\min}$ and maximum $f_i^{\max}$ of each objective function $f_i(x)$ under the given constraints for $\alpha = 0$ and $\alpha = 1$.

Step 1*: (Membership functions) Elicit a membership function from the DM for each of the objective functions.

Step 2*: (Initialization) Ask the DM to select the initial values of α $(0 \le \alpha \le 1)$ and set the initial reference membership values $\bar{\mu}_i = 1$, $i = 1, \ldots, k$.

Step 3: (M-α-Pareto optimal solution) For the degree α and the reference membership values specified by the DM, solve the corresponding augmented minimax problem to obtain the (local) M-α-Pareto optimal solution and the trade-off rates between the membership functions and the degree α.

Step 4*: (Termination or updating) The DM is supplied with the corresponding M-α-Pareto optimal solution and the trade-off rates between the membership functions and the degree α. If the DM is satisfied with the current membership function values of the M-α-Pareto optimal solution and α, stop. Otherwise, the DM must update the reference membership values and/or the degree α by considering the current values of the membership functions and α together with the trade-off rates between the membership functions and the degree α and return to step 3.

Here it should be stressed to the DM that (1) any improvement of one membership function can be achieved only at the expense of at least one of the other membership function for some fixed degree α and (2) the greater value of the degree α gives the worse values of the membership functions for some fixed reference membership values.

In this chapter, nonlinear programming with single or multiple objective functions in a fuzzy environment is briefly discussed on the basis of the author's continuing research works. For further details of linear and nonlinear programming with single or multiple objective functions in a fuzzy environment, including interactive computer programs and some application, the readers might refer to a recently published book on "Fuzzy Sets and Interactive Multiobjective Optimizations" (Sakawa 1993). In addition to this book, recently published books of Lai and Hwang (1994) and edited volumes of Kacprzyk and Orlovski (1987), Verdegay and Delgado (1989), Slowinski and Teghem (1990), and Delgado, Kacprzyk, Verdegay and Vila (1994) would be very useful for interested readers.

Acknowledgments

The author wishes to thank B. Werners and H. Yano for their helpful comments on the first draft of this chapter.

References

Avriel, M. (1976). *Nonlinear Programming: Analysis and Method*, Prentice-Hall, Englewood.

Bazarra, M.S., Sherali, H.D. and Shetty, C.M. (1993). *Nonlinear Programming: Theory and Algorithms*, 2nd Edition, John Wiley & Sons, New York.

Bellman, R.E. and Zadeh, L.A. (1970), Decision making in a fuzzy environment, it Management Science, 17, 141-164.

Buckley, J.J. (1983). Fuzzy programming and the Pareto optimal set, *Fuzzy Sets and Systems*, 10, 56-64.

Choo, E.U. and Atkins, D.R. (1983). Proper efficiency in nonconvex multicriteria programming, *Mathematics of Operations Research*, 8, 467-470.

Delgado, M., Kacprzyk, J., Verdegay J.-L. and Vila, M.A. (eds.) (1994). *Fuzzy Optimization: Recent Advances*, Physica-Verlag, Heidelberg.

Dubois, D. and Prade, H. (1978). Operations on fuzzy numbers, *International Journal of Systems Science*, 9, 613-626.

Feng, Y. (1983). A method using fuzzy mathematics to solve vectormaximum problems, *Fuzzy Sets and Systems*, 9, 129-136.

Fiacco, A.V. (1983). *Introduction to Sensitivity and Stability Analysis in Nonlinear Programming*, Academic Press, New York.

Haimes, Y.Y. and Chankong, V. (1979). Kuhn-Tucker multipliers as trade-offs in multiobjective decision-making analysis; *Automatica*, 15, 59-72.

Kacprzyk, J. and Orlovski, S.A. (eds.) (1987). *Optimization Models Using Fuzzy Sets and Possibility Theory*, D. Reidel Publishing Company, Dordrecht.

Lai, Y.J. and Hwang, C.L. (1994). *Fuzzy Multiple Objective Decision Making*, Springer-Verlag, Berlin.

Luenberger, D.G. (1984). *Linear and Nonlinear Programming*, Second Edition, Addison-Wesley, California.

Orlovski, S.A. (1984). Multiobjective programming problems with fuzzy parameters, *Control and Cybernetics*, 13, 175-183.

Sakawa, M. (1983). Interactive computer programs for fuzzy linear programming with multiple objectives, *International Journal of Man-Machine Studies*, 18, 489-503.

Sakawa, M. (1984a). Interactive fuzzy decision making for multiobjective nonlinear programming problems, in: Grauer, M. and Wierzbicki, A. P. (Eds), *Interactive Decision Analysis*, Springer-Verlag, 105-112.

Sakawa, M. (1984b). Interactive fuzzy goal programming for multiobjective nonlinear programming problems and its applications to water quality management, *Control and Cybernetics*, 13, 217-228.

Sakawa, M. (1986). Interactive multiobjective decision-making using fuzzy satisficing intervals and its application to an urban water resources system, *Large Scale Systems*, 10, 203-213.

Sakawa, M. (1993). *Fuzzy Sets and Interactive Multiobjective Optimization*, Plenum Press, New York.

Sakawa, M. and Yano, H. (1984). An interactive fuzzy satisficing method using penalty scalarizing problems, *Proceedings of International Computer Symposium (ICS'84)*, Tamkang University, Taiwan, 1122-1129.

Sakawa, M. and Yano, H. (1985a). Interactive fuzzy decision-making for multiobjective nonlinear programming using reference membership intervals, *International Journal of Man-Machine Studies*, 23, 407-421.

Sakawa, M. and Yano, H. (1985b). An interactive fuzzy satisficing method using augmented minimax problems and its application to environmental systems, *IEEE Trans. Systems, Man, and Cybernetics*, SMC-15, 720-729.

Sakawa, M. and Yano, H. (1986a). An interactive fuzzy decisionmaking method using constraint problems, *IEEE Transactions on Systems, Man, and Cybernetics*, SMC-16, 179-182.

Sakawa, M. and Yano, H. (1986b). An interactive method for multiobjective nonlinear programming problems with fuzzy parameters, in: Trappl, R. (Ed.), *Cybernetics and Systems '86*, D. Reidel Publishing Company, Dordrecht, 607-614.

Sakawa, M. and Yano, H. (1986c). Interactive fuzzy decision making for multiobjective nonlinear programming using augmented minimax problems, *Fuzzy Sets and Systems*, 20, 31-43.

Sakawa, M. and Yano, H. (1987). An interactive satisficing method for multiobjective nonlinear programming problems with fuzzy parameters, in: Kacprzyk, J. and Orlovski, S.A. (Eds), *Optimization Models Using Fuzzy Sets and Possibility Theory*, D. Reidel Publishing Company, Dordrecht, 258-271.

Sakawa, M. and Yano, H. (1989a). Interactive decision making for multiobjective nonlinear programming problems with fuzzy parameters, *Fuzzy Sets and Systems*, 29, 325-326.

Sakawa, M. and Yano, H. (1989b). An interactive fuzzy satisficing method for multiobjective nonlinear programming problems with fuzzy parameters, *Fuzzy Sets and Systems*, 30, 221-238.

Sakawa, M., Yumine, T. and Yano, H. (1987). An interactive fuzzy satisficing method for multiobjective nonlinear programming problems, in: Bezdek, J.C.

(Ed.), *Analysis of Fuzzy Information*, Volume III, CRC Press. Inc., Florida, 273-283.

Slowinski, R. and Teghem, J. (1990). *Stochastic Versus Fuzzy Approaches to Multiobjective Mathematical Programming under Uncertainty*, Kluwer Academic Publishers, Dordrecht.

Sommer, G. and Pollastschek, M.A. (1978). A fuzzy programming approach to an air pollution regulation problem, in: Trappl, R., Klir, G.J. and Ricciardi, L. (Eds), *Progress in Cybernetics and Systems Research*, 303-323.

Steuer, R.E. and Choo, E.U. (1983). An interactive weighted Tchebycheff procedure for multiple objective programming, *Mathematical Programming*, 26, 326-344.

Verdegay, J.-L. and Delgado, M. (Eds.) (1989). *The Interface between Artificial Intelligence and Operations Research in Fuzzy Environment*, Verlag TÜV Rheinland, Köln.

Wierzbicki, A.P. (1980) The use of reference objectives in multiobjective optimization, in: G. Fandel and T. Gal (Eds.), *Multiple Criteria Decision Making: Theory and Application*, Springer-Verlag, 468-486.

Zadeh, L.A. (1965). Fuzzy sets, *Information and Control*, 8, 338-353.

Zimmermann, H.-J. (1976) Description and optimization of fuzzy systems, *International Journal of General Systems*, 2, 209-215.

Zimmermann, H.-J. (1978) Fuzzy programming and linear programming with several objective functions, *Fuzzy Sets and Systems*, 1, 45-55.

8 DISCRETE FUZZY OPTIMIZATION

Stefan Chanas

Dorota Kuchta

Abstract: The chapter presents selected problems and algorithms of fuzzy discrete optimization. The problems discussed constitute fuzzy equivalents of the following crisp ones: the maximal and cheapest flow and the shortest route problem in a network, the transportation and assignment problems, the CPM and PERT method, a selection of scheduling problems, the set covering problem, the knapsack problem as well as the general 0-1 programming problem.

8.1 INTRODUCTION

Generally, the problem of fuzzy discrete optimization can be formulated in the following way:

$$\begin{aligned} & f(x;C) \to \max(\min) \\ & x \in X \end{aligned}$$

where $C=(C_1,C_2,\ldots,C_r)$ is a vector of fuzzy parameters, whose components are usual fuzzy numbers. X is a fuzzy set in a countable or finite universe of feasible solutions U. The elements of the universe U can be of various nature. They can be vectors of integer components, permutations of a certain set, paths connecting two fixed vertices in a distance network, an so on. Of course, one of the elements C or X of the problem can be unfuzzy. However, if we want to speak of a fuzzy problem of discrete optimization, they cannot both be unfuzzy at the same time. The above formulation is not a complete definition of the problem. It determines the problem in an exact way only when additionally the notions of feasibility and optimality of the problem are defined. These notions can be defined in various ways, thus even one problem may have several different formulations.

In this paper we have made a survey of fuzzy discrete optimization problems, stressing above all the problem formulations themselves and the assumed solution concepts. In some cases, but not very numerous, we also quote the solution algorithms — if they seem interesting and concern important problems of fuzzy discrete optimization.

The survey presented here is not complete. In fact, it was not the aim of this paper to present such a survey. Our intention was rather to indicate various applications of fuzzy sets also in the domain of those optimization problems, which in the classic sense are considered to be discrete optimization problems.

Among those problems the best known and the most representative ones are those connected with optimization on networks as well as with scheduling. Fuzzy counterparts of these problems have an important place in the present paper.

In section 8.2 we present selected problems connected with optimization on graphs. We discuss the problem of a maximal flow in a network with fuzzy arc capacities, stressing the discrete case, i.e. one with the integrality condition imposed on the arc flows. We also present the problem of the cheapest flow with fuzzy arc capacities and a fuzzy constraint on the flow value. Moreover, the problem of the shortest route in a network with fuzzy lengths of the arcs has been discussed too. The proposals and remarks presented in connection with the latter problem are important because they concern in fact also other problems of fuzzy optimization on distance networks, such as the travelling salesman, the minimal spanning tree and the linear ordering problems.

Section 8.3 is devoted to the transportation and assignment problems with fuzzy parameters. The classical transportation problem does not cause any problems if we require its solution, the cheapest shipping schedule, to be integer. It is because this problem possesses, like the problem of the cheapest flow in a network, a so called integrality property. This property consists in the fact that if the coefficients of the problem are integer, then among the optimal solutions of the problem there exists an integer one, and it is determined by classical algorithms. The fuzzy transportation problem does not possess this property and for this reason the integer case of this problem requires a special treatment.

A separate section, numbered 8.4, is devoted to the techniques of network planning, with the assumption that the duration times of the activities in the project are given in an imprecise form of fuzzy numbers. Two approaches are possible here. One of them consists in a direct generalisation of the CPM (Critical Path Method), by replacing the usual numerical operations with operations on fuzzy numbers. A relatively large number of papers have been devoted to this approach. The proposals of the authors differ from one to another in principle only in the use of different definitions of operations on fuzzy numbers (the aim is to reduce the effect of the rapidly progressing fuzziness in the formula calculating the latest dates of events) and in different definitions of the notion of criticality of a path and activities in the project. Here we present in fact only one proposal. In the second approach presented here (suggested by Chanas, 1987), the fuzzy estimates of the duration times of the activities in the project are used only as a basis for generating (deriving) a certain type of probability distribution for the activities duration times. The further procedure is identical as the one in the classical PERT method. That is why the second approach has been termed fuzzy PERT.

Section 8.5 is devoted to scheduling problems. Scheduling problems comprise a very large class of problems, which considerably differ one from another in the formulation itself. We discuss some of the fuzzy scheduling problems studied in the literature, confining ourselves to the formulations themselves and indicating where corresponding algorithms can be found.

The last section, numbered 8.6, contains a short discussion of several other selected fuzzy discrete optimization problems, which seemed interesting, but which could not be included in any of the problem classes presented before. We write there about the fuzzy set covering problem, about a certain case of the knapsack problem and about the general 0-1 linear programming problem.

8.2 FUZZY OPTIMIZATION IN GRAPHS AND NETWORKS

From among classic mathematical programming problems, a class of network optimization problems may be singled out. A network as the main element of a problem description is a common feature of all problems in this class. This fact is utilised to a considerable degree in construction of special solution algorithms which are more efficient than the general mathematical programming methods that could also be used to solve the problems. For example the following problems: the max-flow problem, the min-cost flow problem, the shortest route problem and others, can also be stated as linear programming problems. But, applying the simplex method to solve them would be very inefficient. That is why some special "network" algorithms have been developed separately for those problems. There is a similar situation among fuzzy mathematical programming problems. Also here a class of fuzzy network optimization problems can be distinguished, and, as in the classic case, the network description of problems can be utilised to construct solution algorithms.

In the following part of this section we review some problems of this type and present for some of them solution algorithms based on a network representation of these problems.

8.2.1 Max-flow Problem

Let $S=\langle N,A\rangle$ be a directed graph, where N denotes the set of vertices and $A\subseteq N\times N$ is the set of arcs. Two vertices, a source $s\in N$ and a sink $t\in N$, are specified in the network S. A fuzzy interval C_{ij} is associated with each arc $(i,j)\in A$. It is assumed that C_{ij} is a fuzzy interval number of the form $C_{ij}=\left(0,\bar{c}_{ij},0,\beta_{ij}\right)_{L_{ij}-R_{ij}}$. The C_{ij} is a non-sharply defined capacity of arc $(i,j)\in A$ or, more precisely, a fuzzy interval of feasible flow values in arc $(i,j)\in A$. The fuzzy max-flow problem consists in finding a v-value flow x_v from the source s to the sink t, $x_v = \left\{ x_{ij} \in R \middle| (i,j) \in A \right\}$, such that

$$v \,\hat{\in}\, G \tag{1}$$

$$\sum_i x_{ij} - \sum_k x_{jk} = \begin{cases} -v & \text{for} \quad j = s \\ 0 & \text{for} \quad j \neq s,t \\ v & \text{for} \quad j = t \end{cases} \tag{2}$$

$$x_{ij} \,\hat{\in}\, C_{ij}, \quad (i,j) \in A \tag{3}$$

where G is a fuzzy goal being a fuzzy interval of flow values accepted by the decision-maker (the expression $a \,\hat{\in}\, A$ means: "a belongs to A to the highest possible degree"). The higher the value of the membership function $\mu_G(v)$, the higher is the degree to which the decision maker accepts the flow value v. Consequently, it seems natural to assume the following form of G: $G = \left(\underline{v}, +\infty, \alpha_G, 0\right)_{L_G - R_G}$.

Verbally, problem (1)-(3) can be stated as follows: find a v-value flow x_v fulfilling the conservation constraints (2) and satisfying the fuzzy goal (1) and fuzzy capacity constraints (3) to the maximum degree.

Using Bellman-Zadeh's (1970) approach we define a fuzzy set D in the set X of flows x_v satisfying the conservation condition (2), by the following membership function

$$\mu_D(x_v) = \mu_G(v) \wedge \mu_C(x_v) \tag{4}$$

where

$$\mu_C(x_v) = \bigwedge\nolimits_{(i,j)\in A} \mu_{C_{ij}}(x_{ij}) \tag{5}$$

and „$\wedge$" stands for the minimum operation.

It is natural to consider a flow x_v which belongs to D to the maximum degree, called the maximizing solution, as a proper final choice. Thus, problem (1)-(3) can be reduced, according to the Bellman-Zadeh's methodology, to the following mathematical problem:

$$\text{find } x_v \text{, such that}$$
$$\mu_D(x_v) \to \max, \quad \text{subject to (2)} \tag{6}$$

The optimal solution of problem (6) does not have to be an integer one. For this reason Chanas and Kolodziejczyk (1982, 1984) called this problem a real-valued max-flow problem with fuzzy capacity constraints. In Chanas and Kolodziejczyk (1984) an effective algorithm solving the real case of problem (6) was proposed. Furthermore, Chanas (1989) reduced the analysis of problem (1)-(3), also in the real case, to solving a parametric problem of maximal flow in a network (with a parameter in the constraints on the arcs capacity). In the latter case apart from the maximizing solution, i.e. the solution of problem (6), a fuzzy solution of the problem is given in a analytic form.

As we confine ourselves here to problems of fuzzy discrete optimization, we will present after Chanas and Kolodziejczyk (1986) an algorithm solving problem (6) with the additional condition of the integrality of the flow x_v, i.e. the following one:

$$x_{ij} \in I,\ (i,j) \in A \tag{7}$$

In the description of the algorithm it is assumed — what is natural, if we confine ourselves to integer flows — that coefficients $\bar{c}_{ij}$ i $\underline{v}$ in fuzzy interval numbers C_{ij} i G are integer.

Algorithm (Chanas and Kolodziejczyk, 1986):

Step1: Find a maximum flow x_v in the classic sense in the network S, assuming the arc capacities to be equal to $\bar{c}_{ij}$, $(i,j) \in A$. Evidently, $\mu_C(x_v) = 1$. If $\mu_G(x_v) = 1$, then STOP — x_v is a maximizing solution. Otherwise, go to Step 2.

Step 2: Let x_v be a current flow. Determine a path r leading from s to t and maximizing the value of the expression

$$\mu^r(x_v) = \wedge_{(i,j)\in\vec{r}}\, \mu_{C_{ij}}(x_{ij}+1) \wedge \wedge_{(i,j)\in\overleftarrow{r}}\, \mu_{C_{ij}}(x_{ij}-1) \rightarrow \max \tag{8}$$

where $\vec{r}$ and $\overleftarrow{r}$ are sets of the forward and backward arcs in the path r, respectively. Additionally, it is assumed that if $x_{ij} = 0$, then $\mu_{ij}(x_{ij}-1) = -1$. If $\mu_D(x_v) > \mu^r(x_v)$, then STOP; x_v is a maximizing solution. Otherwise, go to Step 3.

Step 3: $x_v := x_{v+1}$, where x_{v+1} is the v+1-value flow obtained by increasing the flow x_v with a unit on the path r (adding a unit flow to the forward arcs and subtracting the same value from the backward arcs). Go to Step 2.

To determine the path in Step 2, the shortest path algorithm of Dijkstra (see e.g. Boffey, 1982) can be easily adapted.

8.2.2 Min-cost Flow Problem

Let $S = \langle N, A \rangle$ stand, as before, for a directed network with a distinguished source $s \in N$ and sink $t \in N$. The fuzzy interval of feasible flows on an arc has now a more general form $C_{ij} = \left(\underline{c}_{ij}, \bar{c}_{ij}, \alpha_{ij}, \beta_{ij}\right)_{L-R}$, $(i,j) \in A$. However, we assume that this time all fuzzy intervals C_{ij}, $(i,j) \in A$, are fuzzy interval numbers of the same type. The fuzzy interval V of admissible (desirable) flow values, $V = (\underline{v}, \bar{v}, \alpha_V, \beta_V)_{L-R}$, being also of type L-R, is fixed too. With each arc there is associated a real number d_{ij}, a unit cost of the flow in the arc $(i,j) \in A$. The fuzzy min-cost flow problem consists in finding a v-value flow x_v from the source s to the sink t, $x_v = \left\{x_{ij} \in R \middle| (i,j) \in A\right\}$, such that

$$d(x_v) = \sum\nolimits_{(i,j)\in A} d_{ij} x_{ij} \mathrel{\hat{\in}} G \tag{9}$$

$$\sum_i x_{ij} - \sum_k x_{jk} = \begin{cases} -v & \text{for} \quad j = s \\ 0 & \text{for} \quad j \neq s,t \\ v & \text{for} \quad j = t \end{cases} \tag{10}$$

$$v \mathrel{\hat{\in}} V \tag{11}$$

$$x_{ij} \mathrel{\hat{\in}} C_{ij}, \quad (i,j) \in A \tag{12}$$

where G is a fuzzy goal, being now a fuzzy interval of the values of the total cost accepted by the decision-maker. The higher the value of the membership function $\mu_G(d)$, the higher is the degree to which the decision maker accepts the total flow cost d. It is natural to assume that G is now a fuzzy interval of the form $G = \left(-\infty, \bar{d}, 0, \beta_G\right)_{L_G - R_G}$.

If to the set of arcs A we add the arc (t,s) with a fuzzy capacity $C_{ij} = V_{ij}$, we can then replace problem (9)-(12) with a fuzzy min-cost circulation flow problem. It consists in determining a flow $x = \left\{x_{ij} \in R \middle| (i,j) \in A \cup (t,s)\right\}$ in the extended network S fulfilling the following conditions:

$$d(x) = \sum\nolimits_{(i,j)\in A} d_{ij} x_{ij} \mathrel{\hat{\in}} G \tag{13}$$

$$\sum_i x_{ij} - \sum_k x_{jk} = 0, \ j \in N \tag{14}$$

$$x_{ij} \mathrel{\hat{\in}} C_{ij}, \quad (i,j) \in A \cup (t,s) \tag{15}$$

The flow value on the arc (t,s) is equal to the value of the flow from s to t, i.e. $x_{ts} = v$.

After the approach of Bellman-Zadeh has been applied, the determination of a maximizing solution of problem (13)-(15) is reduced to solving the following mathematical programming problem:

$$\mu_D(x) = \mu_G(d(x)) \wedge \mu_C(x) \to \max \tag{16}$$

where

$$\mu_C(x) = \wedge_{(i,j)\in A\cup(t,s)} \mu_{C_{ij}}(x_{ij}) \tag{17}$$

Of course the solution of problem (16) does not have to be integer. Chanas and Machaj (1985) (see also Chanas, 1989) showed that if all the fuzzy interval numbers C_{ij}, $(i,j) \in A \cup (t,s)$ are of the same type, then problem (16) can be replaced with a parametric problem of the cheapest flow in a network with the arc capacities being dependent in a linear way on a parameter, and that problem can be solved by means of the algorithm proposed in the same paper. In that case, apart from the maximizing solution we receive also, in an analytic form, a membership function of the fuzzy decision D. Chanas and Machaj (1985) present also an algorithm determining the optimal integer solution of problem (16), if the real optimal solution has already been found. In this algorithm, which we will not quote here, the so called property of integrality of the cheapest flow in a network plays a substantial role. This property consists in the fact that if the coefficients of the problem are integer, then among the optimal solutions of the problem there are integer ones.

An algorithm for the transportation problem, which we present in section 8.3, can also be easily adopted (through generalisation) to solve problem (16) with the additional condition of the integrality of the solution. In this case no prior knowledge of the real solution is needed. This generalisation is easy, because the transportation problem can be treated as a special case of the problem of the cheapest flow in a network.

8.2.3 The Shortest Route Problem

In this section we will analyse the problem of the shortest route in a network with fuzzy lengths of arcs, given as fuzzy numbers. The importance of the results, remarks and comments presented here is broader, because they can be directly applied to other problems of fuzzy network optimization, such as the salesman problem or the minimal spanning tree problem in a network with fuzzy lengths of the arcs. All these problems have one common feature: the evaluation of the solution (of the path, spanning tree, route of the salesman, etc.) is a fuzzy number being the sum of several fuzzy numbers (lengths of arcs).

Let $S = \langle N, A \rangle$ stand as before for a directed network, in which to each arc $(i,j) \in A$ there has been assigned a fuzzy number $L_{ij} = \left(\underline{l}_{ij}, \bar{l}_{ij}, \alpha_{ij}, \beta_{ij}\right)_{L-R}$, defining its length. The problem consists in determining the "shortest route" between a fixed couple of vertices $i, j \in N$. It is not by accident that the formulation "shortest route" has been put between quotation marks. As the length of a fuzzy route is a fuzzy number (being the sum of fuzzy lengths of the arcs), the notion of the shortest route needs a separate, precise definition. Authors dealing with the fuzzy problem of the shortest route do not always provide such a definition.

On Dubois and Prade approach

Dubois and Prade (1978, 1980) and Prade (1980) touch several times upon the problem of choosing a shortest path. Their idea consists in an appropriate modification of classical algorithms destined for solving problems with crisp arc lengths. They propose to substitute in the Ford algorithm (for acyclic networks) and in that of Floyd (in a more general case) for the operations "+" and "$\wedge$" the respective fuzzy extended operations "$\tilde{+}$" and "$\tilde{\wedge}$". The realisation of the algorithms with the operations changed in this way yields for a fixed couple of vertices $i, j \in N$ a fuzzy number D_{ij}, being a fuzzy equivalent of the length of the shortest path between i and j. Because the $\tilde{\wedge}$ operation for several fuzzy numbers does not necessarily yield one of those numbers, it is possible that no path has fuzzy length D_{ij}. For this reason the authors do not make use of the parts in Ford's and Floyd's algorithms serving merely to identify an optimal path (the labelling procedure in Ford's algorithm and the procedure of calculating the table of predecessors in Floyd's algorithm). Having the fuzzy quantity D_{ij} defined, Dubois and Prade introduce the notion of the criticality of a path, defining for each path $p \in P(i, j)$ the value of its criticality as equal to $hgt(L_p \cap D_{ij}) = Poss(L_p = D_{ij})$, where L_p is the fuzzy length of the path p (i.e. the extended sum of the fuzzy lengths of the arcs of p). The value of $hgt(L_p \cap D_{ij})$ can be treated (as stated in Prade, 1980) as a membership of path p in the "fuzzy set of the shortest paths between i and j". It is easy to notice that it follows from the properties of fuzzy numbers and from those of the extended operations on fuzzy numbers that there always exists a path $\bar{p} \in P(i, j)$ such that $hgt(L_{\bar{p}} \cap D_{ij}) = 1$. To determine this path it is enough to apply one of the classical algorithms to the network $S = \langle N, A \rangle$ with crisp lengths of arcs, equal to the modal values m_{ij} of fuzzy numbers L_{ij}, $\mu_{L_{ij}}(m_{ij}) = 1$.

Let us come back to the fuzzy number D_{ij}. This number can be treated as a possibility distribution of the shortest distance between vertices i and j. $\mu_{D_{ij}}(d)$ determines the degree to which it is possible that the shortest distance between vertices i and j is equal to d, $d \in R^+$. For different values of $d \in R^+$ this distance can be generated by different routes — that is why we cannot link the fuzzy number D_{ij} with only one route. Lin and Chern (1993) considered the problem of the shortest route treating it only as the problem of determining the shortest distance between two specified nodes in a network. They applied fuzzy linear programming approach to determine the fuzzy distance D_{ij} with triangular fuzzy arc lengths. Basing themselves on this result, they propose an algorithm for finding one most vital arc in a network, i.e. the arc whose removal from the network causes a maximal increase of the shortest distance between the vertices in question.

On Chanas and Kamburowski approach

Chanas and Kamburowski (1983) applied the Orlovski approach to define a concept of the solution of the fuzzy shortest route problem. We will shortly present this approach. Let us assume that there is given a fuzzy preference (order) relation R on the set FN of fuzzy numbers, with a membership function μ_R,

$\mu_R: FN \times FN \to [0,1]$. $\mu_R(A,B)$, $A, B \in FN$, is interpreted as the degree to which the preference $A \leq B$ is true. Having defined the relation R one can construct the fuzzy strict preference relation R^s as follows (Orlovsky, 1978):

$$\mu_{R^s}(A,B) = \max\{\mu_R(A,B) - \mu_R(B,A),\ 0\} \tag{18}$$

Using the Orlovsky approach, the fuzzy solution of the problem in question is understood as a fuzzy set in the set $P(i,j)$ of the routes which are non dominated with respect to the fuzzy relation R, defined by the membership function

$$\mu_{ND}(p) = 1 - \max_{d \in P(i,j)} \mu_{R^s}(L_d, L_p),\ \ p \in P(i,j) \tag{19}$$

It is of course natural to assume that the best (optimal) solution is the route which belongs to the fuzzy set *ND* to the highest degree. Particularly interesting are those cases in which the highest value of the membership function μ_{ND} equals one. A route $\overline{p} \in P(i,j)$ for which $\mu_{ND}(\overline{p}) = 1$ is called an unfuzzy non dominated solution. The set of unfuzzy non dominated solutions is denoted with *UND*.

If we assume that $\mu_R(A,B) = Poss(A \leq B)$, then, making use of the results presented in Chanas and Zielinski (1997), we can conclude the following theorem:

$$p \in UND \Leftrightarrow \sum_{(k,l)\in p} \underline{l}_{kl} \leq \bar{l}_{\min} \tag{20}$$

where $\bar{l}_{\min}$ stands for the length of the shortest route in a network $S = \langle N, A \rangle$, in which the arcs have been assigned lengths equal to $\bar{l}_{ij}$, $(i,j) \in A$. Thus the value $\bar{l}_{\min}$ and the corresponding route (which is also an element of the set *UND*) can be determined with any classical algorithm solving the shortest route problem.

We will now assume a more complex preference relation introduced by Kolodziejczyk (1986). The membership function of this relation has the following form:

$$\mu(A,B) = \frac{d(A^L \tilde{\vee} B^L, A^L) + d(A^R \tilde{\vee} B^R, A^R) + d(A \cap B, \varnothing)}{d(A^L, B^L) + d(A^R, B^R) + 2d(A \cap B, \varnothing)} \tag{21}$$

where

$$d(A,B) =^{df} \int_R |\mu_A(x) - \mu_B(x)| dx \tag{22}$$

is the measure of the Hamming distance between fuzzy sets, and

$$\mu_{A^L}(x) =^{df} \begin{cases} \mu_A(x) & for\ x \leq z \\ 1 & for\ x \geq z \end{cases} \tag{23}$$

$$\mu_{A^P}(x) =^{df} \begin{cases} \mu_A(x) & for\ x \geq z \\ 1 & for\ x \leq z \end{cases} \tag{24}$$

where z is any number such that $\mu_A(z) = 1$.

Chanas (1987) showed that the unfuzzy non dominated solution of the fuzzy shortest route problem with respect to fuzzy relation (21) is the shortest route in the network $S = \langle N, A \rangle$, in which the arcs have been assigned lengths equal to generative expected values (see Chanas and Nowakowski, 1988) of the fuzzy numbers L_{ij}, $GE(L_{ij})$.

The generative expected value of a fuzzy number $A = (\underline{a}, \overline{a}, \alpha_A, \beta_A)_{L-R}$ coincides with the value of the ranking function on fuzzy numbers, introduced by Yager (1981), and can be defined with the following general formula:

$$GE(A) = \frac{1}{2} \int_0^1 \left(\underline{a} - L^{-1}(t)\alpha_A + \overline{a} + R^{-1}(t)\beta_A \right) dt \tag{25}$$

If A is a triangular fuzzy number, i.e. $\underline{a} = \overline{a}$ and $L(y) = R(y) = \max\{1 - y, 0\}$, $y \in R^+$, then

$$GE(A) = a + \frac{\overline{a} - \underline{a} + \alpha_A + \beta_A}{4} \tag{26}$$

8.3 FUZZY TRANSPORTATION AND ASSIGNMENT PROBLEM

In this section we discuss two fuzzy formulations of the transportation problem: one with fuzziness in the costs coefficients and one with fuzziness in the demand and supply values as well as in the formulation of the goal. In the second case we quote the corresponding algorithm. As far as the fuzzy assignment problem is concerned, the fuzziness concerns the coefficients of the problem and the formulation of the goal. We discuss three different formulations of the problem, together with the corresponding algorithms.

8.3.1 Integer Transportation Problem with Fuzzy Coefficients in the Objective Function

The transportation problem with fuzzy coefficients in the objective function (whose other coefficients, i.e. the demand and supply values, are crisp numbers) can be applied to the frequent cases when the transportation costs are unstable or known only in an inexact form, but the demand and supply is known exactly, determined e.g. by production plans and contracts. Chanas and Kuchta (1996) proposed an algorithm for the following form of the problem:

$$\sum_{i=1}^{n} \sum_{j=1}^{m} \tilde{c}_{ij} x_{ij} \to \min \qquad \underline{x} \in X \tag{27}$$

where:

$$\underline{x} \in X \Leftrightarrow \begin{cases} \sum_{j=1}^{m} x_{ij} = a_i \ (i = 1,\dots,n); \\ \sum_{i=1}^{n} x_{ij} = b_j \ (j = 1,\dots,m); \\ x_{ij} \geq 0 \text{ and integer } (i = 1,\dots,n;\ j = 1,\dots,m) \end{cases}$$

with a_i and b_j $(i = 1,\dots,n, j = 1,\dots,m)$ being given non-negative integer numbers and $\tilde{c}_{ij}$ $(i = 1,\dots,n,\ j = 1,\dots,m)$ given fuzzy numbers of the L-L type (L is a given shape function): $\tilde{c}_{ij} = (\underline{c}_{ij}, \overline{c}_{ij}, \alpha_{ij}, \beta_{ij})_{L-L}$.

Chanas and Kuchta (1996) define the solution of the above problem in the following way[1]:

Definition 1: For two given numbers t_1 and t_2 such that $0 \leq t_1 \leq t_2 \leq 1$, the solution of problem (27) is a fuzzy set in X with the following membership function:

$$\mu_{FS}(\underline{x}) = \left|\left\{t \in (0,1]:\ \underline{x} \text{ is an efficient solution of problem (28) for } \theta = L^{-1}(t)\right\}\right|$$

where |.| stands for the geometric measure and problem (28) is the following one:

$$f_1(\underline{x}) = \sum_{i=1}^{n}\sum_{j=1}^{m}(p_{ij}^1 + r_{ij}^1 \cdot \theta)\cdot x_{ij} \rightarrow \min$$

$$f_2(\underline{x}) = \sum_{i=1}^{n}\sum_{j=1}^{m}(p_{ij}^2 + r_{ij}^2 \cdot \theta)\cdot x_{ij} \rightarrow \min \tag{28}$$

$$\underline{x} \in X,\ \theta \in [0, L^{-1}(0)),$$

where

$$p_{ij}^1 = \underline{c}_{ij} + t_1(\overline{c}_{ij} - \underline{c}_{ij});\ r_{ij}^1 = t_1(\alpha_{ij} - \beta_{ij}) - \alpha_{ij};$$

$$p_{ij}^2 = \underline{c}_{ij} + t_2(\overline{c}_{ij} - \underline{c}_{ij});\ r_{ij}^2 = t_2(\alpha_{ij} - \beta_{ij}) - \alpha_{ij};$$

The authors present an algorithm solving problem (27) according to Definition 1 in the sense of finding the values of μ_{FS} for all basic solutions of (27). This algorithm includes an algorithm for the parametric bicriteria problem (28).

8.3.2 Integer Transportation Problem with Fuzzy Demand and Supply Values and a Fuzzy Goal

This time we assume the supply and demand values to be fuzzy. The transportation costs are supposed to be crisp, but the goal itself is fuzzy. In fact, it can be said that we look for a compromise between two opposite goals: that of minimizing the costs and that of satisfying the demand to a satisfactory degree. The problem, in the form for which an algorithm finding a solution exists, is formulated as follows:

$$
\begin{aligned}
& c(\underline{x}) = \sum_{i=1}^{n}\sum_{j=1}^{m} c_{ij}x_{ij} \rightarrow \widetilde{\min} \\
& \sum_{j=1}^{m} x_{ij} \cong A_i,\ i = 1,\dots,n \\
& \sum_{i=1}^{n} x_{ij} \cong B_j,\ j = 1,\dots,m \\
& x_{ij} \geq 0 \text{ and integer},\ i = 1,\dots,n,\ j = 1,\dots,m
\end{aligned}
\tag{29}
$$

The notation is analogous to the previous section. A_i and B_j are fuzzy numbers of the following form:

$A_i = (\underline{a}_i, \overline{a}_i, \alpha_{A_i}, \beta_{A_i})_{L_i - R_i}$, $B_i = (\underline{b}_i, \overline{b}_i, \alpha_{B_j}, \beta_{B_j})_{S_j - \Gamma_j}$, $i = 1,\dots,n,\ j = 1,\dots,m.$

The unit transportation costs c_{ij}, $i = 1,\dots,n,\ j = 1,\dots,m,$ are assumed to be crisp numbers. The fuzzy number representing the fuzzy goal is denoted by G and takes on the following form:

$$G = (-\infty, c_0, 0, \beta_G)_{L_G - R_G}$$

The solution of the above problem is defined in the following way:

Definition 2: Let $\mu_D(\underline{x}) = \min\{\mu_C(\underline{x}), \mu_G(\underline{x})\}$, where:

$$\mu_C(\underline{x}) = \min\left\{\mu_{A_i}\left(\sum_{j=1}^{n} x_{ij}\right)(i = 1,\dots,m),\ \mu_{B_j}\left(\sum_{i=1}^{m} x_{ij}\right)(j = 1,\dots,n)\right\}$$

$$\mu_G(\underline{x}) = \mu_G(c(x)) = \mu_G\left(\sum_{i=1}^{m}\sum_{j=1}^{n} c_{ij}x_{ij}\right).$$

The solution of problem (29) is such $\underline{x}$ for which function $\mu_D(\underline{x})$ attains the maximal value. If this maximal value is zero, we say that problem (29) is infeasible.

This problem has been considered by Chanas and others (1984) and by Chanas and Kuchta (1997). In each of the papers an algorithm solving problem (29) has been proposed. However, the algorithm from Chanas and others (1984) solves only a special case of (29) and is more difficult to implement. That is why we will concentrate on the algorithm from Chanas and Kuchta (1997).

In order to present the algorithm solving problem (29), we need the following notation:

Definition 3: Let A be an arbitrary interval. The symbol $[A]$ denotes the widest interval having integer ends and contained in A, i.e. $[A]=[a,b]$, where

$$
\begin{aligned}
a &= \min\{t \mid t \in A,\ t \text{ integer}\}, \\
b &= \max\{t \mid t \in A,\ t \text{ integer}\}.
\end{aligned}
$$

We will use the following auxiliary problem (it is defined for any fixed $\lambda>0$):

$$\begin{aligned}
& c(\underline{x}) \rightarrow \min \\
& \sum_{j=1}^{n} x_{ij} \in \left[A_i^{\lambda}\right],\ i = 1,\dots,m \\
& \sum_{i=1}^{m} x_{ij} \in \left[B_j^{\lambda}\right],\ j = 1,\dots,n \\
& x_{ij} \geq 0 \text{ and integer},\ \ i = 1,\dots,m,\ j = 1,\dots,n
\end{aligned} \tag{30}$$

and the following notion of minimal extension:

Definition 4: Problem (30) for $\lambda=\lambda_1$ is a minimal extension of problem (30) for $\lambda=\lambda_2$ when problem (30) for $\lambda=\lambda_1$ is identical to problem (30) for $\lambda=\lambda^*$, where

$$\lambda^* = \max\left\{\max_{1\leq i\leq m,\ t\notin A_i^{\lambda_2}} \mu_{A_i}(t),\ \max_{1\leq j\leq n,\ t\notin B_j^{\lambda_2}} \mu_{B_j}(t)\right\},$$

where t is integer.

Before we proceed to the presentation of the algorithm itself, let us remark that for a fixed λ problem (30) can be transformed to the classical crisp transportation problem (see Chanas and others, 1993) and thus solved by any classical transportation algorithm. And here is the algorithm solving problem (29):

Step 1: Set λ (1):=0, λ (2):=1;

Step 2: Solve problem (30) for $\lambda=\lambda\ (1)$. If the problem is feasible and $c(x(\lambda\ (1))\in G^{\lambda\ (1)}$, then go to step 3. Otherwise STOP — problem (29) is infeasible ($\mu_D(x) = 0$ for all x)

Step 3: Solve problem (30) for $\lambda=\lambda$ (2). If the problem is feasible and $c(x(\lambda\ (2))\in G^{\lambda\ (2)}$, then STOP — $x(\lambda\ (2))$ is the optimal solution of problem (29) and $\mu_D(x(\lambda(2))) = 1$. Otherwise go to step 4.

Step 4: Set λ (half):=(λ (1)+ λ (2))/2 and go to step 5.

Step 5: Solve problem (30) for $\lambda = \lambda\,(half)$. If the problem is infeasible, then set $\lambda\,(2) := \lambda\,(half)$ and go to step 6. Otherwise three cases are possible:

(i) $\mu_G(x(\lambda\,(half)) = \mu_C(x(\lambda\,(half))$; then $x(\lambda\,(half))$ is an optimal solution of problem (29). STOP.

(ii) $\mu_G(x(\lambda\,(half)) > \mu_C(x(\lambda\,(half))$; then set $\lambda\,(1) := \mu_C(x(\lambda\,(half)))$ and go to step 6.

(iii) $\mu_G(x(\lambda(half)) < \mu_C(x(\lambda(half))$; then set $\lambda(2) := \mu_C(x(\lambda(half)))$ or, if $\lambda(2) = \mu_C(x(\lambda(half)))$, then $\lambda(2) := \lambda(half)$. Go to step 6.

Step 6: If $\lambda(2) - \lambda(1) > \varepsilon$, then go to step 4. Otherwise check whether problem (30) for $\lambda = \lambda(1)$ is a minimal extension of problem (30) for $\lambda = \lambda(2)$. If not, then go to step 4. Otherwise STOP - one of the solutions, $x(\lambda(1))$ or $x(\lambda(2))$, is the optimal solution of problem (29). If $\mu_G(x(\lambda(2) < \mu_C(x(\lambda(2)$, then $x(\lambda(1))$ is an optimal solution of problem (29).

Number ε is given by the user. It seems that it should not be greater than 0.1 and not smaller than 0.05.

8.3.3 Fuzzy Assignment Problem

Let us begin this section with a general presentation of the fuzzy assignment problem.

A fuzzy assignment problem implies that a matrix $\tilde{A} = \left[\tilde{a}_{ij}\right]_{i=1,\ldots,n}^{j=1,\ldots,m}$ is given, whose elements represent fuzzy evaluations of individual assignments, e.g. fuzzy evaluations of the effectiveness of the *i*th worker on the *j*th machine. The membership functions μ_{ij} of the numbers $\tilde{a}_{ij}$ are defined on a space U of grades in which we evaluate the effectiveness of the workers on individual machines.

The solution of the fuzzy assignment problem is a permutation $P = (p_1, p_2, \ldots, p_n)$ of the set $\{1,2,\ldots,n\}$. The interpretation of this solution is that the *i*th (i=1,2,...,n) worker is to work on the p_i-th machine. The choice of the permutation depends of course on the criterion assumed .

There are three formulations of the fuzzy assignment problem for which there exist algorithms solving it. The first formulation is the following one:

$$\sum_{i=1}^{n} \tilde{a}_{ip_i} \xrightarrow{P} \max$$

where the numbers $\tilde{a}_{ij}$ are fuzzy numbers of the type L-L (L is a shape function).

This problem can be solved by transforming the fuzzy assignment problem to the integer transportation problem with fuzzy coefficients in the objective function (this transformation is well known, the resulting transportation problem will have supply and demand values equal to one and its objective function coefficients will be $\tilde{a}_{ij}$) and applying the corresponding fuzzy transportation algorithm.

Now we will present two other formulations of the assignment problem, considered by Chanas and Kokalanov (1980). In these forms of the assignment problem it is assumed that the space U is finite ($U = \{u_1, u_2, \ldots, u_m\}$) and that a fuzzy number G, with a membership function μ_G defined on U, represents the fuzzy goal — the desired overall effectiveness.

With these assumptions the following problem is solved:

$$\max_{u \in U}\left[\left(\mu_{1p_1}(u) * \mu_{1p_1}(u) * \ldots * \mu_{np_n}(u)\right) * \mu_G(u)\right] \xrightarrow{p} \max$$

where the operation * can stand for the algebraic product or for the intersection of fuzzy numbers.

This problem consists in finding such a permutation which gives a fairly satisfactory overall effectiveness (which is measured by function μ_G) with the individual effectiveness being possibly close to each other.

The algorithms (one for each of the possible meanings of the operation *) make use of two known algorithms for the crisp classical assignment problem and bottleneck assignment problem.

In both crisp problems the elements of the matrix A are crisp ($A = \left[a_{ij}\right]_{i=1,\ldots,n}^{j=1,\ldots,m}$).

The classical assignment problem is the following one:

$$\sum_{i=1}^{n} a_{ip_i} \xrightarrow{p} \max \tag{31}$$

Algorithms solving this problem are well known.

The bottleneck assignment problem is the following one:

$$\min_{i=1,2,\ldots,n} a_{ip_i} \xrightarrow{p} \max \tag{32}$$

An algorithm for this problem is presented in Slominski (1977).

Let us return to the fuzzy assignment problem:

*Algorithm for * denoting the algebraic product:*

Step 1: For each $k = 1,2\ldots,m$ solve problem (31) assuming $a_{ij} = \log \mu_{ij}(u_k)$. Denote the respective solutions with P_k;

Step 2: Among the permutations P_k ($k = 1,2\ldots,m$) choose permutation $P = P_r$ such that P_r satisfies the following condition:

$$w(u_r)\mu_G(u_r) = \max_{k=1,2,\ldots,m}\left[w(u_k)\mu_G(u_k)\right]$$

where

$$w(u_k) = \exp\left[\sum_{i=1}^{n} \log \mu_{ip_i}(u_k)\right] \quad (k = 1,2,\ldots,m)$$

*Algorithm for * denoting the intersection operation:*

Step 1: For each k = 1,2...,m solve problem (32) assuming $a_{ij} = \mu_{ij}(u_k)$. Denote the respective solutions with P_k;

Step 2: Among the permutations P_k (k = 1,2...,m) choose permutation $P = P_r$ such that P_r satisfies the following condition:

$$\min[z(u_r), \mu_G(u_r)] = \max_{k=1,2,\ldots,m} \min[z(u_k), \mu_G(u_k)]$$

where

$$z(u_k) = \min_{i=1,2,\ldots,n} \mu_{ip_i}(u_k) \ (k = 1,2,\ldots,m)$$

8.4 FUZZY NETWORK PLANNING

The network planning problem of a project in which the activity duration times are given in the imprecise form of fuzzy numbers has been the subject of a very rich collection of papers, e.g.: Chanas and Kamburowski (1981), Kamburowski (1983), Chanas (1987), Buckley (1989), Rommelfanger (1994), Chang and others (1995). In addition to the temporal analysis of the project with fuzzy activity durations Hapke and Slowinski (1993,1995), Hapke at al. (1993,1997), have considered constrained resources of different categories, multiple performing modes of activities and multiple criteria of project performance.

Here we have to confine ourselves to the presentation of the most important results included in the first three papers from the above mentioned ones.

In the following we assume that a project can be represented as a directed compact and acyclic network $S = \langle N, A \rangle$. Assume that $N = \{1,2,\ldots,n\}$ is the set of nodes (events), 1 — a single start node, n — a single terminal node, and $A \subset N \times N$ is the set of arcs (activities). The events of the project are of the „conjunction" type, i.e. all the activities starting from a fixed node i are to be executed and their execution can be initiated at the time when all the activities entering node i are completed. For convenience we assume that the events are labelled from 1 to n in such a way that $i<j$ for each activity $(i, j) \in A$. There is a fuzzy number associated with each activity $(i, j) \in A$: $T_{ij} = (\underline{t}_{ij}, \bar{t}_{ij}, \alpha_{ij}, \beta_{ij})_{L-R}$, the fuzzy duration time of activity (i, j).

8.4.1 A Generalization of the CPM Method — Chanas and Kamburowski Approach

Determination of the earliest time at which a project can be completed is one of the main problems of network analysis. A natural approach to this problem consists in a direct extension of Ford's algorithm used in *CPM* (Critical Path Method) by replacing the addition and maximum operations in the algorithm with proper extended operations, i.e.

$$T_i = \tilde{\vee}_{k \in P(i)} (T_k \tilde{+} T_{ki}), \quad i = 2,3,\ldots,n \tag{33}$$

where $P(i) = \{k \in N | (k,i) \in A\}$ and T_1 is the crisp number equal to zero, $T_1 = (0,0,0,0)_{L-R}$. T_i is the earliest (fuzzy) time at which event i may occur, i=1,2,...,n. If in representing the duration times in the project we limit ourselves to fuzzy numbers of the same type *L-R*, then the addition operation in the above formula can be carried out easily and exactly. On the other hand, it is not possible to carry out the operation $\tilde{\vee}$ in an exact way — the result of this operation does not have to be a number of the *L-R* type. Dubois and Prade (1978) proposed the following approximate formula for the operation $\tilde{\vee}$ executed on fuzzy numbers with a single modal value $A = (a,a,\alpha_A,\beta_A)_{L-R}$ and $B = (b,b,\alpha_B,\beta_B)_{L-R}$:

$$A \tilde{\vee} B = \begin{cases} (a \vee b, a \vee b, \alpha_A \wedge \alpha_B, \beta_A \vee \beta_B)_{L-R} & \text{if } a \approx b \\ (b,b,\alpha_B,\beta_B)_{L-R} & \text{if } a << b \end{cases} \tag{34}$$

A disadvantage of the above formula is the use of relations $a \approx b$ and $a<<b$, which are not defined unequivocally. Its results can be far from being exact.

In Chanas and Kamburowski (1981) another way of applying formula (33) is presented. In the method given there the properties of the r-cuts of fuzzy numbers are utilised. The r-cut of fuzzy number T_{ij}, $r \in (0,1]$, is the interval of the following form:

$$T_{ij}^r = \left[\underline{t}_{ij}^r, \bar{t}_{ij}^r\right] = \left[\underline{t}_{ij} - L^{-1}(r)\alpha_A, \bar{t}_{ij} + R^{-1}(r)\beta_A\right] \tag{35}$$

Since the extended operations preserve the interval operations on the r-cuts of the arguments, the following formula gives the precise result for any $r \in (0,1]$:

$$T_i^r = \left[\underline{t}_i^r, \bar{t}_i^r\right] = \left[\vee_{k \in P(k)} (\underline{t}_k^r + \underline{t}_{ki}^r), \vee_{k \in P(k)} (\bar{t}_k^r + \bar{t}_{ki}^r)\right] \quad i = 2,3,\ldots,n \tag{36}$$

In order to be able, in a general case, to identify the membership functions μ_{T_i}, $i = 1,2,\ldots,n$ in a precise way, one has to apply formula (36) for many values of r. It seems that knowledge of the r-cuts of T_i for a few chosen values of r (e.g., r=0.1, 0.2, ...,0.9, 1) will be sufficient for the decision maker.

The formula (33) is a natural generalisation of the one used in *CPM* to determine the earliest times of events. On the other hand, it does not seem sensible to use a second formula used in *CPM* — the one which is executed "backwards" and determines the latest times of events, needed in order to find the slacks of the activities and by means of these slacks the critical activities. The reason is that in this formula we start from the earliest project completion time, which in our case is a "strongly" fuzzy number. A direct use of the formula would cause a considerable increase in the range of fuzziness of the times being calculated. For this reason Kamburowski (1983) proposed a trick allowing to avoid this disadvantage (see also Chanas, 1982). He suggested a formula calculating, also backwards from the end of the network, the "earliest" times of events, in which again only generalised operations of addition and maximum are used and there is no need to introduce further operations, those of subtracting and minimum, which do not preserve the number types. Moreover, the calculations according to this formula start with a crisp number equal zero. And here is this formula:

$$\overline{T}_i = \tilde{\vee}_{k \in S(i)} (\overline{T}_k \tilde{-} T_{ik}), \quad i = n-1, n-2, \ldots, 1 \tag{37}$$

where $S(i) = \{k \in N | (i,k) \in A\}$ and $\overline{T}_n$ is the crisp number equal to zero, $\overline{T}_n = (0,0,0,0)_{L-R}$. The formula (37) can be executed, just like earlier formula (33), by computing r-cuts of $\overline{T}_i$. It is true that the times $\overline{T}_i$, $i = 1,2,\ldots,n$, themselves are not useful for the decision maker, but they can be used in calculating the criticality degree of events and activities, according to the following definitions.

Definition 5: Fuzzy set NC on the set of vertices N with the membership function

$$\mu_{NC}(i) = Poss\left(T_i \tilde{+} \overline{T}_i \text{ } is \text{ } T_n\right) \tag{38}$$

is called fuzzy set of critical events.

Definition 6: Fuzzy set AC on the set of activities A with the membership function

$$\mu_{AC}(i) = Poss\left(T_i \tilde{+} T_{ij} \tilde{+} \overline{T}_i \text{ } is \text{ } T_n\right) \tag{39}$$

is called fuzzy set of critical activities

Definition 7: Fuzzy set PC on the set of paths $P(1,n)$ with the membership function

$$\mu_{PC}(p) = Poss\left(T_p \text{ } is \text{ } T_n\right) \tag{40}$$

is called fuzzy set of critical paths in a project.

8.4.2 Fuzzy PERT

The approach presented above could be called fuzzy *CPM*, because it is a direct extension of the *CPM* method to the case of fuzzy activity times. Chanas (1987) proposed a certain approach, which here will be called fuzzy *PERT*, because it makes a substantial use of the *PERT* method, destined for analysing projects with random activity times. What does this approach consist in? For the authors of the first version of the *PERT* method (Malcom and others, 1959) it was natural to assume the beta distribution for the activity duration times, because this distribution seemed to "suit" in the best way the three estimations $a \le m \le b$ given by experts for each activity (the optimistic, the most probable and the pessimistic duration of an activity). The formulae for the expected values and variances are a mere consequence of this assumption:

$$t_e = \frac{1}{3}\left[2m + \frac{1}{2}(a+b)\right], \quad \sigma^2 = \frac{(b-a)^2}{36}, \tag{41}$$

The mean duration times and variances of activities calculated by means of the above formulae are used in calculating an (approximate) expected value and variance of the project completion time, where by the way the *CPM* method is applied. By reference to the limit theorems of the probability theory, it is assumed that the project completion time is a random variable with a normal distribution.

Let us return now to the approach proposed by Chanas (1987). Following a similar way of thinking as above, let us assume that we have at our disposal estimations of the activities duration times, formulated on the base of experts opinion in the form of fuzzy numbers T_{ij}, $(i,j) \in A$. The next natural step consists in replacing each fuzzy activity duration time T_{ij}, $(i,j) \in A$, with a random variable of a probability distribution which would in some sense coincide with this duration time. The following random variable, $X_{T_{ij}}$, may be associated with T_{ij} (see Chanas and Nowakowski, 1988):

$$\begin{aligned} X_{T_{ij}} &= \underline{t}_{ij}^T + S\left(\bar{t}_{ij}^T - \underline{t}_{ij}^T\right) = \\ &= \underline{t}_{ij} - L^{-1}(T)\alpha_{ij} + S\left[\bar{t}_{ij} - \underline{t}_{ij} + L^{-1}(T)\alpha_{ij} + R^{-1}(T)\beta_{ij}\right] \end{aligned} \tag{42}$$

where T and S are independent random variables uniformly distributed over (0,1] interval. The mean value and variance of the random variable $X_{T_{ij}}$, called in Chanas and Nowakowski (1988) respectively the generative mean value $GE(T_{ij})$ and the generative variance $GVar(T_{ij})$ of the fuzzy number T_{ij}, in case of T_{ij} being triangular fuzzy numbers, i.e. $T_{ij} = (t_{ij}, t_{ij}, \alpha_{ij}, \beta_{ij})_{L-R}$, where $L(y) = R(y) = \max\{1-y, 0\}$, $y \in R^+$, are expressed by the following formulae:

$$GE(T_{ij}) = t_{ij} + \frac{\beta_{ij} - \alpha_{ij}}{4} \tag{43}$$

$$GVar(T_{ij}) = \frac{7\alpha_{ij}^2 + 7\beta_{ij}^2 + 2\alpha_{ij}\beta_{ij}}{144} \tag{44}$$

It is of course possible to link other random values with the fuzzy activity duration times T_{ij}, which would lead to other formulae for the expected value and variance, i.e. another version of the fuzzy *PERT* method.

8.5 FUZZY SCHEDULING ON MACHINES

Scheduling problems constitute a large class of problems in which different jobs are to sequenced and scheduled on various machines, additional resources are to be assigned to the jobs, while a set of constraints has to be observed and certain objectives attained. This class is so rich that we had to confine ourselves to a survey of formulations of the problem, only indicating where corresponding algorithms can be found.

8.5.1 The Fuzzy n-job m-machine Flow Shop Problem

The n-job m-machine flow shop problem consists in processing n jobs through m machines. All jobs pass through machine 1 first, than machine 2 and so on up to machine m. A job is processed on one machine at a time without preemption, and a machine processes no more than one job at a time. The processing times p_{ij} (i=1,...,n, j=1,...,m) of the ith job on the jth machine are given. The due dates dd_i (i=1,...,n) for each job may be given too. Let us denote with C_{ij} (i=1,...,n, j=1,...,m) the completion time of the i th job on the j th machine, and with C_i (i=1,...,n) the completion time of the i th job on the last machine, With the fixed order of processing the jobs through the machines, we have $C_i = C_{im}$ (i=1,...,n).

The goal is to sequence and schedule the jobs such that some criterion is optimized. This criterion may be for example the mean flow time (MFT) or the makespan (M), defined as follows:

$$MFT = \sum_{i=1}^{n} C_i / n, \; M = \max_{i=1,\dots,n} C_i \tag{45}$$

Another criterion may be the minimization of the tardiness, which can be stated in two different ways:

- as minimization of the maximal tardiness, which gives the following objective function:

$$\max_{i=1,\dots,n} (\max(C_i - dd_i, 0)) \to \min$$

- as minimization of the total tardiness, which gives the following objective function:

$$\sum_{i=1}^{n} \max(C_i - dd_i, 0) \to \min .$$

For other criteria used in the flow shop, see e.g. Ishibuchi and others, 1994 b.

An algorithm solving this problem in the crisp case with objective functions (45) is given by Cambwell and others (1970). It consists in formulating a series of auxiliary two machine flowshop problems and solving each of them with the well known algorithm of Johnson (1954). The best solution from among those of the auxiliary problems is accepted as the solution of the initial problem.

We have been talking about the crisp flow shop problem so far. Now we will turn to its fuzzy counterpart.

McCahon and Lee (1992) consider a fuzzy version of the n-job m-machine flow shop problem with objectives (45), in which the processing times are trapezoidal fuzzy numbers $\tilde{p}_{ij}$ (i=1,...,n, j=1,...,m). In this case the mean flow time MFT and the makespan M are fuzzy numbers too

The authors adapt to the fuzzy case the algorithm from Cambwell and others (1970) with the following changes:

- the addition and subtraction of real numbers are replaced with the corresponding fuzzy operations;

- whenever it is to decide which one of two given fuzzy numbers has a higher value, the Lee-Li method is used. This method consists in comparing the generalised mean values (*GMV*) of the respective fuzzy numbers and choosing the fuzzy number with a higher *GMV*. If the two *GMV*s happen to be equal, the spreads *SPR* of the two fuzzy numbers in question are determined and the fuzzy number with a lower spread is chosen as the higher one.

The *GMV* and *SPR* for a given fuzzy number $\tilde{A}$ are defined as follows (*S* stands for the support of $\tilde{A}$, μ_A for its membership function):

$$GMV(\tilde{A}) = \frac{\int_S x\,\mu_A(x)dx}{\int_S \mu_A(x)dx}$$

$$SPR(\tilde{A}) = \left(\frac{\int_S x^2\,\mu_A(x)dx}{\int_S \mu_A(x)dx} - [GMV(A)]^2 \right)^{1/2}$$

The same fuzzy scheduling problem is considered by Tsujumira and Seung (1993), who use another method of ranking fuzzy numbers.

Another way of introducing fuzziness into the *n*-job *m*-machine flowshop problem consists in fuzzifying not the processing times, but the due dates dd_i (i=1,...,n) of individual jobs. This approach is considered by Ishibuchi and others (1994 a and b)

The due dates $\tilde{dd}_i$ (i=1,...,n) are considered to be fuzzy numbers, representing the satisfaction of a decision maker with the completion time of, respectively, the i-th job. These fuzzy numbers, or rather their membership functions μ_{dd_i}, usually take on the following form (dd_i^L and dd_i^U are given positive real numbers):

$$\mu_{dd_i} = \begin{cases} 1 \text{ if } C_i \le dd_i^L \\ 1-(C_i - dd_i^L)/(dd_i^U - dd_i^L) \text{ if } dd_i^L < C_i < dd_i^U \\ 0 \text{ if } dd_i^U \le C_i \end{cases} \quad (46)$$

However, they can also be trapezoidal fuzzy numbers (in a just-in-time production system the earliness of a job is not satisfactory either). Some extensions to non-linear fuzzy numbers have also been considered by the mentioned authors.

Two criteria are considered in the fuzzy flowshop with fuzzy due dates. These criteria constitute a natural generalisation of the tardiness criteria from the crisp case.

- maximization of the minimum grade of satisfaction, which gives the following objective function:

$$\min(\mu_{dd_i}(C_i): i = 1,\ldots,n) \to \max$$

- maximization of the total grade of satisfaction, which gives the following objective function:

$$\sum_{i=1}^{n} \mu_{dd_i}(C_i) \to \max$$

The authors consider also some modifications of the objective functions, whose aim is to improve the performance of solution algorithms. One such modification consists in introducing a tardiness penalty.

The authors present and test several heuristic algorithms solving the problem: multi-start descent, simulated annealing, taboo search, genetic algorithms and combinations of algorithms of different types.

8.5.2 The Fuzzy Open Shop Problem (with Fuzzy Due Dates)

Like the flow shop problem, the n-job m-machine open shop problem consists in processing n jobs through m machines, but here the order in which each of the jobs is to pass through the individual machines is not specified. Like in the flow shop problem, a job is processed on one machine at a time, and a machine processes no more than one job at a time, but preemptions are allowed. Additionally, we assume that it is possible to control the speed of each machines — either once for the whole set of jobs or jobwise.

Like in section 8.5.1, the due dates $\widetilde{dd}_i$ (i=1,...,n) of individual jobs are assumed to be fuzzy numbers, representing the satisfaction of the decision maker with the completion time of, respectively, the i-th job.

In an analogous way as for the flow shop problem, different objective functions connected with the due dates may be considered. Here we confine ourselves to the maximization of the minimum grade of satisfaction, which is equal to

$$\min(\mu_{dd_i}(C_i) : i = 1,\ldots,n) \quad (47)$$

Additionally, the cost of speeding up the machines is also included in the objective function.

Ishii and others (1992) consider this problem in the two machines version. They assume that the speed of each of the machines is controllable, but is set once for the whole set of n jobs. They give an algorithm solving the problem with the following objective function:

$$-g_0(\min(\mu_{dd_i}(C_i) : i = 1,\ldots,n)) + g_1 s_1 + g_2 s_2 \to \min$$

where g_0 evaluates the gain at a unit increase of minimal satisfaction of the decision maker and $g_1(g_2)$ is the cost to speed up the processing of the first (second) machine respectively.

Han and others (1994) treat the one machine version of the problem, allowing the speed of the machines to be controllable jobwise.

8.5.3 The Fuzzy Identical Machines Maximum Lateness Problem

This problem consists in assigning n jobs to m identical parallel type machines. The processing time p_i (i=1,...,n) of each job is given (it does not depend on the machine being used). Moreover, each job has a due date $\widetilde{dd}_{ij}$ (i=1,...,n; j=1,...,m), given in the form of a fuzzy number, representing the satisfaction of the decision maker with the completion time of, respectively, the i-th job on the j-th machine. The due date for a given job depends on the machine being used, which can for example correspond to the situation when the machines are located at various distances from the final receiver.

Ishi and others (1992) present an algorithm solving this problem, i.e. finding an optimal schedule maximizing the minimum satisfaction among all job completion times.

8.5.4 General Fuzzy Job Shop Problems

A job shop problem can be much more complicated than the problems treated above. It can involve technological constraints of the production facilities (machines), certain production requirements in terms of both quantity and quality, various time constraints, precedence relations, more or less strict due dates, data about the vulnerability of machines to failures etc., and all of these constraints and data can be of fuzzy type. The set of machine operations for each job can be completely or partially ordered, possibly in a different way for each job (let us remind that e.g. in the flow shop problem it is completely ordered and is identical for each job). There may be significantly more than just one objective, and these implicit or explicit objectives can be also fuzzy. Very often such problems are solved in interactive procedures which can generate several feasible schedules, out of which the decision maker can choose those satisfying in the best way his or her most important objectives.

To give a more precise example, let us suppose that we have n jobs and m machines. For each job i we have a set $Q_i = \{o^i_p\}_{p=r}^{P}$ of operations which are to be performed on this job, each of the operations corresponding to one of the m machines. Let us suppose that in a certain moment of time operation o^i_p, where i,p are certain integers (i belongs to the interval $[1,n]$, p to the interval $[1,m]$), terminates. We have to make the following decisions:

- what is the next operation from the set Q_i to be performed on the i-th job
- what is the next job i_1 ($i_1 \in [1,n]$) to be performed on machine p;

Rules used in making decisions of the first type are called **routing rules** and those used in making decisions of the second type are called **dispatching rules**.

This problem in its fuzzy version is considered by Custodio and others (1994). The authors admit fuzzy characteristics (general and current, i.e. changing with time) of the jobs and of the machines. These characteristics constitute the fuzzy criteria on the base of which the two decisions mentioned above are made at appropriate moments. The fuzzy characteristics of the machines include: the number of jobs waiting to be processed on a given machine, the vulnerability of the machine to failures, the transportation times, etc. The fuzzy characteristics of the jobs include: processing times, waiting times, slack times, priorities of jobs, etc.

Let $\{A_t\}_{t=1}^{T}$ be the set of alternative choices. The decision criteria (derived from the fuzzy characteristics presented above, e.g. minimizing the vulnerability to failures or minimizing the slack times) are represented by fuzzy sets C_l (l=1,..L) with corresponding membership functions μ_{C_l}, whose domain is the set of alternative choices. For each pair of criteria a fuzzy number a_{lh} (l,h=1,...,L) is defined representing the dominance relation between the criteria l and h. Then for each l=1,..L the weight w_l is calculated (using the geometric mean technique, see (Buckley 1985)), being also a fuzzy number and representing the degree of importance of the criterion C_l. Finally a decision $A^* \in \{A_t\}_{t=1}^{T}$ satisfying the following condition is selected:

$$\min_{l=1,.,L}\left(\mu_{C_l}(A^*)\right)^{w_l} = \max_{t=1,..T}\min_{l=1,.,L}\left(\mu_{C_l}(A_t)\right)^{w_l}$$

where the power operation is defined according to the extension principle of Zadeh.

In Dubois and Prade (1982), Dubois (1989) and Dubois and others (1994) the jobs have fuzzy time windows in which they are to be processed. These fuzzy time windows follow for example from the earliest possible processing times (pessimistic and optimiztic) and from the due dates (for each job there is a due date assuring the highest satisfaction and a due date until which the job has to be finished no matter what). Moreover, the sets Q_i (i=1,..,n) are partially ordered, which means that certain operations have strictly defined predecessors and/or successors. Other technological constraints can be taken into account too.

The routing and dispatching decisions to be taken are formulated pairwise, i.e. in the form: "Is it preferable to choose alternative A_{t_1} or A_{t_2} (t_1,t_2 = 1,...T)?". There are several fuzzy rules which are taken into account in making the decisions. An example of such a rule is: „If the processing time of the operation corresponding to the alternative A_{t_1} is significantly smaller than that corresponding to the alternative A_{t_2}, choose A_{t_1}". The authors introduce also a measure of possibility of operation A_{t_2} succeeding A_{t_1}. This measure, based on the time windows constraints, sometimes indicates that a certain order of operations is not possible (because with this order no feasible schedule could be constructed), sometimes says that a certain order is more possible than another one ("more possible" means: giving a schedule with more satisfactory starting and completion times of the individual jobs) etc. Consequently, another rule says: "If it is significantly more possible to schedule A_{t_2} after A_{t_1} then A_{t_3}, choose A_{t_2} to follow A_{t_1}". In case of conflicting rules, certain attributes of the latter (a restrictive rule, an advising rule, an index indicating the rule importance) or some (fuzzy) metarules help to decide which rule(s) to obey.

Slany (1996) goes further in this direction, proposing a complex fuzzy scheduling system, whose efficiency has been tested on practical examples. Other latest results can be found in Dubois and others (1995), Hapke and Slowinski (1996) as well as Hapke and others (1997).

Sometimes the job shop problems considered in the literature are not fuzzy per se, but the non-fuzzy criteria or rules occurring in them are aggregated into fuzzy criteria (rules), which makes it possible to take all of them in account. Such approach is presented by Grabot and Geneste (1994).

8.6 OTHER SELECTED PROBLEMS

In this section we consider several fuzzy discrete optimization problems which we could not include in any of the previous ones: the fuzzy set covering problem, a certain verison of fuzzy knapsack problem and the general fuzzy linear 0-1 programming problem.

8.6.1 Fuzzy Set Covering Problem

The fuzzy set covering problem, considered by Zimmermann (1991), is formulated using the following notation.

A classical (non-fuzzy) finite set $S=\{1,...,m\}$ is given, as well as n fuzzy sets $\tilde{S}_1$, $\tilde{S}_2,..., \tilde{S}_n$. The membership functions ###$_j$ of, respectively, the sets $\tilde{S}_j$ ($j=1,...,n$), are defined on set S ($\mu_j : S \rightarrow [0,1]$).

Definition 8: A fuzzy covering $\tilde{\wp}$ of set S is a fuzzy set $\tilde{\wp}$ with membership function $\mu_\wp : \{1,...,n\} \rightarrow [0,1]$, where $\mu_\wp(j)$ denotes the degree of membership of set $\tilde{S}_j$ in the fuzzy covering $\tilde{\wp}$.

Let f_j ($j=1,...,n$) be real-valued functions defined on the interval [0,1] and denoting the cost connected with using set $\tilde{S}_j$ in a fuzzy covering $\tilde{\wp}$ with a given membership degree in this covering. Let b_i ($i=1,...,m$) be given numbers from the interval [0,1] expressing the desired minimal degree of covering of the element i of set S. The fuzzy set covering problem can be formulated in two following ways:

Formulation 1:
Find $\mu_\wp(j)$ for $j = 1,...,n$, such that:

$$\sum_{j=1}^{n} f_j\left(\mu_\wp(j)\right) \rightarrow \min, \quad \max_{1 \le j \le n} \min(\mu_j(i), \mu_\wp(j)) \ge b_i, \; i = 1,...,m.$$

Formulation 2:
Find $\mu_\wp(j)$ for $j = 1,...,n$, such that:

$$\max_{1\le j\le n} f_j\left(\mu_{\wp}(j)\right) \to \min\,, \quad \max_{1\le j\le n} \min(\mu_j(i), \mu_{\wp}(j)) \ge b_i\,, \; i = 1,\ldots,m\,.$$

The fuzzy set covering problem in the 1. formulation is NP-hard, but for its second formulation there exists, under some not very constraining assumptions about functions f_j (j=1,...,n), a polynomial solution algorithm, presented in Zimmermann (1984).

8.6.2 Fuzzy Multiple Choice Knapsack Problem

In this section we will refer only to one type of fuzzy numbers: to the triangular ones. For a triangular fuzzy number $\tilde{a}$ the symbols a^L and a^R will stand, respectively, for the lower and upper bound of its support.

The following definition of a partial order between fuzzy numbers $\tilde{a}$ and $\tilde{b}$ will be needed:

Definition 9: $\tilde{a} \prec \tilde{b} \Leftrightarrow \tilde{a} \prec_L \tilde{b}$ and $\tilde{a} \prec_R \tilde{b}$

where $\tilde{a} \prec_R \tilde{b} \Leftrightarrow P\left(\tilde{a} \le \tilde{b}\right) = 1$ and $\tilde{a} \prec_L \tilde{b} \Leftrightarrow P\left(-\tilde{b} \le -\tilde{a}\right) = 1$

with $$P\left(\tilde{a} \le \tilde{b}\right) = \frac{\int_{a^L}^{a^R} \min\{\mu_a(x), \mu_b(x)\} dx}{\int_{a^L}^{a^R} \mu_a(x) dx}$$

The multiple choice knapsack problem, considered by Okada and Gen (1994), is formulated in the following way:

$$\sum_{t=1}^{p} \sum_{j=1}^{n_t} \tilde{c}_{tj}\, x_{tj} \to \min$$

such that

$$\sum_{t=1}^{p} \sum_{j=1}^{n_t} \tilde{a}_{tj}\, x_{tj} \le \tilde{d}$$

$$\sum_{j=1}^{n_t} x_{tj} = 1, \quad t = 1,2,\ldots,p$$

$$x_{tj} \in \{0,1\}, \quad j = 1,2,\ldots,n_t\,,\; t = 1,2,\ldots,p$$

where the coefficients $\tilde{c}_{tj}, \tilde{a}_{tj}, \tilde{d}$ $(t = 1,2,\ldots,p;\; j = 1,2,\ldots,n_t)$ are non-negative triangular fuzzy numbers. They are assumed to fulfil the following, additional conditions about reciprocal comparability within groups corresponding to the mutually disjoint variables ($\{x_{tj}\}_{j=1}^{n_t}$ for $t = 1,2,\ldots,p$):

$$\tilde{c}_{ti} \succ \tilde{c}_{tj} \text{ or } \tilde{c}_{ti} \prec \tilde{c}_{tj} \text{ for any } i \neq j, \quad t = 1,2,\ldots,p$$

and

$$\tilde{a}_{ti} \succ \tilde{a}_{tj} \text{ or } \tilde{a}_{ti} \prec \tilde{a}_{tj} \text{ for any } i \neq j, \quad t = 1,2,\ldots,p$$

Okada and Gen (1994) propose an approximate algorithm solving the problem formulated above. This algorithm constitutes a modification and extension of a greedy algorithm for the non-fuzzy multiple choice knapsack problem proposed by Gen (1990). In the course of the algorithm the objective function is improved (in the sense of the order relation defined in Definition 9). Finally the decision maker is presented with a collection of solutions, whose values of the objectives functions cannot be improved by the algorithm and are incomparable in the sense of the above order relation.

8.6.3 General Fuzzy 0-1 Linear Problems

Zimmermann and Pollatschek (1984) propose an algorithm solving the general fuzzy 0-1 linear problem. The formulation of a fuzzy problem does not have to distinguish between fuzzy objectives and fuzzy constraints: the objective in a fuzzy problem is to maximize the satisfaction of both the constraints and the objective functions. For this reason the problem is formulated in the following way:

$$\min_{i=1,\ldots,m} \left\{ \mu_i \left(\sum_{j=1}^{n} a_{ij} x_j \right) \right\} \to \max \tag{48}$$

$$x_j \in \{0,1\}, \quad j = 1,2,\ldots,n;$$

where the membership functions μ_i are defined on the set of real numbers and take on the following form:

$$\mu_i(x) = \begin{cases} 1 \text{ if } x \leq d_i \\ 1 - \left(\dfrac{x - d_i}{e_i} \right) \text{ if } d_i \leq x \leq d_i + e_i \\ 0 \text{ if } x \geq d_i + e_i \end{cases}$$

The coefficients a_{ij} (i=1,...,m; j=1,....,n) and c_i, d_i (i=1,...,m) are real numbers.

Solving the following crisp linear 0-1 programming problem gives the solution of the fuzzy one (the value of the objective function of problem (48) will be equal to λ):

$$\lambda \to \max$$
$$\lambda \le 1 - \frac{\sum_{j=1}^{n} a_{ij} x_j - d_i}{e_i}, \quad i = 1,\dots,m \tag{49}$$
$$x_j \in \{0,1\}, \quad j = 1,\dots,n; \quad 0 \le \lambda \le 1$$

A branch-and-bound algorithm solving the above problem and presented by Zimmermann and Pollatschek (1984) is based on (Taja, 1975). Here we will present its general scheme.

For the sake of simplicity let us denote: $\bar{a}_{ij} = \frac{a_{ij}}{e_i}$; $\bar{d}_i = \frac{d_i}{e_i}$. J will denote an ordered set containing the indices of those variables x_i which have already been assigned a value. If this value is 1, the index i is contained in J with the sign "+", if this value is zero, the index i has in J the sign "-". The order of the elements will be determined by the algorithm: each new element will be added on the right hand side of the current sequence.

Step 1: Set L= -1; J:=∅. Create the first node of the tree, corresponding to the set J;

Step 2: Calculate

$$BL(J) := \min_{i=1,\dots,m} \left[\left(\bar{d}_i - \sum_{\substack{j \in J, \\ j>0}} \bar{a}_{ij} \right) - \sum_{j \in \{1,2\dots,n\} - J} \min\left(0, \bar{a}_{ij}\right) \right] \tag{50}$$

If $BL(J) \le L$, go to Step 3; otherwise go to step 5;

Step 3: Fanthom the current branch (i.e. mark it as terminated) and backtrack, that is look for a node on a branch which has not been fanthomed yet[2]. If there is such a node, update the set J and go to Step 2, otherwise go to Step 4;

Step 4: STOP. If L= -1, problem (48) is infeasible (λ=0 in problem (49)), otherwise the current solution is the solution of problem (48) and λ=L+1;

Step 5: Check if the following statement is true: For each $j \in \{1,2,\dots,n\}$-J $BL(J \cup \{+j\}) \le L$ and $BL(J \cup \{-j\}) \le L$. If so, go to Step 3;

Step 6: Check if the following statement is true: For each $j \in \{1,2,\dots,n\}$-J $BL(J \cup \{+j\}) > L$ and $BL(J \cup \{-j\}) > L$. If so, go to Step 9;

Step 7: For each $j \in \{1,2,...,n\}-J$ do:
If $BL(J \cup \{+j\}) > L$ and $BL(J \cup \{-j\}) \leq L$ then create two new nodes, emanating from the current one, corresponding to the sets $J \cup \{+j\}$ and $J \cup \{-j\}$, go on to the first one and set $J := J \cup \{+j\}$;

Step 8: For each $j \in \{1,2,...,n\}-J$ do:
If $BL(J \cup \{+j\}) \leq L$ and $BL(J \cup \{-j\}) > L$ then create two new nodes, emanating from the current one, corresponding to the sets $J \cup \{+j\}$ and $J \cup \{-j\}$, go on to the second one and set $J := J \cup \{-j\}$. Go to step 10;

Step 9: Select $j \in \{1,2,...,n\}-J$ for which $|BL(J \cup \{+j\}) - BL(J \cup \{-j\})|$ is minimal;
If such j is not unique, select the one for which $\max\{BL(J \cup \{+j\}), BL(J \cup \{-j\})\}$ is minimal. If j is still not unique, select one j from among those fulfilling the two previous criteria using the following, third criterion: "$\min\{BL(J \cup \{+j\}), BL(J \cup \{-j\})\}$ is minimal". For j selected in this way create two new nodes, emanating from the current one, corresponding to the sets $J \cup \{+j\}$ and $J \cup \{-j\}$. If $BL(J \cup \{+j\}) > BL(J \cup \{-j\})$, go on to the node corresponding to the set $J \cup \{+j\}$ and set $J := J \cup \{+j\}$, otherwise go to the second node and set $J := J \cup \{-j\}$;

Step 10: If J contains less than n elements, go to step 2. Otherwise a new current solution $\{x_j\}_{j=1}^{n}$ has been found. Calculate the new value of L corresponding to this solution:

$$L = \min_{i=1,...,m} \left[\bar{d}_i - \sum_{j=1}^{n} \bar{a}_{ij} x_j \right]$$

If $L=0$, go to Step 4, otherwise go to Step 3.

Notes

1. In fact, the definition introduced in that paper is somewhat more general. It constitutes an extension — to the fuzzy case — of any concept of the solution of a transportation problem with interval coefficients in the objective function. However, the proposed algorithm is limited to the special case we quote here.
2. Backtracking is described in detail by Taja (1975).

References:

Bellman R.E. and Zadeh L.A. (1970), Decision-making in a fuzzy environment. *Management Sci.*, 17, 141-164.
Boffey T.B. (1982), *Graph Theory in Operations Research*, Macmillan, London.
Buckley J.J. (1985), Fuzzy Hierarchical Analysis, *Fuzzy Sets and Systems*, 17, 233-247.

Buckley J.J. (1989), Fuzzy PERT, in: *Applications of Fuzzy Set Methodologies in Industrial Engineering*, ed. by Evans G.W., Karwowski W. and Wilhelm M.R., Elsevier; Amsterdam-Oxford-New York-Tokyo.

Campbell H.G., Dudek R.A. and Smith M.L. (1970), A heuristic algorithm for the *n*-job *m*-machine sequencing problem, *Management Science* 16, 630-637.

Chang I.S., Tsujimura Y., Gen M. and Tozawa T. (1995), An efficient approach for large scale project planning based on fuzzy Delphi method, *Fuzzy Sets and Systems*, 76, 277-288.

Chanas S. (1982), Fuzzy sets in few classical operational research problems, in: Gupta M.M. and Sanches E. (eds), *Approximate Reasoning in Decision Analysis*, North-Holland Publishing Comp., 351-363.

Chanas S. (1987), Fuzzy optimization in networks, in: Kacprzyk J. and. Orlovski S.A (eds.), *Optimization Models Using Fuzy Sets and Possibility Theory*, D. Reidel Publishing Comp., 303-327.

Chanas S. (1989), Parametric techniques in fuzzy linear programming problems, in Verdegay J.L and Delgado M. (eds.), *The Interface between Artificial Intelligence and Operations Research in Fuzzy Environment*, Verlag TUV Rheinland, ISR Series 95, 105-116.

Chanas S., Delgado M., Verdegay J.L. and Vila M.A. (1993), Interval and fuzzy extensions of classical transportation problems, *Transportation Planning and Technology* 17, 203-218.

Chanas S. and Kamburowski J. (1981), The use of fuzzy variables in PERT, *Fuzzy Sets and Systems*, 1, 11-19.

Chanas S. and Kokalanov M. (1980), Assignement problem with fuzzy estimates of effectiveness, *Zastosowania Matematyki (Applicationes Mathematicae)*, XVII, 87-94.

Chanas S. and Kolodziejczyk W. (1982), Maximum flow in a network with fuzzy arc capacities, *Fuzzy Sets and Systems*, 8, 165-173.

Chanas S. and Kolodziejczyk W. (1984), Real-valued flows in a network with fuzzy arc capacities, *Fuzzy Sets and Systems*, 13, 139-151.

Chanas S., Kolodziejczyk W. and Machaj A. (1984), A fuzzy approach to the transportation problem, *Fuzzy Sets and Systems* 13, 211-221.

Chanas S. and Kolodziejczyk W. (1986), Integer flows in network with fuzzy capacity constraints, *Networks*, 16, 17-31.

Chanas S. and Kuchta D. (1996), A concept of the optimal solution of the transportation problem with fuzzy cost coefficients, *Fuzzy Sets and Systems*, 82, 299-305.

Chanas S. and Kuchta D. (1997), Fuzzy Integer Transportation Problem, to appear *in Fuzzy Sets and Systems*.

Chanas S. and Machaj, A. (1985), A parametric version of the out-of-kilter method and its application to the fuzzy network flow problem, *Wissenschaftliche Berichte der Technischen Hochschule Leipzig*, Heft 17, 8-18.

Chanas S. and Nowakowski, M. (1988), Single simulation of fuzzy variable, *Fuzzy Sets and Systems*, 25, 43-57.

Chanas S. and Zielinski P. (1997), Unfuzzy Non Dominated Solutions in the Linear Programming Problem with Fuzzy Coefficients in the Objective Functions, *The Journal of Fuzzy Mathematics*, 5, 115-131.

Custódio L.,M.,M., Sentieiro J.J.S., Bispo C.F.G. (1994), Production Planning and Scheduling Using a Fuzzy Decision System, *IEEE Transactions on Robotics and Automation*, 10, 160-168.

Dubois D. (1989), Fuzzy Knowledge in an Artificial Intelligence System for Job-Shop Scheduling, in: *Applications of Fuzzy Set Methodologies in Industrial Engineering*, ed. by Evans G.W., Karwowski W., Wilhelm M.R.; Elsevier; Amsterdam-Oxford-New York-Tokyo.

Dubois D., Fargier H., Prade H. (1994), The Use of Fuzzy Constraints in Job-Shop Scheduling, Proceedings of the Conference "*Moving Towards Experts Systems Globally in the 21st Century*", 10-14. Jan 1994, 483-492.

Dubois D., Fargier H., Prade H. (1995), Fuzzy Constraints in Job-Shop Scheduling; *J. of Intelligent Manufacturing*, 6, 215-234.

Dubois D. and Prade H. (1978), Shortest path algorithms with fuzzy data, (in French), *R.A.I.R.O. Op. Res.*, 12, 214-227.

Dubois D. and Prade H. (1980), *Fuzzy Sets and Systems: Theory and Applications*, Academic Press, New York.

Dubois D. and Prade H. (1982), The Advantages of Fuzzy Approach in OR/MS Demonstrated on Two Examples of Resource Allocation Problems, *Progress in Cybernetics and Systems Research*, Hemisphere, Washington, 8, 491-497.

Gen M. (1990)., Reliability Optimization by 0-1 Programming for a System With Several Failure Modes; in: Rai S., Agraval D.P., Eds., *Distributed Computing Network Reliability* (IEEE Comp., Soc. Press, Los Angeles), 252-256.

Grabot B. and Geneste L. (1994), Dispatching Rules in Scheduling: a Fuzzy Approach, *International Journal of Production Research*, 32, 903-915.

Han S., Ishii H., Fujii S. (1994), One machine scheduling problem with fuzzy due dates, *European Journal of Operational Research*, 79, 1-12.

Hapke M. and Slowinski R. (1993), A DSS ressource-constrained project scheduling under uncertainty, *Journal of Decision Systems*, 2, 111-128.

Hapke M., Jaszkiewicz A. and Slowinski R. (1994), Fuzzy project scheduling system for software development, *Fuzzy Sets and Systems*, 67, 101-117.

Hapke M., and Slowinski R. (1996), Fuzzy priority heuristic for project scheduling, *Fuzzy Sets and Systems*, 83, 291-299.

Hapke M., Jaszkiewicz A. and Slowinski R. (1997), Fuzzy project scheduling with multiple criteria, in: *Proc. 6th IEEE International Conference on Fuzzy Systems, FUZZ-IEEE'97*, Barcelona, 1277-1282.

Ishii H. (1992), Fuzzy combinatorial optimization, Japanese *Journal of Fuzzy Theory and Systems*, 4, 111-122.

Ishii H., Tada M., Masuda T. (1992), Two scheduling problems with fuzzy due dates, *Fuzzy Sets and Systems* 46, 339-347.

Ishibuchi H., Yamatoto N., Misaki S., Tanaka H. (1994), Local search algorithms for flowshop scheduling with fuzzy due dates, *International Journal of Production Economics*, 33, 53-66.

Ishibuchi H., Yamatoto N., Murata T., Tanaka H. (1994), Genetic algorithms and neighbourhood search algorithms for fuzzy flowshop scheduling problems, *Fuzzy Sets and Systems* 67, 81-100.

Johnson S.M. (1954), Optimal two- and three-stage production schedules with setup times included, *Naval Research Logistics Quarterly* 1, 61-68.

Kamburowski J. (1983), Fuzzy activity duration times in critical path analysis, *Inter. Symp. on Project Management*, New Delhi, 194-199.

Kolodziejczyk W. (1986), Orlovsky's concept of decision making with fuzzy preference relation — further results, *Fuzzy Sets and Systems* 19, 11-20.

Lin K.-C. and Chern M.-S. (1993), The fuzzy shortest path problem and its most vital arcs, *Fuzzy Sets and Systems*, 58, 343-353.

Malcom D.D., Roseboom J.H., Clarc C.I. and Fazar W. (1959), Application of technique for research and development program evaluation, *Operations Research*, 6, 649-669.

McCahon C.S., Lee E.S. (1992), Fuzzy job sequencing for a flow shop, *European Journal of Operational Research* 62, 294-301.

Okada S., Gen M. (1994), Fuzzy Multiple Choice Knapsack Problem, *Fuzzy Sets and Systems*, 67, 71-80.

Orlovsky S.A. (1978), Decision-making with a fuzzy preference relation, *Fuzzy Sets and Systems*, 1, 155-167.

Prade H. (1980), Operations research with fuzzy data, in: P.P. Wang and S.K. Chang (eds.), *Fuzzy Sets. Theory and Application to Policy Analysis and Information Systems*, Plenum Press, New York, 155-170.

Rommelfanger H.J. (1994), Network analysis and information flow in fuzzy environment, *Fuzzy Sets and Systems*, 67, 119-128.

Slany W. (1996), Scheduling as a fuzzy multiple criteria optimization problem, *Fuzzy Sets and Systems*, 78, 197-222.

Slominski L. (1977), An efficient approach to the bottleneck assignement problem, *Bull. Acad. Polon. Sci., Ser. sci. tech.*, 25, 17-23.

Taja A.H. (1975), *Integer Programming*, Academic Press, New York, 87-94;

Tsujumira Y., Seung Hun Park, In Seong Chang, and Gen M. (1993), An effective method for solving flow shop scheduling problems with fuzzy processing times, *Computers & Industrial Engineering*, 25, 239-242.

Zimmermann H.J. (1987). *Fuzzy Sets, Decision Making and Expert Systems*. Kluwer, Dordrecht.

Zimmermann H.-J., Pollatschek M.A. (1984), Fuzzy 0-1 Linear Programs, *TIMS/Studies in the Management Sciences*, 20, 133-145.

Zimmermann K. (1984), On Max-separable Optimization Problems, *Annals of Discrete Mathematics*, 19.

Zimmermann K. (1991), Fuzzy Set Covering Problem, *International Journal of General Systems*, 20, 127-131.

9 FUZZY DYNAMIC PROGRAMMING

Augustine O. Esogbue
Janusz Kacprzyk

Abstract: We provide a brief review of basic problem classes and developments of fuzzy dynamic programming which is a promising tool for dealing with multistage decision making and optimization problems under fuzziness (under fuzzy constraints on decisions made and fuzzy goals on states attained). We discuss cases of a deterministic, stochastic, and fuzzy state transitions, and of the fixed and specified, implicitly given, fuzzy, and infinite termination times.

9.1 INTRODUCTION

Multistage decision making (DM) problems are omnipresent in the real world. They may be exemplified by seeking an optimal amount of funds to be invested in a number of preselected portfolios over a number of years to attain the highest total return on investment, seeking optimal intermediate stops to fly from one city, called the origin, to another, the destination, to minimize the total flight time, or travel cost, etc.

Their essence, for the purposes of our discussion, may be summarized as follows. First, the problem is dynamic in that some of its characteristics, exemplified by its output, evolve in time as a result of their present values and some decision(s) made. There are some stages, exemplified by time (temporal) instants, discrete or continuous, at which a decision is to be made. These stages can also be locations (spatial), e.g., geographical points (as in the example of seeking the best set of intermediate stops to fly from one city to another) but, to be more specific, we will consider here temporal stages only.

The decision made at a given stage and state induces a change in the output of the process considered (e.g. in the total return on investment or total flight time), hence has clearly an impact on what will happen in the future.

The performance of the process is measured over some planning horizon, and is expressed by an aggregate of partial scores which expresses the performance of the particular stage decision. This aggregate may take on various forms as, e.g, from a pessimistic, "safety-first" minimum to an optimistic maximum, through all intermediate cases exemplified by an (weighted) average. An optimal sequence of decisions or controls at the consecutive stages over the planning horizon (policy) assumed is then sought.

Dynamic programming is a powerful formal apparatus for dealing with a large spectrum of multistage DM problems. Since its inception in the mid-1950's (cf. Bellman, 1957), dynamic programming has become a standard tool in many fields including operations research, systems analysis, engineering, data analysis, control, computer science, etc.

In real world problems, we are faced however with imperfect information (data). We have to deal with, e.g., uncertain, imprecise, vague, etc. data available. In modeling and analyzing problems of this genre, previous works tended to equate all aspects of imperfect information with uncertainty (of a random character). Thus, a multitude of probabilistic models were proposed. This was also the case with the use of dynamic programming.

However, no simple and adequate formal apparatus for handling imprecise data – which may stem, e.g., from the use of natural language by humans – was available until the mid-1960's when Zadeh (1965) proposed fuzzy sets theory.

And, indeed, dynamic programming has been one of the earliest general techniques to which fuzzy sets theory has been applied (Chang, 1969; Bellman and Zadeh, 1970; Esogbue and Ramesh, 1970). These pioneering works have been followed by numerous contributions of both a foundational and applied character. They have been reviewed in many surveys as, e.g., by Esogbue and Bellman (1984), Esogbue, Fedrizzi and Kacprzyk (1988), Kacprzyk (1994), Kacprzyk and Esogbue (1996) as well as in Kacprzyk (1983b, 1997).

Recently, one may notice an increased interest in fuzzy dynamic programming which is certainly implied by more and more applications (cf. Kacprzyk, 1997) some of which will be mentioned in this chapter too.

The purpose of this chapter is to provide a short, readable, and up-to-date survey of fuzzy dynamic programming which would include a review of main problem classes, foundations, developments, and more relevant applications. Due to lack of space, we will assume that the reader is familiar with conventional (nonfuzzy) dynamic programming, though only a basic knowledge is required. To those readers who have not yet been exposed to dynamic programming, we can recommend any book on operations research, control, decision analysis, etc. Such books are available from virtually all major publishers.

Our discussion will basically proceed in a broadly perceived context of decision analysis and operations research, though we have to be aware of the fact that (fuzzy) dynamic programming is a more general problem solving technique which goes well beyond these two fields. For instance, it is crucial for control with applications in all aspects of engineering and medicine, data analysis, computer science, economics, etc.

In Section 9.2 we present the essence of Bellman and Zadeh's (1970) approach to decision making under fuzziness which is our point of departure. Next, we present formulations of various dynamic programming problem classes: in Section 9.3 the basic case of a fixed and specified termination time, in Section 9.4 the case of a fuzzy termination time, and in Sections 9.5 and 9.6 the cases of an implicit and infinite termination times. In Section 9.7 stochastic multistage decision making under fuzzy criteria is examined, while in Section 9.8 the computational complexity of fuzzy dynamic programming is briefly discussed. In Section 9.9 we review main applications, and in Section 9.10 give some concluding remarks.

9.2 MULTISTAGE DECISION MAKING UNDER FUZZINESS

In this section we will briefly present a general framework for the analysis of multistage decision making under fuzziness which is due to Bellman and Zadeh (1970), and which is very convenient for our purposes. First, however, we will discuss its point of departure, that framework for one-stage decision making which is also due to Bellman and Zadeh (1970).

9.2.1 Multistage Decision Making under Fuzziness – Bellman and Zadeh's Approach

Virtually all works related to decision making under fuzziness, including the multistage decision making case, have as a point of departure Bellman and Zadeh's (1970) framework. Its basic elements are: a fuzzy goal G in X, a fuzzy constraint C in X, and a fuzzy decision D in X; X is a (nonfuzzy) space of options (alternatives, variants, decisions, ...).

To show the essence of the approach, assume the situation as in Figure 1. Suppose that the fuzzy goal G is "x should be much more than 6" and the fuzzy constraint C is "x should be about 5", and both are defined as fuzzy sets with piecewise linear membership functions.

The fuzzy constraint C = "about 6" is to be understood as: the numbers between 5 and 7 are certainly *about 6*, those between 2 and 5, and 7 and 10 are *about 6* to some extent (between 0 and 1), and those below 2 and above 10 are certainly not *about 6*; and similarly for the fuzzy goal G = "much more than 6". Notice that the above interpretation of a fuzzy constraint and a fuzzy goal is clearly implied by an aspiration-level-based attitude which is convenient in the context considered.

Evidently, the piecewise linear form of the membership functions is here assumed only for the sake of simplicity, though it is often convenient in decision making to deal with aspiration levels.

Notice that if $f : X \to R$ is a conventional objective (performance) function, then

$$\mu_G(x) = \frac{f(x)}{\sup_{x \in X} f(x)} \tag{1}$$

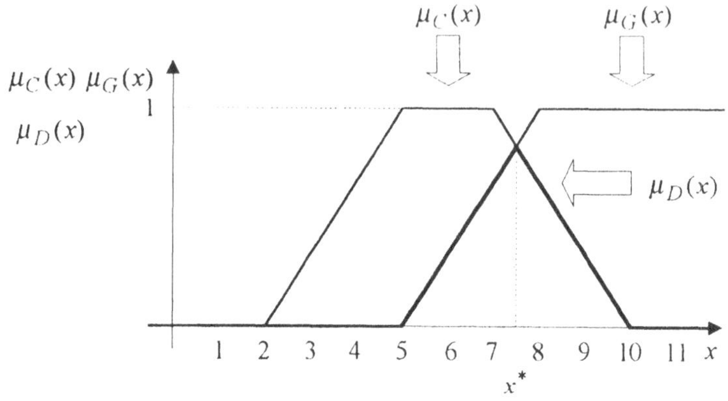

Figure 1 Fuzzy goal, fuzzy constraint, fuzzy decision, and the optimal (maximizing) decision

is a plausible choice provided that $0 \neq \sup_{x \in X} f(x) < \infty$; in this perspective, the fuzzy decision making framework considered may therefore be viewed as a generalization of the conventional (nonfuzzy) one.

Basically, an objective of the form:

"satisfy C and attain G"

will lead to the fuzzy decision

$$\mu_D(x) = \mu_C(x) \wedge \mu_G(x), \forall x \in X \tag{2}$$

which expresses the "goodness" of an $x \in X$ as a solution to the decision making problem considered: from 1 for definitely good (perfect) to 0 for definitely bad (unacceptable), through all intermediate values. The "$a \wedge b = \min(a, b)$" operation is commonly used and is assumed throughout this paper. However, it is by no means the only choice, and may be replaced by, e.g., a t-norm, weighted average or any suitable operation - see Kacprzyk (1983b, 1997) for more detail.

In our example shown in 1, the membership function of the (min-type) fuzzy decision is given in heavy line, and should be interpreted as follows. The set of possible options is the interval $[5, 10]$ because $\mu_D(x) > 0$, for $5 \leq x \leq 10$. The other options, i.e. $x < 5$ and $x > 10$ are impossible since $\mu_D(x) = 0$. However, not all the options belonging to the interval $[5, 10]$ are equally satisfactory (or preferable). The value of $\mu_D(x) \in [0, 1]$ may be meant as the degree of satisfaction from the choice of a particular $x \in X$, from 0 for full dissatisfaction (impossibility of x) to 1 for full satisfaction, though all intermediate values; thus, the higher the value of $\mu_D(x)$, the higher the satisfaction from x.

For an optimal (nonfuzzy) solution to this problem, an $x \in X$ such that

$$\mu_D(x^*) = \sup_{x \in X} \mu_D(x) = \sup_{x \in X} (\mu_C(x) \wedge \mu_G(x)) \tag{3}$$

is a natural (but not the only possible – cf. Kacprzyk, 1997) choice. Notice that in Figure 1, $\mu_D(x) < 1$ which indicates some discrepancy between C and G.

An analogous reasoning may be performed for the case of multiple fuzzy constraints and fuzzy goals, even if defined in different spaces.

Suppose that the fuzzy constraint C is defined as a fuzzy set in $X = \{x\}$, and the fuzzy goal G is defined as a fuzzy set in $Y = \{y\}$. Moreover, suppose that a function $f : X \longrightarrow Y$, $y = f(x)$, is known. Typically, X and Y may be sets of decisions and their outcomes, respectively.

Now, the *induced fuzzy goal* G' in X generated by the given fuzzy goal G in Y is defined as

$$\mu_{G'}(x) = \mu_G[f(x)], \qquad \text{for each } x \in X \tag{4}$$

Notice that by introducing the induced fuzzy goal, both G' and C are defined as fuzzy sets in the same space X which is evidently a *sine qua non* condition.

And now the (min-type) fuzzy decision is

$$\mu_D(x) = \mu_{G'}(x) \wedge \mu_C(x) = \mu_G[f(x)] \wedge \mu_C(x), \qquad \text{for each } x \in X \tag{5}$$

To generalize, if we have n fuzzy constraints defined in X, $C_1, \ldots, C_n$, m fuzzy goals defined in Y, $G_1, \ldots, G_m$, and a function $y = f(x)$, then the (min-type) fuzzy decision is

$$\begin{aligned} \mu_D(x) = \mu_{G_1}[f(x)] \wedge \ldots \wedge \mu_{G_n}[f(x)] \wedge \\ \wedge \mu_{C_1}(x) \wedge \ldots \wedge \mu_{C_m}(x) \end{aligned} \tag{6}$$

for each $x \in X$.

Now we will proceed to the multistage decision making case within the above general framework.

9.2.2 Essentials of Multistage Decision Making (Control) under Fuzziness – Bellman and Zadeh's Approach

We assume and employ here a general framework for dealing with multistage decision making (control) under fuzziness as given in Bellman and Zadeh (1970) and Kacprzyk (1983b, 1997). First, we assume that the dynamics of the problem considered is equated with a deterministic dynamic system described by a state transition equation

$$x_{t+1} = f(x_t, u_t), \qquad t = 0, 1, \ldots \tag{7}$$

where $x_t, x_{t+1} \in X = \{x\} = \{s_1, \ldots, s_n\}$ are the states (equated here, for simplicity but without loss of generality, with outputs) at time (stage) t and $t + 1$, respectively, and $u \in U = \{u\} = \{c_1, \ldots, c_m\}$ is the decision (control or input) at t; the state and decision spaces, X and U, are assumed finite throughout this chapter.

Recall that a stage need not be a time instant though this is assumed here for a better intuitive appeal.

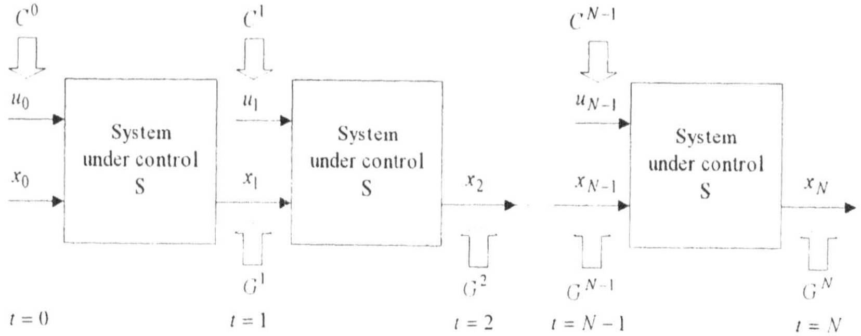

Figure 2 A general framework for multistage decision making (control) under fuzziness

The essence of multistage decision making under fuzziness may then be depicted as in Figure 2. Basically, we start from an initial state at stage (time) $t = 0$, x_0, make a decision at $t = 0$, u_0, attain a state at time $t = 1$, x_1, make a decision u_1, ... Finally, being at stage $t = N - 1$ in state x_{N-1} we make decision u_{N-1} and attain the final state x_N.

The state transitions from state x_t to x_{t+1} under decision u_t are given by (7), the the consecutive decisions applied u_t are subjected to fuzzy constraints C^t, and on the states attained x_{t+1} fuzzy goals G^{t+1} are imposed, $t = 0, 1, \ldots, N-1$.

The performance of the multistage decision making process is evaluated by the fuzzy decision which, assumed to be a decomposable fuzzy set in $U \times X \times \ldots \times U \times X$, is

$$
\begin{aligned}
&\mu_D(u_0, \ldots, u_{N-1} \mid x_0) = \\
&= \mu_C^0(u_0) \wedge \mu_{G^1}(x_1) \wedge \mu_{C^{N-1}}(u_{N-1}) \wedge \mu_{G^N}(x_N) = \\
&= \bigwedge_{t=0}^{N-1} [\mu_{C^t}(u_t) \wedge \mu_{G^{t+1}}(x_{t+1})] \qquad (8)
\end{aligned}
$$

where $x_0 \in X$ is an initial state, the x_t's are given by (7), N is some termination time which is fixed and specified in the basic case.

The problem is to find and optimal sequence of decisions $u_0^*, \ldots, u_{N-1}^*$ such that

$$
\begin{aligned}
&\mu_D(u_0^*, \ldots, u_{N-1}^* \mid x_0) = \\
&= \max_{u_0, \ldots, u_{N-1}} [\mu_C^0(u_0) \wedge \mu_{G^1}(x_1) \wedge \mu_{C^{N-1}}(u_{N-1}) \wedge \mu_{G^N}(x_N)] = \\
&= \max_{u_0, \ldots, u_{N-1}} \bigwedge_{t=0}^{N-1} [\mu_{C^t}(u_t) \wedge \mu_{G^{t+1}}(x_{t+1})] \qquad (9)
\end{aligned}
$$

For simplicity, it is often assumed (also here, if not otherwise specified) that at each stage t, fuzzy constraints are given, $\mu_{C^0}(u_0), \ldots, \mu_{C^{N-1}}(u_{N-1})$, and a fuzzy goal, $\mu_{G^N}(x_N)$, is only imposed on the final state x_N.

Then, the fuzzy decision is

$$\mu_D(u_0, \ldots, u_{N-1} \mid x_0) = \\ = \mu_C^0(u_0) \wedge \ldots \wedge \mu_{C^{N-1}}(u_{N-1}) \wedge \mu_{G^N}(x_N) = \tag{10}$$

and the problem is to find $u_0^*, \ldots, u_{N-1}^*$ such that

$$\mu_D(u_0^*, \ldots, u_{N-1}^* \mid x_0) = \\ = \max_{u_0, \ldots, u_{N-1}} [\mu_C^0(u_0) \wedge \ldots \wedge \mu_{C^{N-1}}(u_{N-1}) \wedge \mu_{G^N}(x_N)] \tag{11}$$

It may readily be seen that the above general problem formulation may be viewed as a starting point for numerous extensions. A convenient classification of problem classes, especially the discrete models, may be done using the following basis:

- The type of the termination time: fixed and specified in advance, fuzzy, implicitly given (by entering a termination set of states), infinite
- The type of the dynamic system: deterministic, stochastic, fuzzy, and fuzzy-stochastic.
- The type of objective function: cost minimization, profit/benefit maximization, and fuzzy criterion set based satisfactory degree maximization.

In virtually all of the foregoing cases a dynamic-programming-type algorithm can be devised. These algorithms will be presented in the sequel.

9.3 FUZZY DYNAMIC PROGRAMMING FOR MULTISTAGE DECISION MAKING WITH A FIXED AND SPECIFIED TERMINATION TIME

The case of a fixed and specified (in advance) termination time is basic. In fact, it provides a suitable point of departure for virtually all fuzzy dynamic programming model extensions. We will therefore discuss it now in some detail considering consecutively a deterministic, stochastic, fuzzy dynamic system, and fuzzy criterion set dynamic program.

9.3.1 The Case of a Deterministic Dynamic System

A deterministic system is described by its state transition equation (7), i.e. $x_{t+1} = f(x_t, u_t); x_t, x_{t+1} \in X = \{s_1, \ldots, s_n\}, u_t \in U = \{c_1, \ldots, c_m\}; t = 0, 1, \ldots, N-1; x_0 \in X$ is the initial state, and $N < \infty$ is a fixed and specified termination time (planning horizon).

The fuzzy constraints are $\mu_{C^0}(u_0), \ldots, \mu_{C^{N-1}}(u_{N-1})$ and the fuzzy goal is $\mu_{G^N}(x_N)$. The fuzzy decision is

$$\mu_D(u_0, \ldots, u_{N-1} \mid x_0) = \\ = \mu_{C^0}(u_0) \wedge \ldots \wedge \mu_{C^{N-1}}(u_{N-1}) \wedge \mu_{G^N}(x_N) \tag{12}$$

and the problem is to find $u_0^*, \ldots, u_{N-1}^*$ such that

$$\begin{aligned}\mu_D(u_0^*, \ldots, u_{N-1}^* \mid x_0) &= \\ &= \max_{u_0, \ldots, u_{N-1}} [\mu_{C^0}(u_0) \wedge \ldots \wedge \mu_{C^{N-1}}(u_{N-1}) \wedge \mu_{G^N}(x_N)] \qquad (13)\end{aligned}$$

It is easy to see that only the two last terms in this problem, i.e.,

$$\mu_{C^{N-1}}(u_{N-1}) \wedge \mu_{G^N}(f(x_{N-1}, u_{N-1}))$$

depend on u_{N-1} (and not on other decisions), hence (13) can be rewritten as

$$\begin{aligned}&\max_{u_0, \ldots, u_{N-1}} [\mu_{C^0}(u_0) \wedge \ldots \wedge \mu_{C^{N-1}}(u_{N-1}) \wedge \mu_{G^N}(x_N)] = \\ &= \max_{u_0, \ldots, u_{N-2}} [\mu_{C^0}(u_0) \wedge \ldots \wedge \mu_{C^{N-2}}(u_{N-2}) \wedge \\ &\qquad \wedge \max_{u_{N-1}} (\mu_{C^{N-1}}(u_{N-1}) \wedge \mu_{G^N}(f(x_{N-1}, u_{N-1})))] \qquad (14)\end{aligned}$$

On repeating this backward iteration for $u_{N-2}, \ldots, u_0$, we obtain the set of dynamic programming recurrence equations

$$\begin{cases} \mu_{G^{N-i}}(x_{N-i}) = \max_{u_{N-i}} [\mu_{C^{N-i}}(u_{N-i}) \wedge \mu_{G^{N-i+1}}(x_{N-i+1})] \\ x_{N-i+1} = f(x_{N-i}, u_{N-i}); i = 1, \ldots, N \end{cases} \qquad (15)$$

where $\mu_{G^{N-i}}(.)$ is a fuzzy goal at $t = N - i$ induced by a fuzzy goal at $t = N - i + 1$.

An optimal sequence of decisions sought, $u_0^*, \ldots, u_{N-1}^*$, is given by the successive maximizing values of u_{N-i} in (15). It is convenient to represent the solution, u_t^*, by an optimal policy $a_t^* : X \longrightarrow U$, such that $u_t^* = a_t^*(x_t), t = 0, 1, \ldots, N-1$, i.e. relating an optimal decision to the current state.

9.3.2 The Case of a Stochastic Dynamic System

The stochastic system is assumed to be a Markov chain whose temporal evolution is described by a conditional probability

$$p(x_{t+1} \mid x_t, u_t)$$

such that $x_t, x_{t+1} \in X = \{s_1, \ldots, s_n\}$, $u_t \in U = \{c_1, \ldots, c_m\}$, $x_0 \in X$ is an initial state, $t = 0, 1, \ldots, N-1$, and $N < \infty$ is a fixed and specified termination time.

We have now the following two problem formulations:

- due to Bellman and Zadeh (1970): find an optimal sequence of decisions $u_0^*, \ldots, u_{N-1}^*$ to maximize the probability of attainment of the fuzzy goal subject to the fuzzy constraints, i.e.

$$\begin{aligned}\mu_D(u_0^*, \ldots, u_{N-1}^* \mid x_0) &= \\ &= \max_{u_0, \ldots, u_{N-1}} [\mu_{C^0}(u_0) \wedge \ldots \wedge \mu_{C^{N-1}}(u_{N-1}) \wedge E\mu_{G^N}(x_N)] \qquad (16)\end{aligned}$$

where the fuzzy goal is viewed to be a fuzzy event in X whose (nonfuzzy) probability is (cf. Zadeh, 1968)

$$E\mu_{G^N}(x_N) = \sum_{x_N \in X} p(x_N \mid x_{N-1}, u_{N-1}) \cdot \mu_{G^N}(x_N) \tag{17}$$

- due to Kacprzyk and Staniewski (1980): find an optimal sequence of decisions $u_0^*, \ldots, u_{N-1}^*$ to maximize the expectation of the fuzzy decision's membership function, i.e.

$$\begin{aligned} &\mu_D(u_0^*, \ldots, u_{N-1}^* \mid x_0) = \\ &= \max_{u_0, \ldots, u_{N-1}} E[\mu_{C^0}(u_0) \wedge \ldots \wedge \mu_{C^{N-1}}(u_{N-1}) \wedge \mu_{G^N}(x_N)] \end{aligned} \tag{18}$$

and these formulations are clearly not equivalent.

9.3.2.1 Bellman and Zadeh's approach. Since in (16) $\mu_{C^{N-1}}(u_{N-1}) \wedge E\mu_{G^N}[f(x_{N-1}, u_{N-1})]$ depend only on u_{N-1}, the next two right-most terms depend only on u_{N-2}, etc., the structure of (16) is essentially the same as that of (14), and the set of fuzzy dynamic programming recurrence equations is

$$\begin{cases} \mu_{G^{N-i}}(x_{N-i}) = \max_{u_{N-i}} [\mu_{C^{N-i}}(u_{N-i}) \wedge E\mu_{G^{N-i+1}}(x_{N-i+1})] \\ E\mu_{G^{N-i+1}}(x_{N-i+1}) = \sum_{x_{N-i+1} \in X} p(x_{N-i+1} \mid x_{N-1}, u_{N-1}) \times \\ \qquad \times \mu_{G^{N-i+1}}(x_{N-i+1}); \qquad i = 1, \ldots, N \end{cases} \tag{19}$$

and we consecutively obtain u_{N-i}^* or, in fact, optimal policies a_{N-i}^* such that $u_{N-i}^* = a_{N-i}^*(x_{N-i}), i = 1, \ldots, N$.

9.3.2.2 Kacprzyk and Staniewski's approach. To solve problem (18), we first introduce a sequence of functions $h_i : X \times \times_{j=1}^{i} U \longrightarrow [0,1]$ and $g_j : X \times \times_{l=1}^{j+1} U \longrightarrow [0,1]$; $i = 0, 1, \ldots, N$; $j = 1, \ldots, N-1$; such that

$$\begin{cases} h_N(x_N, u_0, \ldots, u_{N-1}) = \mu_{C^0}(u_0) \wedge \ldots \wedge \mu_{C^{N-1}}(u_{N-1}) \wedge \\ \qquad \wedge \mu_D(u_0, \ldots, u_{N-1} \mid x_0) \\ \ldots\ldots\ldots\ldots\ldots\ldots\ldots\ldots\ldots\ldots \\ g_k(x_k, u_0, \ldots, u_k) = \sum_{i=1}^{n} h_{k+1}(s_i, u_0, \ldots, u_k)\, p(s_i \mid x_k, u_k) \\ h_k(x_k, u_0, \ldots, u_{k-1}) = \max_{u_k} g_k(x_k, u_0, \ldots, u_k) \\ \ldots\ldots\ldots\ldots\ldots\ldots\ldots\ldots\ldots\ldots \\ h_0(x_0) = \max_{u_0} g_0(x_0, u_0) \end{cases} \tag{20}$$

Basically, if the consecutive decisions and states are $u_0, \ldots, u_j$ and $x_0, \ldots, x_j$, respectively, then g_j is the expected value of $\mu_D(. \mid x_0)$ provided that the next decisions are optimal, i.e., $u_{j+1}^*, \ldots, u_{N-1}^*$.

It can be shown (Kacprzyk and Staniewski, 1980; Kacprzyk, 1983b, 1997) that there exist functions $w_k : X \times \times_{i=1}^{k} U \longrightarrow U$ such that $h_k(x_k, u_0, \ldots, u_{k-1}) =$

$g_k(x_k, u_0, \ldots, u_{k-1}, w_k(x_k, u_0, \ldots, u_{k-1}))$. Then, as shown by Kacprzyk and Staniewski (1980), an optimal policy sought, a_t^*, $t = 0, 1, \ldots, N-1$, is given by

$$\begin{cases} a_0^* = w_0(x_0) \\ a_1^*(x_0, x_1) = w_1(x_1, a_0^*(x_0)) \\ \cdots\cdots\cdots\cdots\cdots\cdots\cdots\cdots\cdots\cdots\cdots\cdots \\ a_{N-1}^*(x_0, \ldots, x_{N-1}) = \\ \qquad = w_{N-1}(x_{N-1}, b_{N-2}^*(x_0, \ldots, x_{N-2}, \ldots, b_0^*(x_0)) \ldots) \end{cases} \tag{21}$$

and it depends now not only on the current state but also on the trajectory.

Needless to say that the solution of this formulation is more difficult than of that due to Bellman and Zadeh.

9.3.3 The Case of a Fuzzy Dynamic System

In this challenging case the system is fuzzy and its dynamics is governed by a state transitions equation

$$X_{t+1} = F(X_t, U_t), \qquad t = 0, 1, \ldots \tag{22}$$

where X_t, X_{t+1} are fuzzy states at time (stage) t and $t+1$, and U_t is a fuzzy decision at t, characterized by their membership functions $\mu_{X_t}(x_t), \mu_{X_{t+1}}(x_{t+1})$, and $\mu_{U_t}(u_t)$, respectively; (22) is equivalent to a conditioned fuzzy set $\mu_{X_{t+1}}(x_{t+1} \mid x_t, u_t)$ (cf. Kacprzyk, 1983b, 1997).

Notice that the fuzzy state transition equation (22) is just a general form which can be given as, e.g., a set of IF–THEN rules or a neural network.

Baldwin and Pilsworth (1982) proposed a dynamic programming scheme. First, for each $t = 0, 1, \ldots, N-1$ a fuzzy relation $\mu_{R^t}(u_t, x_{t+1}) = \mu_{C^t}(u_t) \wedge \mu_{G^{t+1}}(x_{t+1})$ is constructed. The degree to which U_t and X_{t+1} satisfy C^t and G^{t+1} is

$$\begin{aligned} &\mu_T(u_t, \mu_{R^t}(u_t, x_{t+1}), x_{t+1}) = \\ &= \max_{u_t}[(\mu_{U_t}(u_t) \wedge \mu_{C^t}(u_t)) \wedge \max_{x_{t+1}}(\mu_{X_{t+1}}(x_{t+1}) \wedge \mu_{G^{t+1}}(x_{t+1}))] \end{aligned} \tag{23}$$

Assuming now, for simplicity and clarity but without loss of generality, that the fuzzy constraints are imposed at $t = 0, 1, \ldots, N-1$ but the fuzzy goal is imposed at $t = N$ only, we obtain the fuzzy decision

$$\begin{aligned} &\mu_D(U_0, \ldots, U_{N-1} \mid X_0) = \\ &= \max_{u_0}[\mu_{U_0}(u_0) \wedge \mu_{C^0}(u_0)) \wedge \ldots \wedge \max_{u_{N-1}}((\mu_{U_{N-1}}(u_{N-1}) \wedge \\ &\quad \wedge \mu_{C^{N-1}}(u_{N-1})) \wedge \max_{x_N}(\mu_{X_N}(x_N) \wedge \mu_{G^N}(x_N))] \end{aligned} \tag{24}$$

and we seek an optimal sequence of fuzzy decisions $U_0^*, \ldots, U_{N-1}^*$ such that

$$\begin{aligned} &\mu_D(U_0^*, \ldots, U_{N-1}^* \mid X_0) = \max_{U_0, \ldots, U_{N-1}} \mu_D(u_0, \ldots, U_{N-1} \mid X_0) = \\ &= \max_{U_0} \max_{u_0}[\mu_{U_0}(u_0) \wedge \mu_{C^0}(u_0)) \wedge \ldots \end{aligned}$$

$$\ldots \wedge \max_{U_{N-1}((\max_{u_{N-1}}} (\mu_{U_{N-1}}(u_{N-1}) \wedge \mu_{C^{N-1}}(u_{N-1}) \wedge$$
$$\wedge \max_{X_N \max_{x_N}} (\mu_{X_N}(x_N) \wedge \mu_{G^N}(x_N)))] \tag{25}$$

hence the set of dynamic programming recurrence equations is

$$\begin{cases} \mu_{\overline{G}^N}(X_N) = \max_{x_N}[\mu_{X_N}(x_N) \wedge \mu_{G^N}(x_N)] \\ \mu_{\overline{G}^{N-i}}(X_{N-i}) = \max_{U_{N-i}[(\max_{u_{N-i}}} (\mu_{U_{N-i}}(u_{N-i}) \wedge \mu_{C^{N-i}}(u_{N-i}))\wedge \\ \qquad \wedge \mu_{\overline{G}^{N-i+1}}(X_{N-i+1})] \\ \mu_{X_{N-i+1}}(x_{N-i+1}) = \max_{x_{N-i}[\max_{u_{N-i}}} (\mu_{U_{N-i}}(u_{N-i})\wedge \\ \qquad \wedge \mu_{X_{N-i+1}}(x_{N-i+1} \mid x_{N-i}, u_{N-i})] \wedge \mu_{X_{N-i}}(x_{N-i})) \\ i = 1, \ldots, N-1 \end{cases} \tag{26}$$

In principle, this may be solved though a prohibitive difficulty is that $\mu_{\overline{G}^{N-i}}(X_{N-i})$ must be specified for all the possible X_{N-i}'s, and the maximization is to proceed over all the possible U_{N-i}'s. As the number of both of them may be very high (theoretically infinite), Baldwin and Pilsworth (1982) predefine some (sufficiently small) number of reference (standard) fuzzy states and fuzzy decisions. Then, they redefine their problem formulation in terms of the reference fuzzy states and fuzzy decisions to finally make (26) solvable. For details, see the source (Baldwin and Pilsworth, 1982) or a readable coverage in Kacprzyk (1983b, 1997) and Zimmermann (1987). The idea of using a small number of reference fuzzy sets is not new (cf. Kacprzyk and Staniewski, 1982), and has also been recently advocated for solving larger fuzzy dynamic programming problems (cf. Kacprzyk, 1993a, b).

Experience with Baldwin and Pilsworth's (1982) approach is often not encouraging (cf. Zimmermann, 1987) and an earlier and simple branch-and-bound approach by Kacprzyk (1979) may be a better choice.

9.4 FUZZY DYNAMIC PROGRAMMING FOR MULTISTAGE DECISION MAKING WITH A FUZZY TERMINATION TIME

In a number of real life scenarios one cannot (or is not willing to) precisely specify the termination time of a process as is the case in the usual decision analysis and programming models. In this case, it may be more adequate to assume a fuzzy termination time as *more or less 5 years, a couple of days, ten years or so ....* This idea appeared in Fung and Fu (1977) and Kacprzyk (1977).

Let $R = \{0, 1, \ldots, K-1, K, K+1, \ldots, N\}$ be the set of decision making stages. At each $t \in R$ we have a fuzzy constraint $\mu_{C^t}(u_t)$, and a fuzzy goal $\mu_{G^v}(x_v)$, $v \in R$, is imposed on the final state. The fuzzy termination time is given by $\mu_T(v), v \in R$, which can be viewed as a degree of how preferable v is as the termination time.

The fuzzy decision is now (Kacprzyk, 1977, 1978b, c)

$$\mu_D(u_0, \ldots, u_{v-1} \mid x_0) = \mu_{C^0}(u_0) \wedge \ldots \wedge \mu_{C^{v-1}}(u_{v-1}) \wedge \mu_T(v) \cdot \mu_{G^v}(x_v) \tag{27}$$

and the problem is to find an optimal termination time v^* and an optimal sequence of decisions $u_0^*, \ldots, u_{v^*-1}^*$ such that

$$\mu_D(u_0^*, \ldots u_{v^*-1}^* \mid x_0) = \\ = \max_{v, u_0, \ldots, u_{v-1}} [\mu_{C^0}(u_0) \wedge \ldots \wedge \mu_{C^{v-1}}(u_{v-1}) \wedge \mu_T(v) \cdot \mu_{G^v}(x_v)] \quad (28)$$

9.4.1 The Case of a Deterministic Dynamic System

Problem (28) was formulated and solved by Kacprzyk (1977, 1978b). However, Stein (1980) presented a computationally more efficient model and solution.

9.4.1.1 Kacprzyk's approach. In Kacprzyk's (1977, 1978c) formulation the set of possible termination times is $\{v \in R : \mu_T(v) > 0\} = \{K, K+1, \ldots, N\} \subseteq R$, hence an optimal sequence of decisions is $u_0^*, \ldots, u_{K-2}^*, u_{K-1}^*, \ldots, u_{v^*-1}^*$.

The part $u_{K-1}^*, u_K^*, \ldots, u_{v^*-1}^*$ is determined by solving

$$\begin{cases} \mu_{G^{v-i}}(x_{v-i}, v) = \max_{u_{v-i}} [\mu_{C^{v-i}}(u_{v-i}) \wedge \mu_{G^{v-i+1}}(x_{v-i+1}, v)] \\ x_{v-i+1} = f(x_{v-i}, u_{v-i}) \\ i = 1, \ldots, v-i+1; v = K, K+1, \ldots, N-1 \end{cases} \quad (29)$$

where $\mu_{G^v}(x_v, v) = \mu_T(v)\mu_{G^v}(x_v)$.

An optimal termination time v^* is then found by the maximizing v in

$$\mu_{G^{K-1}}(x_{K-1}) = \max_v \mu_{G^{K-1}}(x_{K-1}, v) \quad (30)$$

The part $u_0^*, \ldots, u_{K-2}^*$ is then determined by solving

$$\begin{cases} \mu_{G^{K-i-1}}(x_{K-i-1}) = \max_{u_{K-i-1}} [\mu_{C^{K-i-1}}(u_{K-i-1}) \wedge \mu_{G^{K-i}}(x_{K-i})] \\ x_{K-i} = f(x_{K-i-1}, u_{K-i-1}); i = 1, \ldots, K-1 \end{cases} \quad (31)$$

9.4.1.2 Stein's approach. Stein (1980) presented a computationally more efficient dynamic programming approach whose idea is as follows. At $t = N-i, i \in \{1, \ldots, N-1\}$, we can either stop and attain $\mu_{\overline{G}^{N-i}}(x_{N-i}) = \mu_T(N-i)\mu_{G^{N-i}}(x_{N-i})$, or apply u_{N-i} and attain $\mu_{C^{N-i}}(u_{N-i}(u_{N-i}) \wedge \mu_{G^{N-i+1}}(x_{N-i+1})$. The better alternative should evidently be chosen, and this is repeated for $t = N-i-1, N-i-2, \ldots, 0$. The set of recurrence equations is therefore

$$\begin{cases} \mu_{G^{N-i}}(x_{N-i}) = \mu_{\overline{G}_{N-i}}(x_{N-i}) \wedge \\ \qquad \wedge \max_{u_{N-i}} [\mu_{C^{N-i}}(u_{N-i}) \wedge \mu_{G^{N-i+1}}(x_{N-i+1})] \\ x_{N-i+1} = f(x_{N-i}, u_{N-i}); i = 1, 2, \ldots, N \end{cases} \quad (32)$$

and an optimal termination time is such a $t = N-i$ at which the terminating decision, $u_{v^*-1}^*$, occurs, i.e., when

$$\mu_{G^{N-i}}(x_{N-i}) > \max_{u_{N-i}} [\mu_{C^{N-i}}(u_{N-i}) \wedge \mu_{G^{N-i+1}}(x_{N-i+1})] \quad (33)$$

9.4.2 The Case of a Stochastic Dynamic System

The case with a fuzzy termination time and a stochastic system was first formulated and solved in Kacprzyk (1978b, c) by combining the elements of Sections 3.2.1 and 4.1.

The problem is to find an optimal termination time v^* and an optimal sequence of decisions $u_0^*, \ldots, u_{v^*-1}^*$ such that

$$\begin{aligned}&\mu_D(u_0^*, \ldots, u_{v^*-1}^* \mid x_0) = \\ &= \max_{v, u_0, \ldots, u_{N-1}} [\mu_{C^0}(u_0) \wedge \ldots \wedge \mu_{C^{v-1}}(u_{v-1}) \wedge E\mu_{G^v}(x_v)]\end{aligned} \tag{34}$$

where $\mu_{G^v}(x_v) = \mu_T(v)\,\mu_{G^v}(x_v)$; moreover, $\{v : \mu_T(v) > 0\} = \{K, K+1, \ldots, N\}$.

As in Section 3.2.1, we determine v^* and $u_{K-1}^*, u_K^*, \ldots, u_{v^*-1}^*$ by solving

$$\begin{cases} \mu_{G^{v-i}}(x_{v-i}) = \max_{u_{v-i}}[\mu_{C^{v-i}}(u_{v-i}) \wedge E\mu_{G^{v-i+1}}(x_{v-i+1})] \\ E\mu_{G^{v-i+1}}(x_{v-i+1}) = \sum_{x_{v-i+1} \in X} p(x_{v-i+1}) \mid x_{v-1}, u_{v-1} \times \\ \qquad \times\, \mu_{G^{v-i+1}}(x_{v-i+1}); i = 1, \ldots, N \end{cases} \tag{35}$$

and v^* is given by the maximizing v in

$$\mu_{G^{K-1}}(x_{K-1}) = \max_v \mu_{G^{K-1}}(x_{K-1}, v) \tag{36}$$

The remaining part $u_{K-2}^*, u_{K-3}^*, \ldots, u_0^*$ is obtained by solving

$$\begin{cases} \mu_{G^{K-1-i}}(x_{K-1-i}) = \max_{u_{K-1-i}}[\mu_{C^{K-1-i}}(u_{K-1-i}) \wedge E\mu_{G^{K-i}}(x_{K-i})] \\ E\mu_{G^{K-i}}(x_{K-i}) = \sum_{x_{K-i} \in X} p(x_{K-i} \mid x_{K-1-i}, u_{K-1-i}) \times \\ \qquad \times\, \mu_{G^{K-1-i}}(x_{K-1-i}); i = 1, \ldots, K-1 \end{cases} \tag{37}$$

In the later Stein's (1980) formulation the problem is solved by the following set of recurrence equations

$$\begin{cases} \mu_{G^{N-i}}(x_{N-i}) = \mu_{\overline{G}_{N-i}}(x_{N-i}) \wedge \\ \qquad \wedge \max_{u_{N-i}}[\mu_{C^{N-i}}(u_{N-i}) \wedge E\mu_{G^{N-i+1}}(x_{N-i+1})] \\ E\mu_{G^{N-i+1}}(x_{N-i+1}) = \sum_{x_{K-i} \in X} p(x_{K-i+1} \mid x_{N-i}, u_{N-i}) \times \\ \qquad \times\, \mu_{G^{N-i+1}}(f(x_{N-i}, u_{N-i})); i = 1, 2, \ldots, N \end{cases} \tag{38}$$

where v^* occurs when

$$\mu_{G^{N-i}}(x_{N-i}) > \max_{u_{N-i}}[\mu_{C^{N-i}}(u_{N-i}) \wedge E\mu_{G^{N-i+1}}(x_{N-i+1})] \tag{39}$$

9.4.3 Remarks on the Case of a Fuzzy Dynamic System

To obtain a dynamic programming problem formulation for a fuzzy termination time and a fuzzy system, we fix some (finite and relatively small) number of reference fuzzy states (and possibly decisions), and obtain an auxiliary approximate system whose state transitions are of a deterministic system type (cf. Section 3.3 or Kacprzyk, 1983b and Kacprzyk and Staniewski, 1982). Then, Stein's approach can be employed. In many cases, however, a simple Kacprzyk's (1983b) branch-and-bound algorithm is a better choice.

9.5 MULTISTAGE DECISION MAKING WITH AN IMPLICITLY SPECIFIED TERMINATION TIME

The process terminates now when the state enters for the first time a termination set of states $W = \{s_{p+1}, s_{p+2}, \ldots, s_n\} \subset X$. The problem is to determine an optimal sequence of decisions $u_0^*, \ldots, u_{\overline{N}-1}^*$ such that

$$\begin{aligned}\mu_D(u_0^*, \ldots, u_{\overline{N}-1}^* \mid x_0) &= \\ &= \max_{u_0,\ldots,u_{\overline{N}-1}} [\mu_C(u_0 \mid x_0) \wedge \ldots \wedge \mu_C(u_{\overline{N}-1} \mid x_{\overline{N}-1}) \wedge \mu_{G^{\overline{N}}}(x_{\overline{N}})] \qquad (40)\end{aligned}$$

where $x_0, \ldots, x_{\overline{N}-1} \in X \setminus W$, and $x_{\overline{N}} \in W$; we seek an optimal stationary strategy in fact.

The solution of (40) may proceed by using:

- Bellman and Zadeh's (1970) iterative approach,
- Komolov's *et et al.* (1979) graph-theoretic approach, and
- Kacprzyk's (1978a. b) branch-and-bound approach,

and the first one is somehow related to dynamic programming. For details we refer the reader to, e.g., Kacprzyk's (1983b, 1997) books.

9.6 MULTISTAGE DECISION MAKING WITH AN INFINITE TERMINATION TIME

In all the problems considered so far the termination time (planning horizon) was assumed finite. The solution process required some iterations over the consecutive stages. This may be justified when the number of stages is not too high, and – above all – when the process itself exhibits a sufficient variability over time. However, this is not always the case. Then it may expedient to assume an infinite termination time and seek an optimal stationary strategy.

In the fuzzy setting, the multistage DM problem with an infinite termination time was first formulated and solved by Kacprzyk and Staniewski (1982, 1983) – see also Kacprzyk's (1983b, 1997) books.

For the deterministic dynamic system (7), the fuzzy decision is

$$\begin{aligned}\mu_D(u_0, u_1, \ldots \mid x_0) &= \\ &= \mu_C(u_0 \mid x_o) \wedge \mu_G(x_1) \wedge \mu_C(u_1 \mid x_1) \wedge \mu_G(x_2) \wedge \ldots = \\ &= \lim_{N\to\infty} \bigwedge_{t=0}^{N} [\mu_C(u_t \mid x_t) \wedge \mu_G(x_{t+1})] \qquad (41)\end{aligned}$$

The problem becomes to find an optimal stationary strategy $a^*_\infty = (a^*, a^*, \ldots)$ such that

$$\mu_D(a^*_\infty \mid x_0) = \max_{a_\infty} \mu_D(a_\infty \mid x_0) =$$
$$= \max_{a_\infty} \lim_{N\to\infty} \bigwedge_{t=0}^{N} [\mu_C(a(x_t) \mid x_t) \wedge \mu_G(x_{t+1})] \tag{42}$$

As shown in Kacprzyk and Staniewski (1983), problem (42) may be solved in a finite number of steps by using a policy iteration algorithm whose essence is a step-by-step improvement of stationary policies.

A policy iteration type algorithm was also proposed for the stochastic system (the most challenging case!) by Kacprzyk, Safteruk, and Staniewski (1981), and for a fuzzy system by Kacprzyk and Staniewski (1982).

9.7 STOCHASTIC MULTISTAGE DECISION MAKING UNDER FUZZY CRITERIA

9.7.1 Fuzzy Criterion Sets and Fuzzy Dynamic Programming

Let us now turn our attention to a new and potent dimension of fuzzy decision making but specifically fuzzy dynamic programming first introduced in Liu and Esogbue (1996) with extensions given by Esogbue and Liu (1996). Additional studies and applications may be found in Esogbue and Liu (1997a, b, c). In this section, we present the framework for fuzzy criterion set and fuzzy criterion dynamic programming which is a general tool for dealing with many decision and control situations arising in many fields but exemplified by the stochastic reservoir operation and stochastic inventory control models of operations research and engineering.

The distinguishing feature of fuzzy criterion dynamic programming is multistage decision making under fuzzy criteria and, ipso facto, fuzzy objective function. Specifically, the objective is expressed as the maximization of the expected fuzzy criterion function of the product of fuzzy criterion sets. Because of the central role played by this concept in our development, we define it in the sequel.

9.7.1.1 Fuzzy criterion sets and fuzzy criterion functions. Let X be a collection of elements denoted generally by x with the fuzzy set A in X the set of ordered pairs, defined as

$$A = \{(x, \mu_A(x)) | x \in X\} \tag{43}$$

where $\mu_A(x)$ is called the membership function of x in A. We say that A is a *fuzzy criterion set* if A is the set of all *satisfactory* elements and $\mu_A(x)$ is the *satisfactory degree* of x. In this presentation, we call $\mu_A(x)$ the *fuzzy criterion function*.

As in the basic theory of fuzzy sets, we define an *α-level set* as the subset of elements that belong to the fuzzy criterion set A at least to the degree α:

$$A_\alpha = \{x \in X | \mu_A(x) \geq \alpha\}. \tag{44}$$

A fuzzy criterion set A is said to be *unimodal* if its fuzzy criterion function $\mu_A(x)$ is unimodal. For additional concepts employed in the exposition of fuzzy criterion dynamic programming see Liu and Esogbue (1996).

9.7.1.2 Fuzzy criterion dynamic programming. To motivate our model, consider an inventory system. If the demand for goods in this system must be satisfied regardless of the physical constraints of the warehouse or reservoir, then the state of the system can be described by the *imaginary* inventory level. When this imaginary inventory level is less than the dead inventory level (dead storage), then we cannot fulfill demand. In this case the difference represents the shortage quantity. When the imaginary inventory level is greater than the largest physical storage, the amount exceeds the capacity of the warehouse or reservoir, and the difference represents the degree of exceeding the level or, for the reservoir problem, when flooding would result. Usually, there exists a best state at which we define the value of fuzzy criterion to be 1. When the inventory level deviates from the best state, the fuzzy criterion value decreases. Thus, the set of all satisfactory states is a *fuzzy criterion set* whose fuzzy criterion function is the *satisfactory degree of elements*.

For a given N-stage decision process, inventory control process or reservoir operation problem, let us define $A_1, A_2, \ldots, A_N$ as the fuzzy criterion sets of satisfactory states with fuzzy criterion functions $\mu_1, \mu_2, \ldots, \mu_N$ at stages $1, 2, \ldots, N$, respectively, on the real line R. Assume that $\lambda_1, \lambda_2, \ldots, \lambda_N$ are coefficients of convex combination representing the relative importance among $A_1, A_2, \ldots, A_N$. Let x_i, d_i and ξ_i be the state, decision and stochastic variable respectively at stage i, then the state transition equation has the following form:

$$x_{i+1} = x_i + d_i + \xi_i, \quad i = 1, 2, \ldots, N. \tag{45}$$

Our problem is then to control this system such that the states over all stages are satisfactory, i.e., at the stage n, the objective is to maximize the expected fuzzy criterion function of the product $A_n \otimes A_{n+1} \otimes \ldots \otimes A_N$.

Based on the fuzzy criterion set operations, the expected fuzzy criterion function J_n of product $A_n \otimes A_{n+1} \otimes \ldots \otimes A_N$ is

$$J_n(\mathbf{x}; \mathbf{d}) = \sum_{i=n}^{N} \gamma_i^{(n)} \int_R \mu_i(x_i + d_i + \xi_i) d\Phi_i(\xi_i) \tag{46}$$

where $\mathbf{x}$ is a state vector, $\mathbf{d}$ is a decision vector, and

$$\gamma_i^{(n)} = \lambda_i / \sum_{j=n}^{N} \lambda_j \tag{47}$$

For the quantities d_i and ξ_i, positivity implies that it is an input, negativity implies that it is an output. We also mention that $\gamma_n^{(n)} : \gamma_{n+1}^{(n)} : \cdots : \gamma_N^{(n)}$ are simply coefficients of convex combination and $\gamma_n^{(n)} : \gamma_{n+1}^{(n)} : \cdots : \gamma_N^{(n)} = \lambda_n : \lambda_{n+1} : \cdots : \lambda_N$.

Let us now introduce the fuzzy criterion dynamic programming model associated with problem (46) as follows:

$$\begin{aligned} f_N(x) &= \sup_{d \in D_N} L_N(x+d) \\ f_n(x) &= \sup_{d \in D_n} \Big\{ \theta_n L_n(x+d) \\ &\quad + (1-\theta_n) \int_R f_{n+1}(x+d+\xi) d\Phi_n(\xi) \Big\} \\ n &\leq N-1 \end{aligned} \tag{48}$$

where

$$L_n(y) = \int_R \mu_n(y+\xi) d\Phi_n(\xi), \tag{49}$$

and $\theta_n = \gamma_n^{(n)}$, $D_n = [q_n, Q_n]$ is a set of feasible policies, q_n and Q_n are not necessarily finite and positive.

We also make the following assumptions:

(A$_1$) The quantities $\xi_1, \xi_2, \ldots, \xi_N$ are independently stochastic variables with distributions $\Phi_1, \Phi_2, \ldots, \Phi_N$, respectively, and $E|\xi_i| < +\infty$, where E denotes the expected operator.

(A$_2$) The fuzzy criterion functions $\mu_i(x)$ of fuzzy criterion sets A_i: $R \to [0,1]$ are continuous almost everywhere for all i, and $\lim_{x \to \pm\infty} \mu_i(x) = 0$.

(A$_3$) At least one of the following conditions holds,

1. μ_i are continuous functions for all i;
2. Φ_i are continuous distributions for all i.

Additionally, we mention the following facts:

1. When $D_n = R^+$ and the support of the distribution Φ_n is R^-, the equation (48) is a standard inventory model. In this case, the control is to order commodities from outside and the stochastic variables are quantities of demand.

2. When D_n is a closed interval on R^- and the support of the distribution Φ_n is R^+, the equation (48) is a reservoir operation model. Meanwhile, the control is to release water from the reservoir and the stochastic variables are quantities of inflow.

3. $f_n(x)$ is the expected fuzzy criterion function of the product $A_n \otimes A_{n+1} \otimes \cdots \times A_N$ in Euclidean space R^{N-n+1} for any n.

9.7.1.3 Basic theorem. Consider the following convolution operator,

$$H(y) = \int_R h(y+\xi)d\Phi(\xi) \tag{50}$$

where h is an integrable and bounded functional. Usually, the functional $H(y)$ is not necessarily continuous, but we have the following result.

Lemma 1 *If h is a measurable bounded functional which can have at most a countable number of discontinuities, then so is H. In particular, H is a continuous functional if at least one of the following conditions holds:*

1. *h is a continuous functional;*

2. *Φ is a continuous distribution.*

Liu and Esogbue (1996) establish and prove the following theorem which is fundamental in fuzzy criterion dynamic programming.

Theorem 1 *Assume (A_1), (A_2) and (A_3) for all stages, then the dynamic programming equation (48) defines a sequence of continuous functions. Moreover, there exists a Borel function $\widehat{d_n}(x)$ such that the supremum in (48) is attained for any x if D_n can be restricted to a compact set for any n.*

They consider the following infinite horizon problem which is to maximize

$$J_n(\mathbf{x};\mathbf{d}) = \sum_{i=n}^{\infty} \gamma_i^{(n)} \int_R \mu_i(x_i + d_i + \xi_i)d\Phi_i(\xi_i) \tag{51}$$

where $\gamma_i^{(n)} = \lambda_i/(\lambda_n + \lambda_{n+1} + \cdots)$ and $\lambda_1 + \lambda_2 + \cdots = 1$. Liu and Esogbue then develop the dynamic programming equation associated with problem (51) as follows:

$$\begin{aligned} f_n(x) &= \sup_{d \in D_n} \Big\{ \theta_n L_n(x+d) \\ &\quad + (1-\theta_n) \int_R f_{n+1}(x+d+\xi)d\Phi_n(\xi) \Big\} \end{aligned} \tag{52}$$

where $\theta_n = \gamma_n^{(n)}$. In this section, we will suppose that all $\lambda_i > 0$. This implies that

$$0 < \theta_n < 1, \quad \forall n. \tag{53}$$

They proposed and provide the proofs of the following lemmas useful in establishing the existence and uniqueness theorems for the solution to (52):

Lemma 2 *Assume (A_1), (A_2) and (A_3) for all stages, then the relations*

$$\begin{aligned} W_n(x) &= \theta_n L_n(x+d_n) \\ &\quad + (1-\theta_n) \int_R W_{n+1}(x+d_n+\xi)d\Phi_n(\xi) \end{aligned} \tag{54}$$

define a bounded sequence of continuous functions. Moreover, W_n is explicitly defined by

$$W_n(x) = \sum_{i=n}^{\infty} \gamma_i^{(n)} L_{n,i}(x + d_n + \cdots + d_i) \tag{55}$$

where

$$L_{n,i}(y) = \int_R \cdots \int_R \mu_i(y + \xi_n + \cdots + \xi_i) d\Phi_n(\xi_n) \cdots d\Phi_i(\xi_i) \tag{56}$$

and d_n are any given feasible policies in D_n, respectively.

We will also need the following lemma which we state without proof:

Lemma 3 *Consider*

$$t(p) = \sup_{d \in D} \left\{ \theta h(p,d) + (1-\theta) \int_R f(p,d,r) d\Phi(r) \right\} \tag{57}$$

and

$$T(p) = \sup_{d \in D} \left\{ \theta H(p,d) + (1-\theta) \int_R F(p,d,r) d\Phi(r) \right\} \tag{58}$$

where Φ is a probability measure, h, H, f and F are integrable and bounded and $0 \leq \theta \leq 1$. Then, for any given p and $\epsilon > 0$, there exists a point $d^0 \in D$ such that

$$\begin{aligned} |t(p) - T(p)| \leq\ & \theta |h(p,d^0) - H(p,d^0)| \\ & + (1-\theta) \int_R \left[|f(p,d^0,r) \right. \\ & \left. - F(p,d^0,r)| d\Phi(r) \right] + \epsilon. \end{aligned} \tag{59}$$

Liu and Esogbue (1996) then prove the following existence and uniqueness theorems associated with solutions to their fuzzy dynamic programming functional equation.

Theorem 2 [Existence Theorem] *We consider*

$$\begin{aligned} W_n^0(x) &= W_n(x) \\ W_n^{k+1}(x) &= \sup_{d \in D_n} \left\{ \theta_n L_n(x+d) \right. \\ & \left. + (1-\theta_n) \int_R W_{n+1}^k (x + d + \xi) d\Phi_n(\xi) \right\}. \end{aligned} \tag{60}$$

Then we have

$$W_n^{k+1} \geq W_n^k \quad , \qquad k = 0, 1, 2, \ldots \tag{61}$$

and

$$W_n^k \uparrow W_n^*. \tag{62}$$

Moreover, the limit $W_n^*(x)$ *is a continuous bounded solution to (52) if we assume* (A_1), (A_2) *and* (A_3).

Theorem 3 [Uniqueness Theorem] *Assume* (A_1), (A_2) *and* (A_3) *for all stages, then there is one, and only one, bounded solution to (52).*

9.7.2 Optimal Control Policy and Additional Studies in Fuzzy Criterion Dynamic Programming

The operating characteristics of the resultant control policy, as well as various specific applications to open inventory networks and systems, multidimensional versions, and stochastic joint reservoir operations are developed in Esogbue and Liu (1997a, b, c). Additional studies and applications are in progress.

9.8 COMPUTATIONAL COMPLEXITY ANALYSIS OF FUZZY DYNAMIC PROGRAMS

The development of efficient algorithms for processing various aspects of fuzzy dynamic programs is an important research area in dynamic programming which is hardly addressed in the works of most authors. We have begun to fill this gap in our work. This is exemplified by the research of Esogbue and Warsi with regards to nonserial dynamic programming but especially the contribution of Esogbue (1997) with regards to fuzzy dynamic programs. In this section, we illustrate this concern by summarizing the results of the computational complexity analysis provided by Esogbue using the two algorithms of Kacprzyk and Stein for the case of fuzzy termination time for a decision making problem in a fuzzy environment. These models were discussed earlier in this chapter.

9.8.1 Computational Analysis of Kacprzyk's Model

We now examine the space and the computational complexities of the algorithm proposed by Kacprzyk. For storage complexity S, we assume that each variable (computer word) takes a unit of storage space. We may then calculate the demand for the storage tables during computations ignoring all input tables and intermediate variables created during the course of processing.

For a performance profile of an algorithm, we consider the computational complexity T, as a function of the basic operations. Apparently, comparison, mod operation, assignment and arithmetic operations may be considered as such basic operations. We further assume that the total number of all above operations performed is roughly proportional to the number of comparisons. Hence the comparison constitutes the basic operation and T of an algorithm is approximated by the number of comparisons.

Using the above framework, Esogbue showed that the dynamic programming approach presented by Kacprzyk requires $N(N+1)/2$ iterations while the one proposed by Stein requires only N and hence it is computationally more efficient. Let us focus here on the time and space complexity analysis for both.

In this model consisting of n equations and u_{N-1} controls each assuming m values, and time $t = N-1, \ldots, K.$, the total number of operations involved is

$$n(2mt - 2mK - 2m - t + K) + n(N - K + 1) + (K - 1)n(2m - 1). \quad (63)$$

If we let $m = K = t$, we can see that this is of the order $O(2K^3)$.

9.8.2 Stein's Approach

The basic dynamic programming formulation of the problem as proposed by Stein is the same except the structure of the recurrence equations that requires N only iterations of the optimizing process as opposed to $N(N+1)/2$ required in Kacprzyk's algorithm.

For this model, the space and the time complexities are of order $O(n)$ and $O(mn)$, respectively. Essentially, the time complexity is of order $O((2m-1)(N-K))$. This is of order $O(K^2)$. The total storage demand is $n(N-K+1)(N-K+2)/2 + 2n + n(K-1)$. This is of order $O(K)$.

It was pointed out that the difference between the earliest and the latest possible termination times is an important factor in the computational burden of this optimization process.

9.8.3 Comparative Summary

Esogbue concludes his analysis by noting that the orders of time and space complexities of the two algorithms clearly indicate that the model proposed by Stein is computationally more efficient, taking up $O(K)$ memory spaces in $O(K^2)$ operations, as opposed to $O(K^2)$ memory spaces in $O(2K^3)$ operations in Kacprzyk's model. Thus the superiority is exhibited both from the space and time complexity considerations.

In the case of a stochastic system, it was shown that the computational burden and storage requirements are identical for both algorithms if, in the dynamic programming formulation, the objective is to maximize the probability of attainment of the fuzzy goal G subject to non fuzzy constraints.

As a final remark, Esogbue notes that, although the fuzziness of the termination time influences the computational aspect of the optimizing process, it does not considerably decrease the computational efficiency of the algorithms when compared with the case of non-fuzzy termination time.

9.9 BRIEF REVIEW OF APPLICATIONS OF FUZZY DYNAMIC PROGRAMMING

In this section, we present some of the more relevant applications of fuzzy dynamic programming which have been proposed in widely perceived areas of operations research and systems analysis. Clearly, this review is not exhaustive. However, we have attempted to include most of the more representative examples of areas and approaches.

These applications basically fall into two main categories:

- applications related to more general (standard) problems,
- applications related to more specific problems.

In the first group, we should first mention a pioneering work on a general resource allocation problem with fuzzy goals and constraints by Esogbue and Ramesh (1970). This represents the first reported effort detailing the computational schema and algorithms for fuzzy dynamic programs. Another problem class of a universal relevance is data clustering, and the application of fuzzy dynamic programming was proposed here by Esogbue (1986), and Esogbue and Bellman (1981). These applications involve environmental and water pollution control modeling. There exist several applications which are directed at some mathematical programming problems. For example, Hussein and Abo-Sinna (1993) apply a hybrid fuzzy dynamic programming approach to determine a set of efficient solutions of the multiobjective mathematical programming problem. Narshima Sastry, Tiwari, and Sastry (1993), on the other hand, solved a generalized multiple goal fuzzy control (optimization) problem by using fuzzy dynamic programming.

In the second group, we should first mention Esogbue's (1983) work on a general research and development (R&D) control problem in which some limited funds are to be optimally used to attain certain goals distributed over time. This problem may be a prototype for a much wider class of applications. Second, in a series of papers, Kacprzyk and Straszak (1981, 1982a, b, 1984) proposed a fuzzy dynamic programming model for determining socioeconomic regional development strategies which take into account limited resources, requirements on the effectiveness and stability of development, some objective and subjective aspects, etc.

Relevant medical applications have been proposed, and one should mention here Esogbue and Elder's (1977) work on a fuzzy dynamic programming model of medical diagnosis and Esogbue's (1985) approach to intra-operative anesthesia administrations which is useful for planning and control. In fact, Esogbue's (1983) work on R&D is much concerned with the cancer research appropriation process which seeks an optimal resource allocation strategy such that the probability/possibility of attaining a breakthrough in cancer containment is maximized.

A fruitful area of research is energy systems. Among more relevant works there, we should first of all mention Su and Hsu (1991) who applied fuzzy dynamic programming to the unit commitment of a power system, and practically implemented the results obtained for Taiwan's power system. Huang, Lin and Huang (1992) presented a fuzzy dynamic programming model for the determination of an optimal schedule for the outage starting times of power generators for maintenance over the one year planning horizon for a Taiwanese power company. Sugianto and Mielczarski (1995a, b, 1996) consider the optimization of a spare parts inventory in a power generation plant by balancing requirements on power availability and reduction of spare parts' costs.

Water resources systems comprise an important field in which fuzzy dynamic programming models have found applications. One of the earliest applications in this involve the work of Esogbue and Bellman (1981) where an application to the analysis of water resources related data was presented. Esogbue and Ahipo (1982a, b) utilize fuzzy DP tools in developing some effectiveness measures in water resources planning. Esogbue, Theologidu, and Guo (1992) presented an application to an extremely relevant problem of flood control, while Esogbue (1996) generalizes the models to disaster control planning. An application of fuzzy dynamic programming to the open inventory problem is reported in Esogbue and Liu (1997c) while its application to the multi-reservoir operation problem, with both fuzzy and stochastic parameters, is discussed by Esogbue and Liu (1997b).

In chemical engineering we cite the work of Krasławski, Górak and Vogenpohl (1989) who proposed an application of fuzzy dynamic programming to the determination of distillation sequence (sequence of distillation columns) for a mixture of some number of components. An important problem involving the determination of a best itinerary for the transportation of hazardous waste loads in a given road structure was considered by Klein (1991) taking into account the safety of the road segments and intersection, distance, driving difficulty, dangers related to population density, etc.

Yuan and Wu (1991) proposed an application of fuzzy dynamic programming to real-time control of a transportation system in a flexible manufacturing environment, more particularly involving the so-called autonomous guided vehicles (AGVs) which are driverless, programmable vehicles that can move around the factory.

This concludes our brief analysis of the main applications of fuzzy dynamic programming in widely perceived operations research related areas. For further details, as well as their applications to other fields, we refer the interested reader to the original sources as well as Kacprzyk's (1997) book.

9.10 CONCLUDING REMARKS

In this chapter we have presented, first, a readable exposition of the idea of fuzzy dynamic programming. Then, a brief survey of main problem classes has been provided. A brief summary of the major applications in a variety of fields related to broadly perceived operations research is also given.

However, because of space limitations and the scope of this short review, we have not discussed an extremely important problem of computational issues of fuzzy dynamic programming models. It must be noted however that these models, analogous to their nonfuzzy counterparts, exhibit the so-called but over-publicized "curse of dimensionality". This may prohibit their brute force use in large-scale real-world problems without first ensuring adroit formulation and employing recent computational devices including parallel computing. For discussion of these issues, we refer the interested reader to Esogbue (1991a, b, 1997), and to some "trickery" to overcome this difficulties – to Kacprzyk's (1997) book.

For other surveys of fuzzy dynamic programming and its applications, we refer the reader to Esogbue and Bellman (1984), a follow up study by Kacprzyk and Esogbue (1996), and then the fundamental presentation by Kacprzyk (1983a, 1997).

References

Baldwin, J.F. and Pilsworth, B.W. (1982). Dynamic programming for fuzzy systems with fuzzy environment. *J. Math. Anal. and Appls.*, 85, 1–23.

Bellman, R.E. (1957). *Dynamic Programming.* Princeton Univ. Press, Princeton, NJ.

Bellman, R.E. and Zadeh, L.A. (1970). Decision-making in a fuzzy environment. *Management Sci.*, 17, 141–164.

Chang, S.S.L. (1969b). Fuzzy dynamic programming and the decision making process. *Proc. 3rd Princeton Conf. on Inf. Sci. and Syst.* (Princeton, NJ, USA), 200–203.

Esogbue A.O. (1983) Dynamic programming, fuzzy sets and the modelling of R&D management control systems. *IEEE Transactions on Systems, Man, and Cybernetics* SMC-13: 18–30.

Esogbue A.O. (1984) Some novel applications of fuzzy dynamic programming. *Proceedings of IEEE Systems, Man, and Cybernetics Conference* (Bombay, India).

Esogbue A.O. (1985) A fuzzy dynamic programming model of intra-operative anesthesia administration. In J. Kacprzyk and R.R. Yager (Eds.): *Management Decision Support Systems using Fuzzy Sets and Possibility Theory.* Verlag TÜV Rheinland, Cologne, 155–161.

Esogbue A.O. (1986) Optimal clustering of fuzzy data via fuzzy dynamic programming. *Fuzzy Sets and Systems* 18: 283–298.

Esogbue A.O. (1989) *Dynamic Programming for Optimal Water Resources Systems Analysis.* Prentice-Hall, Englewood Cliffs, NJ.

Esogbue A.O. (1991a) Computational and data acquisition issues in fuzzy dynamic programming. *Proceedings of ORSA/TIMS Conference* (Anaheim, CA).

Esogbue A.O. (1991b) Computational aspects and applications of a branch and bound algorithm for fuzzy multistage decision processes. *Computers and Mathematics with Applications* 21, 117–127.

Esogbue, A.O. (1995) On the computational complexity of fuzzy dynamic programs. *Proc. 6th International Association World Congress,* Sao Paolo, Brazil, July, 1995, 217-220.

Esogbue A.O. and Z.M. Ahipo (1982a) Fuzzy sets in water resources planning. In R.R. Yager (Eds.): *Recent Developments in Fuzzy Sets and Possibility Theory,* Pergamon Press, New York, 450–465.

Esogbue A.O. and Z.M. Ahipo (1982b) A fuzzy sets model for measuring the effectiveness of public participation in water resources planning. *Water Resources Bulletin* 18: 451–456.

Esogbue A.O. and R.E. Bellman (1981) A fuzzy dynamic programming algorithm for clustering non-quantitative data arising in water pollution control planning. *Proceedings of Third International Conference on Mathematical Modelling* (Los Angeles, CA).

Esogbue A.O. and R.E. Bellman (1984) Fuzzy dynamic programming and its extensions. *TIMS/Studies in the Management Sciences* 20: 147–167.

Esogbue A.O. and R.C. Elder (1983) Measurement and valuation of a fuzzy mathematical model for medical diagnosis. *Fuzzy Sets and Systems* 10: 223–242.

Esogbue A.O., M. Fedrizzi and J. Kacprzyk (1988) Fuzzy dynamic programming with stochastic systems. In J. Kacprzyk and M. Fedrizzi (Eds.): *Combining Fuzzy Imprecision with Probabilistic Uncertainty in Decision Making.* Springer–Verlag, Berlin/New York, 266–285.

Esogbue A.O. and B. Liu (1996) On fuzzy criterion set dynamic programming. *Proc. 1996 Biennial Conference of the North American Fuzzy Information Processing Society, Berkeley,* CA, June 19-22, 1996, 70-74.

Esogbue A.O. and B. Liu (1997a) On stochastic multistage decision making under fuzzy criteria. *Proc. International Conference on Fuzzy Logic and Applications,* Zichron-Ya'akov, Israel, May 11-14, 1997.

Esogbue A.O. and B. Liu (1997b) Stochastic multi-joint reservior operation via fuzzy criterion dynamic programming. *Proc. 7th International Fuzzy Systems Association (IFSA) World Congress,* Prague, Czech Republic, June 25-29, 1997.

Esogbue A.O. and B. Liu (1997c) Fuzzy criterion dynamic programming and optimal control of multi-dimensional open inventory networks. *Proc. Sixth IEEE International Conference on Fuzzy Systems,* Barcelona, Spain, July 1-5, 1997.

Esogbue A.O. and V. Ramesh (1970) Dynamic programming and fuzzy allocation processes. *Technical Memo No. 202,* Dept. of Operations Research, Case Western University, Cleveland, OH.

Esogbue A.O., M. Theologidu and K. Guo (1992) On the application of fuzzy sets theory to the optimal flood control problem arising in water resources systems. *Fuzzy Sets and Systems* 48: 155–172.

Fung, L.W. and Fu, K.S. (1977). Characterization of a class of fuzzy optimal control problems. In M.M. Gupta, G.N. Saridis and B.R. Gaines (Eds.): *Fuzzy Automata and Decision Processes.* North-Holland, Amsterdam, 205–220.

Huang C.J., Lin, C.E. and Huang, C.L. (1992) Fuzzy approach for generator maintenance scheduling. *Electric Power Systems Research,* 24, 31–38.

Hussein, M.L. and Abo-Sinna, M.A. 1993) Decomposition of multiobjective programming problems by hybrid fuzzy dynamic programming. *Fuzzy Sets and Systems,* 60, 25–32, 1993.

Kacprzyk, J. (1977). Control of a nonfuzzy system in a fuzzy environment with fuzzy termination time. *Systems Sci.,* 3, 325–341.

Kacprzyk, J. (1978a). A branch-and-bound algorithm for the multistage control of a nonfuzzy system in a fuzzy environment. *Control and Cybern.,* 7, 51–64.

Kacprzyk, J. (1978b). Control of a stochastic system in a fuzzy environment with fuzzy termination time. *Systems Sci.,* 4, 291–300.

Kacprzyk, J. (1978c). Decision making in a fuzzy environment with fuzzy termination time. Fuzzy *Sets and Systems,* 1, 169–179.

Kacprzyk, J. (1979). A branch-and-bound algorithm for the multistage control of a fuzzy system in a fuzzy environment. *Kybernetes,* 8, 139–147.

Kacprzyk, J. (1983a). A generalization of fuzzy multistage decision making and control via linguistic quantifiers. Int. *J. of Control,* 38, 1249–1270.

Kacprzyk, J. (1983b). Multistage Decision-Making under Fuzziness. Verlag TÜV Rheinland, Cologne.

Kacprzyk, J. (1993). Interpolative reasoning in optimal fuzzy control, *Proc. of Second IEEE International Conference on Fuzzy Systems (FUZZ-IEEE '93),* Vol. II, 1259–1263.

Kacprzyk, J. (1994a). Fuzzy dynamic programming – basic issues, In M. Delgado, J. Kacprzyk, J.-L. Verdegay and M.A. Vila (Eds.): *Fuzzy Optimization: Recent Advances,* Physica-Verlag, Heidelberg, 321–331.

Kacprzyk, J. (1997). *Multistage Fuzzy Control: A Model-Based Approach to Control and Decision Making.* Wiley, Chichester.

Kacprzyk, J. and Esogbue, A.O. (1996). Fuzzy dynamic programming: Main developments and applications. *Fuzzy Sets and Systems,* 81:1, 31-45.

Kacprzyk, J., Safteruk, K. and Staniewski, P. (1981). On the control of stochastic systems in a fuzzy environment over infinite horizon, *Systems Sci.,* 7, 121–131.

Kacprzyk, J. and Staniewski P. (1980). A new approach to the control of stochastic systems in a fuzzy environment, *Arch. Automat. i Telemech.,* XXV, 443–444.

Kacprzyk, J. and Staniewski, P. (1982). Long-term inventory policy-making through fuzzy decision-making models. *Fuzzy Sets and Systems,* 8, 117–132.

Kacprzyk, J. and Staniewski, P. (1983). Control of a deterministic system in a fuzzy environment over an infinite planning horizon, Fuzzy Sets and Systems , 10, 291–298.

Kacprzyk J. and A. Straszak (1981) Application of fuzzy decision -making models for determining optimal policies in 'stable' integrated regional development. In P.P. Wang and S.K. Chang (Eds.): *Fuzzy Sets Theory and Applications to Policy Analysis and Information Systems,* Plenum Press, New York, 321–328.

Kacprzyk J. and A. Straszak (1982a) A fuzzy approach to the stability of integrated regional development. In G.E. Lasker (Ed.): *Applied Systems and Cybernetics,* Vol. 6, Pergamon Press, New York, 2997–3004.

Kacprzyk J. and A. Straszak (1982b) Determination of 'stable' regional development trajectories via a fuzzy decision-making model. In R.R. Yager (Ed.): *Recent Developments in Fuzzy Sets and Possibility Theory,* Pergamon Press, New York, 531–541

Kacprzyk J. and A. Straszak (1984) Determination of stable trajectories for integrated regional development using fuzzy decision models. *IEEE Transactions on Systems, Man and Cybernetics* SMC-14: 310–313.

Klein, C.M. (1991). A model for the transportation of hazardous waste. *Decision Sciences,* 22, 1091–1108.

Komolov, S.V. et al. (1979). On the problem of optimal control of a finite automaton with fuzzy constraints and fuzzy goal (in Russian), *Kybernetika* (Kiev), 6, 30–34.

Krasławski, A., A. Górak and A. Vogelpohl (1989) Fuzzy dynamic programming in the synthesis of distillation column systems. *Computers and Chemical Engineering,* 13, 611–618.

Liu, B. and A.O. Esogbue (1996) Fuzzy criterion set and fuzzy criterion dynamic programming. *Journal of Mathematical Analysis and Applications,* 199:1, 293-311.

Narshima Sastry, V., Tiwari, R.N., and Sastry, K.S. (1993) Dynamic programming approach to multiple objective control problems having deterministic or fuzzy goals. *Fuzzy Sets and Systems,* 57, 195–202, 1993.

Stein, W.E. (1980). Optimal stopping in a fuzzy environment, *Fuzzy Sets and Systems,* 3, 252–259.

Sugianto, L.F. and W., Mielczarski (1995a) A fuzzy logic approach to optimize inventory. In G.F. Forsyth and M. Ali (Eds.): *Proceedings of Eighth International Conference on Industrial and Engineering Applications of Artificial Intelligence and Expert Systems,* Gordon and Breach (Australia), 419–424.

Sugianto, L.F. and W. Mielczarski (1995b) Control of power station inventory by optimal allocation process with fuzzy logic approach. *Proceedings of Control '95 Conference* (Australia), Vol. 2, 529–533.

Yuan, Y. and Z. Wu (1991) Algorithm of fuzzy dynamic programming in AGV scheduling. *Proceedings of the International Conference on Computer Integrated Manufacturing - ICCIM '91,* 405-408.

Zadeh, L.A. (1968). Probability measures of fuzzy events, *J. of Math. Anal. and Appls.*, 23, 421–427.

Zimmermann, H.-J. (1987). *Fuzzy Sets, Decision Making, and Expert Systems.* Kluwer, Dordrecht.

III Statistics and Data Analysis

10 FUZZY SET-THEORETIC METHODS IN STATISTICS

Jörg Gebhardt
María Angeles Gil
Rudolf Kruse

Abstract: In this chapter we have gathered some of the approaches which have been introduced in the literature to deal with fuzzy statistical data, as well as some methods to solve inferential (in particular, parameter estimation and hypothesis testing) and decision problems from them.

10.1 INTRODUCTION

Statistical problems are commonly focused on drawing inferences or making decisions regarding a population. Inferences and decisions are assumed to be based on the information supplied by a *random experiment* associated with that population and, occasionally, on additional information about the experiment.

To arrive at a statistical inference or decision in the presence of absolute certainty and precision with respect to all relevant elements, becomes quite often impossible.

Since Fuzzy Set Theory was introduced, several studies have been developed to establishing and handling concepts that combine notions from both Fuzzy Set and Probability Theories.

The aim of this Chapter is to examine some models and methods to manage statistical problems involving fuzziness in elements or stages of the random experiment.

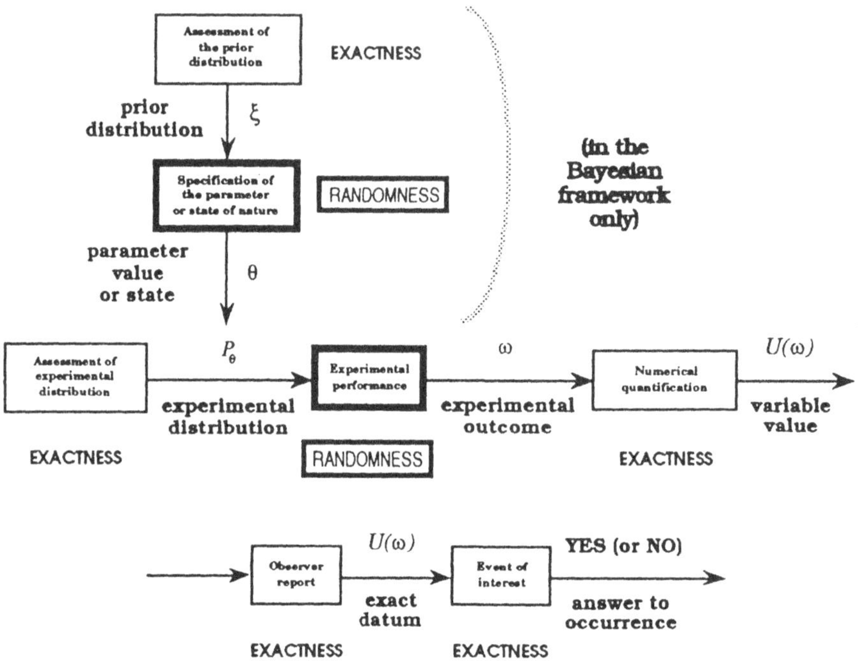

Figure 1 Elements and stages in a random experiment involving the observation of a random variable

10.2 FUZZINESS IN RANDOM EXPERIMENTS

To characterize a random experiment we need to identify all *experimental outcomes*, to define all *events of interest* associated with the experiment, and to assess *probabilities* to these events (and, if needed, to the population parameter values or states of nature).

Random experiments are often assumed to involve the observation of a *random variable* U defined on the space Ω of possible experimental outcomes. Figure 1 schematically represents the mechanism of such a random experiment.

The mathematical model used to describe the random experiment in Figure 1 is given by the induced probability space $(\mathbb{R}, \mathcal{B}_1, P_\theta), \theta \in \Theta$ (or, alternatively, by $(U(\Omega), \mathcal{B}_{U(\Omega)}, P_\theta))$, $\mathcal{B}_1$ being the Borel σ-field on $\mathbb{R}$ ($\mathcal{B}_{U(\Omega)}$ the Borel σ-field on $U(\Omega)$) and P_θ is the probability measure induced by the variable on that σ-field and often depending on a certain parameter or state θ with unknown value which ranges on the space Θ. When a Bayesian framework is considered, θ is assumed to behave as a random variable with (prior) distribution ξ.

In Statistics one traditionally assumes that the experimental performance (and the parameter value or state specification, in a Bayesian setting) are accomplished under randomness, whereas the remaining stages in the experiment are carried out under absolutely certain and well-defined conditions.

However, in practice fuzziness can arise in some of these remaining stages, that is, in the assessment of the experimental (and/or prior) distribution, in the quantification process of the random variable, in the observer report of the variable value and/or in the definition of events of interest associated with the experiment (see, for instance, [Kahneman; Slovic; Tversky 1985], [Zimmer 1986, 1990], [Walley 1991] and [Thomas 1995]).

In that sense, limitations sometimes appear when assessing exact probabilities, so that the available information about probabilities is more properly described in terms of imprecise propositions stating that a set of experimental results (or a subset of the parameter space) is *"highly probable"*, or *"unlikely"*, or *"more or less probable"*. On the other hand, the quantification process in the random variable can associate with each experimental outcome an imprecise value, like *"a very small number"*, *"a high number"*, and other; in the same way, although the quantification above was numerical, the observer can imprecisely report the variable value. Finally, imprecise events of interest associated with the experiment can be defined (for instance, the event being *"very tall"* associated with the experiment of measuring the height of people in a given population) so that, even if variable value is exactly known, one cannot conclude whether these events occur or not, but rather one prefers to determine what their "degree" of occurrence is.

Fuzziness could be also involved in getting the experimental outcome (or the parameter value of the experimental distribution), but that fact would not affect the development and application of Probability Theory, since these elements are not forced to be real or well-defined elements in that theory. To deal with the preceding situations some proper mathematical models have been proposed in the literature on fuzzy sets.

Regarding the assessment of **fuzzy probabilities** we can see, for instance, papers from Zadeh [1975, 1984], Negoita and Ralescu [1975], Nguyen [1979], Dubois and Prade [1980], Jain and Agogino [1988], Rappoport; Wallsten; Cox [1987], Utkin [1993], and more recently Ralescu [1995a], among others. Nevertheless, there are still many open questions in connection with this topic.

For the presence of fuzziness in the quantification process characterizing the random variable, the notion of **fuzzy random variable** (see, for instance, Kwakernaak [1978], Ralescu [1982], Puri and Ralescu [1985, 1986], and Kruse and Meyer [1987]) provides us with a valuable model very manageable in the probabilistic framework (cf. [Kruse and Meyer 1987], [Negoita and Ralescu 1987], [Kruse; Gebhardt; Klawonn 1994], and [López-Díaz 1996]).

Finally, the concept of **fuzzy information** (Zadeh [1978], Okuda; Tanaka; Asai [1978], Tanaka; Okuda; Asai [1979]) can suitably formalize either experimental data or events of interest involving fuzziness.

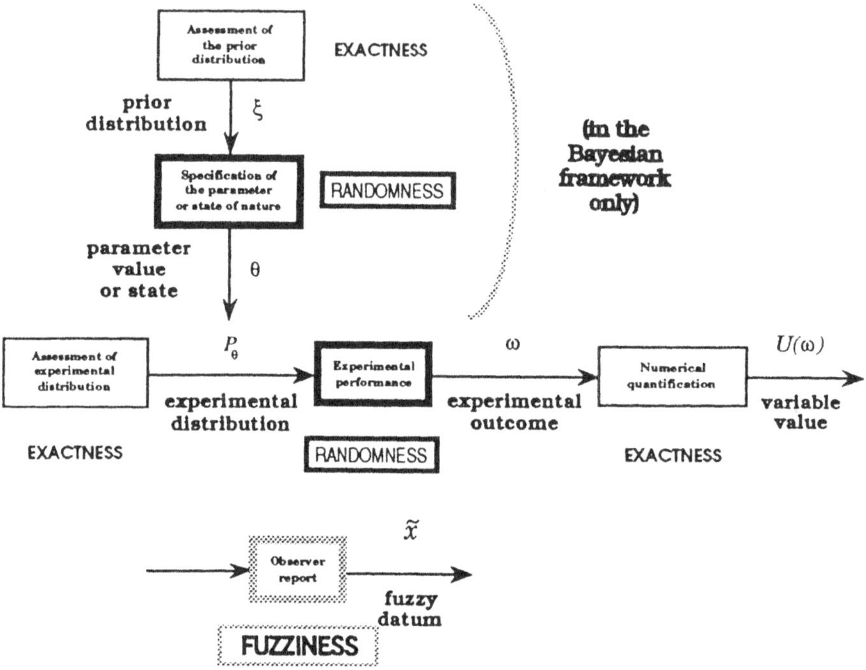

Figure 2 Elements and stages in a random experiment involving fuzziness in the the observed report of the random variable value

The studies which are going to be developed in the next two sections, are focused on statistical problems concerning experiments where the variable value is assumed to be exact, but the information reported from the observer to the statistician is expressed in terms of a fuzzy datum (Figure 2).

This type of uncertainty in the observer report is not herein assumed to be due to a possible experimental error (that the statistician would usually identify with a random error being normally distributed around an exact datum), or to the fact that the reported datum is either grouped or the statistician only knows its image through a transformation (so that one could properly characterize it by means of an incomplete datum).

Sometimes, the uncertainty is intrinsically due to the definition of the reported data, whose values or set of possible values are ill-defined. In that way, data like *"moderately expensive"* to indicate the price of a given item, or *"more or less tall"* to describe the height of a person, or *"warm"* in a report on the average temperature of a zone, etc., involve lexical imprecision so they cannot easily be expressed as classical subsets of the sample space for the associated experiment and, hence, as exact or grouped data. When the statistician receives data of this type, one of the most appropriate ways to characterize these data is to describe them by means of fuzzy subsets of the sample space.

In practice, the management of the imprecision in the observer report has been usually based either on the "rounding" or on the "grouping" of data. However, sometimes the information available for the statistician is not precise enough to accept rounding as a proper approximation, but it is not so imprecise that he has to be compelled to express it in terms of a more or less wide class of values (grouped data).

Two different approaches to deal with fuzzy data will be next presented. Purposes of the problems these approaches try to model and solve are close, but essentially distinct. Thus, the first approach is considered to model and solve problems regarding either real- or fuzzy-valued parameters by using fuzzy-valued statistics, and is based on the concept of fuzzy random variable, whereas the second approach concerns inferential problems regarding real-valued parameters of the experimental distribution by using real-valued statistics and is based on the concept of fuzzy information system.

10.3 AN APPROACH TO FUZZY-VALUED STATISTICS WITH FUZZY EXPERIMENTAL DATA

Whenever an extension of probability theory to non-standard data is going to be established, the fundamental aim is to provide an appropriate concept of a generalized random variable that allows us to verify the validity of essential limit theorems such as the strong law of large numbers and the central limit theorem. In the case of set-valued data, the generalized random variable is a *random set* (see [Matheron 1975], [Kendall 1974], [Stoyan; Kendall; Mecke 1987]), for which a strong law of large numbers was proved in Artstein and Vitale [1975]. With regard to fuzzy data and the basic notion of a *fuzzy random variable* ([Kwakernaak 1978], [Puri and Ralescu 1986]), an analogous theorem can be formulated (see Ralescu [1982], Kruse [1982], Miyakoshi and Shimbo [1984], Klement; Puri; Ralescu [1986], Kruse and Meyer [1987]). It supports the development of a fuzzy probability theory and by this to lay down the concepts for mathematical statistics on fuzzy sets. The monographs [Kruse and Meyer 1987] and [Bandemer and Näther 1992] describe the theoretical and practical methods of fuzzy statistics in much detail. For comparable discussions and alternative approaches we can mention, for instance, the studies of Kandel [1979], Hirota [1981], Czogała and Hirota [1986], Tanaka [1987], Gil [1988] and Viertl [1992].

The background of fuzzy statistics can be distinguished from two different viewpoints of modelling imperfect information using fuzzy sets. The first one regards a fuzzy datum as an existing object, for example a physical grey scale picture. Therefore this view is called the *physical interpretation* of fuzzy data. The second view, the *epistemic interpretation*, applies fuzzy data to imperfectly specify a value that is existing and precise, but not measurable with exactitude under the given observation conditions. Thus the first view does not examine real-valued data, but objects that are more complex. In the most simple case we need multivalued data, as they turn up in the field of random sets.

We restrict ourselves to consider the second view, namely the extension of traditional probability theory and mathematical statistics from the treatment of real-valued crisp data to handling fuzzy data in their epistemic interpretation as *possibility distributions* (Zadeh [1978], Dubois and Prade [1988]). The statistical analysis of this sort of data was first studied by Kwakernaak [1978]. An extensive investigation of relevant aspects of statistical inference in the presence of possibilistic data can be found in Kruse and Meyer [1987]. The corresponding methods have been incorporated into the software tool SOLD (Statistics on Linguistic Data) that offers several operations for analyzing fuzzy random samples (see [Kruse 1987], [Kruse and Gebhardt 1989]).

We now introduce the concept of a fuzzy random variable and outline how to use it for the development of a theory of fuzzy probability and fuzzy statistics. Additionally, we show some implementation aspects and features of the mentioned software tool SOLD.

10.3.1 Fuzzy Random Variables

Introducing the concept of a *fuzzy random variable* means that we deal with situations, in which two different types of uncertainty appear simultaneously, namely randomness and possibility.

Randomness refers to the description of a random experiment by a probability space $(\Omega, \mathcal{A}, P)$, where Ω is the set of all possible outcomes of this experiment, $\mathcal{A}$ is a σ-field of subsets of Ω (the set of possible events), and the set function P, defined on $\mathcal{A}$, is a probability measure. We assume that the whole information that is relevant for further analysis of any outcome of the random experiment can be expressed with the aid of a real number, so that we can specify a mapping $U : \Omega \to \mathbb{R}$, which assigns to each outcome in Ω its random value in $\mathbb{R}$. U is called a *random variable* and is expected to be measurable with respect to the σ-field $\mathcal{A}$ and the Borel σ-field $\mathcal{B}_1$ of the real line.

Possibility as a second kind of uncertainty in our description of a random experiment has to be involved whenever we are not in the position to fix the random values $U(\omega)$ as crisp numbers in $\mathbb{R}$, but only to imperfectly specify these values by possibility distributions on $\mathbb{R}$. In this case, the random variable $U : \Omega \to \mathbb{R}$ changes into a *fuzzy random variable* $X : \Omega \to \mathcal{F}(\mathbb{R})$ with $\mathcal{F}(\mathbb{R}) = \{\tilde{x} \mid \mu_{\tilde{x}} : \mathbb{R} \to [0, 1]\}$ denoting the class of all fuzzy subsets (unnormalized possibility distributions) of the real numbers.

A fuzzy random variable $X : \Omega \to \mathcal{F}(\mathbb{R})$ is interpreted as a (fuzzy) perception of an inaccessible usual random variable $U_0 : \Omega \to \mathbb{R}$, which is called the *original* of X. The basic idea is to assume that the considered random experiment is characterized by U_0, but the available description of its attached random values $U_0(\omega)$ is imperfect in the sense that their most specific specification is the possibility distribution $X_\omega = X(\omega)$. In this case, for any $r \in \mathbb{R}$, the value $X_\omega(r)$ quantifies the *degree of possibility* with which the proposition $U_0(\omega) = r$ is regarded as being true.

More particularly, $X_\omega(r) = 0$ means that there is no supporting evidence for the possibility of truth of $U_0(\omega) = r$, whereas $X_\omega(r) = 1$ means that there is no evidence against the possibility of truth of $U_0(\omega) = r$, so that this proposition is fully possible, and $X_\omega(r) \in (0, 1)$ reflects that there is evidence that supports the truth of the proposition as well as evidence that contradicts it, based on a set of competing contexts for the specification of $U_0(\omega)$.

Recent research activities in *possibility theory* have delivered a variety of different approaches to the semantic background of a degree of possibility, similar to the several interpretations that have been proposed with respect to the meaning of subjective probabilities (Shafer [1976], Nguyen [1978], Kampé de Fériet [1982], Wang [1983], Dubois; Moral; Prade [1993]). A quite promising way of interpreting a possibility distribution $X_\omega : \mathbb{R} \to [0, 1]$ is that of viewing X_ω in terms of the *context approach* (Gebhardt [1992], Gebhardt and Kruse [1993, 1995]). It is very important to provide such semantical underpinnings in order to obtain a well-founded concept of a fuzzy random variable. On the other hand, the following results are presented in a way that makes it sufficient for the reader to confine to the intuitive view of a possibility distribution X_ω as a *gradual constraint* on the set $\mathbb{R}$ of possible values [Zadeh 1978].

The concept of a fuzzy random variable is a reasonable extension of the concept of a usual random variable in the many practical applications of random experiments where the implicit assumption of data precision seems to be an inappropriate simplification rather than an adequate modelling of the real physical conditions. Considering possibility distributions allows us to involve *uncertainty* (due to the probabilities of occurrence of competing specification contexts) as well as *imprecision* (due to the context-dependent set-valued specifications of $U_0(\omega)$). For this reason a frequent case in applications, namely using error intervals instead of crisp points for measuring $U_0(\omega)$, is covered by the concept of a fuzzy random variable.

Note that fuzzy random variables describe situations where the uncertainty and imprecision in observing a random value $U_0(\omega)$ is functionally dependent of the respective outcome ω. If observation conditions are not influenced by the random experiment, so that for any $\omega_1, \omega_2 \in \Omega$, the equality of random values $U_0(\omega_1)$ and $U_0(\omega_2)$ does not imply that their imperfect specifications with the aid of possibility functions are the same, then theoretical considerations in fuzzy statistics become much simpler, since in this case we do not need anymore a concept of a fuzzy random variable. It suffices to generalize operations of traditional statistical inference for crisp data to operations on possibility distributions using the well-known *extension principle* [Zadeh 1975]. We reconsider this topic later.

After the semantical underpinnings and aims of the concept of a fuzzy random variable have been clarified, we will now present its full formal definition and show how to use it for a probability theory based on fuzzy sets.

Let $\mathcal{F}_N(\mathbb{R}) = \{\tilde{x} \mid \mu_{\tilde{x}} : \mathbb{R} \to [0,1] \mid (\exists x)(\mu_{\tilde{x}}(x) = 1)\}$ be the *class of all normal fuzzy sets of the real line*. Moreover, let $\mathcal{F}_c(\mathbb{R})$ denote the *class of all upper semicontinuous fuzzy sets* $\tilde{x} \in \mathcal{F}_N(\mathbb{R})$, which means that for all $\alpha \in (0,1]$, the α-cuts $(\tilde{x})_\alpha = \{x \in \mathbb{R} \mid \mu_{\tilde{x}}(x) \geq \alpha\}$ are closed and bounded.

Definition 1 Let $(\Omega, \mathcal{A}, P)$ be a probability space. A function $X : \Omega \to \mathcal{F}_c(\mathbb{R})$ is called a *fuzzy random variable* , iff

$$\begin{aligned} \underline{X}_\alpha &: \Omega \to \mathbb{R}, \quad \omega \mapsto \inf(X(\omega))_\alpha \text{ and} \\ \overline{X}_\alpha &: \Omega \to \mathbb{R}, \quad \omega \mapsto \sup(X(\omega))_\alpha \end{aligned}$$

are $\mathcal{A}$–$\mathcal{B}_1$–measurable for all $\alpha \in (0,1)$,with $\mathcal{B}_1$ being the Borel σ–field of $\mathbb{R}$.

The notion of a *fuzzy random variable* and the related notion of a *probabilistic set* were introduced by several authors in different ways. From a formal viewpoint, our definition is similar to that of Kwakernaak [1978] and Miyakoshi and Shimbo [1984]. For other work we refer to [Klement; Puri; Ralescu 1986]. They considered fuzzy random variables whose values are fuzzy subsets of $\mathbb{R}^n$, or, more generally, of a Banach space. This approach involves distances on spaces of fuzzy sets and measurability of random elements valued in a metric space.

10.3.2 Fuzzy Probability Theory

Our generalization of concepts of traditional probability theory to a fuzzy probability theory is based on the idea that a fuzzy random variable is considered as a (fuzzy) perception of an inaccessible usual random variable $U_0 : \Omega \to \mathbb{R}$, which we referred to as the unknown *original* of X.

Let $\mathcal{U} = \{U \mid U : \Omega \to \mathbb{R}$ and U is $\mathcal{A}$-$\mathcal{B}_1$–measurable$\}$ be the set of all one–dimensional random variables w.r.t. $(\Omega, \mathcal{A}, P)$.

If only fuzzy data are available, then it is of course not possible to identify one of the candidates in $\mathcal{U}$ as the true original of X, but we can evaluate the degree of possibility $\mathrm{Orig}_X(U)$ of the truth of the statement "U is the original of X", determined by the following possibility distribution Orig_X on $\mathcal{U}$:

$$\mathrm{Orig}_X : \mathcal{U} \to [0,1], \; U \mapsto \inf_{\omega \in \Omega}\{X_\omega(U(\omega))\}.$$

The definition of Orig_X shows relationships to random set theory (see Matheron [1975]) in the way that for all $\alpha \in [0,1]$, $[\mathrm{Orig}_X]_\alpha$ coincides with the set of all selectors of the random set $X_\alpha : \Omega \mapsto \mathcal{B}_1$, $X_\alpha(\omega) = [X_\omega]_\alpha$.

Zadeh's extension principle, which can be justified by the context approach mentioned above (Gebhardt [1993]), helps us to define fuzzifications of well–known probability theoretical notions. As an example consider the generalization of characteristic parameters of crisp random variables to fuzzy random variables:

If $\gamma(U)$ is a characteristic of a crisp random variable $U : \Omega \to \mathbb{R}$, then

$$\gamma(X) : \mathbb{R} \to [0,1],\ t \mapsto \sup_{\substack{U \in \mathcal{U},\\ \gamma(U)=t}} \inf_{\omega \in \Omega} \{X_\omega(U(\omega))\}$$

turns out to be the corresponding characteristic of a fuzzy random variable. For example, expected value and variance of a fuzzy random variable are defined as follows:

Definition 2 *(a)* $E(X) : \mathbb{R} \to [0,1],\ t \mapsto \sup\{\mathrm{Orig}_X(U) \mid U \in \mathcal{U},\ E|U| < \infty,\ E(U) = t\}$ is called the *expected value of* X. *(b)* $Var(X) : \mathbb{R} \to [0,1]$, $t \mapsto \sup\{\mathrm{Orig}_X(U) \mid U \in \mathcal{U},\ E|U - E(U)|^2 < \infty, E(U - E(U))^2 = t\}$ is the *variance of* X.

In a similar way, other notions of probability theory and descriptive statistics can be generalized to fuzzy data. Based on the semantically well-founded concept of a fuzzy random variable, the fuzzification step is quite simple, since it only refers to an appropriate application of the extension principle. The main theoretical problem consists in finding simplifications that support the development of efficient algorithms for calculations in fuzzy statistics. It turns out that the *horizontal representation* of fuzzy sets and possibility distributions by using the family of their α-cuts is more appropriate than fixing on the *vertical representation* that attaches a membership degree or a degree of possibility to each element of the domain of the respective fuzzy set or possibility distribution.

The horizontal representation has the advantage that it reduces operations on fuzzy sets and possibility distributions to operations on α-cuts (Kruse; Gebhardt; Klawonn [1994]). Nevertheless many algorithms for efficient computations in fuzzy statistics require deeper theoretical effort. For more details, see [Kruse and Meyer 1987].

Complexity problems may also result from non-trivial structures of probability spaces. Tractability therefore often means that one has to confine oneself to the consideration of finite probability spaces or to use appropriate approximation techniques (Kruse and Meyer [1987]).

The following theorem shows that a convenient representation of the fuzzy expected value as one example for a characteristic of a fuzzy random variable X is derived under certain restrictions on X.

Theorem 1 *Let* $X : \Omega \to \mathcal{F}_c(\mathbb{R})$ *be a finite fuzzy random variable such that* $X(\Omega) = \{\tilde{x}_1, \tilde{x}_2, \ldots, \tilde{x}_n\}$ *and* $p_i = P\Big(\{\omega \in \Omega \mid X_\omega = \tilde{x}_i\}\Big)$, $i = 1, 2, \ldots, n$. *Then,*

$$\left\{ \left[\sum_{i=1}^{n} p_i \cdot \inf(\tilde{x}_i)_\alpha, \sum_{i=1}^{n} p_i \cdot \sup(\tilde{x}_i)_\alpha \right] \right\}_{\alpha \in (0,1]}$$

is an α*-cut representation of* $E(coX)$, *where* $coX : \Omega \to \mathcal{F}_c(\mathbb{R})$ *is defined by* $(coX)(\omega) = co(X_\omega)$ *with* $co(X_\omega)$ *denoting the convex hull of* X_ω.

10.3.3 Fuzzy Statistics

When fuzzy sets are chosen to be applied in mathematical statistics we have to consider two conceptual different approaches. The first one strictly refers to the concept of a fuzzy random variable. It assumes that, given a generic $X : \Omega \to \mathcal{F}_c(\mathbb{R})$ and a fuzzy random sample $X_1, \ldots, X_n$ i.i.d. F_X, the realization of an underlying random experiment is formalized by a tuple $(\tilde{x}_1, \ldots, \tilde{x}_n) \in \left[\mathcal{F}_c(\mathbb{R})\right]^n$ of fuzzy-valued outcomes. Kruse and Meyer [1987] verified that all important limit theorems (e.g., the strong law of large numbers, the central limit theorem, and the theorem of Gliwenko–Cantelli) remain valid in the more general context of fuzzy random variables. From this it follows that the extension of mathematical statistics from crisp to fuzzy data is well-founded. As an example for the generalization of an essential theorem we present a fuzzy data version of the strong law of large numbers. More general versions can be found in [Klement; Puri; Ralescu 1986], [Meyer 1987] and [Kruse and Meyer 1987].

Theorem 2 *Let $\{X_i\}_{i\in\mathbb{N}}$ be an i.i.d.-sequence on the probability space $(\Omega, \mathcal{A}, P)$ with the generic fuzzy random variable $X : \Omega \to \mathcal{F}_c(\mathbb{R})$. Let $E(|(\underline{X_i})_0|) < \infty$ and $E(|(\overline{X_i})_0|) < \infty$. Then there exists a zero set N such that*

$$(\forall \omega \in \Omega \setminus N)\left(\left\{\frac{1}{n}\sum_{i=1}^{n} X_i(\omega)\right\}_{n\in\mathbb{N}} \xrightarrow{d_\infty} E(X)\right),$$

where

$$d_\infty : \left[\mathcal{F}_c(\mathbb{R})\right]^2 \to \mathbb{R}_0^+,\ d_\infty(\tilde{x}, \tilde{x}') = \sup_{\alpha\in(0,1]} d_H((\tilde{x})_\alpha, (\tilde{x}')_\alpha)$$

denotes the "generalized Hausdorff pseudometric w.r.t. $\mathcal{F}_c(\mathbb{R})$", and

$$d_H : (2^{\mathbb{R}}\backslash\{\emptyset\})^2 \to \mathbb{R}_0^+,\ d_H(A, B) = \max\left\{\sup_{a\in A}\inf_{b\in B}|a-b|, \sup_{b\in B}\inf_{a\in A}|a-b|\right\},$$

the "Hausdorff pseudometric w.r.t. $\mathbb{R}$".

The second approach to statistics with fuzzy data does not base on the concept of a fuzzy random variable, but rather on the presupposition that there is a generic random variable $U : \Omega \to \mathbb{R}$, a crisp random sample $U_1, \ldots, U_n$ i.i.d. F_U, and a corresponding realization $(u_1, \ldots, u_n)$, imperfectly specified by $(\tilde{x}_1, \ldots, \tilde{x}_n) \in [\mathcal{F}_c(\mathbb{R})]^n$.

Furthermore let $U_1, \ldots, U_n$ i.i.d. F_U and $T(u_1, \ldots, u_n)$ be a realization of a statistical function $T(U_1, \ldots, U_n)$. The task is to calculate the corresponding fuzzy statistical function $T[\tilde{x}_1, \ldots, \tilde{x}_n]$.

As an example consider the problem of computing fuzzy parameter tests:

Suppose that F_U depends on a parameter $\gamma \in \Gamma$ of a predefined parameter space $\Gamma \subseteq \mathbb{R}^k$, $k \in \mathbb{N}$. Let $\mathcal{D}$ be a class of distribution functions, $F_U \in \mathcal{D}$, $D : \Gamma \to \mathcal{D}$ a mapping, and $\Gamma_0, \Gamma_1 \subseteq \Gamma$ two disjoint sets of parameters.

A function $\Phi : \mathbb{R}^n \to \{0, 1\}$ is called non-randomized parameter test for $(\delta, \Gamma_0, \Gamma_1)$ with respect to D based on a given significance level $\delta \in (0, 1)$, null hypothesis $H_0 : \gamma \in \Gamma_0$, and alternative hypothesis $H_1 : \gamma \in \Gamma_1$, iff Φ is $\mathcal{A}$–$\mathcal{B}_1$–measurable and $(\gamma \in \Gamma_0)(\mathrm{E}_\gamma(\Phi(U_1,\ldots,U_n)) \leq \delta)$ holds for $U_1,\ldots,U_n$ i.i.d. F_U.

By application of the extension principle we obtain the corresponding fuzzy parameter test, where the calculation of $\Phi[\tilde{x}_1,\ldots,\tilde{x}_n]$ often turns out to be a time consuming task. For this reason we present one of the simple extensions, which is the *fuzzy chi-square test.*

Theorem 3 *Let $\mathcal{N}$ be the class of all normal distributions $N(\mu,\sigma^2)$ and $U : \Omega \to I\!R$ a $N(\mu_0,\hat{\sigma}^2)$-distributed random variable with given expected value μ_0, but unknown $\hat{\sigma} \in \Gamma = I\!R^+$.*
Define $D : \Gamma \to \mathcal{N}$, $D(\sigma) = N(\mu_0,\sigma^2)$, $\Gamma_0 = \{\sigma_0\}$, $\Gamma_1 = \Gamma\backslash\Gamma_0$, and choose $U_1,\ldots,U_n$ i.i.d. F_U and $\delta \in (0,1)$.
Suppose $\Phi : I\!R^n \to \{0,1\}$ to be the non-randomized double-sided chi-square test for $(\delta,\Gamma_0,\Gamma_1)$ with respect to D.
If $(\tilde{x}_1,\ldots,\tilde{x}_n) \in \left[\mathcal{F}_c(I\!R)\right]^n$, then $\Phi[\tilde{x}_1,\ldots,\tilde{x}_n]$ is the realization of the corresponding fuzzy chi-square test. For $\alpha \in (0,1]$ we obtain:

$$\left(\Phi[\tilde{x}_1,\ldots,\tilde{x}_n]\right)_\alpha = \begin{cases} \{0\}, & \text{iff } (\tilde{x}_1,\ldots,\tilde{x}_n] > \sigma_0^2\chi^2_{\frac{\delta}{2}}(n) \\ & \quad \wedge S_\alpha[\tilde{x}_1,\ldots,\tilde{x}_n] < \sigma_0^2\chi^2_{1-\frac{\delta}{2}}(n), \\ \{1\}, & \text{iff } S_\alpha[\tilde{x}_1,\ldots,\tilde{x}_n] \leq \sigma_0^2\chi^2_{\frac{\delta}{2}}(n) \\ & \quad \wedge I_\alpha[\tilde{x}_1,\ldots,\tilde{x}_n] \geq \sigma_0^2\chi^2_{1-\frac{\delta}{2}}(n), \\ \{0,1\}, & \text{otherwise} \end{cases}$$

where $\chi^2_{\frac{\delta}{2}}(n)$ denotes the $\frac{\delta}{2}$-quantile of the chi-square distribution with n degrees of freedom, and

$$I_\alpha[\tilde{x}_1,\ldots,\tilde{x}_n] = \sum_{\substack{i=1 \\ \mu\leq\inf(\tilde{x}_i)_\alpha}}^{n} (\inf(\tilde{x}_i)_\alpha - \mu)^2 + \sum_{\substack{i=1 \\ \mu\geq\sup(\tilde{x}_i)_\alpha}}^{n} (\mu - \sup(\tilde{x}_i)_\alpha)^2,$$

$$S_\alpha[\tilde{x}_1,\ldots,\tilde{x}_n] = \sum_{i=1}^{n} \max\{(\inf(\tilde{x}_i)_\alpha - \mu)^2\ ,\ (\sup(\tilde{x}_i)_\alpha - \mu)^2\}.$$

10.3.4 The SOLD-System - An Implementation

As an example for the application of many of the concepts, methods, and results discussed here, we briefly present the software tool SOLD (Statistics On Linguistic Data) (Kruse and Gebhardt [1989]), that supports the modelling and statistical analysis of linguistic data, which are representable by fuzzy sets.

An application of the SOLD system consists of two steps, which have to be considered separately with regard to their underlying concepts.

In the first step (*specification phase*) SOLD enables its user to create an application environment (e.g. to analyze weather data), that consists of a finite set of *attributes* (e.g. *clouding, temperature, precipitation*) with their *domains* (intervals of real numbers, e.g. [0, 100] for the clouding of the sky in %). For each attribute A the user states several (possibly parameterized) *elementary linguis-*

tic values (e.g. *cloudy* or *approximately*(75) as fuzzy degrees of the clouding of the sky) and defines for all of these values w the fuzzy sets $\tilde{x}_w$, that shall be associated with them. For this reason SOLD provides 15 different classes of parameterized fuzzy sets of $I\!R$ (e.g. triangular, rectangular, trapezoidal, Gaussian, and exponential functions) as well as 16 logical and arithmetical operators (*and, or, not*, +, −, *, /, **) and functions (e.g. *exp, log, min, max*), that are generalized to fuzzy sets using the extension principle.

The application of context-free generic grammars G_A permits the combination of elementary linguistic values by logic operators (*and, or, not*) and linguistic hedges (*very, considerable*) to increase or decrease the specificity of fuzzy data. By this, formal languages $L(G_A)$ are obtained, which consist of the linguistic expressions that are permitted to describe the values of the attributes A (e.g. *cloudless or fair* as a linguistic expression with respect to the attribute *clouding*).

In the second step (*analysis phase*) the application environments created in the specification phase can be applied to describe realizations of random samples by tuples of linguistic expressions. Since the random samples consist of existing numeric values, that generally cannot be observed exactly, the fuzzy sets, which are related to the particular linguistic expressions, are interpreted epistemically as possibility distributions.

The SOLD system allows to determine convex fuzzy estimators for several characteristic parameters of the generic random variables for the considered attributes (e.g. for the expected value, variance, p-quantile, and range). In addition SOLD calculates fuzzy estimates for the unknown parameters of several classes of given distributions and also determines fuzzy tests for one- or two-valued hypotheses with regard to the parameters of normally distributed random variables.

The algorithms incorporated in this tool are based on the original results about fuzzy statistics that were presented in the monograph [Kruse and Meyer 1987]. For reasons of efficiency, in SOLD only fuzzy sets of the classes $\mathcal{F}_{D_k}(I\!R)$ are employed, namely the subclasses of $\mathcal{F}_N(I\!R)$ that consist of the fuzzy sets with membership degrees out of $\{0, 1/k, \ldots, 1\}$, and α–cuts that are representable as the union of a finite number of closed intervals. In this case the operations to be performed can be reduced to the α-cuts of the involved fuzzy sets (Kruse; Gebhardt; Klawonn [1994]). Nevertheless the simplification achieved by this restriction does not guarantee that we gain an efficient implementation, since operations on α-cuts are not equivalent to elementary interval arithmetics. The difficulties that arise can be recognized already in the following example of determining a fuzzy estimator for the variance.

Let $U : \Omega \to I\!R$ be a random variable defined with respect to a probability space $(\Omega, \mathcal{A}, P)$ and F_U its distribution function. By a realization $(u_1, \ldots, u_n) \in I\!R^n$ of a random sample $(U_1, \ldots, U_n)$ with random variables $U_1, \ldots, U_n : \Omega \to I\!R$, $n \geq 2$, that are completely independent and equally distributed according to F_U, the parameter $\mathrm{Var}(U)$ can be estimated with the help of the variance of the random sample, defined as

$$S_n(U_1,\ldots,U_n) \;=\; \frac{1}{n-1}\left(\sum_{i=1}^{n}\left(U_i - \frac{1}{n}\sum_{j=1}^{n}U_j\right)^2\right).$$

$S_n(U_1,\ldots,U_n)$ is an unbiased, consistent estimator for $\mathrm{Var}(U)$.

If $(\tilde{x}_1,\ldots,\tilde{x}_n) \in [\mathcal{F}_{D_k}(\mathbb{R})]^n$ is the specification of a fuzzy observation of $(u_1,\ldots,u_n)$ for a given $k \in \mathbb{N}$, then by applying the extension principle we obtain the following fuzzy estimator for $\mathrm{Var}(U)$:

$$\begin{aligned}\hat{S}_n : [\mathcal{F}_{D_k}(\mathbb{R})]^n \;&\to\; [\mathcal{F}_{D_k}(\mathbb{R})],\\ \hat{S}_n(\tilde{x}_1,\ldots,\tilde{x}_n)(y) \;&=\; \sup\big\{\min\{\mu_{\tilde{x}_1}(x_1),\ldots,\mu_{\tilde{x}_n}(x_n)\} \mid \\ &\qquad (x_1,\ldots,x_n)\in\mathbb{R}^n \wedge S_n(x_1,\ldots,x_n)=y\big\}.\end{aligned}$$

For $\alpha \in (0,1]$, this leads to the α-cuts

$$\begin{aligned}&\left[\hat{S}_n(\tilde{x}_1,\ldots,\tilde{x}_n)\right]_\alpha\\ &= S_n\left((\tilde{x}_1)_\alpha,\ldots,(\tilde{x}_n)_\alpha\right)\\ &= \left\{y \mid \exists\,(x_1,\ldots,x_n)\in\prod_{i=1}^{n}(\tilde{x}_i)_\alpha : S_n(x_1,\ldots,x_n)=y\right\}\\ &= \left\{y \mid \exists\,(x_1,\ldots,x_n)\in\prod_{i=1}^{n}(\tilde{x}_i)_\alpha : \frac{1}{n-1}\sum_{i=1}^{n}\left(x_i-\frac{1}{n}\sum_{j=1}^{n}x_j\right)^2 = y\right\}.\end{aligned}$$

It is

$$S_n\left((\tilde{x}_1)_\alpha,\ldots,(\tilde{x}_n)_\alpha\right) \;\subseteq\; \frac{1}{n-1}\sum_{i=1}^{n}\left((\tilde{x}_i)_\alpha - \frac{1}{n}\sum_{j=1}^{n}(\tilde{x}_j)_\alpha\right)^2,$$

and equality does not hold in general, so that $S_n\left((\tilde{x}_1)_\alpha,\ldots,(\tilde{x}_n)_\alpha\right)$ cannot be determined by elementary interval arithmetics.

Therefore the creation of SOLD had to be preceded by further mathematical considerations, that were helpful to the development of efficient algorithms for the calculation of fuzzy estimators. Some results can be found in [Kruse and Meyer 1987], [Kruse and Gebhardt 1990].

The fuzzy set ν calculated during the analysis phase by statistical inference with regard to an attribute A (e.g. fuzzy estimation for the variance of the *temperature*) is not transformed back to a linguistic expression by SOLD, as might be expected at first glance. The fundamental problem consists in the fact that in general no $w \in L(G_A)$ can be found, for which $\nu \equiv \mu_w$ holds. Consequently one is left to a *linguistic approximation* of ν, i.e. to find those linguistic expressions w of $L(G_A)$, whose interpretations μ_w approximate the fuzzy set ν under consideration as accurately as possible. The *distance* between two fuzzy sets is measured with the help of the *generalized Hausdorff pseudo-metric* d_∞.

The aim of this linguistic approximation is to determine a $w_{\text{opt}} \in L(G_A)$ that satisfies

$$\forall w \in L(G_A): \qquad d_\infty(\mu_{w_{\text{opt}}}, \nu) \leq d_\infty(\mu_w, \nu).$$

Since this optimization problem in general is very difficult and can lead to unsatisfactory approximations, if $L(G_A)$ is chosen unfavourably (Hausdorff distance too large or linguistic expressions too complicated), SOLD uses the language $L(G_A)$ only to name the fuzzy data that appear in the random samples related to A in an expressive way. SOLD calculates the Hausdorff distance $d_\infty(\mu_w, \nu)$ between ν and a fuzzy set μ_w, provided by the user as a linguistic expression $w \in L(G_A)$, that turns out to be suitable, but does not carry out a linguistic approximation by itself, since the resulting linguistic expression would not be very useful in order to *make a decision* in consequence of the statistical inference.

10.4 AN APPROACH TO REAL-VALUED STATISTICS WITH FUZZY EXPERIMENTAL DATA

The second mathematical approach to model this type of imprecise data, is considered to deal with statistical inferential problems regarding the experimental distribution of the random experiment (or a parameter of its distribution), on the basis of the information supplied by the available fuzzy data. Conclusions from methods to solve the problems will be crisp.

This approach means an intermediate one between the rounding and the grouping of data, so that we can consider ideas for the cases of exact and grouped data when the available experimental ones are fuzzy. Nevertheless, we have to take into account that such a consideration entails two essential inconveniences. Thus, unlike those two situations the set of all available fuzzy data is not a partition of the sample space $U(\Omega)$ by means of subsets in $\mathcal{B}_{U(\Omega)}$. On the other hand, since data cannot be identified with elements in $\mathcal{B}_{U(\Omega)}$, the assessment of probabilities to them cannot be directly carried out according to Probability Theory. To simplify notations we will refer too $U(\Omega)$ as X throughout this section.

The model was suggested by Okuda; Tanaka; Asai [1978], and Tanaka; Okuda; Asai [1979]. This model avoids the inconveniences we have just pointed out, by combining mathematical concepts in Fuzzy Set and Probability Theories. The basic elements in the model are given in the following definitions (stated in [Okuda; Tanaka; Asai 1978], [Tanaka; Okuda; Asai 1979]):

Definition 3 Given a random experiment, $(X, \mathcal{B}_X, P_\theta), \theta \in \Theta$, a fuzzy subset $\tilde{x}$ of X, characterized by a Borel-measurable membership function $\mu_{\tilde{x}} : X \to [0,1]$, is called *fuzzy information* associated with the random experiment.

Definition 4 Given a random experiment, $(X, \mathcal{B}_X, P_\theta), \theta \in \Theta$, a *fuzzy information system* associated with the random experiment is a fuzzy partition $\mathcal{X}$ of X by means of measurable fuzzy subsets of X, that is, a class of measurable fuzzy subsets of X satisfying the *orthogonality condition* given by

$$\sum_{\tilde{x} \in \mathcal{X}} \mu_{\tilde{x}}(x) = 1$$

for all $x \in X$.

The probability measure to be used is that defined by Zadeh [1968], that is,

Definition 5 Given a random experiment, $(X, \mathcal{B}_X, P_\theta), \theta \in \Theta$, and the fuzzy information $\tilde{x}$ associated with it, the *probability of $\tilde{x}$ induced by the experimental distribution P_θ*, is given by the expected value of the membership function $\mu_{\tilde{x}}$ with respect to P_θ, that is,

$$\mathcal{P}_\theta(\tilde{x}) = \int_X \mu_{\tilde{x}}(x)\, dP_\theta(x).$$

We are now going to examine the management and analysis of fuzzy experimental data by means of some procedures of statistical information and inference. The analysis will be based on the preceding model, that is: fuzzy experimental data will be formalized in terms of the notion of fuzzy information associated with the experiment; the set of available fuzzy data will be assumed, whenever required, to be a fuzzy information system; and the assessment of probabilities to fuzzy data will be given in terms of Zadeh's probabilistic definition. To this respect, it should be emphasized that

The structure of fuzzy partition in Definition 4 is the counterpart to the structure of classical partition in the case of grouped data. Actually, to assume the first structure does not mean a strong constraint for the analysis to be developed. Thus, on one hand some of those studies do not need to use the orthogonality condition (e.g., those concerning procedures for point estimation). On the other hand, to assume such a condition in other studies (like those concerning testing hypothesis) will not entail a loss of generality, since it will be possible to construct a fuzzy partition associated with the experiment and providing us with the same inferences that the set of available data. Finally, if the orthogonality condition is not assumed in the remaining studies (Bayesian statistical decision with fuzzy data in Section 10.5, the definition of notions to be introduced can be properly revised without appreciably affecting basic results.

It should be remarked that the probability of a fuzzy event, as intended by Zadeh [1968], is a new concept (not a result derived from Probability Theory), although it can be justified by means of two arguments. In accordance with the simplest one, the induced probability of a fuzzy datum is the most immediate extension of that for grouped data (by simply replacing the indicator function of a grouped datum by the membership function of a fuzzy datum). The second argument is based on a generalization of the traditional definition of random experiment, that ignores σ-additivity and other regularity conditions (Le Cam [1964, 1986]). That generalization defines a *single stage experiment* as a triplet $(X, \mathcal{E}_X, \mathcal{P}_\theta), \theta \in \Theta$ (with a structure weaker than that for $(X, \mathcal{B}_X, P_\theta), \theta \in \Theta$), in which: $\mathcal{B}_X$ has been replaced by a set $\mathcal{E}_X$ of bounded numerical functions from X (in practice, the set of bounded observable random variables), $\mathcal{E}_X$ being a vector lattice for the usual operations (sum, product by real numbers, pointwise supremum and infimum) containing the indicator of X, and being complete for the norm $\|\eta\| = \sup\{|f(x)|, x \in X\}$; and P_θ is replaced by $\mathcal{P}_\theta$, which is a normalized linear functional on $\mathcal{E}_X$, associating with each $\eta \in \mathcal{E}_X$ the expected value with respect to P_θ (i.e., $\mathcal{P}_\theta(\eta) = \int_X \eta(x)\, dP_\theta(x)$). Consequently, and because whatever the fuzzy information $\tilde{x}$ associated with the (traditional) experiment $(X, \mathcal{B}_X, P_\theta), \theta \in \Theta$, may be, we have that $\mu_{\tilde{x}} \in \mathcal{E}_X$, then Definition 5 is equivalent to state that $\mathcal{P}_\theta(\tilde{x}) = \mathcal{P}_\theta(\mu_{\tilde{x}})$. Other arguments can be found in [Thomas 1995].

The formalization of a random experiment with fuzzy data in terms of concepts in Definitions 3 to 5 allows us to guarantee that $\sum_{\tilde{x} \in \mathcal{X}} \mathcal{P}_\theta(\tilde{x}) = 1$. Hence, we can establish a model for this situation given by a new probability space, the sample space being $\mathcal{X}$, the σ-field of events being the σ-field generated by $\mathcal{X}$, $\sigma(\mathcal{X})$, and the probability measure being that induced by P_θ on that σ-field according to Zadeh's probabilistic definition. By using this model we can immediately apply many procedures to analyze complete or grouped data to analyze fuzzy data. For purposes of the mathematical development of those procedures, the use of the new probability space allows us viewing the analysis of fuzzy data as a particular case of the analysis of (nonfuzzy) complete or grouped data. Nevertheless, from a conceptual viewpoint the analysis of fuzzy data actually means an extension of the nonfuzzy case.

To properly present and develop some of these procedures, a model for the management of samples of fuzzy data being consistent with that for individual fuzzy data, is required. The model we will use is based on concepts introduced by Corral and Gil [1984] and Gil; López; Gil [1984].

Suppose that $(X^n, \mathcal{B}_{X^n}, P_\theta), \theta \in \Theta$, is the probability space associated with a simple random sample of size n (which involves the observation of n independent and identically distributed random variables). Assume that the common distribution is that of the random experiment $(X, \mathcal{B}_X, P_\theta), \theta \in \Theta$, and that the available observation reports from that experiment are fuzzy. The basic elements in the associated sampling are presented in the following definitions (Corral and Gil [1984] and Gil; López; Gil [1984]):

Definition 6 Given the random experiment, $(X, \mathcal{B}_X, P_\theta), \theta \in \Theta$, and the fuzzy information associated with it $\tilde{x}_i, i = 1, 2, \ldots, n$, the n-tuple $(\tilde{x}_1, \tilde{x}_2, \ldots, \tilde{x}_n)$ which represents the algebraic product of its compounds (that is, the fuzzy information associated with $(X^n, \mathcal{B}_{X^n}, P_\theta), \theta \in \Theta$, and such that for all sample $(x_1, x_2, \ldots, x_n) \in X^n$,

$$\mu_{(\tilde{x}_1, \tilde{x}_2, \ldots, \tilde{x}_n)}(x_1, x_2, \ldots, x_n) = \mu_{\tilde{x}_1}(x_1) \times \mu_{\tilde{x}_2}(x_2) \times \ldots \times \mu_{\tilde{x}_n}(x_n)$$

is called *sample fuzzy information of size n associated with the experiment.*

Definition 7 If $\mathcal{X}$ is a fuzzy information system associated with the random experiment $(X, \mathcal{B}_X, P_\theta), \theta \in \Theta$, the fuzzy information system $\mathcal{X}^n$ associated with $(X^n, \mathcal{B}_{X^n}, P_\theta), \theta \in \Theta$, and whose elements are the n-tuples $(\tilde{x}_1, \tilde{x}_2, \ldots, \tilde{x}_n)$, $\tilde{x}_i \in \mathcal{X}$ is called *fuzzy random sample of size n from $\mathcal{X}$*.

The preceding definitions, and especially Definition 6, could be introduced in a more general way. In fact, the development of most of the procedures that can be immediately applied to the case of fuzzy data only requires that $\mu_{(\tilde{x}_1, \tilde{x}_2, \ldots, \tilde{x}_n)}(x_1, x_2, \ldots, x_n) = f(\mu_{\tilde{x}_1}(x_1), \mu_{\tilde{x}_2}(x_2), \ldots, \mu_{\tilde{x}_n}(x_n), \tilde{x}_1, \tilde{x}_2, \ldots, \tilde{x}_n)$ (cf. [French 1984]), where f is a function taking on values on $[0, 1]$ and satisfying some plausible conditions. Nevertheless, one of the most operative a proper functions f is that given by the product of the first n components (for a more detailed study, see [Römer and Kandel 1995]). In particular when both, Definition 6 and Zadeh's probabilistic definition, are considered, then the stochastic independence of (nonfuzzy) sample observations implies the "induced independence" of the obtained fuzzy data $\tilde{x}_1$, $\tilde{x}_2$, ..., and $\tilde{x}_n$, that is, $\mathcal{P}_\theta(\tilde{x}_1, \tilde{x}_2, \ldots, \tilde{x}_n) = \mathcal{P}_\theta(\tilde{x}_1) \times \mathcal{P}_\theta(\tilde{x}_2) \times \ldots \times \mathcal{P}_\theta(\tilde{x}_n)$.

When we consider a random experiment and a collection of data from it, the main purpose of Statistics is the use of the information contained in those data to draw inferences or making decisions on the experimental distribution.

Studies on classical inferential problems of statistical estimation and hypothesis testing have provided us with a large number of procedures to solve them. The structure of many of those procedures for the case of exact data allows us to immediately accomplish their extension to the case of fuzzy experimental data. However, the incorporation of membership functions can entail in practice some computational inconveniences. Among the inferential procedures whose theoretical extension is easy to be carried out, we can remark the maximum likelihood principle of point estimation and the likelihood ratio test for testing hypotheses. Both of them make use of the log-likelihood and score functions.

We are next going to examine the above mentioned methods of estimation and testing, as well as to analyze their properties and computational behavior.

10.4.1 Point Estimation from Fuzzy Data. Maximum Likelihood Principle and an Operative Approximation

The aim of *the point estimation problem* of the parameter θ on the basis of fuzzy experimental data from $(X, \mathcal{B}_X, P_\theta), \theta \in \Theta$, is to make use of the information contained in these data to determine a single value to be employed as an estimate of the unknown value of θ, that will be assumed to be a real number. One of the most remarkable techniques of parameter point estimation in the exact case is the method of maximum likelihood introduced by Fisher [1922]. The extension of that method to the case in which data reported from the observer to the statistician are fuzzy does not force the set of available data to be a fuzzy information system. In this way, if we consider a sample fuzzy information $(\tilde{x}_1, \tilde{x}_2, \ldots, \tilde{x}_n)$ from the experiment, we can state (see [Gil; Corral; Gil 1988], [Gil and Casals 1988], and [Gil; Corral; Casals 1989])

Definition 8 If we define the *likelihood function for* $(\tilde{x}_1, \tilde{x}_2, \ldots, \tilde{x}_n)$ as the function $\mathcal{L}(\tilde{x}_1, \tilde{x}_2, \ldots, \tilde{x}_n; \cdot) : \Theta \rightarrow [0, 1]$ such that

$$\mathcal{L}(\tilde{x}_1, \tilde{x}_2, \ldots, \tilde{x}_n; \theta) = \mathcal{P}_\theta(\tilde{x}_1) \times \mathcal{P}_\theta(\tilde{x}_2) \times \ldots \times \mathcal{P}_\theta(\tilde{x}_n)$$

$$= \int_{X^n} \mu_{(\tilde{x}_1, \tilde{x}_2, \ldots, \tilde{x}_n)}(x_1, x_2, \ldots, x_n) L(x_1, x_2, \ldots, x_n; \theta) d\lambda(x_1) d\lambda(x_2) \ldots d\lambda(x_n)$$

for all $\theta \in \Theta$, $L(x_1, x_2, \ldots, x_n; \theta)$ being the likelihood function of θ with respect to a σ-finite measure λ on $(X^n, \mathcal{B}_{X^n})$, then the value $\widehat{\theta}(\tilde{x}_1, \tilde{x}_2, \ldots, \tilde{x}_n) \in \Theta$, if it exists, satisfying that

$$\mathcal{L}(\tilde{x}_1, \tilde{x}_2, \ldots, \tilde{x}_n; \widehat{\theta}(\tilde{x}_1, \tilde{x}_2, \ldots, \tilde{x}_n)) = \sup_{\theta \in \Theta} \mathcal{L}(\tilde{x}_1, \tilde{x}_2, \ldots, \tilde{x}_n; \theta)$$

is called the *maximum likelihood estimate of* θ for the sample fuzzy information $(\tilde{x}_1, \tilde{x}_2, \ldots, \tilde{x}_n)$. Given a fuzzy random sample $\mathcal{X}^n$ from the fuzzy information system $\mathcal{X}$ associated with the experiment $(X, \mathcal{B}_X, P_\theta), \theta \in \Theta$, the *estimator* $\widehat{\theta}(\mathcal{X}^n)$ *of* θ which allocates to the sample fuzzy information $(\tilde{x}_1, \tilde{x}_2, \ldots, \tilde{x}_n) \in \mathcal{X}^n$, the value of a maximum likelihood estimate $\widehat{\theta}(\tilde{x}_1, \tilde{x}_2, \ldots, \tilde{x}_n)$ of θ for that sample information is referred to as a *maximum likelihood estimator of* θ for $\mathcal{X}^n$.

The maximum likelihood principle for fuzzy data preserves the equivariance property by one-to-one transformations in the exact case. Thus,

Proposition 1 *Let* $(X, \mathcal{B}_X, P_\theta), \theta \in \Theta$, *be an experiment and let* $g : \Theta \rightarrow \Phi$ *be a one-to-one transformation, where* $\Phi = g(\Theta) \subseteq R$. *If* $(\tilde{x}_1, \tilde{x}_2, \ldots, \tilde{x}_n)$ *is a sample fuzzy information from that experiment and* $\theta(\tilde{x}_1, \tilde{x}_2, \ldots, \tilde{x}_n)$ *is a maximum likelihood estimate of* θ *for* $(\tilde{x}_1, \tilde{x}_2, \ldots, \tilde{x}_n)$, *then* $g(\theta(\tilde{x}_1, \tilde{x}_2, \ldots, \tilde{x}_n))$ *is a maximum likelihood estimate of* $\phi = g(\theta)$ *for* $(\tilde{x}_1, \tilde{x}_2, \ldots, \tilde{x}_n)$.

Before examining other interesting properties of the maximum likelihood principle with fuzzy data, we have to note that the application of that principle only makes use of the "relative" degree of membership of each x to each $\tilde{x}_i$, so that it is a "scale invariant" procedure with respect to membership functions of fuzzy data, that is,

Proposition 2 *The maximum likelihood estimate of a parameter for a sample fuzzy information is scale invariant with respect to the membership functions of its components.*

The assumption of orthogonality for the set of available fuzzy data, is not a requirement to introduce the maximum likelihood method. However, the implications from that condition are necessary in the study of most of the properties of the estimators supplied by this method. Nevertheless, and because of the above mentioned scale invariance with respect to the membership functions of fuzzy data, one can assert that: whatever the set of available fuzzy data may be, it is possible to construct a set containing fuzzy data being "equivalent" (the equivalence intended in the sense of supplying the same estimate) and satisfying the orthogonality condition.

The maximum likelihood principle when the parameter is vector-valued can be introduced in an analogous way. Since the maximum likelihood principle is directly based on the induced distribution of samples, rather than on the values of the component observations, most of the properties in the exact case can be extended to the case of samples of fuzzy data whenever we consider a fuzzy information system $\mathcal{X}$ and the induced probability space $(\mathcal{X}, \sigma(\mathcal{X}), \mathcal{P}_\theta)$ (or, more precisely, $\mathcal{X}^n$ and the induced space $(\mathcal{X}^n, \sigma(\mathcal{X}^n), \mathcal{P}_\theta)$).

In that sense, the property indicating that the maximum likelihood estimators are the best asymptotically normal ones under very general conditions, is preserved in the fuzzy case. To formalize this assertion, consider now the following "regularity conditions": Θ is a real subset containing an open interval, the true value, θ_0, of the parameter belongs to; whatever the fuzzy information $\tilde{x} \in \mathcal{X}$ may be, the induced probability $\mathcal{P}_\theta(\tilde{x})$ is twice differentiable for every $\theta \in \Theta$; whatever $\tilde{x} \in \mathcal{X}$ may be, we have that $P_\theta(supp(\tilde{x})) > 0$, for every $\theta \in \Theta$; and the Fisher information function associated with the fuzzy information system $\mathcal{X}$ in θ_0 is positive, where *the Fisher information function of θ associated with the fuzzy information system $\mathcal{X}$* is defined as the mapping $I_{\mathcal{X}}^F : \Theta \to [0, +\infty)$, assigning to each $\theta \in \Theta$ the value

$$I_{\mathcal{X}}^F(\theta) = \sum_{\tilde{x} \in \mathcal{X}} \mathcal{P}_\theta(\tilde{x})[\frac{\partial}{\partial \theta} log \mathcal{P}_\theta(\tilde{x})]^2.$$

Then,

Theorem 4 *Under the preceding regularity conditions, there exists a sequence of solutions of the (induced) likelihood equation, $\frac{\partial}{\partial \theta}\mathcal{L}(\tilde{x}_1, \tilde{x}_2, \ldots, \tilde{x}_n; \theta) = 0$, $\{\theta_n(\tilde{x}_1, \tilde{x}_2, \ldots, \tilde{x}_n)\}_n$, such that the sequence of random variables $\{\theta_n(\mathcal{X}^n)\}_n$ almost surely $[\mathcal{P}_{\theta_0}]$ converges to θ_0. In addition, if $\frac{\partial^2}{\partial \theta^2}\mathcal{P}_\theta(\tilde{x})$ is continuous at θ, uniformly in $\tilde{x} \in \mathcal{X}$, then the sequence $\{\theta_n(\mathcal{X}^n)\}_n$ will be from a given sample size a sequence of maximum likelihood estimators asymptotically normal distributed according to a $N(\theta_0, nI_{\mathcal{X}}^F(\theta_0))$, as n tends to ∞.*

The second part in Theorem 4 also guarantees that the maximum likelihood estimators are asymptotically efficient and unbiased, under the assumed conditions. In spite of these valuable properties and although the application of the maximum likelihood principle to some examples is very simple, to determine solutions in most of situations becomes too complex. In particular, when we try to apply it to the case of grouped data (see, for instance, [Zacks 1971], [Carter and Myers 1974], [Cramér 1974], [Sundberg 1974], [McDonald and Ransom 1979], and [Schader und Schmid 1988]), many difficulties arise and it becomes necessary to look for approximated solutions under some plausible conditions on the grouping of data.

A possible way to avoid the inconveniences we have just pointed out, is that of approximating the membership functions characterizing the fuzzy data so that the likelihood and the log-likelihood functions for samples of fuzzy observations are manageable. That way has been briefly examined in a previous paper (Gil; Corral; Casals [1989]). Another possible way is to construct alternative principles coinciding with the maximum likelihood one if available data are exact, being operative and providing us with estimates that are close to the maximum likelihood ones under reasonable conditions on the membership functions of the fuzzy data.

We now present a method satisfying the preceding requirements, which is in part based on the fact that, in the exact case, ease in searching for the solutions of the maximum likelihood equation is due to the employment of the log-likelihood function and the exponentiallity of most of the experimental distributions, whereas the induced distributions in the fuzzy case are not of the exponential type. The alternative suggested method is introduced ([Corral and Gil 1984], [Gil; Corral; Gil 1988], and [Gil; Corral; Casals 1989]) as follows:

Let $(X, \mathcal{B}_X, P_\theta), \theta \in \Theta$, be an experiment in which $\{P_\theta, \theta \in \Theta\}$ is a parametric family of probability measures dominated by the counting or the Lebesgue measure (that we will generically denote by λ). Assume that the set $X^n = \{(x_1, x_2, \ldots, x_n) | L(x_1, x_2, \ldots, x_n; \theta) > 0\}$ does not depend on θ. If we consider the sample fuzzy information $(\tilde{x}_1, \tilde{x}_2, \ldots, \tilde{x}_n)$ from the experiment, then the value $\theta^\star(\tilde{x}_1, \tilde{x}_2, \ldots, \tilde{x}_n) \in \Theta$, if it exists, such that

$$\mathcal{I}(\tilde{x}_1, \tilde{x}_2, \ldots, \tilde{x}_n; \theta^\star(\tilde{x}_1, \tilde{x}_2, \ldots, \tilde{x}_n)) = \inf_{\theta \in \Theta} \mathcal{I}(\tilde{x}_1, \tilde{x}_2, \ldots, \tilde{x}_n; \theta)$$

with

$$\mathcal{I}(\tilde{x}_1, \tilde{x}_2, \ldots, \tilde{x}_n; \theta) = -\int_{X^n} \mu_{|(\tilde{x}_1, \tilde{x}_2, \ldots, \tilde{x}_n)|}(x_1, x_2, \ldots, x_n)$$
$$log L(x_1, x_2, \ldots, x_n; \theta) d\lambda(x_1) d\lambda(x_2) \ldots d\lambda(x_n)$$

which is the Kerridge inaccuracy between the membership function of the "standardized form" (Saaty [1974]) of the sample fuzzy information $(\tilde{x}_1, \tilde{x}_2, \ldots, \tilde{x}_n)$, that will be denoted by $\mu_{|(\tilde{x}_1, \tilde{x}_2, \ldots, \tilde{x}_n)|}(\cdot)$, and the likelihood function of θ, $L(\cdot; \theta)$, is called *minimum inaccuracy estimate of* θ *for* $(\tilde{x}_1, \tilde{x}_2, \ldots, \tilde{x}_n)$.

From the point of view of Statistical Information Theory, there is a very simple structural connection between the principles above stated. Thus, whereas the second principle is based on the minimization of Kerridge's inaccuracy [Kerridge 1961] between two special "distributions", the first one would be equivalent to that based on minimizing the inaccuracy measure of order $\alpha = 2$ [Rathie 1971] between those distributions.

It would be now useful to analyze some properties that can considerably simplify computations. In this respect, the minimum inaccuracy ones satisfy properties of equivariance by one-to-one transformations and scale invariance with respect to the membership functions of fuzzy data (see [Gil; Corral; Casals 1989]).

Obviously, the assumption of the orthogonality condition for the set of available fuzzy data is not a requirement to introduce the minimum inaccuracy method, and the implications from that condition in the properties we next examine. On the other hand, and because of the above mentioned invariance property, the introduction of the minimum inaccuracy principle does not require the standardization process (Saaty [1974]) of the sample fuzzy information. Actually, the only reason to consider such a standardization is the ability to express the procedure in terms of a well-known measure in Statistical Information Theory, like the Kerridge inaccuracy.

We are now going to give sufficient conditions for the existence of minimum inaccuracy estimates, and conditions to construct a procedure to obtain minimum inaccuracy estimates for fuzzy samples from efficient estimates from exact samples.

Theorem 5 *Let $(X, \mathcal{B}_X, P_\theta), \theta \in \Theta$, be a random experiment in which $\{P_\theta, \theta \in \Theta\}$ is a parametric family of probability measures dominated by the counting or the Lebesgue measure, λ. Assume that the experiment satisfies the following regularity conditions: i) Θ is a real interval which is not a singleton; ii) the set $X^n = \{(x_1, x_2, \ldots, x_n) | L(x_1, x_2, \ldots, x_n; \theta) > 0\}$ does not depend on θ; iii) P_θ is associated with a parametric distribution function which is regular with respect to all its second θ-derivatives in Θ. Suppose that the sample fuzzy information $(\tilde{x}_1, \tilde{x}_2, \ldots, \tilde{x}_n)$ satisfies the following regularity conditions: iv) $\lambda(\tilde{x}_1, \tilde{x}_2, \ldots, \tilde{x}_n) = \int_{X^n} \mu_{(\tilde{x}_1, \tilde{x}_2, \ldots, \tilde{x}_n)|}(x_1, x_2, \ldots, x_n) d\lambda(x_1) d\lambda(x_2) \ldots d\lambda(x_n) < \infty$ and $\mathcal{I}(\tilde{x}_1, \tilde{x}_2, \ldots, \tilde{x}_n; \theta) < \infty$, for all $\theta \in \Theta$; v) the product function $H(\cdot) = \mu_{(\tilde{x}_1, \tilde{x}_2, \ldots, \tilde{x}_n)}(\cdot) log L(\cdot; \theta)$ is "regular" with respect to all its first and second θ-derivatives in Θ, in the sense that*

$$\frac{\partial}{\partial \theta} \mathcal{I}(\tilde{x}_1, \tilde{x}_2, \ldots, \tilde{x}_n; \theta) = - \int_{X^n} \mu_{|(\tilde{x}_1, \tilde{x}_2, \ldots, \tilde{x}_n)|}(x_1, x_2, \ldots, x_n)$$

$$\frac{\partial}{\partial \theta} log L(x_1, x_2, \ldots, x_n; \theta) d\lambda(x_1) d\lambda(x_2) \ldots d\lambda(x_n),$$

and

$$\frac{\partial^2}{\partial \theta^2} \mathcal{I}(\tilde{x}_1, \tilde{x}_2, \ldots, \tilde{x}_n; \theta) = - \int_{X^n} \mu_{|(\tilde{x}_1, \tilde{x}_2, \ldots, \tilde{x}_n)|}(x_1, x_2, \ldots, x_n)$$

$$\frac{\partial^2}{\partial \theta^2} log L(x_1, x_2, \ldots, x_n; \theta) d\lambda(x_1) d\lambda(x_2) \ldots d\lambda(x_n).$$

Under the regularity conditions i) - v), if there is an estimator of θ for the simple random sample $(X^n, \mathcal{B}_{X^n}, P_\theta), \theta \in \Theta$, whose variance attains the Fréchet-Cramér-Rao bound, then the inaccuracy equation, $\frac{\partial}{\partial \theta} \mathcal{I}(\tilde{x}_1, \tilde{x}_2, \ldots, \tilde{x}_n; \theta) = 0$, *admits a solution minimizing the inaccuracy $\mathcal{I}(\tilde{x}_1, \tilde{x}_2, \ldots, \tilde{x}_n; \theta)$ with respect to θ in Θ.*

Theorem 6 *Under the regularity conditions i) - v) in Theorem 5, let $T(X^n)$ be an estimator of θ for the simple random sample $(X^n, \mathcal{B}_{X^n}, P_\theta), \theta \in \Theta$, attaining the Fréchet-Cramér-Rao lower bound for the variance, and whose expected value is given by $E_\theta(T) = h(\theta)$, (h being a one-to-one real-valued function on Θ). Then, for the sample fuzzy information $(\tilde{x}_1, \tilde{x}_2, \ldots, \tilde{x}_n)$ the inaccuracy equation admits a unique solution minimizing the inaccuracy $\mathcal{I}(\tilde{x}_1, \tilde{x}_2, \ldots, \tilde{x}_n; \theta)$ and* *taking on the value* $\theta^\star(\tilde{x}_1, \tilde{x}_2, \ldots, \tilde{x}_n) \in \Theta$ *such that*

$$h(\theta^\star(\tilde{x}_1, \tilde{x}_2, \ldots, \tilde{x}_n)) = \int_{X^n} \mu_{|(\tilde{x}_1, \tilde{x}_2, \ldots, \tilde{x}_n)|}(x_1, x_2, \ldots, x_n)$$
$$T(x_1, x_2, \ldots, x_n) d\lambda(x_1) d\lambda(x_2) \ldots d\lambda(x_n).$$

In consequence, we can deduce the next relationship between the minimum inaccuracy estimates for fuzzy data and the maximum likelihood ones for exact data.

Corollary 1 *Under the regularity conditions i) - v) in Theorem 5, let* $T(X^n)$ *be an unbiased and efficient estimator of* θ *for the simple random sample* $(X^n, \mathcal{B}_{X^n}, P_\theta), \theta \in \Theta$ *(and, hence, it is the maximum likelihood estimator of* θ *for this random sample). Then, for* $(\tilde{x}_1, \tilde{x}_2, \ldots, \tilde{x}_n)$ *there is a unique minimum inaccuracy estimate given by the value*

$$\theta^\star(\tilde{x}_1, \tilde{x}_2, \ldots, \tilde{x}_n) = \int_{X^n} \mu_{|(\tilde{x}_1, \tilde{x}_2, \ldots, \tilde{x}_n)|}(x_1, x_2, \ldots, x_n)$$
$$T(x_1, x_2, \ldots, x_n) d\lambda(x_1) d\lambda(x_2) \ldots d\lambda(x_n).$$

It should be emphasized that if $(\tilde{x}_1, \tilde{x}_2, \ldots, \tilde{x}_n)$ is the algebraic product of $\tilde{x}_1$, $\tilde{x}_2$, ..., and $\tilde{x}_n$, and the statistic $T(X^n)$ is such that $T((x_1, x_2, \ldots, x_n) = \sum_{i=1}^n k(n) W(x_i)$, then the minimum inaccuracy estimate $\theta^\star(\tilde{x}_1, \tilde{x}_2, \ldots, \tilde{x}_n)$ satisfies that

$$h(\theta^\star(\tilde{x}_1, \tilde{x}_2, \ldots, \tilde{x}_n)) = \sum_{i=1}^n k(n) \int_X \mu_{|\tilde{x}_i|}(x) W(x) d\lambda(x).$$

Finally, in the study of properties of the minimum inaccuracy principle, we are going to analyze the property which formalizes the purpose for which that principle was introduced: to approximate (under certain conditions) the maximum likelihood method in the case of fuzzy data, and to coincide with it in the case of exact data. The last assertion is obvious, but the first one requires to add some assumptions on the involved fuzzy data. Thus,

Proposition 3 *Let* $(X, \mathcal{B}_X, P_\theta), \theta \in \Theta$, *be a random experiment in which* $\{P_\theta, \theta \in \Theta\}$ *is a parametric family of probability measures dominated by the counting or the Lebesgue measure,* λ. *Let* $\mathcal{X}$ *be a fuzzy information system associated with the experiment. Assume that: i*) if* $\{P_\theta, \theta \in \Theta\}$ *is dominated by the counting (or the Lebesgue) measure, then the sets* $supp(\tilde{x}), \tilde{x} \in \mathcal{X}$, *are discrete sets of consecutive points (or intervals) in* X, *being not necessarily disjoint, and such that* $\bigcup_{\tilde{x} \in \mathcal{X}} supp(\tilde{x}) = X$; *ii*) we agree that whenever* $n(\tilde{x}) = 0$ *(with* $n(\tilde{x})$ = *absolute frequency of* $\tilde{x}$ *in the sample information* $(\tilde{x}_1, \tilde{x}_2, \ldots, \tilde{x}_n)$*) implies that* $[\lambda(\tilde{x})]^{n(\tilde{x})} = 1$ *and* $n(\tilde{x})\mathcal{I}(\tilde{x}; \theta) = 0$, *whatever* $\lambda(\tilde{x})$ *and* $\mathcal{I}(\tilde{x}; \theta)$ *(finite or infinite) may be, then* $\lambda(\tilde{x}_1, \tilde{x}_2, \ldots, \tilde{x}_n) = \prod_{\tilde{x} \in \mathcal{X}} [\lambda(\tilde{x})]^{n(\tilde{x})} < \infty$ *and* $\mathcal{I}(\tilde{x}_1, \tilde{x}_2, \ldots, \tilde{x}_n; \theta) = \sum_{\tilde{x} \in \mathcal{X}} n(\tilde{x})\mathcal{I}(\tilde{x}; \theta) < \infty$; *iii*) if* $\lambda(X) = \infty$, *then for every reported datum* $\tilde{x} \in \mathcal{X}$ *such that* $\lambda(\tilde{x}) = \infty$, *we have that* $n(\tilde{x}) = 0$; *iv*)* $n(\tilde{x})$

decreases with respect to $\lambda(supp(\tilde{x}))$. Under conditions i) - v) and i) - iv*), the minimum inaccuracy method supplies an estimate of θ for the sample fuzzy information $(\tilde{x}_1, \tilde{x}_2, \ldots, \tilde{x}_n)$), which is a "good approximation" of the maximum likelihood one for fuzzy data.*

The extension of the minimum inaccuracy procedure and its properties to the case of vector-valued parameters could be developed in an analogous way.

10.4.2 Testing Hypothesis with Fuzzy Data. Likelihood Ratio Test

The aim of the problem of *(nonrandomized) testing (parameter) statistical hypothesis* on the basis of fuzzy experimental data from $(X, \mathcal{B}_X, P_\theta), \theta \in \Theta$, is to make use of the information contained in fuzzy data to conclude whether or not a given hypothesis about the experimental distribution could be accepted. One of the most useful methods in the construction of hypothesis tests, in the exact case, is the likelihood ratio one (Neyman and Pearson [1928, 1933]). This method has asymptotically convenient properties, under the conditions guaranteeing the existence of asymptotically normal maximum likelihood estimates, and is closely related to that estimation procedure when the test concerns a composite hypothesis. Assume that data reported from the observer to the statistician on the experiment $(X, \mathcal{B}_X, P_\theta), \theta \in \Theta$, are fuzzy, and one can construct a fuzzy information system, $\mathcal{X}$. If we consider a fuzzy random sample $\mathcal{X}^n$ from $\mathcal{X}$ and the probability space $(\mathcal{X}^n, \sigma(\mathcal{X}^n), P_\theta)$, we can state (see [Gil and Casals 1988]) that for testing $H_0 : \theta \in \Theta_0 \subset \Theta$ against $H_1 : \theta \in \Theta - \Theta_0$, on the basis of the fuzzy random sample $\mathcal{X}^n$, a nonrandomized test whose critical region is given by

$$\{(\tilde{x}_1, \tilde{x}_2, \ldots, \tilde{x}_n) \in \mathcal{X}^n | \Lambda(\tilde{x}_1, \tilde{x}_2, \ldots, \tilde{x}_n) < c\}$$

with

$$\Lambda(\tilde{x}_1, \tilde{x}_2, \ldots, \tilde{x}_n) = \frac{\sup_{\theta \in \Theta_0} \mathcal{L}(\tilde{x}_1, \tilde{x}_2, \ldots, \tilde{x}_n; \theta)}{\sup_{\theta \in \Theta} \mathcal{L}(\tilde{x}_1, \tilde{x}_2, \ldots, \tilde{x}_n; \theta)}$$

for some constant c, is called a *likelihood ratio test from $\mathcal{X}^n$*.

In many texts in Mathematical Statistics, one can find valuable references on testing hypothesis with exact data (see, for instance, [Wilks 1962], [Fourgeaud et Fuchs 1972], [Rohatgi 1976], [Lehmann 1986]).

An immediate property of the likelihood ratio test is the scale invariance with respect to the membership functions of fuzzy data. Consequently, the assumption of orthogonality does not mean a constraint for this procedure.

On the other hand, the likelihood ratio test is directly based on the induced distribution of samples, so that most of the properties in the exact case can be extended to the case of samples of fuzzy data whenever we consider the fuzzy random sample $\mathcal{X}^n$ and the induced probability space. In that sense, we have to point out that the asymptotic distribution of the statistic $\Lambda(\mathcal{X}^n)$, under regularity conditions, is preserved in the fuzzy case which allows us to determine the critical value of that statistic for each significance level.

Assume that the regularity conditions are satisfied, that is: Θ is a set in a euclidean space, $\mathbb{R}^k$ $(k \in N)$ containing an open set the true parameter value belongs to; for any fuzzy datum $\tilde{x} \in \mathcal{X}$ the induced probability $\mathcal{P}_\theta(\tilde{x})$ is twice differentiable with respect to all the components of $\theta \in \Theta$; for every $\tilde{x} \in \mathcal{X}$ we have that $P_\theta(supp(\tilde{x})) > 0$, whatever $\theta \in \Theta$ may be; and the Fisher information function for $\mathcal{X}$ in the true parameter value is positive. Then, we have that

Theorem 7 *For testing the null hypothesis* $H_0 : \theta \in \Theta_0 = \{(\theta_1, \dots, \theta_k) \in \Theta | \theta_1 = \theta_1^0, \dots, \theta_s = \theta_s^0\} (s \leq k)$ *against* $H_1 : \theta \in \Theta - \Theta_0$, *on the basis of the fuzzy random sample* $\mathcal{X}^n$, *the test rejecting* H_0 *for the sample fuzzy information* $(\tilde{x}_1, \tilde{x}_2, \dots, \tilde{x}_n)$ *whenever* $-2log\Lambda(\tilde{x}_1, \tilde{x}_2, \dots, \tilde{x}_n) > c_\alpha$, *with* $c_\alpha = 100(1-\alpha)$*th percentile of the chi-square with s degrees of freedom is a test having size approximately equal to* α, *as n tends to* ∞ *(more precisely, the distribution of* $-2log\Lambda(\mathcal{X}^n)$ *is approximately* χ^2_s, *as n tends to* ∞*).*

Main troubles in applying the likelihood ratio method to the case of fuzzy experimental data, arise in the process of obtaining the values in $\Lambda(\mathcal{X}^n)$, since this process forces us to compute the maximum likelihood estimates of θ in Θ_0 and in Θ for the available sample fuzzy information. By properly combining Proposition 1 and Theorem 7, it is possible to construct an approximation to the likelihood ratio test with fuzzy data which leads to good results when conditions i*) - iv*) in Proposition 1 are satisfied. This approximation is simply based on replacing the statistic $\Lambda(\mathcal{X}^n)$ by the statistic $\Lambda^*(\mathcal{X}^n)$, associating with the sample fuzzy information $(\tilde{x}_1, \tilde{x}_2, \dots, \tilde{x}_n)$ the value

$$\Lambda^*(\tilde{x}_1, \tilde{x}_2, \dots, \tilde{x}_n) = \frac{\mathcal{L}(\tilde{x}_1, \tilde{x}_2, \dots, \tilde{x}_n; \theta_0^*(\tilde{x}_1, \tilde{x}_2, \dots, \tilde{x}_n))}{\mathcal{L}(\tilde{x}_1, \tilde{x}_2, \dots, \tilde{x}_n; \theta^*(\tilde{x}_1, \tilde{x}_2, \dots, \tilde{x}_n)},$$

where $\theta_0^*(\tilde{x}_1, \tilde{x}_2, \dots, \tilde{x}_n)$ and $\theta^*(\tilde{x}_1, \tilde{x}_2, \dots, \tilde{x}_n)$ are the minimum inaccuracy estimates of θ in Θ_0 and in Θ, respectively. In particular, when the null hypothesis to test by means of the likelihood procedure is a simple one, we have that $-2log\Lambda(\mathcal{X}^n) \geq -2log\Lambda^*(\mathcal{X}^n)$, whence the rejection of the hypothesis by the approximate test ensures the rejection by means of the extension in Theorem 7.

An approach different to the preceding ones has been given by Manton; Woodbury; Tolley [1994], by stating a complex model to deal with high dimensional fuzzy data in statistical data analysis. Additional related studies can be found in the literature. Among the most recent ones, we can refer to [Viertl 1987], the papers in [Kacprzyk and Fedrizzi 1988], [Anisimov 1989], [Fruhwirthschnatter 1992], [Cai 1993], [Casals 1993], [Watanabe and Imaizumi 1993], [Casals and Gil 1994], [Römer and Kandel 1995], and [Römer; Kandel; Backer 1995].

Other studies connecting Probability Theory and Fuzzy Logic that should be mention in this section are those relating Random Set and Fuzzy Set Theories (see, for instance, [Goodman 1982], [Goodman and Nguyen 1985], [Wang 1982, 1989], [Wang and Sanchez 1983], [Gil 1992], and those relating Probability and

Possibility Theories (see, for instance, [Dubois and Prade 1986, 1989, 1993], and [Delgado and Moral 1987]).

10.5 APPROACHES TO MODELING FUZZY-BAYES STATISTICAL DECISION MAKING

In a decision problem, a decision maker has to make a choice from the set of all possible *actions*), and consequences of choosing a decision are assumed to depend on some uncertainties, which are referred to as *states*. In a decision problem under uncertainty the decision maker is assumed to be unable to control and know the true state (that means the state that actually occurs) before making the decision.

In a statistical decision problem under uncertainty,to obtain further information about the true state, the decision maker can perform an experiment $(X, \mathcal{B}_X, P_\theta)$, where the experimental distribution, P_θ, depends on the true state $\theta \in \Theta$.

Fuzziness could be involved in elements of a statistical decision problem. We are now going to analyze two approaches to deal with problems in which either the sample information or the quantification of consequences are assumed to be fuzzy.

10.5.1 An Approach to Modeling Statistical Decision Making with Fuzzy Data

Indeed, the model and objectives in the approach to solve statistical inferential problems involving fuzzy data, could also be considered to solve statistical decision problems when the sample information is fuzzy. In that sense, Tanaka; Okuda; Asai [1979], Gil; López; Gil [1984], and Gil [1987] have examined statistical decision problems in a Bayesian setting and involving fuzzy experimental data, so that the elements of the problems are the following:

- the state space, Θ;
- the action space, A;
- the utility function, u, which is assumed to be a real-valued function on $\Theta \times A$ and quantifying consequences in accordance with decision maker's preferences;
- the sample information supplied, by a fuzzy information system $\mathcal{X}$ associated with $(X, \mathcal{B}_X, P_\theta), \theta \in \Theta$;
- the prior distribution ξ defined on a measurable space $(\Theta, \mathcal{C})$ associated with Θ.

In this new framework, Bayes' principle of choice among actions could be easily applied by considering the *expected utility* of an action $a \in A$ *with respect to the distribution* π on $(\Theta, \mathcal{C})$ given (see [Tanaka; Okuda; Asai 1979], [Gil;

López; Gil 1984], and [Gil 1987]) by

$$u(\pi, a) = \int_{\Theta} u(\theta, a)\, d\pi(\theta),$$

and an action $a^{\pi} \in A$ such that $u(\pi, a^{\pi}) = \max_{a \in A} u(\pi, a)$ is called a *Bayes action with respect to* π. In particular, when π is the prior distribution, ξ, a^{π} is referred to as a *prior Bayes action*, and when π is the (induced) posterior distribution given $\tilde{x} \in \mathcal{X}$,$\xi_{\tilde{x}}$, then a^{π} is referred to as a *posterior Bayes action* given $\tilde{x}$.

The induced posterior distribution $\xi_{\tilde{x}}$ is intended (cf. [Okuda; Tanaka; Asai 1978], Tanaka; Okuda; Asai 1979], [Gil 1987]) as follows:

$$d\xi_{\tilde{x}}(\theta) = \frac{\mathcal{P}_{\theta}(\tilde{x})\, d\xi(\theta)}{\int_{\Theta} \mathcal{P}_{\theta}(\tilde{x})\, d\xi(\theta)}.$$

One of the basic modes of Bayes decision analysis of a statistical problem with fuzzy experimental data, will be the *extensive form of Bayes decision analysis* (the name being the same used for the analogue nonfuzzy mode - see [Raiffa and Schlaifer 1961], [Berger 1993], [Pratt; Raiffa; Schlaiffer 1995] -). This form proceeds as follows: if the observer report from experimentation leads to the fuzzy information $\tilde{x}$, a posterior Bayes action $a^{\xi_{\tilde{x}}}$ is choosen, and the value associated with the decision problem can then be measured by means of the *expected terminal utility* associated with $\mathcal{X}$ given by

$$V(\mathcal{X}) = \sum_{\tilde{x} \in \mathcal{X}} \mathcal{P}(\tilde{x}) \int_{\Theta} u(\theta, a^{\xi_{\tilde{x}}}) d\xi_{\tilde{x}}(\theta),$$

where $\mathcal{P}(\tilde{x}) = \int_{\Theta} \mathcal{P}_{\theta}(\tilde{x})\, d\xi(\theta)$ is the induced marginal probability of $\tilde{x}$.

On the other hand, we can consider *decision rules* based on $\mathcal{X}$, $d \in \mathcal{D}_{\mathcal{X}}$, d mapping $\mathcal{X}$ into A, and so that two decision rules $d, d' \in \mathcal{D}_{\mathcal{X}}$ are considered to be equivalent when $d(\tilde{x}) = d'(\tilde{x})$ for all $\tilde{x} \in \mathcal{X}$. Thus, if $d \in \mathcal{D}_{\mathcal{X}}$, the *normal form of Bayes decision analysis* would proceed by considering that the *expected utility* of decision d with respect to the probability distribution π on $(\theta, \mathcal{C})$, given by

$$u(\pi, d(\mathcal{X})) = \int_{\Theta} \sum_{\tilde{x} \in \mathcal{X}} u(\theta, d(\tilde{x}))\, \mathcal{P}_{\theta}(\tilde{x})\, d\pi(\theta)$$

and a decision rule $d^{\pi} \in \mathcal{D}_{\mathcal{X}}$ is said to be a *Bayes decision rule with respect to* π for the decision problem, if and only if

$$u(\pi, d^{\pi}(\mathcal{X})) = \max_{d \in \mathcal{D}_{\mathcal{X}}} u(\pi, d(\mathcal{X})).$$

The value associated with the decision problem could now be measured by the *greatest expected utility* associated with $\mathcal{X}$ given by

$$W(\mathcal{X}) = u(\xi, d^{\xi}(\mathcal{X})).$$

Under quite general conditions, the extensive and normal forms of Bayes analysis of a decision problem with fuzzy experimental data are equivalent. Thus,

Theorem 8 *Consider a decision problem in a Bayesian framework, with state space* Θ*, and action space* A*. Let* $(\Theta, \mathcal{C})$ *be a measurable space associated with* Θ*, and let* ξ *be a prior distribution. Let* u *be a real-valued utility function associated with this problem. Then, if the order of integration and summation in* $V(\mathcal{X})$ *and* $W(\mathcal{X})$ *can be reversed, we have that* $V(\mathcal{X}) = W(\mathcal{X})$.

For operational purposes, the extensive form is the most convenient and the simplest to be used, whenever they coincide. However, when the class of decision rules is restricted to a subset of $\mathcal{D}_{\mathcal{X}}$, the normal form could be needed.

10.5.2 An Approach to Modeling Statistical Decision Making with Fuzzy Utilities

On the other hand, another problem of statistical decision making with fuzzy elements has been examined in the literature of fuzzy decision analysis (see, for instance, [Watson; Weiss, Donnell 1979], [Freeling 1980], [Dubois and Prade 1982], [Whalen 1984], and [Gil and Jain 1992]): the statistical decision problem with fuzzy-valued consequences.

Utility Theory deals with the development of the assessment of utilities, or "values" of consequences which are assumed to be stated in agreement with decision maker's preferences. In the literature on decision problems (see, for instance, [Berger 1993]), it is sometimes argued that the the utilities are often very vague or even nonunique.

Gil and López-Díaz [1996] have recently developed a general mathematical model (extending those in the referred previous studies in case of real-valued probabilities) to properly deal with the problem above considered. The main purpose of this model is to avoid the decision maker is compeled to evaluate utilities in a numerical scale. To this purpose, the scale of measurement is first enlarged from $\mathbb{R}$ to $\mathcal{F}_{cc}(\mathbb{R})$(set of normalized convex fuzzy sets of $\mathbb{R}$, whose level sets and support closure are compact).

The elements of a statistical decision problem with fuzzy utilities are the following:

- the state space, Θ;
- the action space, A;
- the utility function, $\mathcal{U}$, which is assumed to be a fuzzy-valued function on $\Theta \times A$ and quantifying consequences in accordance with decision maker's preferences;
- the sample information supplied, by a random experiment $(X, \mathcal{B}_X, P_\theta), \theta \in \Theta$;

- the prior distribution ξ defined on a measurable space $(\Theta, \mathcal{C})$ associated with Θ.

Then, the (fuzzy) utility function will be formalized (Gil and López-Díaz [1996]) as follows:

Definition 9 A *fuzzy utility function* for the decision problem is a mapping $\mathcal{U} : \Theta \times A \to \mathcal{F}_{cc}(\mathbb{R})$, such that

1. for each action $a \in A$, the projection $\mathcal{U}(.,a) : \Theta \to \mathcal{F}_{cc}(R)$, is an integrably bounded fuzzy random variable, for which the real-valued mapping $\mathrm{cl}(\mathrm{supp}(\mathcal{U}(.,a)))$ is also integrably bounded, and its fuzzy expected value [Puri and Ralescu 1986] with respect to ξ is a fuzzy number, the *fuzzy expected utility* that will be denoted by $\widetilde{E}^{\pi}[\mathcal{U}(\theta,a)]$;

2. whatever actions $a, a' \in A$ may be, a is considered to be *preferred or indifferent to* a', if and only if

$$V_L^{\lambda}(\widetilde{E}^{\pi}[\mathcal{U}(\theta,a)]) \geq V_L^{\lambda}(\widetilde{E}^{\pi}[\mathcal{U}(\theta,a')]),$$

where (see [Campos and González 1989])

$$V_L^{\lambda}(\widetilde{A}) = \int_0^1 [\lambda a_{\alpha 2} + (1-\lambda) a_{\alpha 1}]\, dL(\alpha),$$

$\widetilde{A}_\alpha = [a_{\alpha 1}, a_{\alpha 2}]$, $\lambda \in [0,1]$, and L being the Lebesgue measure on the unit interval $[0,1]$.

The notion of fuzzy random variable as intended by [Puri and Ralescu 1986], (see also [Klement; Puri; Ralescu 1986]) differs from the one introduced by Kwakernaak [1978] and considered in Section 10.3 (see also [Ralescu 1995b]). The essential difference in that Kwakernaak's one are based on the measurability of real-valued functions (inf and sup of α-cuts), whereas the othe one is based on the measurability of set-valued functions, the α-cut functions).

Then, on the basis of the preceding definition, we can conclude that

Theorem 9 *Given a decision problem with consequence space $\mathcal{R}$, any set of axioms guaranteeing the existence of a bounded real-valued utility function on $\mathcal{R}$, being unique up to an increasing linear transformation and agreeing with the preference relation on the set of probability distributions (lotteries) on $\mathcal{R}$, $\mathcal{P}$, through the expected utilities, also guarantees the existence of a class of fuzzy utility functions on $\mathcal{R}$ agreeing with the preference relation through the fuzzy expected utilities.*

Dubois and Prade [1995] have recently developed an axiomatic approach to the qualitative quantification of utilities, as a counterpart to von Neumann and Morgenstern [1953] expected utility theory in the setting of Possibility Theory. Nevertheless, in Dubois and Prade's approach, "lotteries" are assumed to be *possibility* distributions on $\mathcal{R}$, which means a definitely different problem to what is considered in this subsection.

On the other hand, it should be emphasized that the nonuniqueness of a fuzzy utility function for a given set of axioms, should be viewed as an advantage with respect to the traditional case, rather than an inconvenience, since as a consequence the decision maker does not feel forced to assess utilities in a unique way (up to an increasing linear transformation).

It should be also remarked that the decision problem with fuzzy consequences only reduces to a problem with real-valued ones, in stages where decisions have to be made. However, there are stages in this problem whose outcomes will not be reduced to crisp ones. These stages are those concerning the computation of some useful values like fuzzy expected utilities, values associated with experimentation, and so on.

The ideas of Bayes' principle of choice among actions without and with experimentation can be easility extended as follows:

Consider a decision problem in a Bayesian framework, with state space Θ, and action space A. Let $(\Theta, \mathcal{C})$ be a measurable space associated with Θ, and let π be a probability distribution on $\mathcal{C}$. Then, (see [Gil and López-Díaz 1996]) the *fuzzy expected utility* of action $a \in A$ *with respect to* π is given by

$$\widetilde{\mathcal{U}}(\pi, a) = \widetilde{E}^{\pi}[\mathcal{U}(\theta, a)],$$

and an action (assuming there exists) $a^{\pi} \in A$ is said to be a *Bayes action with respect to* π for the decision problem with fuzzy utilities, if and only if

$$V_L^{\lambda}(\widetilde{\mathcal{U}}(\pi, a^{\pi})) \geq V_L^{\lambda}(\widetilde{\mathcal{U}}(\pi, a))$$

for all $a \in A$.

As in the approach to modeling statistical decision making with fuzzy data, one of the basic modes of Bayes decision analysis of a statistical problem with fuzzy utilities and involving experimentation, will be the *extensive form of Bayes decision analysis.* This form proceeds as follows: if experimentation from $(X, \mathcal{B}_X, P_\theta)$ leads to the outcome x, a posterior Bayes action a_x^{ξ} is choosen, and the value associated with the decision problem can be then measured by means of a *fuzzy expected terminal utility* associated with $(X, \mathcal{B}_X, P_\theta)$ given by

$$\widetilde{\mathcal{V}}(X) = \widetilde{E}^{P}[\widetilde{E}^{\xi_x}[\widetilde{\mathcal{U}}(\theta, a_x^{\xi})]] = \widetilde{E}^{P}[\widetilde{\mathcal{U}}(\xi_x, a_x^{\xi})],$$

P being the marginal distribution of $(X, \mathcal{B}_X, P_\theta)$.

On the other hand, we can consider *decision rules* based on X, $d \in \mathcal{D}_X$, d mapping X into A, and so that two decision rules $d, d' \in \mathcal{D}_X$ are considered to be equivalent when $d(x) = d'(x)$ a. s. $[P_\theta]$ for all $\theta \in \Theta$. Then, the *normal form of Bayes decision analysis* would proceed as follows (see [Gil and López-Díaz 1996]): if $d \in D_x$, the *fuzzy expected utility* of decision d with respect to π is given by

$$\widetilde{\mathcal{U}}(\pi, d(X)) = \widetilde{E}^{\pi}[\widetilde{E}^{P_\theta}[\mathcal{U}(\theta, d(x))]],$$

and a decision rule (assuming there exists) $d^{\pi} \in \mathcal{D}_X$ is said to be a *Bayes decision rule with respect to* π for the decision problem, if and only if for all

$d \in \mathcal{D}_X$

$$V_L^\lambda(\widetilde{\mathcal{U}}(\pi, d^\pi(X))) \geq V_L^\lambda(\widetilde{\mathcal{U}}(\pi, d(X))).$$

The value associated with the decision problem could now be measured by a value of the *greatest fuzzy expected utility* associated with $(X, \mathcal{B}_X, P_\theta)$ is given by

$$\widetilde{\mathcal{W}}(X) = \widetilde{\mathcal{U}}(\xi, d^\xi(X)).$$

Obviously, both, $\widetilde{\mathcal{V}}(X)$ and $\widetilde{\mathcal{W}}(X)$, are not always well-defined, in the sense that each of them are not necessarily given by a unique fuzzy value.

However, $\widetilde{\mathcal{V}}(X)$ and $\widetilde{\mathcal{W}}(X)$ are well-defined, up to an indifference relation, in terms of the ranking ordering by Campos and González. In addition, under the conditions in Definition 9, the extensive and normal forms of analysis of a decision problem with fuzzy utilities could be considered "equivalent", since a comparison between them lead to indifference, that is,

Theorem 10 *Consider a decision problem in a Bayesian framework, with state space Θ, and action space A. Let $(\Theta, \mathcal{C})$ be a measurable space associated with Θ, and let π be a prior distribution on $\mathcal{C}$. Let $\mathcal{U}$ be a fuzzy utility function associated with this problem. Then,*

$$V_L^\lambda(\widetilde{\mathcal{V}}(X)) = V_L^\lambda(\widetilde{\mathcal{W}}(X)).$$

For operational purposes, the extensive form is the most convenient and the simplest to be used. However, when the class of decision rules is restricted to a subset of $\mathcal{D}_X$, the normal form could be needed.

Operative criteria to rank fuzzy numbers, different from Campos and González's one, could be alternatively considered. Nevertheless, it would be useful that such criteria were based on a real-valued function (like V_L^λ) and satisfy properties similar to those required for proving Theorems 9 and 10.

The statistical decision problem examined in the second part of this section could be modeled and managed in other different ways (following, for instance, the ideas in [Lamata 1994], [Dubois and Prade 1995], and [Thomas 1995], in decision problems with no experimentation). Anyway, the model presented in this subsection, operationally generalized the traditional one by means of the use of fuzzy random variables, and it provides a rigourous formalization supplying results like those we have presented.

10.6 CONCLUDING REMARKS

The analyses in Sections 10.3 to 10.5 could be directly particularized to the case in which data received by the statistician are grouped. This particularization would simply consist in replacing the membership functions of the fuzzy data by the indicators of the grouped ones (cf. [Gil and Corral 1987]).

On the other hand, the studies in the preceding sections can also be applied to the case in which the imprecision in data reported from the observer is random. In this way, one can sometimes have enough evidence to assert that for a given value of the random variable the reported information varies from observer to observer, or from outcome to outcome (or with respect to another element in the experiment), but it is not fuzzy for each observer. In those circumstances, data have to be modeled by means of random sets. In general, whenever observer reports are formalized in terms of random sets, the studies in the preceding sections could be particularized by simply replacing membership functions by coverage ones (see [Goodman 1982], [Gil 1992]).

Acknowledgments

The research in this paper has been supported in part by DGES Grant No. PB95-1049. Its financial support is gratefully acknowledged.

References

Anisimov, V.Y. (1989). Parameter estimation in the case of fuzzy information on the observation conditions, *Telecommunications and Radio Engineering*, 44, 86-88.

Artstein, Z. and Vitale, R.A. (1975). A Strong Law of Large Numbers for Random Compact Sets, *Annals of Probability*, 3, 879-882.

Bandemer, H. and Näther, W. (1992). *Fuzzy Data Analysis*, Kluwer, Series B: Mathematical and Statistical Methods, Dordrecht.

Berger, J.O. (1993). *Statistical Decision Theory and Bayesian Analysis*, 3rd ed., Springer-Verlag, New York.

Campos, L.M. and González, A. (1989). A subjective approach for ranking fuzzy numbers, *Fuzzy Sets and Systems*, 29, 145-153.

Carter, W.H. and Myers, R.H. (1974). The effect of grouping on the variance and bias of the maximum likelihood estimator of the Poisson parameter - Some Monte Carlo results, *Journal of American Statistical Association*, 69, 482-484.

Casals, M.R. (1993). Bayesian testing of fuzzy parametric hypotheses from fuzzy information, *R.A.I.R.O.-Recherche Opérationnelle*, 27, 189-199.

Casals, M.R. and Gil, P. (1994). Bayesian sequential test for fuzzy parametric hypotheses from fuzzy information, *Information Sciences*, 80, 283-298.

Cai, K.Y. (1993). Parameter estimations of normal fuzzy variables, *Fuzzy Sets and Systems*, 55, 179-185.

Corral, N. and Gil, M.A. (1984). The minimum inaccuracy fuzzy estimation: an extension of the maximum likelihood principle, *Stochastica*, VIII, 63-81.

Cramér, H. (1974). *Mathematical Methods of Statistics*, Princeton University Press, Princeton.

Czogała, E. and Hirota, K. (1986). *Probabilistic Sets: Fuzzy and Stochastic Approach to Decision Control and Recognition Processes.* Verlag TÜV Rheinland, Köln.

Delgado, M. and Moral, S. (1987). On the concept of possibility-probability consistency, *Fuzzy Sets and Systems*, 21, 311-318.

Dubois, D. and Prade, H. (1980). *Fuzzy Sets and Systems. Theory and Applications*, Academic Press, New York.

Dubois, D. and Prade, H. (1982). The use of Fuzzy Numbers in Decision Analysis, in: Gupta, M.M. and Sanchez, E. (Eds.), *Fuzzy Information and Decision Processes*, North-Holland, Amsterdam, 309-321.

Dubois, D. and Prade, H. (1986). Fuzzy sets and statistical data, *European Journal of Operational Research*, 25, 345-356.

D. Dubois and H. Prade (1988). *Possibility Theory*, Plenum Press, New York

Dubois, D. and Prade, H. (1989). Fuzzy sets, probability and measurement, *European Journal of Operational Research*, 40, 135-154.

Dubois, D. and Prade, H. (1993). Fuzzy sets and probability: misunderstandings, bridges and gaps, *Proc. 2nd IEEE Int. Conf. on Fuzzy Systems* (San Francisco), 1059-1068.

D. Dubois and S. Moral and H. Prade (1993). A Semantics for Possibility Theory Based on Likelihoods, *Tech. Rep. CEC-Esprit III BRA 6156 DRUMS II.*

Dubois, D. and Prade, H. (1995). A qualitative decision theory based on possibility theory, *Proc. 14th Int. Joint Conf. on Art. Intell.* (Montreal), 1924-1930.

Fisher, R.A. (1922). On the mathematical foundations of theoretical statistics, *Philosophical Transactions of the Royal Society of London*, Series A, 222, 309-368.

Fourgeaud, G. et Fuchs, A. (1972). *Statistique*, Dunod, Paris.

Freeling, A.N.S. (1980). Fuzzy Sets and Decision Analysis, *IEEE Transactions on Systems, Man and Cybernetics*, 10, 341-354.

French, S. (1984). Fuzzy Decision Analysis: Some Criticisms, in: Zimmermann, H.-J., Zadeh, L.A. andGaines, B.R. (Eds.), *Fuzzy Sets and Decision Analysis*, North-Holland, Amsterdam, 29-44.

Fruhwirthschnatter, S. (1992). On statistical inference for fuzzy data with applications to descriptive statistics, *Fuzzy Sets and Systems*, 50, 143-165.

Gebhardt, J. (1992). On the Epistemic View of Fuzzy Statistics. in: Bandemer, H. (Ed.), *Modelling Uncertain Data*, Series: Mathematical Research - Akademie-Verlag, Vol. 68, 136-141.

Gebhardt, J. and Kruse, R. (1993). A New Approach to Semantic Aspects of Possibilistic Reasoning, in: Clarke, M., Kruse, R. and Moral, S. (Eds.), *Symbolic and Quantitative Approaches to Reasoning and Uncertainty*, Lecture Notes in Computer Science, 747 - Springer-Verlag, 151-160.

Gebhardt J. and Kruse, R. (1995). Learning possibilistic networks from data, *Proc. 5th Int. Workshop on Artificial Intelligence and Statistics* (Fort Lauderdale), 233-244.

Gil, M.A., López, M.T. and Gil, P. (1984). Comparison between fuzzy information systems, *Kybernetes*, 13, 245-251.

Gil, M.A. (1987). Fuzziness and loss of information in statistical problems, *IEEE Transactions on Systems, Man and Cybernetics*, 17, 1016-1025.

Gil, M.A. and Corral, N. (1987). The Minimum inaccuracy principle in estimating population parameters from grouped data, *Kybernetes*, 16, 43-49.

Gil, M.A. (1988). On the Loss of Information due to Fuzziness in Experimental Observations, *Ann. Inst. Statist. Math.*, 40, 627–639.

Gil, M.A. and Casals, M.R. (1988). An operative extension of the likelihood ratio test from fuzzy data, *Statistical Papers*, 29, 191-203.

Gil, M.A., Corral, N. and Gil, P. (1988). The minimum inaccuracy estimates in χ^2 tests for goodness of fit with fuzzy observations, *Journal of Statistical Planning and Inference*, 19, 95-115.

Gil, M.A., Corral, N. and Casals, M.R. (1989). The likelihood ratio test for goodness of fit with fuzzy experimental observations, *IEEE Transactions on Systems, Man and Cybernetics*, 19, 771-779.

Gil, M.A. (1992). A note on the connection between fuzzy numbers and random intervals, *Statistics and Probability Letters*, 13, 311-319.

Gil, M.A. and Jain, P. (1992). Comparison of experiments in statistical decision problems with fuzzy utilities, *IEEE Transactions on Systems, Man and Cybernetics*, 22, 662-670.

Gil, M.A. and López-Díaz, M. (1996). Fundamentals and Bayesian analyses of decision problems with fuzzy-valued utilities, *International Journal of Approximate Reasoning*, 15,203, 224.

Goodman, I.R. (1982). Fuzzy sets as equivalence classes of possibility random sets, in: Yager, R.R. (Ed.), *Fuzzy Set and Possibility Theory: Recent Development*, Pergamon, Oxford, 327-343.

Goodman, I.R. and Nguyen, H.T. (1985). *Uncertainty Models for Knowledge Based Systems*, North-Holland, Amsterdam.

Hirota, K. (1981). Concepts of Probabilistic Sets, *Fuzzy Sets and Systems*, 5, pp 31-46.

Jain, P. and Agogino, A. (1988). Calibration of fuzzy linguistic variables for expert systems, in: Tipnis, V.A. and Patton, E.M. (Eds.), *Computers in Engineering 1988* - Vol. 1, American Society of Mechanical Engineering, New York, 313-318.

Kacprzyk, J. and Fedrizzi, M. (Eds.) (1988). *Combining fuzzy imprecision with probabilistic uncertainty in decision making*, Lecture Notes in Economics and Mathematical Systems, 310, Springer-Verlag, Berlin.

Kahneman, D., Slovic, P. and Tversky, A. (1985). *Judgement under Uncertainty: Heuristics and Biases*, Cambridge University Press, Cambridge, MA.

Kampé de Fériet, J. (1982). Interpretation of Membership Functions of Fuzzy Sets in Terms of Plausibility and Belief, in: Gupta, M.M. and Sanchez, E. (Eds.), *Fuzzy Information and Decision Processes*, North-Holland, Amsterdam, 13-98.

Kandel, A. (1979). On Fuzzy Statistics, in: Gupta, M.M., Ragade, R.K. and Yager, R.R. (Eds.), *Advances in Fuzzy Set Theory and Applications*, North-Holland, Amsterdam, 181-200.

Kendall, D.G. (1974). Foundations of a Theory of Random Set, in: Harding, E.F. and Kendall, D.G., (Eds.), *Stochastic Geometry*, John Wiley, New York, 322-376.

Kerridge, D.F. (1961). Inaccuracy and inference, *Journal of the Royal Statistical Society*, Series B, 23, 184-194.

Klement, E.P., Puri, M.L. and Ralescu, D.A. (1986). Limit Theorems for Fuzzy Random Variables. *Proceedings of the Royal Society of London - Series A*, 19, 171-182.

Kruse, R. (1982). The Strong Law of Large Numbers for Fuzzy Random Variables, *Information Sciences*, 12, 53-57.

Kruse, R. and Meyer, K.D. (1987). *Statistics with Vague Data*, Reidel Publ. Co., Dordrecht.

Kruse, R. (1987). On a Software Tool for Statistics with Linguistic Data *Fuzzy Sets and Systems*, 24, 377-383.

Kruse, R. and Gebhardt, J. (1989). On a Dialog System for Modelling and Statistical Analysis of Linguistic Data, *Proc. IFSA Congress* (Seattle), 157-160.

Kruse, R. and Gebhardt, J. (1990). Some New Aspects of Testing Hypothesis in Fuzzy Statistics. *Proc. NAFIPS'90 Conference* (Toronto), 185-187.

Kruse, R., Gebhardt, J. and Klawonn, F. (1994). *Foundations of Fuzzy Systems*, John Wiley, New York.

Kwakernaak, H. (1978). Fuzzy random variables. Part I, *Information Sciences*, 15, 1-29.

Lamata, M.T. (1994). A model of decision with linguistic knowledge, *Mathware & Soft Computing*, 1, 253-263.

Le Cam, L. (1964). Sufficiency and approximate sufficiency, *Ann. Math. Statist.*, 35, 1419-1455.

Le Cam, L. (1986). *Asymptotic Methods in Statistical Decision Theory*, Springer-Verlag, New York.

Lehmann, E.L. (1986). *Testing Statistical Hypotheses*, John Wiley, New York.

López-Díaz, M. (1996). *Medibilidad e integración de variables aleatorias difusas. Aplicación a problemas de decisión*, PhD Thesis, Universidad de Oviedo.

Manton, K.G., Woodbury, M.A. and Tolley, H.D. (1994). *Statistical applications using fuzzy sets*, John Wiley, New York.

Matheron, G. (1975). *Random Sets and Integral Geometry*, John Wiley, New York

McDonald, J.B. and Ransom, M.R. (1979). Alternative parameter estimators based upon grouped data, *Communications in Statistics - Theory and Methods*, 8, 899-917.

Meyer, K.D. (1987). *Grenzwerte zum Schatzen von Parametern unscharfer Zufallsvariablen*, PhD Thesis, University of Braunschweig.

Miyakoshi, M. and Shimbo, M. (1984). A Strong Law of Large Numbers for Fuzzy Random Variables, *Fuzzy Sets and Systems*, 12, 133-142.

Negoita, C.V. and Ralescu, D.A. (1975). *Applications of fuzzy sets to systems analysis*, John Wiley, New York.

Negoita, C.V. and Ralescu, D.A. (1987). *Simulation, Knowledge-based Computing, and Fuzzy Statistics*, Van Nostrand Reinhold, New York.

Neyman, J. and Pearson, E.S. (1928). On the use and interpretation of certain test criteria for purposes of statistical inference, *Biometrika*, 20A, Part I, 175-240, Part II, 263-294.

Neyman, J. and Pearson, E.S. (1933). On the problem of the most efficient tests of statistical hypotheses, *Transactions of the Royal Society of London, Series A*, 231, 289-337.

Nguyen, H.T. (1978). On Random Sets and Belief Functions, *Journal of Mathematical Analysis and Applications*, 65, 531-542.

Nguyen, H.T. (1979). Some mathematical tools for linguistic probabilities, *Fuzzy Sets and Systems*, 2, 53-65.

Okuda, T., Tanaka, H. and Asai, K. (1978). A formulation of fuzzy decision problems with fuzzy information, using probability measures of fuzzy events, *Information and Control*, 38, 135-147.

Pratt, J.W., Raiffa, H. and Schlaifer, R. (1995). *Introduction to Statistical Decision Theory*, The MIT Press, Cambridge.

Puri, M.L. and Ralescu, D.A. (1985). The concept of normality for fuzzy random variables, *Annals of Probability*, 13, 1373-1379.

Puri, M.L. and Ralescu, D.A. (1986). Fuzzy random variables, *Journal of Mathematical Analyysis and Applications*, 114, 409-422.

Raiffa, H. and Schlaifer, R. (1961). *Applied Statistical Decision Theory*, Graduate School of Business, Harvard University, Boston.

Ralescu, D.A. (1982). Fuzzy Logic and Statistical Estimation, *Proc. 2nd World Conf. on Mathematics at the Service of Man* (Canarias), 605–606.

Ralescu, D.A. (1995a). Fuzzy probabilities and their applications to statistical inference, in: Bouchon-Meunier, b. Yager, R.R. and Zadeh, L.A. (Eds.), *Advances in Intelligent Computing*, Lecture Notes in Computer Science, 945, 217-222.

Ralescu, D.A. (1995b). Fuzzy random variables revisited, *Proc. IFES'95 and Fuzzy IEEE Joint Conf.* (Yokohama), Vol. 2, 993-1000.

Rappoport, A., Wallsten, T.S. and Cox, J.A. (1987). Direct and indirect scaling of membership functions of probability phrases, *Mathematical Modelling*, 9, 397-418.

Rathie, P.N. (1971). On some new measures of uncertainty, inaccuracy and information and their characterizations, *Kybernetika*, 7, 394-403.

Rohatgi, V.K. (1976). *An Introduction to Probability Theory and Mathematical Statistics*, New York: John Wiley.

Römer, C. and Kandel, A. (1995). Statistical tests for fuzzy data, *Fuzzy Sets and Systems*, 72, 1-26.

Römer, C., Kandel, A. and Backer, E. (1995). Fuzzy partitionns of the sample space and fuzzy parameter hypotheses, *IEEE Transactions on Systems, Man and Cybernetics*, 25, 1314-1322.

Saaty, T.L. (1974). Measuring the fuzziness of sets, *Journal of Cybernetics*, 4, 53-61.

Schader, M. und Schmid, F. (1988). Maximum-likelihood-schdtzung aus gruppierten data - eine —bersicht, *OR Spektrum*, 10, 1-12.

Shafer, G. (1976). *A Mathematical Theory of Evidence.* Princeton University Press, Princeton.

Stoyan, D., Kendall, W.S. and Mecke, J. (1987). *Stochastic Geometry and its Applications*, John Wiley, New York.

Sundberg, R. (1974). Maximum likelihood theory for incomplete data from an exponential family, *Scandinavian Journal of Statistics*, 1, 49-58.

Tanaka, H., Okuda, T. and Asai, K. (1979). Fuzzy information and decision in statistical model, in: Gupta, M.M., Ragade, R.K. and Yager, R.R. (Eds.), *Advances in Fuzzy Sets Theory and Applications*, North-Holland, Amsterdam, 303-320.

Tanaka, H. (1987). Fuzzy Data Analysis by Possibilistic Linear Models, *Fuzzy Sets and Systems*, 24, 363–375

Thomas, S.F. (1995). *Fuzziness and Probability*, ACG Press, Wichita, Kansas.

Utkin, L.V. (1993). Uncertainty importance of sytem components by fuzzy and interval probability, *Microelectronics and Reliability*, 33, 1357-1364.

Vajda, I. (1973). χ^2-divergence and generalized Fisher's information, *Trans. 6th Prague Conf. Inform. Theory, Stat. Dec. Func., Rand. Proc.*, 873-886.

Viertl, R. (1987). Is it necessary to develop a fuzzy Bayesian inference, in: *Probability and Bayesian Statistics*, Plenum Press, New York, 471-475.

Viertl, R. (1992). On Statistical Inference Based on Non-precise Data, in: Bandemer, H. (Ed.), *Modelling Uncertain Data*, Series: Mathematical Research. - Akademie-Verlag, Vol 68, 121–130

Von Neumann, J. and Morgenstern, 0. (1953). *Theory of Games and Economic Behavior*, 3rd ed., Princeton University Press, Princeton.

Walley, P. (1991). *Statistical Reasoning with Imprecise Probabilities*, Chapman and Hall, London.

Wang, P.Z. (1983). From the Fuzzy Statistics to the Falling Random Subsets in: *Advances in Fuzzy Sets Possibility and Applications*, Plenum Press, New York.

Wang, P.Z. and Sanchez, E. (1983). Hyperfields and random sets, *Proc. of the IFAC Symposium* (Marseille), 335-339.

Wang, P.Z. (1987). Random sets in fuzzy set theory, in: Singh, M.G., *Systems & Control Encyclopedia: Theory, Tecnology, Applications*, Pergamon Press, New York, 3945-3947.

Watanabe, N. and Imaizumi, T. (1993). A fuzzy statistical test of fuzzy hypotheses, *Fuzzy Sets and Systems*, 53, 167-178.

Watson, S.R., Weiss, J.J. and Donnell, M.L. (1979). Fuzzy Decision Analysis, *IEEE Transactions on Systems, Man and Cybernetics*, 9, 1-9.

Whalen, T. (1984). Decisionmaking under uncertainty with various assumptions about available information, *IEEE Transactions on Systems, Man and Cybernetics*, 14, 888-900.

Wilks, S.S. (1962). *Mathematical Statistics*, John Wiley, New York.

Zacks, S. (1971). *The Theory of Statistical Inference*, John Wiley, New York.

Zadeh, L.A. (1968). Probability measures of fuzzy events, *Journal of Mathematical Analysis and Applications*, 23, 421-427.

Zadeh, L.A. (1975). The concept of a linguistic variable and its application to approximate reasoning, *Information Sciences*, Part 1: 8, 199-249; Part 2: 8, 301-353; Part 3: 9, 43-80.

Zadeh, L.A. (1978). Fuzzy sets as a basis for a theory of possibility, *Fuzzy Sets and Systems*, 1, 3-28.

Zadeh, L.A. (1984). Fuzzy probabilities, *Information Processing Management*, 20, 363-372.

Zimmer, A.C. (1986). What uncertainty judgments can tell about the underlying subjective probabilities, *Machine Intelligent Pattern Recognition*, 4, 249-258.

Zimmer, A.C. (1990). A common framework for colloquial quantifiers and probability terms, in: Zetenyi, T. (Ed.), *Fuzzy Sets in Psychology*, North-Holland, Amsterdam, 73-89.

11 FUZZY REGRESSION ANALYSIS

Phil Diamond
Hideo Tanaka

Abstract: Fuzzy regresion analysis gives a fuzzy functional relationship between dependent and independent variables where vagueness is present in some form. The input data may be crisp or fuzzy. This chapter considers two types of fuzzy regression. The first is based on possibilistic concepts and the second upon a least squares approach. However, in both the notion of "best fit" incorporates the optimization of a functional associated with he problem. In possibilistic regression, this functional takes the form of a measure of the spreads of the estimated output, either as a weighted linear sum involving the estimated coefficients in linear regression, or as quadratic form in the case of exponential possibilistic regression. These optimization problems reduce to linear programming. For the least squares approach, the functional to be minimized is an L_2 distance between the observed and estimated outputs. This reduces to a class of quadratic optimization problems and constrained quadratic optimization. The method can incorporate stochastic fuzzy input and fuzzy kriging uses covariances to obtain BLUE estimators.

11.1 INTRODUCTION

Regression Analysis is a fundamental analytic tool in many disciplines. The method gives a crisp relationship between the dependent and independent variables with an estimated variance of measurement error. This functional relationship is based on the given data. If a phenomenon under consideration has not only stochastic variability but is also vague in some sense, it is more natural to seek a fuzzy functional relationship for the given data, which may be fuzzy or crisp. That is to say, a vague phenomenon should be modelled by a fuzzy functional relationship. This is the prime motivation for formulating fuzzy regression analysis models.

Let us review conventional regression analysis based on probability models so as to contrast differences from fuzzy regression analysis to be described later. Generally, regression analysis is a statistical tool for evaluating the relationship of independent variables x_1 ,..., x_n to a single dependent variable y. The observation data are denoted as $(y_j, x_{j1}, ..., x_{jn}) = (y_j, \boldsymbol{x}_j)$, $j = 1, ..., m$ where $\boldsymbol{x}_j = (x_{j1}, ..., x_{jn})^t$ is a column vector.

Let us assume that the regression model is

$$y_i = \beta_1 x_{j1} + \cdots + \beta_n x_{jn} + \epsilon_j, \quad j = 1, ..., m \tag{1}$$

where the β_i are regression coefficients and the ϵ_j observation errors. (1) can be rewritten as the following form

$$\boldsymbol{y} = X\boldsymbol{\beta} + \boldsymbol{\epsilon} \tag{2}$$

where $\boldsymbol{y} = (y_1, ..., y_m)^t$, $\boldsymbol{\beta} = (\beta_1, ..., \beta_n)^t$, $\boldsymbol{\epsilon} = (\epsilon_1, ..., \epsilon_m)^t$ and

$$X = \begin{bmatrix} x_{11}, & \cdots & , x_{1n} \\ \vdots & & \vdots \\ x_{m1}, & \cdots & , x_{mn} \end{bmatrix} \tag{3}$$

The least squares method determines the optimal regression coeficient vector $\boldsymbol{\beta}^*$ that minimizes the sum of squares of observation errors, that is

$$\min_{\boldsymbol{\beta}} \boldsymbol{\epsilon}^t \boldsymbol{\epsilon} = (\boldsymbol{y} - X\boldsymbol{\beta})^t (\boldsymbol{y} - X\boldsymbol{\beta}). \tag{4}$$

By solving (4), the optimal vector $\boldsymbol{\beta}^*$ can be obtained as

$$\boldsymbol{\beta}^* = (X^t X)^{-1} X^t \boldsymbol{y}. \tag{5}$$

Substituting (2) into (5), we find that

$$\boldsymbol{\beta}^* = \boldsymbol{\beta} + (X^t X)^{-1} X^t \boldsymbol{\epsilon}. \tag{6}$$

In general, $\boldsymbol{\epsilon}$ is assumed to be a random vector defined by

$$E(\boldsymbol{\epsilon}) = \boldsymbol{0}, \quad V(\boldsymbol{\epsilon}) = \sigma^2 I \tag{7}$$

where $E(\boldsymbol{\epsilon})$ is an expectation vector, $\boldsymbol{0}$ is the zero vector, $V(\boldsymbol{\epsilon})$ is a covariance matrix, I is an identity matrix and σ^2 is a variance. Since $\boldsymbol{\epsilon}$ is a random vector, $\boldsymbol{\beta}^*$ in (6) is also a random vector. Using (6) and (7), we have

$$E(\boldsymbol{\beta}^*) = E(\boldsymbol{\beta}) + (X^t X)^{-1} X^t E(\boldsymbol{\epsilon}) = \boldsymbol{\beta} \tag{8}$$

which means $\boldsymbol{\beta}$ is a linear unbiased estimator and the variance is given by

$$V(\boldsymbol{\beta}^*) = E((\boldsymbol{\beta}^* - \boldsymbol{\beta})(\boldsymbol{\beta}^* - \boldsymbol{\beta})^t) = \sigma^2 (X^t X)^{-1} \tag{9}$$

Consequently, $\boldsymbol{\beta}^*$ is called the best linear unbiased estimator (BLUE) in the sense that $V(\boldsymbol{\beta}^*) \leq V(\boldsymbol{\beta}')$ where $\boldsymbol{\beta}'$ is any linear unbiased estimator. It should be noted that $\boldsymbol{\beta}^*$ is a random vector denoted as $(E(\boldsymbol{\beta}^*), V(\boldsymbol{\beta}^*)) = (\boldsymbol{\beta}, \sigma^2(X^tX)^{-1})$.

Given an unknown input vector $\boldsymbol{x}_0$, the corresponding output y_0 can be represented as

$$y_0 = \boldsymbol{x}_0^t\boldsymbol{\beta}^* + \epsilon. \tag{10}$$

Here, let y^* be the true output corresponding to $\boldsymbol{x}_0$, that is, $y^* = \boldsymbol{x}_0^t\boldsymbol{\beta}$. Then, the error between y_0 and y^* can be written as

$$y_0 - y^* = \boldsymbol{x}_0^t(\boldsymbol{\beta}^* - \boldsymbol{\beta}) + \epsilon. \tag{11}$$

Then, from (7) and (8), we have

$$E(y_0 - y^*) = 0, \tag{12}$$

$$V(y_0 - y^*) = \sigma^2(\boldsymbol{x}_0^t(X^tX)^{-1}\boldsymbol{x}_0^t + 1). \tag{13}$$

Since σ^2 can be replaced by

$$s^2 = \boldsymbol{\epsilon}^t\boldsymbol{\epsilon}/(m-n), \tag{14}$$

where m and n are respectively the number of data points and independent variables, $\boldsymbol{\beta}^*$ and y_0 can be obtained as the following random variables:

$$(E(\boldsymbol{\beta}^*), V(\boldsymbol{\beta}^*)) = (\boldsymbol{\beta}, s^2(X^tX)^{-1}), \tag{15}$$

$$(E(y_0), V(y_0)) = (\boldsymbol{x}_0^t\boldsymbol{\beta}, s^2((X^tX)^{-1}\boldsymbol{x}_0 + 1)). \tag{16}$$

If we have $\boldsymbol{\epsilon} \sim N(0, \sigma^2 I)$ instead of (7), then (15) and (16) can be replaced by

$$\boldsymbol{\beta}^* \sim N(\boldsymbol{\beta}, s^2(X^tX)^{-1}), \tag{17}$$

$$y_0 \sim N(\boldsymbol{x}_0^t\boldsymbol{\beta}, s^2((X^tX)^{-1}\boldsymbol{x}_0 + 1)). \tag{18}$$

It is well known that the confidence region of $\boldsymbol{\beta}$ and the confidence interval prediction of y_0 can be obtained by using (17) and (18) respectively.

Now let us introduce the concepts of fuzzy regression analysis formulated by many researchers (see the book edited by Kacprzyk and Fedrizzi, 1992). In general, fuzzy regression analysis is a technique to obtain a fuzzy input-output relation that fits the given input-output data in some sense. Thus, a fuzzy regression model for crisp input data $\boldsymbol{x}_j$ is assumed as

$$Y_j = A_1 x_{j1} + \cdots + A_n x_{jn} \tag{19}$$

where A_i, $i = 1, \ldots, n$, are fuzzy numbers. The expression (19) is well defined by the extension principle and is called a fuzzy linear system or a possibilistic linear system (Zadeh, 1995). When input data are fuzzy, a fuzzy regression model is

$$Y_j = A_1 X_{j1} + \cdots + A_n X_{jn}, \tag{20}$$

where X_{ji} is a fuzzy input variable described as a fuzzy number. More generally speaking, a regression model can be written as

$$Y_j = f(\boldsymbol{X}_j, \boldsymbol{A}) \tag{21}$$

where $\boldsymbol{X}_j = (X_{j1}, ..., X_{jn})^t$ and $\boldsymbol{A} = (A_1, ..., A_n)^t$.

The main purpose for formulating fuzzy regression analysis is to explain the given data by a fuzzy input-output relation which can be described as (19), (20) or (21) depending on the given data structures. This is the main point of difference from conventional regression. There are three cases for input-output data to be analyzed: (i) crisp input-output data (no fuzzy data), (ii) crisp input data and fuzzy output data and (iii) fuzzy input-output data. Corresponding to the type of data structures, we have studied several combinations of data structures and models such as (crisp data, fuzzy functional relation), (fuzzy data, crisp functional relation) and (fuzzy data, fuzzy functional relation). Therefore, it is rather hard to classify types of fuzzy regression analyses into clear divisions. Nevertheless, let us try to classify regression techniques into two distinct areas. The first is based on a possibility approach and the second uses least squares ideas and formalism.

11.1.1 Overview of Possibilistic Regression

Possibilistic methods derive new estimators by dealing directly with models formulated in a fuzzy context. Fuzzy regression analysis was first proposed by Tanaka *et al.*(1980) , where a fuzzy linear system was used as a regression model. Since membership functions of fuzzy sets are often described as possibility distributions, this approach is usually called possibilistic regression analysis (Tanaka, 1987,1988,1989), where the fuzzy coefficients are assumed to be noninteractive, that is, determined independently of each other. Because these methods reduce the problem to one of linear programming (LP), some coefficients can become crisp because of the LP solution. In order to deal with interactive possibility distributions of coefficients of the model, quadratic membership functions and exponential possibility distributions have been used to formulate possibilistic regression analysis (Tanaka, 1991, Tanaka, 1995). In these approaches, inclusion relations between the given outputs and the estimated outputs play an important role in formulationg possibilistic regression (PR) because inclusion relations naturally arise in possibility theory (Zadeh, 1977). In (Tanaka, Hayashi, 1987), both possibility and necessity measures are used to evaluate inclusion relations.

When nonlinear regression is involved, the process involves more general mathematical programming (Bardossy, 1990). Since Group Method of Data Handling techniques (GMDH) can deal with nonlinear regression, Fuzzy GMDH (Hayashi, 1990) has been proposed for nonlinear PR. PR was applied by Heshmaty and Kandel to economic forecasting (Heshmaty, 1985). A modified form of PR is proposed in (Savic, 1991) while PR for fuzzy input-output data has been studied in (Sakawa, 1992, Redden, 1996). The properties of PR formulated in Tanaka *et al.*(1982, 1987, 1988, 1989) have been further studied in (Moskowitz,

1993, Reden, 1994), where the degree of fit of the fuzzy linear model is discussed and the effect of outliers also examined. PR based on fuzzy LP has been proposed as a modification of PR (Peters, 1994). Within PR, Minkowski's subtraction has been used to measure interval errors (Inuiguchi, 1993). PR has also been applied to civil engineering problems (Kaneyoshi, 1990) and in ergonomics (Chang, 1996).

11.1.2 Overview of Least Squares Methods

The least squares (LS) approach involves ideas of goodness-of-fit and this requires a notion of distance (Diamond, 1994) between the fuzzy values predicted by a parametric model and the fuzzy data that is actually observed. The principal advantage of these techniques is that the residual gives some idea of the accuracy of the model. There are basically two situations that have been studied. The first has fuzzy data with no assumed error structure in the observations. So, this is approximation with respect to some L_2 norm. However, when an error structure is assumed, the fuzzy data can be regarded as sampled from a fuzzy set valued random variable and the measure of best fit will then involve a variance. In this case, by analogy with crisp regression, an idea of bias can be introduced.

Within the LS approximation model itself, there are several approaches. Diamond considers noninteractive models for crisp input, fuzzy output and for fuzzy input-output (Diamond, 1988, Diamond, 1987). The second paper extends the ideas to multiple LS regression. Both data and output are triangular fuzzy numbers. Celmins applied LS methods for fitting data in the context of symmetric fuzzy numbers appearing as vector elements and having conically dependent membership numbers (Celmins, 1987, Celmins, 1987).

Where the data has an error structure which is assumed in the model, Diamond has used BLUE estimators in kriging problems from mining geostatistics (Diamond, 1989). Very recently, Körner has extended many results to general noninteractive LR fuzzy number data and models (Körner, 1995). This was made possible by using two technical devices: the Steiner centroid (see, for example (Diamond, 1994)) and Fréchet's concept of variance in general metric spaces (Fréchet, 1948). Körner and Näther (Körner, Näther, 1996) and Näther (Näther) also have some results in this direction.

11.2 POSSIBILISTIC REGRESSION

In order to simplify the explanation of PR, consider the linear model with three different types of possibility distributions of coefficients:

- interval coefficients,
- fuzzy coefficients modelled as fuzzy numbers and
- an exponential possibility distribution of coefficients which is similar to a normal distribution in probability.

We will call the corresponding regression analyses respectively interval regression, fuzzy regression and possibilistic regression.

11.2.1 Interval Regression

The simplest version of possibilistic regression analysis is interval regression (Ishibuschi, 1993). The interval regression model is expressed as

$$Y = A_1 x_1 + \cdots + A_n x_n = \boldsymbol{A}\boldsymbol{x}, \tag{22}$$

where x_i is an input variable and A_i is an interval. An interval A_i is denoted as $A_i = (a_{ci}, a_{wi})$ where a_{ci} is a center and $a_{wi} \geq 0$ is a radius. Thus A_i can be written as

$$A_i = \{a | a_{ci} - a_{wi} \leq a \leq a_{ci} + a_{wi}\}. \tag{23}$$

The well-known operations of interval arithmetic yield the interval output as follows :

$$Y = (\boldsymbol{a}_c^t \boldsymbol{x}, \boldsymbol{a}_w^t |\boldsymbol{x}|), \tag{24}$$

where $\boldsymbol{a}_c = (a_{c1}, ..., a_{cn})^t$, $\boldsymbol{a}_w = (a_{w1}, ..., a_{wn})^t$ and$|\boldsymbol{x}| = (|x_1|, ..., |x_n|)^t$.

Let us assume the following in order to formulate interval regression analysis for crisp data.

(i) The data are given as $(y_i, \boldsymbol{x}_j)$, $j = 1, ..., m$.

(ii) The data can be represented by the interval model (22).

(iii) The given output y_j should be included in the estimation interval $Y_j = (\boldsymbol{a}_c^t \boldsymbol{x}_j, \boldsymbol{a}_w^t |\boldsymbol{x}_j|)$,

$$\boldsymbol{a}_c^t \boldsymbol{x}_j - \boldsymbol{a}_w^t |\boldsymbol{x}_j| \leq y_j \leq \boldsymbol{a}_c^t \boldsymbol{x}_j + \boldsymbol{a}_w^t |\boldsymbol{x}_j|, \quad j = 1, ..., m. \tag{25}$$

(iv) The index of the spread of the interval model is defined by

$$J = \sum_{j=1}^{m} \boldsymbol{a}_w^t |\boldsymbol{x}_j|. \tag{26}$$

Interval Regression analysis determines the interval coefficients A_i, $i = 1, ..., n$, that minimize J subject to (25). It is straightforward to show that this leads to the LP problem

$$\min_{\boldsymbol{a}_c, \boldsymbol{a}_w} \quad J = \sum_{j=1}^{m} \boldsymbol{a}_w^t |\boldsymbol{x}_j| \tag{27}$$

$$\textit{subject to} \quad (25) \; \textit{and} \; \boldsymbol{a}_w \geq \boldsymbol{0}.$$

Since interval regression analysis can be reduced to the LP problem (27), constraint conditions for the coefficients can be introduced. For instance, if an input variable, say x_i, has a positive correlation with the output variable, it is advantageous to constrain A_i to be positive. Generally speaking, by introducing expert knowledge suggesting that the interval coefficient A_i should lie in some interval $B_i = (b_{ci}, b_{wi})$, the interval A_i can be estimated within the limit of that knowledge B_i. Thus, we can introduce the following constraint condition.

$$A_i \subset B_i \quad \Leftrightarrow \quad b_{ci} - b_{wi} \leq a_{ci} - a_{wi}, \ \ and \ \ b_{ci} + b_{wi} \geq a_{ci} + a_{wi}. \tag{28}$$

Since A_i is constrained by the expert knowledge B_i, the obtained linear interval regression model appears to be acceptable.

Now, let us study a house price model by interval regression analysis.

House price model : The data in Table 1 were obtained from the pamphlet published by a company which manufactures prefabricated houses.

Input data : x_1 = quality of the construction material , x_2 = area of the first floor (m^2), x_3 = area of the second floor (m^2), x_4 = total number of rooms, x_5 = number of Japanese rooms. For the construction materials, 1=low grade, 2=medium grade, and 3=high grade.

Output data : y = sale price ($\times 10^6$ Yen)

The linear interval model is written as

$$Y = A_0 + A_1x_1 + A_2x_2 + A_3x_3 + A_4x_4 + A_5x_5. \tag{29}$$

Since all coefficients are assumed to be positive, we added the constraint condition on the coefficients $\boldsymbol{a_c} \geq \mathbf{0}$ to the LP problem (27). Solving the LP problem (27) with $\boldsymbol{a_c} \geq \mathbf{0}$, we obtained the optimal interval model :

$$Y^* = (245.167, 37.634)x_1 + (5.833, 0)x_2 + (4.786, 0)x_3. \tag{30}$$

In contrast to this, an ordinary statistical regression analysis without the constraint condition gave the optimal model as

$$y = -112.40 + 236.48x_1 + 9.3568x_2 + 8.2294x_3 - 37.889x_4 - 17.253x_5. \tag{31}$$

In (31), as x_4 (total number of rooms) increases, y (price) decreases. This seems contradictory to intuition. On the other hand, (30) is more in accord with common sense. The estimated interval values are given in Fig. 1, which indicates that sale prices for cheap prefabricated houses are located at the lower ends of the estimation intervals, while sale prices for expensive ones are located at the upper ends of the estimation intervals. This is indicative of the fact that the prices of high-grade products are set higher to reflect their additional values.

Table 1. Data related to prefabricated houses

No.	y_j	x_1	x_2	x_3	x_4	x_5
1	606	1	38.09	36.43	5	1
2	710	1	62.10	26.50	6	1
3	808	1	63.76	44.71	7	1
4	826	1	74.52	38.09	8	1
5	865	1	75.38	41.10	7	2
6	852	2	52.99	26.49	4	2
7	917	2	62.93	26.49	5	2
8	1031	2	72.04	33.12	6	3
9	1092	2	76.12	43.06	7	2
10	1203	2	90.26	42.64	7	2
11	1394	3	85.70	31.33	6	3
12	1420	3	95.27	27.64	6	3
13	1601	3	105.98	27.64	6	3
14	1632	3	79.25	66.81	6	3
15	1699	3	120.50	32.25	6	3

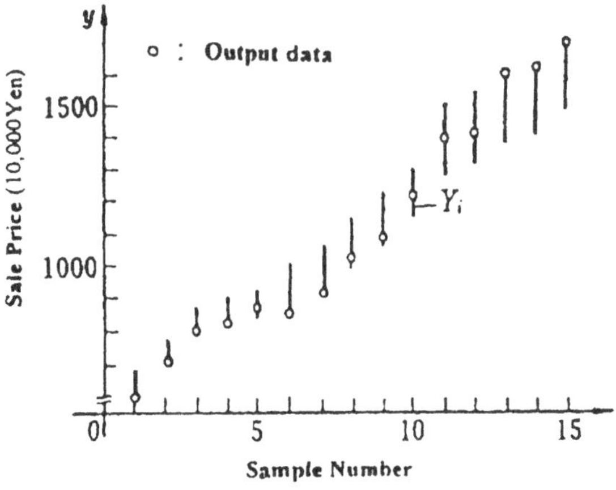

Figure 1 Interval estimation by the houseprice model

When the given outputs are interval but the given inputs are crisp, we can consider two further regression models such as a possibility estimation model and a necessity estimation model. The given data are denoted as $(Y_j, x_{j1}, \dots, x_{jn}) = (Y_j, \boldsymbol{x_j})$ where Y_i is an interval output denoted as (y_j, e_j)

The possibility estimation model and the necessity estimation model are defined respectively as follows:

$$Y^*{}_j = A^*{}_1 x_{j1} + \cdots + A^*{}_n x_{jn}, \tag{32}$$

$$Y_{*j} = A_{*1} x_{j1} + \cdots + A_{*n} x_{jn}. \tag{33}$$

Two regression models are described as follows:

Possibility Regression Model: The problem here is to satisfy

$$Y_j \subset Y^*{}_j, \quad j = 1, ..., m, \tag{34}$$

and to find the interval coefficients $\boldsymbol{A}^* = (\boldsymbol{a}_c^*, \boldsymbol{a}_w^*)$ which minimize the sum of the ranges of the estimation intervals:

$$J^* = \sum_{j=1}^{m} \boldsymbol{a}_w^{*t} |\boldsymbol{x}_j|. \tag{35}$$

Since (34) is equivalent to a system of linear inequalities by (28), the foregoing is an LP problem and is easily solved.

Necessity Regression Model: The problem here is to satisfy

$$Y_{*j} \subset Y_j, \quad j = 1, ..., m, \tag{36}$$

and to find the interval coefficients $\boldsymbol{A}_* = (\boldsymbol{a}_{*c}, \boldsymbol{a}_{*w})$ which maximize the sum of the ranges of the estimation intervals:

$$J_* = \sum_{j=1}^{m} \boldsymbol{a}_{*w}^{t} |\boldsymbol{x}_j|. \tag{37}$$

This is also an LP problem and is solved equally easily. The reason for maximizing J_* is to find the estimation intervals Y_{*j} of a wider range among those satisfying the constraint condition of (36). The necessity estimation intervals Y_{*j} and the possibility estimation intervals Y_j^* have inclusion relations

$$Y_{*j} \subset Y_j \subset Y_j^*. \tag{38}$$

Accordingly, since Y_j^* and Y_{*j} which satisfy (38) are to be found, the formulation is done so that the range of Y_j^* is minimized while the range of Y_{*j} is maximized.

If the given data $(Y_j^o, \boldsymbol{x}_j^o)$, $j = 1, ..., m$ satisfy the linear interval system

$$Y_j^o = A_1^o x_{j1}^o + \cdots + A_n^o x_{jn}^o, \tag{39}$$

the interval vector $\boldsymbol{A}^*$ obtained from the possibility model is the same as the vector $\boldsymbol{A}_*$ obtained by the necessity model, $\boldsymbol{A}^* = \boldsymbol{A}_* = \mathbf{A}_0$ (see Tanaka, 1989). This means that unknown coefficients can be identified by interval regression techniques if the given data satisfy (39).

The features of the linear interval regression model formulated above include the following:

(i) The possibility and necessity estimation models for the interval data are similar to the concepts of upper approximation and lower approximation of rough sets. When the output data are intervals, the possibility and necessity intervals are obtained in the context of incompleteness of data.

(ii) The range of the estimation interval widens as the number of data increases. This is due to the fact that the increased analytical data results in more information and wider possibilities for decision-making. In contrast, an estimation interval in conventional regression analysis diminishes as the number of data increases. Since conventional regression analysis is based on a probability model, objective analysis is done for the large number of data, but linear interval regression analysis is based on a possibility model and is useful for the problem of deciding what is possible.

(iii) Since interval regression analysis can be reduced to LP problems, constraint conditions for the coefficients can be introduced. For instance, it is advantageous to constrain coefficients to be positive if the variables corresponding to those coefficients have a positive correlation with the outputs. In addition, generally speaking, by introducing expert knowledge, interval coefficients can be estimated within the limits of that knowledge. Since it is constrained by the knowledge of the expert, the obtained linear interval regression model appears to be acceptable.

(iv) In general, since the intervals represent partial ignorance, this should also be reflected in the analytical results.

Grinding Model: Nine feed speeds were taken in the experiments, and the surface roughness for each was measured. The interval output data of Table 2 were obtained from the maximum and minimum surface roughness values from the three finishing experiments. The linear interval system is expressed as

$$Y = A_0 + A_1 x + A_2 x^2,$$

and the following results were obtained.

Possibility Estimation Model.

$$A_0^* = (0.236, 0), \quad A_1^* = (-0.007, 0.055), \quad A_2^* = (0.016, 0).$$

Necessity Estimation Model.

$$A_{*0} = (0.477, 0), \quad A_{*1} = (-0.206, 0), \quad A_{*2} = (0.047, 0.002).$$

The estimation intervals and the data for each model are shown in Fig. 2 and 3. The longitudinal lines in these illustrations are the output data Y_j, and the estimation intervals Y_j^* and Y_{*j} are represented by the two curves. As can readily be seen, $Y_{*j} \subset Y_j \subset Y_j^*$.

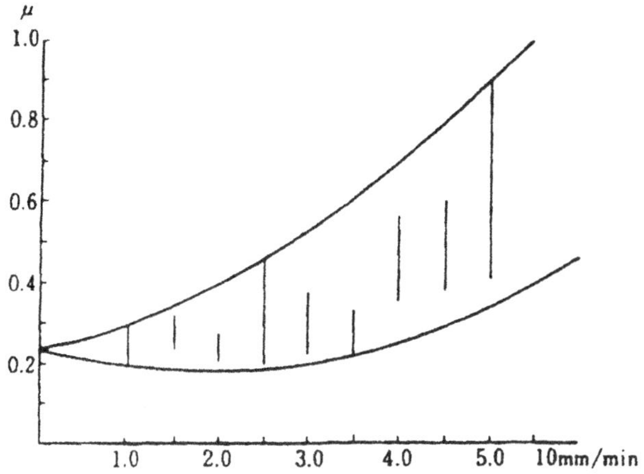

Figure 2 Possibility estimation model

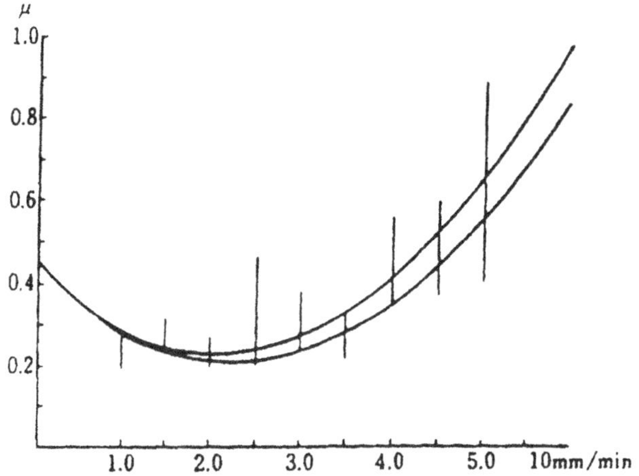

Figure 3 Necessity estimation model

Table 2. Feed speed and surface roughness

No.	Feed speed x (10 mm/min)	Roughness(μ)y_i min. value	max. value	Roughness(μ)Y_i (y_i, e_i)
1	1.0	0.19	0.29	(0.240,0.050)
2	1.5	0.24	0.32	(0.280,0.035)
3	2.0	0.20	0.27	(0.235,0.035)
4	2.5	0.20	0.46	(0.330,0.130)
5	3.0	0.22	0.38	(0.300,0.080)
6	3.5	0.22	0.33	(0.275,0.055)
7	4.0	0.35	0.56	(0.455,0.105)
8	4.5	0.37	0.60	(0.485,0.115)
9	5.0	0.41	0.89	(0.650,0.240)

11.2.2 Fuzzy Regression

In this section, the fuzzy regression model is expressed as

$$Y = A_1 x_1 + \cdots + A_n x_n = \boldsymbol{A}\boldsymbol{x} \tag{40}$$

where for simplicity it is assumed that A_i is a symmetrical fuzzy number. A symmetrical fuzzy number A_i denoted as $(a_i, c_i)_L$ is defined by

$$\mu_{A_i}(z) = L((z - a_i)/c_i) \tag{41}$$

where the reference function $L(z)$ satisfies (i) $L(z) = L(-z)$, (ii) $L(0) = 1$ and (iii) L is strictly decreasing on $[0, +\infty)$. If $L(z) = max(0, 1 - |z|)$, the menbership function of A_i is

$$\mu_{A_i}(z) = 1 - |(z - a_i)/c_i| \tag{42}$$

which is a triangular fuzzy number where a_i is a center value and c_i is a radius. Using the extension principle, the membership function of (40) is obtained for $\boldsymbol{x} \neq \boldsymbol{0}$ as

$$\mu_Y(y) = L((y - \boldsymbol{x}^t\boldsymbol{a})/\boldsymbol{c}^t|\boldsymbol{x}|) \tag{43}$$

for $\boldsymbol{x} = \boldsymbol{0}$ and $y = 0$, $\mu_Y(y) = 1$, and for $\boldsymbol{x} = \boldsymbol{0}$ and $y \neq 0$, $\mu_Y(y) = 0$ (see Tanaka, 1988). The h-level set of Y can be described as the interval

$$Y_h = [\boldsymbol{a}^t\boldsymbol{x} - |L^{-1}(h)|\boldsymbol{c}^t|\boldsymbol{x}|, \boldsymbol{a}^t\boldsymbol{x} + |L^{-1}(h)|\boldsymbol{c}^t|\boldsymbol{x}|] \tag{44}$$

where $Y_h = \{y|\mu_Y(y) \geq h\}$.

Interval techniques used in the section 11.2.1 can be applied to (40) in order to formulate fuzzy regression (Tanaka, 1982,1987,1988,1989). Fuzzy regression for crisp data can be formulated as

$$\min_{a,c} \quad J = \sum_{j=1}^{m} \boldsymbol{c}^t|\boldsymbol{x}_j| \tag{45}$$

$$\textit{subject to } \boldsymbol{c} \geq \boldsymbol{0} \textit{ and}$$

$$\boldsymbol{a}^t\boldsymbol{x}_j - |L^{-1}(h)|\boldsymbol{a}^t\boldsymbol{x}_j \leq y_j \leq \boldsymbol{a}^t\boldsymbol{x}_j + |L^{-1}(h)|\boldsymbol{a}^t\boldsymbol{x}_j, \quad j = 1, ..., m.$$

The fuzzy regression model can be obtained by solving the LP problem (45). When a triangular fuzzy number ($L(z) = \max(0, 1 - |z|)$) is used, $|L^{-1}|$ can be written as

$$|L^{-1}(h)| = 1 - h. \tag{46}$$

Given data $(y_j, \boldsymbol{x}_j)$, $j = 1, ..., m$, there is an optimal solution $\boldsymbol{A}^* = (\boldsymbol{a}^*, \boldsymbol{c}^*)_L$ for $0 \leq h < 1$ in the LP problem (45). If c_i is large enough, it is easy to see that an admissible set exist for (45). When $h = 1$, the data must satisfy

$$y_j = \boldsymbol{a}^t\boldsymbol{x}_j, \quad j = 1, ..., m. \tag{47}$$

Since (47) does not generally hold, $h = 1$ is excluded.

Now let us discuss a degree h which is assumed to be given by an analyst. Let the optimal h-level solution be $\boldsymbol{A}_h = (\boldsymbol{a}_h, \boldsymbol{c}_h)_L$. The optimal solution for h' can be obtained from the optimal h-level solution $\boldsymbol{A}_h$ in the following way (Tanaka, 1988):

$$\boldsymbol{A}_{h'} = (\boldsymbol{a}_h, \frac{L^{-1}(h)}{L^{-1}(h')}\boldsymbol{c}_h)_L \tag{48}$$

From (48), we can see that after we find the $h = 0$ level solution, we can easily obtain the optimal solution for any level. Also, the value of J for any h is

$$J(\boldsymbol{c}_h) = \frac{L^{-1}(0)}{L^{-1}(h)} J(\boldsymbol{c}_0), \tag{49}$$

and if we let $L(x) = 1 - |x|$, we get

$$J(\boldsymbol{c}_h) = (1-h)^{-1} J(\boldsymbol{c}_0). \tag{50}$$

From (48), the relationship between, for example, the solution $\boldsymbol{c}_0$ obtained at $h = 0$ and the solution $\boldsymbol{c}_{0.5}$ obtained at $h = 0.5$ is

$$\boldsymbol{c}_{0.5} = 2\boldsymbol{c}_0. \tag{51}$$

From the above, the degree h can be interpreted in the following manner. If there are enough data, the possibility shown from enough data is sufficient. Thus, $h = 0$ is recommendable. If the given data are considered to include only half, comparing with the idea number of data, $h = 0.5$ is recommendable because of (51). In other words, (51) can be explained as adding the width of the obtainable data to the width of the unobtainable data.

House price model: The given data are shown in Table 1. The fuzzy linear model is shown in (29) where A_i is a triangular fuzzy number denoted as $(a_i, c_i)_L$. The optimal result of (45) for $h = 0.5$ is shown in the following:

$$\left.\begin{array}{lll} A_0^* = (0,0)_L, & A_1^* = (245.17, 37.63)_L & A_2^* = (5.85, 0)_L \\ A_3^* = (4.79, 0)_L & A_4^* = (0,0)_L & A_5^* = (0,0)_L \end{array}\right\} \tag{52}$$

The estimated fuzzy outputs $Y_j^* = \boldsymbol{A}^*\boldsymbol{x}_j$ for the odd-numbered data are shown in Fig. 4 where we can that the given outputs marked by $(\times)$ are included in the 0.5-level sets of Y_j^*.

Fuzzy regression models for fuzzy outputs and crisp inputs can be considered in the same way as described in interval regression. Assuming that two fuzzy regression models are set out as in (32) and (33) where A_i^* and A_{*i} are fuzzy numbers, (34) and (36) can be rewritten as inclusion relations of h-level sets:

$$[Y_j]_h \subset [Y_j^*]_h, \quad [Y_{*j}]_h \subset [Y_j]_h, \quad j = 1, ..., m \tag{53}$$

where $Y_j = (y_j, e_j)_L$, $j = 1, ..., m$ are the given fuzzy outputs. The cost functions of (35) and (37) are written as $J = \sum \boldsymbol{c}^t |\boldsymbol{x}_j|$.

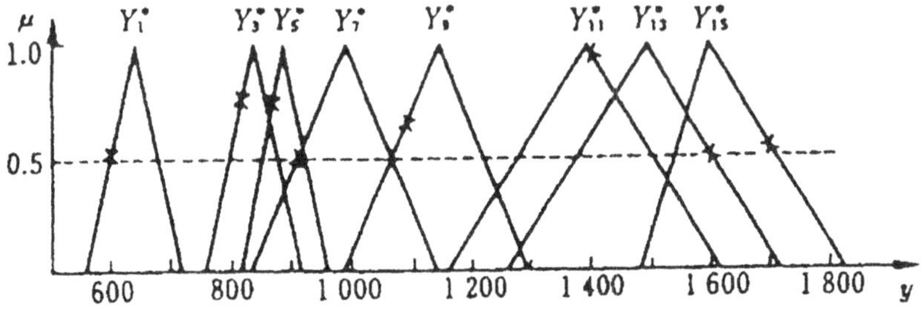

Figure 4 Estimated fuzzy output $Y_j^* = \mathbf{A}^*\mathbf{x}_j$

Possibility Estimation Model for fuzzy regression can be formulated as the following LP problem:

$$\min_{a,c} \quad J = \sum_j \boldsymbol{c}^t|\boldsymbol{x}_j| \tag{54}$$

$$subject \ to \ \ \boldsymbol{c} \geq 0 \ \ and$$

$$y_j + |L^{-1}(h)|e_j \leq \boldsymbol{a}^t\boldsymbol{x}_j + |L^{-1}(h)|\boldsymbol{c}^t|\boldsymbol{x}_j|,$$

$$y_j - |L^{-1}(h)|e_j \geq \boldsymbol{a}^t\boldsymbol{x}_j - |L^{-1}(h)|\boldsymbol{c}^t|\boldsymbol{x}_j|, \ \ j = 1, ..., m.$$

The constraint conditions of (54) stem from $[Y_j]_h \subset [Y_j^*]_h$. Necessity Estimation Model for fuzzy regression can be obtained from the constraint $[Y_{*j}]_h \subset [Y_j]_h$ as follows:

$$\max_{a,c} \quad J = \sum_j \boldsymbol{c}^t|\boldsymbol{x}_j| \tag{55}$$

$$subject \ to \ \ \boldsymbol{c} \geq 0 \ \ and$$

$$y_j + |L^{-1}(h)|e_j \geq \boldsymbol{a}^t\boldsymbol{x}_j + |L^{-1}(h)|\boldsymbol{c}^t|\boldsymbol{x}_j|,$$

$$y_j - |L^{-1}(h)|e_j \leq \boldsymbol{a}^t\boldsymbol{x}_j - |L^{-1}(h)|\boldsymbol{c}^t|\boldsymbol{x}_j|, \ \ j = 1, ..., m.$$

11.2.3 Exponential Possibility Regression

In statistical regression models, the difference between the given outputs and the inferred outputs obtained from the model is taken to be statistical error. In contrast, PR analysis assumes that the gap between the given outputs and the obtained outputs is interpreted by the possibility of the input-output relation. Therefore, exponential possibility linear systems are used as regression models (Tanaka, 1991, 1995). That is,

$$Y = A_1x_1 + \cdots + A_nx_n = \boldsymbol{A}\boldsymbol{x}, \tag{56}$$

where $\boldsymbol{x}$ is a real vector (input vector) and $\boldsymbol{A} = (A_1, ..., A_n)$ is a fuzzy coefficient vector defined by the following exponential possibility distribution with a center vector $\boldsymbol{a}_c$ and a symmetrical positive definite matrix D_A:

$$\Pi_A(\boldsymbol{a}) = \exp\{-(\boldsymbol{a} - \boldsymbol{a}_c)^t D_A^{-1}(\boldsymbol{a} - \boldsymbol{a}_c)\}, \tag{57}$$

whose parametric representation is $\boldsymbol{A} = (\boldsymbol{a}_c, D_A)_e$. Let us calculate the possibility distribution of Y in (56) using the extension principle as follows:

$$\Pi_Y(y) = \max_{\{\boldsymbol{a}|y=\boldsymbol{a}^t\boldsymbol{x}\}} \Pi_A(\boldsymbol{a}). \tag{58}$$

The optimization problem (58) can be written as

$$\min_{\boldsymbol{a}} \quad (\boldsymbol{a} - \boldsymbol{a}_c)^t D_A^{-1}(\boldsymbol{a} - \boldsymbol{a}_c) \tag{59}$$

$$subject \quad to \quad y = \boldsymbol{a}^t\boldsymbol{x}.$$

This optimization problem (59) can be solved by minimizing the following Lagrangian function:

$$L(\boldsymbol{a}, \lambda) = (\boldsymbol{a} - \boldsymbol{a}_c)^t D_A^{-1}(\boldsymbol{a} - \boldsymbol{a}_c) + \lambda(y - \boldsymbol{a}^t\boldsymbol{x}) \tag{60}$$

where λ is a Lagrangian multiplier variable. Since $L(\boldsymbol{a}, \lambda)$ is a quadratic form, the necessary and sufficient conditions for optimality are

$$\partial L/\partial \boldsymbol{a} = \boldsymbol{0}, \quad \partial L/\partial \lambda = 0. \tag{61}$$

From (61), we have the following normal equations:

$$2D_A^{-1}(\boldsymbol{a} - \boldsymbol{a}_c) - \lambda\boldsymbol{x} = 0, \quad y = \boldsymbol{a}^t\boldsymbol{x}. \tag{62}$$

Thus, we have

$$\lambda^* = 2\boldsymbol{x}^t(\boldsymbol{a} - \boldsymbol{a}_c)/\boldsymbol{x}^t D_A\boldsymbol{x} = 2(y - .\boldsymbol{x}^t\boldsymbol{a}_c)/\boldsymbol{x}^t D_A\boldsymbol{x}. \tag{63}$$

Substituting (63) into (62), we have

$$\boldsymbol{a}^* = \boldsymbol{a}_c + D_A\boldsymbol{x}(y - \boldsymbol{x}^t\boldsymbol{a}_c)/\boldsymbol{x}^t D_A\boldsymbol{x}. \tag{64}$$

Finally, substituting (64) into (57) yields

$$\Pi_Y(y) = \exp\{-(y - \boldsymbol{x}^t\boldsymbol{a}_c)^2/\boldsymbol{x}^t D_A\boldsymbol{x}\} \tag{65}$$

whose parametric representation is $Y = (\boldsymbol{x}^t\boldsymbol{a}_c, \boldsymbol{x}^t D_A\boldsymbol{x})_e$. It should be noted that the possibility distribution $\Pi_A(\boldsymbol{a})$ is transformed into the possibility distribution $\Pi_Y(y)$ by the extension principle (58).

Using the above mechanism, let us formulate exponential PR analysis for crisp data (Tanaka, 1995) under the following assumptions being similar to those described in the section 11.2.1.

(i) The data are given as $(y_j, \boldsymbol{x}_j)$, $j = 1, ..., m$, where y_j is an output and $\boldsymbol{x}_j = (x_{j1}, ..., x_{jn})^t$ is an input vector. Assuming $m \gg n$, $E = \{1, ..., n\}$ is the set of subscripts of independent vectors $\{\boldsymbol{x}_1, ..., \boldsymbol{x}_n\}$ that are chosen from the given input vectors.

(ii) The data can be represented by the possibility linear system (56) with the exponential possibility distribution (57).

(iii) Given a threshold h_j, the given output y_j should be included in the h_j-level set of the estimated fuzzy output Y_j, that is

$$[Y_j]_{h_j} = \{y | \Pi_{Y_j}(y) \geq h_j\} \ni y_j. \tag{66}$$

(iv) The cost function is defined by

$$J = \sum_{j=1}^{m} \boldsymbol{x}_j^t D_A \boldsymbol{x}_j. \tag{67}$$

It should be noted that the given data y_j $(j = 1, ..., m)$ can be covered with the h_j-level set of the fuzzy output Y_j $(j = 1, ..., m)$ by the assumption (iii). In other words, the j-th input-output relation is assumed to be possible with the grade h_j. The independent vectors $\{\boldsymbol{x}_1, ..., \boldsymbol{x}_n\}$ in the assumption (i) will be used later for finding a positive definite matrix denoted as $D_A > 0$. The assumption (iv) is the sum of spreads of the estimated outputs.

The optimization problem for finding the possibility distribution $\Pi_A(\boldsymbol{a})$ can be written as

$$\min_{D_A, \boldsymbol{a}_c} \quad J \tag{68}$$

$$\textit{subject to} \quad \Pi_{Y_j}(y_j) \geq h_j \quad \textit{and} \quad D_A > 0$$

where the first constraint condition is the same as (66). The first constraint condition of (68) can be rewritten as

$$\boldsymbol{x}_j^t D_A \boldsymbol{x}_j \geq (y - \boldsymbol{a}_c^t \boldsymbol{x}_j)^2 / (-\log h_j) \tag{69}$$

Unfortunately, it can be seen from (69) and $D_A > 0$ that (68) is a non-linear optimization problem. Thus, in order to transform a non-linear optimization problem into an LP problem, let us take the following procedure :

(I) First, determine the center vector $\boldsymbol{a}_c$ using the interval regression or the conventional regression. If some constraint conditions on coefficients should be introduced, the interval regression is suitable for finding $\boldsymbol{a}_c$ as decribed in the section 11.2.1. Otherwise, the conventional regression will be used.

(II) Second, substitute the obtained center vector $\boldsymbol{a}_c^*$ into (69) and solve the following LP problem :

$$\min_{D_A} \quad J = \sum_{j=1}^{m} \boldsymbol{x}_j^t D_A \boldsymbol{x}_j \tag{70}$$

$$\textit{subject to} \quad \boldsymbol{x}_j^t D_A \boldsymbol{x}_j \geq (y - \boldsymbol{a}_c^{*t} \boldsymbol{x}_j)^2 / (-\log h_j), \quad j = 1, ...m,$$

$$\boldsymbol{x}_i^t D_A \boldsymbol{x}_j = 0 \quad for\ all\ \ i \neq j\ \ and\ \ i, j \in E,$$

where the last constraint called the orthogonal condition ensures that D_A can be obtained as a positive definite matrix.

Let us show $D_A > 0$, using the orthogonal condition. An arbitrary vector $\boldsymbol{z}$ in an n-dimensional space can be represented as

$$\boldsymbol{z} = \lambda_1 \boldsymbol{x}_1 + \cdots + \lambda_n \boldsymbol{x}_n, \tag{71}$$

where $\{\boldsymbol{x}_1, ..., \boldsymbol{x}_n\}$ are independentvectors and λ_i is a real number. From the first constraint, we have in general

$$\boldsymbol{x}_i^t D_A \boldsymbol{x}_i > 0. \tag{72}$$

Thus, it follows from (72) and the orthogonal condition that for an arbitary vector $\boldsymbol{z} \neq \boldsymbol{0}$,

$$\begin{aligned} \boldsymbol{z}^t D_A \boldsymbol{z} &= (\lambda_1 \boldsymbol{x}_1 + \cdots + \lambda_n \boldsymbol{x}_n)^t D_A (\lambda_1 \boldsymbol{x}_1 + \cdots + \lambda_n \boldsymbol{x}_n) \\ &= \sum_{i=1}^{n} \lambda_i^2 \boldsymbol{x}_i^t D_A \boldsymbol{x}_i > 0 \end{aligned} \tag{73}$$

which means that $D_A > 0$.

It is clear that the optimization problem (70) is an LP problem, since $(y_j, \boldsymbol{x}_j)$, $j = 1, ..., m$, and $\boldsymbol{a}_c^*$ are given as numerical numbers. Since (70) can be easily solved by LP, we can obtain the matrix which leads to the interactive possibility distribution of coefficients. This point is defferent from those in the sections 11.2.1 and 11.2.2.

Given an unknown input vector $\boldsymbol{x}_0$, the corresponding fuzzy output Y_0 can be obtained as

$$Y_0 = A\boldsymbol{x}_0, \tag{74}$$

whose possibility distribution $\Pi_{Y_0}(y)$ is

$$\Pi_{Y_0}(y) = \exp\{-(y - \boldsymbol{x}_0^t \boldsymbol{a}_c)^2 (\boldsymbol{x}_0^t D_A \boldsymbol{x}_0)^{-1}\}. \tag{75}$$

In contrast to (17) and (18) in statistical regression analysis, we can have from (57) and (75)

$$\boldsymbol{A} = (\boldsymbol{a}_c, D_A)_e, \tag{76}$$

$$Y_0 = (\boldsymbol{x}_0^t \boldsymbol{a}_c, \boldsymbol{x}_0^t D_A \boldsymbol{x}_0)_e. \tag{77}$$

It can be seen from this that principal difference is that the spread of data is interpreted as statistical error in statistical regression analysis, but as the possibility of coefficients in PR analysis. It should be noted that D_A is directly obtained from the spread of the given data by this method.

Numerical Example: The input-output data are shown in Table 3. The possibility linear system is

$$Y_j = A_0 + A_1 x_1 = \boldsymbol{A}\boldsymbol{x}, \tag{78}$$

where $\boldsymbol{A}$ is a possibility distribution with the center vector $\boldsymbol{a}_c = (a_{0c}, a_{1c})^t$ and the matrix D_A. According to the above procedure (I), we used the conventional regression analysis for determining the center vector $\boldsymbol{a}_c^*$ and obtained

$$\boldsymbol{a}_c^* = (3.7885, 0.4097)^t. \tag{79}$$

Then, we used the procedure (II) to solve the LP problem (70). First, let us discuss the choice of n independent vectors out of the m input vectors. In this example, we have $n = 2$ and $m = 10$. It should be noted that each input vector $\boldsymbol{x}_j$ is represented as $(1, x_j)^t$ where x_j is given in Table 3. Since $n = 2$, two input vectors should be selected from the input data. There are $({}_{10}C_2 = 45)$ candidates for the orthogonal condition. Thus, by solving (70) with $h_j = 0.5$ for all the given data we obtained the optimal values J for 45 candidates, and these are shown in Table 4.

Table 4. Optimal values J with the orthogonal conditions

Orthogonal condition	Optimal value J	Orthogonal condition	Optimal value J
x_1, x_2	1229.21	x_1, x_3	657.68
x_1, x_4	114.23	x_1, x_5	92.37
x_1, x_6	90.87	x_1, x_7	89.72
x_1, x_8	88.07	x_1, x_9	86.51
x_1, x_{10}	86.13	x_2, x_3	1079.80
x_2, x_4	123.94	x_2, x_5	96.96
x_2, x_6	95.44	x_2, x_7	94.23
x_2, x_8	92.41	x_2, x_9	90.59
x_2, x_{10}	90.13	x_3, x_4	135.25
x_3, x_5	107.70	x_3, x_6	104.99
x_3, x_7	102.96	x_3, x_8	100.11
x_3, x_9	97.64	x_3, x_{10}	97.17
x_4, x_5	202.87	x_4, x_6	195.64
x_4, x_7	191.31	x_4, x_8	175.73
x_4, x_9	155.16	x_4, x_{10}	152.09
x_5, x_6	1920.84	x_5, x_7	1180.74
x_5, x_8	810.69	x_5, x_9	353.84
x_5, x_{10}	299.08	x_6, x_7	3217.60
x_6, x_8	1266.18	x_6, x_9	396.58
x_6, x_{10}	337.63	x_7, x_8	1462.82
x_7, x_9	456.42	x_7, x_{10}	394.17
x_8, x_9	774.15	x_8, x_{10}	676.05
x_9, x_{10}	14155.24		

From Table 4, the orthogonal condition of $\{x_1, x_{10}\}$ should be chosen, because this choice leads to the smallest value of J. With this orthogonal condition we obtained

$$D_A^* = \begin{bmatrix} 5.4446 & -0.3769 \\ -0.3769 & 0.0712 \end{bmatrix} > 0. \tag{80}$$

The PR model (75) with a_c^* and D_A^* is shown in Fig. 5 where open circles are the given data. The obtained possibility distribution of the coefficient vector $A^* = (a_c^*, D_A^*)_e$ is shown in Fig. 6. It can be seen from Fig. 5 and Fig. 6 that the spread of data can be well transformed into the possibility distribution of the coefficient vector by this method.

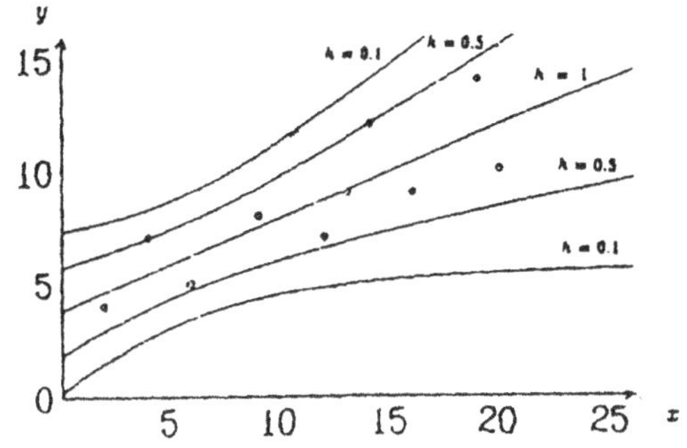

Figure 5 The possibility regression model and the given data

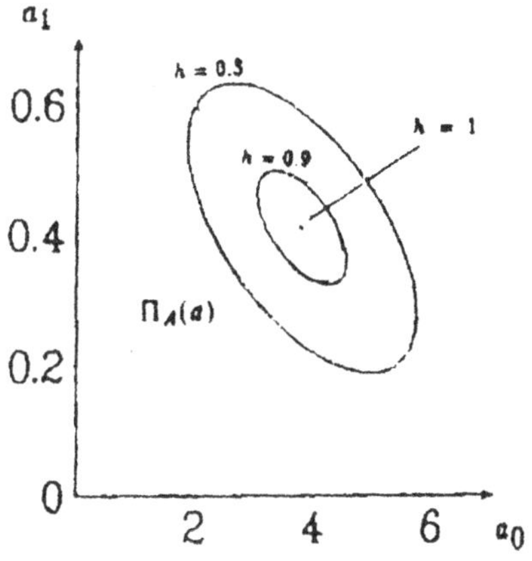

Figure 6 The possibility distribution of coefficient vector **A**

House Price model: PR analysis is applied to the house price model described in section 11.2.1. From the result of the interval regression model (30), the PR model is

$$Y = A_1x_1 + A_2x_2 + A_3x_3. \tag{81}$$

In the procedure (I), the center vector has been already obtained by the interval regression model (30) as

$$\boldsymbol{a}_c^* = (245.167, 5.853, 4.786)^t. \tag{82}$$

In the procedure (II), substitution of (82) into (70) leads to an LP problem. Under $h_j = 0.5$ for all the given data we solved LP problems with 10 different orthogonal conditions. Those results are shown in Table 5. Since $n = 3$, three independent vectors were selected as the orthogonal condition. First, fourth and seventh input vectors in Table 1 were selected as the best combination of independent vectors from Table 5. It should be noted that the data related to x_4 and x_5 in Table 1 were not used, because the PR model was assumed as (81). Using the best orthogonal condition, we obtained the optimal matrix D_A^* as

$$D_A^* = \begin{bmatrix} 5952.900768 & -62.697340 & -74.914836 \\ -62.697340 & 4.513809 & -5.693125 \\ -74.914836 & -5.693125 & 12.678663 \end{bmatrix} > 0, \tag{83}$$

which shows interactive possibility coefficients. Using $\boldsymbol{a}_c^*$ and D_A^*, the estimated fuzzy outputs (75) for samples with even sample numbers are depicted in Fig.7 where open circles are the given outputs.

Exponential PR is very similar to statistical regression in the sense of mathematical structure. Through the comparison between the two analyses, it can be emphasized that the two approaches have different points of view, that is, possibility and probability. Thus, it might be nonsense to simply compare two results obtained by statistical and PR analyses. Generally, since phenomena have many aspects, an analyst should select which aspect of the given phenomenon should be analysed. After that, one will be able to choose which approach will be suitable for the given phenomenon.

In PR analysis, it is emphasized that the spread of the given data is well transformed into the interactive possibility distribution of the coefficient vector by this method. The selection of independent vectors in the orthogonal condition is left for the future study.

Table 5. Optimal value J with the orthgonal conditions

Orthogonal condition (sample number)	Optimal value J
1,4,7	165531
5,10,15	317464
3,8,13	435490
1,2,3	457034
10,11,12	548021
4,9,14	715373
9,12,15	770709
13,14,15	1756514
1,6,11	1842863
5,8,11	6628069

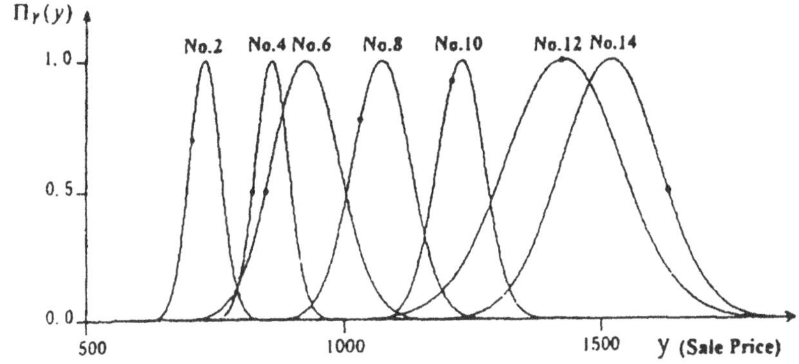

Figure 7 The estimated fuzzy outputs (possibility distributions) with the given crisp outputs

11.3 LEAST SQUARES REGRESSION

There are basically two situations that have been studied:

- Fuzzy input, fuzzy output,
- Numerical input, fuzzy output.

The first typically looks to models of the form

$$\mathbf{Y} = \mathbf{A} + \beta^t \mathbf{X}, \tag{84}$$

where $\mathbf{Y}$, $\mathbf{X}$ are $n-$vectors with fuzzy components, $(\mathbf{Y}_i, \mathbf{X}_i)$, $i = 1, \ldots, m$, are the fuzzy data pairs and the fuzzy $n-$vector $\mathbf{A}$ and crisp $n-$vector β are to be estimated from the data with some concept of least squares best fit. There can be variations on this theme, for example if β is fuzzy. Models of this type have usually not assumed any randomness structure, but this can be introduced if fuzzy random variables are used.

On the other hand, with crisp input the usual model is

$$\mathbf{Y} = \mathbf{A} + \mathbf{x}^t \mathbf{B}, \tag{85}$$

with $\mathbf{Y}$ and bx fuzzy and crisp $n-$vectors respectively. The data pairs are of the form $(\mathbf{Y}_i, \mathbf{x}_i)$, $i = 1, \ldots, m$ and the fuzzy $n-$vectors of parameters $\mathbf{A}$, $\mathbf{B}$ to be determined from some least squares procedure applied to the data.

11.3.1 Celmins Model

Celmins has used a form of least squares method for fitting symmetric fuzzy numbers appearing as vector elements and having conically dependent membership functions (Celmins, 1987),(Celmins, 1987). That is to say, the elements of the vectors are interactive fuzzy numbers. Let $\mathbf{X}$ be the vector of all fuzzy observations of an event. An implicit parametric model is given by

$$\mathbf{F}(\mathbf{X}, \mathbf{A}) = \mathbf{0}, \tag{86}$$

where $\mathbf{F}$ is an $\Re^n$ valued function and $\mathbf{A}$ is a $k-$dimensional vector of fuzzy parameters. In fitting, the data $\mathbf{X}$ do not exactly satisfy (86) but are hopefully close to a set $\mathbf{X} + \mathbf{\Delta_X}$ that does,

$$\mathbf{F}(\mathbf{X} + \mathbf{\Delta_X}, \mathbf{A}) = \mathbf{0}. \tag{87}$$

As a measure of goodness of fit, that is, of small corrections $\mathbf{\Delta_X}$, Celmins uses $W = (1 - \mu_\mathbf{X}(\mathbf{\Delta_X}))^2$. Consequently, he solves the problem of minimizing W subject to the constraint (87).

This is done by locally weighting each component of the residual by information in the fuzzy data. It is assumed that each fuzzy vector $\mathbf{X}$ has a conical membership function

$$\mu_\mathbf{X}(\xi) = \left[1 - ((\xi - \mathbf{X})' P_\mathbf{X} (\xi - \mathbf{X}))^{1/2}\right]^+ ,$$

where $P_\mathbf{X}$ is a diagonal matrix incorporating the spread squares of the components of $\mathbf{X}$. Hence the optimization has elements of a weighted least squares problem. This technique does not fit directly into the classical least squares formulation, which is approximation by projection onto a closed linear subspace or cone. Nor does it give a sum of residuals from which variance might be estimated.

Example 1. The following example is taken from (Celmins, 1987). It involves perforation of plate by projectiles fired orthogonal to the plate and links limit velocity V to nose shape quotient Q. All fuzzy numbers were symmetric triangular, that is, with membership function $\mu(z) = 1 - |(z - m)/s|$, denoted by (m, s). The model used is

$$V = a + bQ, \tag{88}$$

where a, b are symmetric triangular fuzzy parameters to be determined as the solution the optimization problem. Note that bQ is no longer triangular, but drum shaped. Nevertheless, Celmins treats this quantity approximately as a triangular number. The data is shown in Table 6 below. Celmins' fit was

$$V = (819.2,\ 24.5) + (-207.8,\ 226.0)Q$$

Table 6

$Q_i = (q_i, \tau_i)$	$V_i = (v_i, \omega_i)$
(0.000, 0.050)	(841, 50)
(0.052, 0.050)	(846, 50)
(0.165, 0.050)	(803, 50)
(0.232, 0.050)	(795, 50)
(0.431, 0.050)	(772, 100)

11.3.2 Metric Distance

True least squares, that is projection onto a closed linear subspace or cone, can be accomplished with fuzzy data by introducing the ρ_2-metric (Diamond, 1994) on the class of fuzzy sets, or by introducing equivalent metrics on subspaces, like triangular fuzzy numbers, that are especially computationally convenient. Least squares then extends naturally to a number of linear models (Diamond, 1988), (Diamond, 1987) and the optimization problem is very much simpler than that described above. For computational simplicity, the fuzzy data will be considered to lie in a special class.

Let $\mathcal{E}^n$ be the class of all normal, upper semicontinuous fuzzy convex sets on $\Re^n$ with compact support. For brevity, write $\mathcal{E} = \mathcal{E}^1$, the class of fuzzy numbers. In plain language, these are fuzzy sets, nonzero only on a finite interval, whose level sets are closed intervals and with at least one point with membership value 1.

Such fuzzy numbers are characterized by their *support function*. For $u \in \mathcal{E}^n$ define

$$s_u(h,p) = \sup\{< p, a >: \ a \in [u]^h, \ p \in S^{n-1}\},$$

where $[u]^h = \{x \in \Re^n : u(x) \geq h\}$, $0 < h \leq 1$ is the h-level set of u and S^{n-1} is the spherical surface $||x|| = 1$ in $\Re^n$. A metric is defined by

$$\rho_2(u,v) = \left(\int_0^1 \int_{S^{n-1}} (s_u(h,p) - s_v(h,p))^2 \, \mu(dp) \, dh \right)^{1/2},$$

where $\mu(dp)$ is unit measure on S^{n-1}. See (Diamond, 1994) for further details.

An especially convenient form for computation are the class $\mathcal{L}$ of LR-fuzzy numbers. Let $L, R : \Re^+ \to [0,1]$ be fixed left continuous and nonincreasing functions with $L(0) = R(0) = 1$. Then

$$u(x) = \begin{cases} L([m-x]/\ell) & \text{if} \quad x \leq m \\ R([x-m]/r) & \text{if} \quad x > m \end{cases}$$

is a fuzzy number with modal point m and left and right spreads ℓ, $r > 0$ respectively. The symmetric fuzzy numbers $(a,c)_L$, where L is even, are a special case.

Very commonly used in applications are a subset of the LR-fuzzy numbers: $\mathcal{T} \subset \mathcal{L} \subset \mathcal{E}$, the set of *triangular fuzzy numbers*, where $L = R = |1-x|$. That is, $u \triangleq (u^-, u^m, u^+)$ with $u^- \leq u^m \leq u^+$ with membership function $u(x) = 0$ if $x < u^-$ or $x > u^+$, while

$$u(x) = \begin{cases} (x-u^-)/(u^m-u^-), & \text{for } u^- \leq x \leq u^m \\ (u^+-x)/(u^+-u^m), & \text{for } u^m < x \leq u^+. \end{cases}$$

Clearly, both $\mathcal{L}$, $\mathcal{T}$ are linear cones in $\mathcal{E}$, that is, closed under fuzzy addition and scalar multiplication. We give $\mathcal{T}$ a metric $D_2(\cdot,\cdot)$ defined by

$$D_2(u,v)^2 = (u^- - v^-)^2 + (u^m - v^m)^2 + (u^+ - v^+)^2. \tag{89}$$

Then D_2 is an L_2-metric on $\mathcal{T}$, and is in fact an equivalent metric to the ρ_2 described in (Diamond, 1994). Moreover, it is easy to see that $(\mathcal{T}, D_2)$ is a complete metric space. Denote by $\mathcal{P} \subset \mathcal{T}$ the set of triangular fuzzy numbers u with positive support, that is $u^- \geq 0$.

Definition 1. *Let $\mathcal{H}$ be a cone in $\mathcal{T}$ and let $u = (u^-, u^m, u^+) \in \mathcal{T}$. If there exists a fuzzy number $h_0 \in \mathcal{H}$ such that for every $h \in \mathcal{H}$*

$$(h_0^- - h^-)(u^- - h_0^-) + (h_0^m - h^m)(u^m - h_0^m) + (h_0^+ - v^+)(u^+ - h_0^+) \geq 0,$$

then u is is said to be h_0-orthogonal to $\mathcal{H}$.

Theorem 1. (Diamond, 1988) *Let $\mathcal{H}$ be a closed cone in $\mathcal{T}$. For any $u \in \mathcal{T}$ there is a unique triangular fuzzy number h_0 in $\mathcal{H}$ such that $D_2(u, h_0) \leq D_2(u, h)$ for all $h \in \mathcal{H}$. A necessary and sufficient condition for h_0 to be the unique minimizing fuzzy number in $\mathcal{H}$ is that u is h_0-orthogonal to $\mathcal{H}$.*

Remark. The theorem extends to n-vectors of fuzzy numbers $(u_i), (v_i) \in \mathcal{T}^n$ by defining the distance between such to be $\sum_{i=1}^n D_2(u_i, v_i)^2$. The projection theorem is the principle behind systematic least squares regression and kriging of fuzzy triangular data.

A simple closed form for the metric in the class of LR–fuzzy numbers $\mathcal{L}$ has been given by Körner (Körner, 1995). First, define the *Steiner centroid* of a fuzzy set $u \in \mathcal{E}^n$, by

$$\sigma_u = n \cdot \int_0^1 \int_{S^{n-1}} t \cdot s_u(h, p)\, dp\, dha\ .$$

Then

$$d_2(u, v)^2 = d_2(u_0, v_0)^2 + ||\sigma_u - \sigma_v||^2,$$

where $||\cdot||$ is the usual norm in $\Re^n$ and $u_0 = u - \sigma_u$, $v_0 = v - \sigma_v$ are centred fuzzy sets with Steiner points at the origin. Let $L^{(-1)}(h) = \sup\{y \in \Re : L(y) \geq h\}$ and write

$$L_1 = \int_0^1 \left|L^{(-1)}(h)\right| dh/2\ , L_2 = \int_0^1 \left|L^{(-1)}(h)\right|^2 dh/2\ ,$$

with analogous expressions for R_1, R_2. Körner (Körner, 1995) has shown that for any two LR–fuzzy numbers $u = (m_u, \ell_u, r_u)_{LR}$, $v = (m_v, \ell_v, r_v)_{LR}$

$$\sigma_u = m_u + r_u R_1 - \ell_u L_1, \tag{90}$$

$$d_2(u, v)^2 = (m_u - m_v)^2 + R_2(r_u - r_v)^2 + L_2(\ell_u - \ell_v)^2 + 2(m_u - m_v)[(r_u - r_v)R_1 - (\ell_u - \ell_v)L_1]. \tag{91}$$

Note that for these formulas to be well defined, it is necessary that L_i, $R_i < \infty$, $i = 1, 2$. This condition is satisfied since $\mathcal{L} \subset \mathcal{E}$, so that L, R have finite support. For functions with infinite support the condition may still hold - for example, the exponential fuzzy numbers of Section 11.2.3. However, it is untrue for $L(x) = (1 + x)^{-1/2}$.

11.3.2.1 Metric least square models. The metric in $\mathcal{T}$ described above has been exploited by Diamond in a number of simple least squares models (Diamond, 1988, Diamond, 1987).

Case 1: fuzzy input, fuzzy output.

Suppose that observations consist of data pairs X_i, Y_i, $i = 1, 2, \ldots, N$, where $X_i = (X_i^-, X_i^m, X_i^+)$, $Y_i = (Y_i^-, Y_i^m, Y_i^+)$ are triangular fuzzy numbers. Assume that $X \in \mathcal{P}$, that is, $X_i^- \geq 0$, by translation of data if needed. Consider the model $Y = a + bX$, $a, b \in \Re$, to be fitted to the data in the sense of best fit with respect to the D_2-metric. This is the same as minimizing $\sum D_2(a + bX_i, Y_i)^2$. Recall that, if $b \geq 0, bX = (bX^-, bX^m, bX^+)$, while if $b < 0, bX = (-bX^+, bX^m, -bX^-)$. If $b \geq 0$, using (89) we minimize

$$\sum \left((a + bX_i^- - Y_i^-)^2 + (a + bX_i^m - Y_i^m)^2 + (a + bX_i^+ - Y_i^+)^2\right) ,$$

and if $b < 0$, minimize

$$\sum \left((a - bX_i^+ - Y_i^-)^2 + (a + bX_i^m - Y_i^m)^2 + (a - bX_i^- - Y_i^+)^2\right) .$$

Consequently, if a solution to the problem exists for $b \geq 0$, it is given by the solutions a^*, b^* to the equations

$$3Na + b\sum(X_i^- + X_i^m + X_i^+) = \sum(Y_i^- + Y_i^m + Y_i^+), \tag{92}$$

$$\begin{aligned} a\sum(X_i^- + X_i^m + X_i^+) \quad &+ \quad b\sum((X_i^-)^2 + (X_i^m)^2 + (X_i^+)^2) \\ &= \quad \sum(X_i^- Y_i^- + X_i^m Y_i^m + X_i^+ Y_i^+). \end{aligned} \tag{93}$$

A similar pair of equations will hold for $b < 0$ (interchange X^- by $-X^+$ and X^+ by $-X^-$) and the solutions will be written as a_*, b_*. If the data set is not degenerate, that is not sampled all at the one point X, then always $b^* \geq b_*$ (Diamond, 1988).

Definition 2. *The fuzzy data set X_i, Y_i is said to be tight if either $b^* \geq b_* \geq 0$ or $b_* \leq b^* \leq 0$. This gives a sort of compatibility condition between the "spreads" of the fuzzy data and the discernible trend of the regression.*

As an example of violation of this condition, consider the following: $N = 3$, $X_1 = (1, 3/4, 3/4)$, $X_2 = (2, 1, 1)$, $X_3 = (3, 1, 1)$, $Y_1 = (1, 3/4, 3/4)$, $Y_2 = (15/8, 3/2, 3/2)$, $Y_3 = (13/4, 3/2, 3/2)$. Then $b* = 333/225$, $b_* = -21/225$. The spreads of the data points are so large as to swamp trend of the modal points.

Theorem 2. (Diamond, 1988) *The least squares problem has a unique solution if the data is tight. If $b_* \geq 0$ the solution is given by $Y = a^* + b^*X$, and if $b^* \leq 0$ by $Y = a_* + b_*X$.*

Proof. Apply the projection theorem to the cone in $\mathcal{T}^N$ generated by the N-vectors of fuzzy numbers $(\mathbf{1}, \mathbf{1}, \ldots, \mathbf{1}), (X_1, \ldots, X_N)$, where $\mathbf{1} = (1, 0, 0)$. If a minimizing pair a, b exists, then it is unique. But if the data is tight, say

$b^* \geq b_* \geq 0$, only the first of the two systems can arise. Similarly, if $b_* \leq b^* \leq 0$, only the second system is present. □

This model can be mildly generalised to $Y = A + bX, A = (A^-, A^m, A^+) \in \mathcal{T}$ and the details are similar, although a stronger compatibility condition than that of definition 2 is required (Diamond, 1988). We give here the corresponding equations for the case of symmetric triangular fuzzy numbers $\mathcal{S} \subset \mathcal{T}$. Where $A = (a-\alpha, a, a+\alpha$, write $A = (a, \alpha)$ and similarly for $X = (x, \xi)$, $and Y = (y, \eta)$. Write $\widehat{y} = \sum y_i/N$ and similarly for $\widehat{x}$, $\widehat{\eta}$, $\widehat{\xi}$. Then the system for $b \geq 0$ is

$$a = \widehat{y} - b\widehat{x} \tag{94}$$

$$\alpha = \widehat{\eta} - b\widehat{\xi} \tag{95}$$

$$3N\widehat{x}a + 2N\alpha\widehat{\xi} + b\sum(3x_i^2 + 2\xi_i^2) = \sum(3x_iy_i + 2\xi_i\eta_i) \tag{96}$$

A solution to the problem for $b < 0$ is given by leaving (94) unchanged and replacing (95) and (96) by

$$\alpha = \widehat{\eta} + b\widehat{\xi} \tag{97}$$

$$3N\widehat{x}a - 2N\alpha\widehat{\xi} + b\sum(3x_i^2 + 2\xi_i^2) = \sum(3x_iy_i - 2\xi_i\eta_i) \tag{98}$$

Let the solution to the $b \geq 0$ case be b^*, A^*, and for the other, b_*, A_*. A nondegenerate data set (X_i, Y_i) is said to be *coherent* if

$$\sum(\xi_i - \widehat{\xi})(\eta_i - \widehat{\eta}) \geq 0, \tag{99}$$

and either $b_* \geq 0$ or $b^* \leq 0$. Again, if this compatibility condition holds, $b^* \geq b_*$.

Reworking the example of nontight data above, the expression (99) has value $0.5 > 0$ and $b_* = 0.9286$. Thus the data is coherent. The replacement of the crisp constant a by the fuzzy constant A fits the relatively large spreads and the extended model fits the data with a discernible trend. As an example of symmetric incoherent data, artificial to make the effect more obvious, consider $X_1 = (1, 0.1)$, $X_2 = (2, 0.2)$, $X_3 = (3, 0.3)$, $Y_1 = (3, 0.3)$, $Y_2 = (2, 0.2)$, $Y_3 = (1, 0.1)$. Then the expression in (99) is $-0.02 < 0$. Here the trend on the modal points of the data is at odds with the trend on the spreads.

Case 2: numerical input, fuzzy output.

Again for simplicity, consider only the class of symmetric triangular fuzzy numbers $\mathcal{S} \subset \mathcal{T}$, and note that if we write u^-, u^+ as the boundary of the support of $u \in \mathcal{S}$, then $u^m = (u^- + u^+)/2$. A simple metric D_S is defined on $\mathcal{S}$ by

$$D_S(u, v)^2 = (u^- - v^-)^2 + (u^+ - v^+)^2, \; u, v \in \mathcal{S}. \tag{100}$$

If $\mathcal{T}$ is used, the details are very much the same, but the normal equations are more complicated to accommodate the extra parameter introduced by unequal spreads.

Lemma. *The metrics D_2, D_S are equivalent on $\mathcal{S}$,*

$$D_S(u,v) \leq D_2(u,v) \leq \sqrt{3/2}\, D_S(u,v)\ ,\ u,v \in \mathcal{S}.$$

Thus, $(\mathcal{S}, D_S)$ is a complete, separable subspace of $(\mathcal{T}, D_2)$, and the subspace topology is the same as the D_S-metric topology.

Suppose that data pairs $x_i, Y_i, i = 1, 2 \ldots, N$, are observed, where the real numbers $x_i \geq 0$ and $Y_i \in \mathcal{S}$. Consider the model $Y = A + xB$, where $x \in \Re, A, B \in \mathcal{S}$, to be fitted to the data as a best D_2-fit. Solutions are sought in either $\mathcal{C}^* = \{(A + Bx_1, \ldots, A + Bx_N) : B^- \geq 0\}$ or $\mathcal{C}_* = \{(A + Bx_1, \ldots, A + Bx_N) : B^+ \leq 0\}$. In this model, the cones $\mathcal{C}^*$ or $\mathcal{C}_*$ could be thought of as representing positive or negative fuzzy trends. The problem is equivalent to minimizing $\sum D_2(A + Bx_i, Y_i)^2$. If a solution is sought in $\mathcal{C}^*$, this is equivalent to $\sum \left((A^- + B^- x_i - Y_i^-)^2 + (A^+ + B^+ x_i - Y_i^+)^2\right)$, with a similar expression for $\mathcal{C}_*$.

As in the other models, a compatibility condition is needed for the solution to be well-defined. Here, it expresses that the trends of the endpoints of the supports $[Y_i^-, Y_i^+]$ are not opposite. Write $\hat{x} = \sum x_i/N$, $\hat{Y}^{\pm} = \sum Y_i^{\pm}$ and say that the data $\{x_i, Y_i\}$ is *cohesive* if $\sum x_i(Y_i^+ - \hat{Y}^+) \geq \sum x_i(Y_i^- - \hat{Y}^-) \geq 0$. If all inequalities are reversed, then the data $\{2\hat{x} - x_i, Y_i\}$ is cohesive and the model $Y = A + zB, z = 2\hat{x} - x$ fitted.

Theorem 3. (Diamond, 1988) *Let $A, B \in \mathcal{P}$ satisfy the system*

$$A^- = Y^- - B^- \hat{x}, A^+ = Y^+ - B^+ \hat{x}, B^- = \kappa/T^2, B^+ = K/T^2,$$

where

$$\kappa = \sum x_i(Y_i^- - \hat{Y}^-),\ K = \sum x_i(Y_i^+ - \hat{Y}^+),\ T^2 = \sum x_i(x_i - \hat{x}).$$

If the data is cohesive, the fuzzy numbers A, B are well-defined and, if $K \geq \kappa \geq 0$, the model $Y = A + xB$ has a unique solution in the cone $\mathcal{C}^*$. If $K \leq \kappa \leq 0$, then $Y = A + zB$ has a unique solution in the cone $\mathcal{C}_*$.

Example 2. A very simple example is taken from (Diamond, 1988). Data involving student grades Y and family incomes X was fuzzified. Observed fuzzy output is $Y = (y, \underline{\eta}, \overline{\eta})$, where $Y^- = y - \underline{\eta}$, $Y^+ = y + \overline{\eta}$, and similarly fuzzy input $X = (x, \underline{\xi}, \overline{\xi})$ set out in Table 7.

Using the model $Y = a + bX$, a, b crisp, it was found that

$$Y = 1.052 + 0.147X,$$

with a sum of residuals 2.898 as compared with 0.713 for the original crisp numbers on which the fuzzy data were based. The data set is tight positive and replacing (93) and (96) by and replacing (93) and (96) by because $b_* = 0.051 > 0$. With the more general model $Y = A + bX$, A fuzzy symmetric,

$$Y = (1.201, 0.180) + 0.136X,$$

with a sum of residuals 2.413.

Table 7

$Y_i = (y_i, \underline{\eta}_i, \overline{\eta}_i)$	$X_i = (x_i, \underline{\xi}_i, \overline{\xi}_i)$
(4.0,0.6,0.8)	(21.0,4.2,2.1)
(3.0,0.3,0.3)	(15.0,2.25,2.25)
(3.5,0.35,0.35)	(15.0,1.5,2.25)
(2.0,0.4,0.4)	(9.0,1.35,1.35)
(3.0,0.3,0.45)	(12.0,1.2,1.2)
(3.5,0.53,0.7)	(18.0,3.6,1.8)
(2.5,0.25,0.38)	(6.0,0.6,1.2)
(2.5,0.5,0.5)	(12.0,1.8,2.4)

Consider the same data, but with the original crisp x and symmetric fuzzified Y as shown in Table 8.

Table 8

$Y_i = (y_i, \eta_i)$	x_i
(4.0,0.8)	21.0
(3.0,0.3)	15.0
(3.5,0.35)	15.0
(2.0,0.4)	9.0
(3.0,0.45)	12.0
(3.5,0.7)	18.0
(2.5,0.38)	6.0
(2.5,0.5)	12.0

The fitted model is

$$Y = (1.538, 0.223) + x(0.108, 0.019),$$

with residual sum of squares 2.280. The data are cohesive.

Example 3. Now consider Celmins' normal impact data from Section 11.3.1. For the fuzzy-input output model outlined above,

$$V = A + bQ,$$

where the symmetric triangular $A = (840.72, 51)$ and the crisp $b = -166.58$, with residual sum of squares $SSD = 7850.0$. This is quite close to Celmins' computation. The sum of residuals appears large by comparison with Example 2, but so are the parameters, If one were to assume an error structure and proceed as if there were 15 observations (5 observations of mode and left and right foot of the triangular numbers), there is the variance estimate $s^2 = SSD/(15-2) = 603.55$.

In the case of crisp input-fuzzy output,

$$V = A + qB,$$

where the modal q were used as the numerical input, the data failed the cohesiveness test. This was due to the very large increase in spread, despite a negative trend, of the last datum.

11.3.3 Körner's Method

The approach here is to fit LR–fuzzy data Y_i, $\mathbf{X}_i$, $i = 1, \ldots, N$, $y \in \mathcal{L}$, $\mathbf{X} \in \mathcal{L}^n$, to a parametric function $F(\mathbf{X}; \boldsymbol{\beta})$, $\boldsymbol{\beta} = (\beta_1, \ldots, \beta_m)^t$ so as to minimize

$$\sum_{i=1}^{N} d_2 \left(F(\mathbf{X}_i; \boldsymbol{\beta}), Y_i\right)^2 \ ,$$

at some $\widehat{\boldsymbol{\beta}} \in \mathcal{L}^m$. In the particular case where F is linear

$$F(\mathbf{X}; \boldsymbol{\beta}) = \left(f_1(\mathbf{X}), \ldots, f_m(\mathbf{X})\right)^t \boldsymbol{\beta},$$

the existence and uniqueness of a solution $\widehat{\boldsymbol{\beta}}$ is given by the Hilbert space projection theorem (Körner, Näther, 1996).

For simplicity and convenience, assume in what follows that $L = R$. This is done because of the following technical reason: in the linear model below, the Normal Equations solving the least squares problem for positive parameters are very different from those for the negative parameters unless $L = R$. Then the systems of equations are very similar.

The simplest case is $Y = A + bX$, $X \in \mathcal{L}$, where $b \in \Re$, $A \in \mathcal{L}$ are to be fitted. Write

$$X = (X^m, X^-, X^+)_{LR}, \quad Y = (Y^m, Y^-, Y^+)_{LR}, \quad A = (A^m, A^-, A^+)_{LR},$$

and seek parameters which minimize $\sum_{i=1}^{N} d_2(Y_i, A+bX)^2$. As in Section 11.3.2.1 there are two minimization problems, corresponding to $b \geq 0$, $b < 0$. Suppose that the solution to the first is b^*, and to the second b_*. If $b^* \geq 0$, say that it is *admissible* and similarly if $b_* < 0$. If only one is admissible, it is the solution to the fitting problem. If both are admissible, the solution which gives the smaller squared distance is taken as the best fit.

Assuming $b \geq 0$, then from (91), the squared distance $\sum_{i=1}^{N} d_2(Y_i, A + bX)^2$ to be minimized is

$$\sum_{i=1}^{N} \{(Y_i^m - A^m - bX_i^m)^2 + R_2(Y_i^+ - A^+ - bX^+)^2 \tag{101}$$
$$+L_2(Y_i^- - A^- - bX^-)^2$$
$$+2(Y_i^m - A^m - bX_i^m)[R_1(Y_i^+ - A^+ - bX_i^+) - L_1(Y_i^- A^- - bX_i^-)]\},$$

subject to the constraints A^-, $A^+ \geq 0$.

It is straightforward, but lengthy, to derive analogues of the Normal Equations from (101). For some quantities z_i, v_i appearing in the data, write

$$\begin{aligned} \widehat{z} &= 1/n\sum_i z_i, \; s_{zv} = 1/n\sum_i z_i v_i, \; t_{zv} = s_{zv} - \widehat{z}\widehat{v}, & (102)\\ C_{XX} &= t_{X^mX^m} + R_2 t_{X^+X^+} + L_2 t_{X^-X^-} + 2R_1 t_{X^mX^+} & (103)\\ &\quad -2L_1 t_{X^mX^-} \\ C_{XY} &= t_{X^mX^m} + R_2 t_{X^+X^+} + L_2 t_{X^-X^-} + R_1(t_{X^mY^+} + t_{Y^mX^+}) & (104)\\ &\quad -L_1(t_{X^mY^-} + t_{Y^mX^-}) \end{aligned}$$

Then the solutions to the $b \geq 0$ minimization satisfy the equations

$$b^* = \frac{C_{XY}}{C_{XX}}, \qquad A^{*m} = \widehat{Y^m} - b^*\widehat{X^m}, \tag{105}$$

$$A^{*-} = \widehat{Y^-} - b^*\widehat{X^-}, \qquad A^{*+} = \widehat{Y^+} - b^*\widehat{X^+}. \tag{106}$$

The solutions to the $b < 0$ problem satisfy very much the same equations, the last two equations being replaced by

$$A_*^- = \widehat{Y^-} + b_*\widehat{X^+}, \qquad A_*^+ = \widehat{Y^+} + b_*\widehat{X^-},$$

while in the expressions for C_{XX}, C_{XY}, X^- and X^+ are interchanged in the subscripts and some signs are changed.

Example 4. Again consider Celmins' normal impact data in Section 11.3.1. Since all fuzzy numbers are here symmetric triangular, observe that $X^+ = X^-$, $Y^+ = Y^-$ and $L(x) = R(x) = \max\{0, 1-x\}$. Hence, $L_1 = R_1 = 1/4$ and $L_2 = R_2 = 1/6$. Consequently, if we write $X^m = x$, $X^+ = \xi$, $Y^m = y$, $Y^+ = \eta$, equations (101), (102) reduce to

$$C_{XX} = t_{xx} + t_{\xi\xi}/3, \quad C_{XY} = t_{xy} + t_{\xi\eta}/3$$

for the $b \geq 0$ assumption, while for the $b < 0$ case C_{XX} is unchanged but $C_{XY} = t_{xy} - t_{\xi\eta}/3$. From (105), $b^* = -177.72$ and the solution is not admissible. However, $b_* = -177.72$ also and this is admissible. From (106), $A = (842.68, 68.89)$ and so the fitted expression is

$$V = (842.68, 68.89) - 177.72Q.$$

11.3.4 Hukuhara Fitting

Incoherent data may arise in a natural context and any regression methodology should be able to take account of this. For example, spreads may decrease with a negative trend because

- possibly, smaller fuzzy numbers do not have the same absolute spread of imprecision as larger ones;

- if time T is the explanatory variable, it may be possible to measure the Y relatively more precisely for larger T as a aystem evolves;
- measurements may be relatively more precise in higher ranges of the explanatory variable.

Körner (Körner, 1996) has suggested models involving approximations to the Hukuhara difference to resolve the problem of incoherence in some cases. Recall that the Hukuhara difference of fuzzy sets $A \ominus B$ is defined as the solution X to the equation $A = B + X$, when such a solution exists. If it exists it is unique. For example, with symmetric triangular numbers $A = (2, 0.4)$, $B = (0.6, 0.1)$, $A -_H B = (1.4, 0.3)$. However, if $B = (0.6, 0.41)$, no Hukuhara difference exists. The larger spreads of B cannot be accomodated by any fuzzy number X in the equation. However, there is a least squares solution to the problem: find $X = (x, \xi)$ so as to minimize $D_2(A, B + X)^2$. The solution to this problem is

$$\begin{aligned} x &= a - b, \\ \xi &= \begin{cases} \alpha - \beta & \text{if } \alpha > \beta \\ 0 & \text{if } \alpha \leq \beta . \end{cases} \end{aligned}$$

Call this solution $A \ominus_H B = (x, \xi) = X$ the *Least Squares Hukuhara difference* (LSH-difference).

So, this suggests that if minimizing $\sum D_2(Y_i, A + bX_i)^2$ gives no sensible fit because the data are incoherent, instead seek to minimize $\sum D_2(Y_i + bX_i, A)^2$ when there seems to be a negative trend and, "reflecting" $Y \to -Y$, $\sum D_2(bX_i, A + Y_i)^2$ when there appears to be a positive trend.

Definition 3. Let $A, B \in \mathcal{S}$ be symmetric triangular fuzzy numbers. Define $X = A \ominus_H B$ to be that $X \in \mathcal{S}$ which minimizes $D_S(B + X, A)^2$. In particular, write

$$Y = A \ominus_H bX$$

for the Y which minimizes $D_S(Y + bX, A)^2$. Here, D_S is the metric on $\mathcal{S}$ as defined in Section 11.3.2.

Consequently, fitting the model $Y = A \ominus_H bX$ to data (X_i, Y_i), $i = 1, \ldots, N$, comes down to finding parameters b, A which minimize $\sum D_S(Y_i + bX_i, A)^2$. Using the notation $X = (X^-, X^+)$, as in equation (100), $\widehat{X^-} = \sum X_i^- / N$, and so on, obtain the equations

$$A^- = \widehat{Y^-} + b\widehat{X^-} \qquad A^+ = \widehat{Y^+} + b\widehat{X^+} \tag{107}$$

$$\begin{aligned} b\sum((X^-)^2 + (X^+)^2) &= N\widehat{X^-}A^- + N\widehat{X^+}A^+ \\ &\quad - \sum(X^-Y^- + X^+Y^+) . \end{aligned} \tag{108}$$

Substituting (107) in (108) gives

$$b = \frac{t_{X^-Y^-} + t_{X^+Y^+}}{t_{X^-X^-} + t_{X^+X^+}}$$

and because the variance-like terms in the denominator are positive, $b \geq 0$ provided the covariance-like terms in the numerator are nonnegative. In these circumstances, call the data set *Hukuhara coherent negative* and observe that $A \in \mathcal{S}$ is well defined from (107) if $b \geq 0$, since then $A^+ \geq A^-$.

Turning to incoherent data with an evident positive trend, minimize

$$\sum D_S(bX_i, A + Y_i)^2$$

with respect to the parameters. It is easy to show that

$$\begin{aligned} b\widehat{X^+} &= A^+\widehat{Y^+}, \qquad b\widehat{X^-} = A^- + \widehat{Y^-} \\ b\sum((X_i^-)^2 + (X_i^+)^2) &= \sum(X_i^+Y_i^+ + X_i^-Y_i^-) + NA^+\widehat{X^+} + NA^-\widehat{X^-}\ . \end{aligned}$$

So $b \geq 0$ automatically for data lying in $\mathcal{P}$. The fitting will make sense if $A^+ \geq A^-$, and the condition for this is that

$$b(\widehat{X^+} - \widehat{X^-}) \geq \widehat{Y^+} - \widehat{Y^-}\ . \tag{109}$$

If this condition holds, we say that the data is *Hukuhara coherent positive* and fits the model

$$Y = A \oplus_H bX.$$

Both these models can be extended to general LR-fuzzy numbers.

Example 5. Returning to the example of incoherent data for the $Y = A + bX$ model, we find that $b = 1$ and $A^- = 3.6, A^+ = 4.4$. Consequently, the data is H-coherent negative and the best H-fit is

$$Y = (4, 0.4) \oplus_H X.$$

Example 6. Consider the symmetric fuzzy triangular data set $X_1 = (1, 1/4)$, $X_2 = (1.1, 1.0)$, $X_3 = (1.2, 1.1)$, $Y_1 = (1, 3/4)$, $Y_2 = (13/8, 3/2)$, $Y_3 = (7/2, 3/2)$ which appear in (Diamond, 1988). Then the equations give $b = 1.4823$, $A^+ = -0.3023$, $A^- = -0.5200$. Hence, the data is H-coherent positive and the best H-fit is

$$Y = (-0.4112, 0.1089) \oplus_H 1.4823X.$$

11.3.5 Fitting Several Fuzzy Variables

Suppose that data consists of n-vectors of fuzzy explanatory variables $\mathbf{X}_i = (X_{1i}, \ldots, X_{ni})^T$ and Y_i, $i = 1, \ldots, N$. Each component X_{pi} is a triangular fuzzy number $(X_{pi}^m, X_{pi}^-, X_{pi}^+)$ and similarly for Y_i. We discuss least squares fitting of this data to the fuzzy affine function

$$Y = \beta_0 + \boldsymbol{\beta}^T X, \quad \boldsymbol{\beta} = (\beta_1, \ldots, \beta_n), \tag{110}$$

where $\beta_0, \ldots, \beta_n \in \Re$. The fit will be with respect to minimization of

$$r(\beta_0, \boldsymbol{\beta}) = \sum_{i=1}^{N} d_2(\beta_0 + \boldsymbol{\beta}^T \mathbf{X}_i, Y_i)^2. \tag{111}$$

As before, different expressions arise for $r(\beta_0, \boldsymbol{\beta})$ according as some of the β_p are assumed positive or negative. Consequently, to derive equations for the minimizing parameters, we must specify certain cones in which to seek minimizing solutions.

Definition 4. Let e denote the fuzzy number $(1,0,0)$, $\mathbf{E}$ the N–vector with all entries e and X the N–vector of the data with components $\mathbf{X}_i$. Partition the set $\{1,2,\ldots,n\}$ into two exhaustive, mutually exclusive subsets $J(+)$, $J(-)$, one of which may be empty. To each such partition associate a binary multi-index $J=(j_1,\ldots,j_n)$ defined by $j_p=0$ if $p\in J(+)$ and $j_p=1$ if $p\in J(-)$. In particular, write $J_0=(0,0,\ldots,0)$, $J_1=(1,1,\ldots,1)$. Denote by $\mathcal{C}(J)$ the cone in $\mathcal{T}^N$,

$$\mathcal{C}(J)=\left\{\beta_0\mathbf{E}+\boldsymbol{\beta}^T\mathbf{X} : \beta_p\geq 0 \text{ if } j_p=0,\ \beta_p<0 \text{ if } j_p=1\right\}.$$

Call J the *conal index* and $\mathcal{C}(J)$ the cone determined by the index J. For a given conal index J, consider the problem of minimizing $r(\beta_0(J),\boldsymbol{\beta}(J))$ in the cone $\mathcal{C}(J)$. By the Projection Theorem, each such problem is solved by a unique set of parameters $\beta_0(J)$, $\boldsymbol{\beta}(J)$. This J–solution solves the problem of minimizing the distance (110), subject to the constraints

$$\beta_p\geq 0 \quad \text{if} \quad j_p=0,\ p=1,\ldots,n.$$

Thus, each J–problem could be solved by treating it as a constrained quadratic problem with Kuhn–Tucker conditions. In general, this will entail some of the $\beta_p=0$, that is, the corresponding variables are not significant to the fit in $\mathcal{C}(J)$. Let us suppose that all "superfluous" variable have been removed and all explanatory variables are active. Thus the minimum of (110) will not be achieved by setting one or more of the parameters β_p to zero.

Denote by $S(J)$ the system of $n+1$ equations

$$\partial r(\beta_0(J),\boldsymbol{\beta}(J))/\partial\beta_p,\ p=0,1,\ldots,n$$

and write it as

$$S(J)\ :\qquad A(J)\,(\beta_0(J),\boldsymbol{\beta}(J))^T\ =\ \zeta(J).$$

Suppose that $S(J)$ has a solution $\beta_0^*(J),\boldsymbol{\beta}_0^*(J)$ such that $\beta_p(J)>0$ if $j_p=0$ and $\beta_p<0$ if $j_p=1$. That is, the signs of the $\beta(J)$'s are consistent with the conal index J. Then we say that the model $Y=\beta_0+\boldsymbol{\beta}^T\mathbf{X}$ is J–consistent with the data. That is, if the formal solution of the system for unconstrained optimization $S(J)$ s compatible $\beta_0+\boldsymbol{\beta}^T\mathbf{X}$ lying in $\mathcal{C}(J)$.

Example 7. The data in Table 9 relates admission patterns to a hospital with indices of health service availability and indigency, for ten communities in the hospital's catchment area. The data was fuzzified and used in (Diamond, 1987). As in Example 2, $Y^-=y-\underline{\eta}$, $Y^+=y+\overline{\eta}$, and similarly for X_1, X_2. Solutions of $S(J_0)$ were $\beta_0=11.8$, $\overline{\beta_1}=4.1$, $\beta_2=3.0$. The model

$$Y=\beta_0+\beta_1X_1+\beta_2X_2$$

is thus J_0−consistent. Solutions of $S(J_1)$ gave positive values for β_1, β_2, which is incompatible with the possibility of the normal equation solution being in $\mathcal{C}(J_1)$. The other two conal indices were similarly incompatible with their associated systems of equations for $\boldsymbol{\beta}$. The only significant fit is thus

$$Y = 11.8 + 4.1X_1 + 3.0X_2.$$

The SSD was 866.1, with $s^2 = 30.9$.

Table 9

$Y = (y, \underline{\eta}, \overline{\eta})$	$X_1 = (x_1, \underline{\xi_1}, \overline{\xi_1})$	$X_2 = (x_2, \underline{\xi_2}, \overline{xi_2})$
(61.6,6.2,3.1)	(6.0,0.3,0.9)	(6.3,0.9,0.9)
(53.2,2.7,5.3)	(4.4,0.4,0.7)	(5.5,0.8,0.3)
(65.5,9.8,9.8)	(9.1,0.5,0.7)	(3.6,0.2,0.4)
(64.9,3.2,9.8)	(8.1,1.2,1.2)	(5.8,0.8,0.9)
(72.7,3.6,7.3)	(9.7,1.0,1.5)	(6.8,0.3,0.3)
(52.2,2.6,5.2)	(4.8,0.2,0.7)	(7.9,1.2,0.8)
(50.2,2.5,5.0)	(7.6,0.4,1.1)	(4.2,0.2,0.6)
(44.0,2.2,4.4)	(4.4,0.2,0.4)	(6.0,0.6,0.3)
(53.8,8.1,8.1)	(9.1,0.9,0.9)	(2.8,0.1,0.4)
(53.5,8.1,5.4)	(6.7,0.7,0.7)	(6.7,1.0,1.0)

11.3.6 Random Fuzzy Variables

Recall that in Section 11.3.2, the space $\mathcal{E}^n$ was given a metric d_2. Then $(\mathcal{E}^n, \rho_2)$ is a complete separable metric space (Diamond, 1994). This space has a Borel algebra given by the metric topology. Let $(\Omega, \mathcal{A}, P)$ be a probability space. A fuzzy random variable (FRV) is a measurable function $X : X \to \mathcal{E}^n$. That is, an FRV is equivalent to a collection $\{[X]^\alpha : 0 \le \alpha \le 1\}$ of random nonempty compact convex sets, namely the level sets of X. The *expectation* is defined in terms of the level sets, $[E(X)]^\alpha = E([X]^\alpha)$, by means of the Aumann integral

$$\begin{aligned} E([X]^\alpha) &= \int_\Omega [X]^\alpha \, dP \\ &= \left\{ \int_\Omega f \, dP : f \text{ is a } P\text{- measurable selection of } [X]^\alpha \right\}. \end{aligned}$$

From this it is easy to see that when X is integrally bounded, it is indeed true that $[E(X)]^\alpha = E\left([X]^\alpha\right)$, $0 < \alpha \le 1$, as in (Puri, 1986). The *variance* is well-defined by Var $X = E\left(n^{1/2}\rho_2(X, E(X))^2\right)$, while the *covariance* of two FRVs X, Y is given by

$$E\left(\int_0^1 \int_{S^{n-1}} \left(s_X - s_{E(X)}\right)\left(s_Y - s_{E(Y)}\right) \mu(dt)\, dh\right),$$

where $s_X(h, p)$ is the support function of X defined in Section 11.3.2.

Theorem 4. (Körner, 1995) *For any* $u, v \in \mathcal{E}^n$, *and any FRVs* X, Y,

$$\begin{aligned} \rho_2(u,v)^2 &= \rho_2(u-\sigma_u, v-\sigma_v)^2 + \rho_2(\sigma_u,\sigma_v)^2, \\ E(X) &= E(X-\sigma_X) + E(\sigma_X), \\ \mathrm{Var}\,(X) &= \mathrm{Var}\,(X-\sigma_X) + \mathrm{Var}\,(\sigma_X), \\ \mathrm{Cov}\,(X,Y) &= \mathrm{Cov}\,(X-\sigma_X, Y-\sigma_Y) + \mathrm{Cov}\,(\sigma_X,\sigma_Y). \end{aligned}$$

With this result it is possible to develop precise formulas for the variance and covariance of LR–represented FRVs (Körner, 1995). For example, if X, Y are two such and modal points and spreads of a FRV are assumed to be independently distributed,

$$\mathrm{Var}\,(X) = \mathrm{Var}\,(X^m) + R_2\mathrm{Var}\,(X^+) + L_2\mathrm{Var}\,(X^-) \tag{112}$$

$$\begin{aligned} \mathrm{Cov}\,(X,Y) = {} & \mathrm{Cov}\,(X^m,Y^m) + L_2\mathrm{Cov}\,(X^-,Y^-) \\ & + R_2\mathrm{Cov}\,(X^+,Y^+) - L_1(\mathrm{Cov}\,(X^m,Y^-) + \mathrm{Cov}\,(Y^m,X^-)) \\ & - R_1(\mathrm{Cov}\,(X^m,Y^+) + \mathrm{Cov}\,(Y^m,X^+)) \ . \end{aligned} \tag{113}$$

If the FRVs are symmetric, with $L = R$, $X^m = x$, $X^+ = X^- = \xi$, $Y^m = y$, $Y^+ = Y^- = \eta$, these further simplify to

$$\mathrm{Var}\,(X) = \mathrm{Var}\,(x) + 2L_2\mathrm{Var}\,(\xi) \tag{114}$$

$$\begin{aligned} \mathrm{Cov}\,(X,Y) = {} & \mathrm{Cov}\,(x,y) + L_2\mathrm{Cov}\,(\xi,\eta) - 2L_1(\mathrm{Cov}\,(x,\eta) \\ & + \mathrm{Cov}\,(y,\xi)). \end{aligned} \tag{115}$$

11.3.7 Kriging

Linear estimation using L_2–metrics is often called *kriging* and is commonly used in the estimation of naturally occuring quantities, dispersed in space, such as concentrations in a mineral deposit or pollution levels over a region. Diamond considered such estimation for the case where data were $\mathcal{T}$–valued FRVs (Diamond, 1989). The computational problem is reduced to a constrained optimization which is readily solved and is described briefly below.

Let V be a three dimensional region and suppose that $w(x), x \in V$ is a $\mathcal{P}$-valued *regionalised variable.* This is regarded as a realisation of a $\mathcal{P}$-valued random function on V, with the properties that the expectation $E\,W(x)$ exists and is in $\mathcal{P}$; and that the variance is $\mathrm{Var}\;W(x) = E\,D_2(W(x), E\,W(x))^2$, which is a real-valued quantity. This random function is further assumed to be *second order stationary*:

(i) $E\,W(x) \triangleq \omega = (\omega^-, \omega^m, \omega^+)$ exists and is independent of $x \in V$.
(ii) There exist lower, modal and upper covariance functions $C^-(\xi), C^m(\xi), C^+(\xi)$, independent of $x \in V$, such that

$$E\,(W^\beta(x+\xi)W^\beta(x)) - (\omega^\beta)^2 = C^\beta(\xi)\ ,\ \beta \in \mathcal{J} = \{-, m, +\}.$$

These assumptions represent physical homogeneity and the existence of the first two moments of the law underlying the random function. See (Diamond, 1989) for further details and a justification of the results below.

Properties.

1. $C^-(\xi), C^m(\xi), C^+(\xi)$ are each symmetric functions in ξ.
2. Var $W(x) = C^-(0) + C^m(0) + C^+(0)$.
3. $\frac{1}{2}E\, D_2(W(x+\xi), W(x))^2 = \sum_{\beta \in \mathcal{J}}(C^\beta(0) - C^\beta(\xi))$.
4. Given the estimator $W^* = \sum_{i=1}^N \lambda_i W(x_i)$, where $\lambda_i \geq 0$, $i = 1, 2, \ldots, N$, then Var $W^* = \sum_{i,j=1}^N \lambda_i \lambda_j (C^-(x_i - x_j) + C^m(x_i - x_j) + C^+(x_i - x_j))$.

Note that the non-negativity of the λ 's is essential to prevent the appearance of "mixed products" $E(W^-(x_i)W^+(x_j))$, in the sum for Var W^*, as would occur if $\lambda_i\lambda_j$ were negative. This is because, if $W \in \mathcal{P}$, then $-W = (-W^+, -W^m, -W^-)$, and a negative weight has the effect of reversing the spreads of the fuzzy number it multiplies.

With ordinary data that is spatially distributed, kriging under the stationary hypothesis amounts to finding those weights of an estimator that is a combination of data values, which minimize the variance of the estimator. The fuzzy-valued counterpart minimizes the variance of a linear combination of fuzzy data, where the variance is with respect to the D_2-norm in $\mathcal{P}$. Suppose that $w(x)$ is a fuzzy-valued regionalised variable which is a realisation of a second order stationary random function $W(x)$, over a region V. Consider a quantity y_0, fuzzy-valued and associated with the regionalised variable $w(x)$. For example, $y_0 = w(x_0)$, the value taken by w at $x = x_0$, or $y_0 = |V|^{-1} \int_V w(a)\, da$, which may be interpreted as an "average value" of the fuzzy variable $w(x)$ over the whole field V. Here $|V|$ denotes the volume of V while integration is that defined in, for example, (Diamond, 1994). Given N data values $w(x_1), w(x_2), \cdots, w(x_N)$ observed at points $x_i \in V, i = 1, 2, \cdots, N$, we want to estimate y_0. It is convenient to estimate a whole class of estimators into a single problem by considering y_0 as an average value over some region $V_0 \subseteq V$, containing the point x_0. This is certainly justified in the the limit for the case of integrably bounded regionalised variables which are also continuous (see (Diamond, 1994) for further details).

To estimate y_0, consider a weighted average of the data,

$$y_0^* = \sum_{i=1}^N \lambda_i w(x_i).$$

Let the random function corresponding to y_0 be denoted by Y_0. This defines a corresponding estimator $Y_0^* = \sum_{i=1}^N \lambda_i W(x_i)$ for Y_0 from the set of random variables $\{W(x_i) : i = 1, 2, \cdots, N\}$.

The weights λ_i are to be estimated so that the estimator is unbiased, and minimizes the variance $E\left(D_2(Y_0^*, Y_0)^2\right)$, with constraints $\lambda_i \geq 0$. The unbiasedness of Y_0^* means that $E(Y_0^*) = E(Y_0) = E(W(x))$, and this determines the equality constraint $\sum_{i=1}^N \lambda_i = 1$. The quantity to be minimized is

$$E\left(D_2(Y_0^*, Y_0)^2\right) = E\left((Y_0^{*-} - Y_0^-)^2 + (Y_0^{*m} - Y_0^m)^2 + (Y_0^{*+} - Y_0^+)^2\right) .$$

It may be shown that (Diamond, 1989) :

$$\begin{aligned}
E\left(D_2(Y_0^*,Y_0)^2\right) &= \sum_{i,j=1}^{N} \lambda_i\lambda_j\left(C^-(x_i-x_j)+C^m(x_i-x_j)+C^+(x_i-x_j)\right)\\
&- 2\sum_{i=1}^{N}\lambda_i\left(C^-(x_i,V_0)+C^m(x_i,V_0)+C^+(x_i,V_0)\right)\\
&+ C^-(V_0,V_0)+C^m(V_0,V_0)+C^+(V_0,V_0)
\end{aligned}$$

where the various symbols $C(x_i, V_0), C(V_0, V_0)$, etc., involve integrals of the covariance functions C^-, C^m, C^+ over the volume V_0 (Diamond, 1989). It is worth emphasising that without the condition for non-negative weights, the above equation has no meaning.

To minimize this variance, introduce a Lagrange multiplier μ for the equality constraint, and Kuhn-Tucker multipliers $L_1, \ldots, L_N$ corresponding to inequality constraints $\lambda_1 \geq 0, \ldots, \lambda_N \geq 0$.

Theorem 5. (Diamond, 1989) *Let $Y_0^* = \sum_{i=1}^{N} \lambda_i W(x_i)$ be an estimator as above. Suppose the matrix*

$$\Gamma_{ij} = C^-(x_i - x_j) + C^m(x_i - x_j) + C^+(x_i - x_j),\ i,j = 1,\ldots,N,$$

is strictly positive definite. Then there exists a unique linear unbiased $\mathcal{P}$-valued estimator minimizing the variance. The weights satisfy the system

$$\begin{aligned}
\sum_{i=1}^{N}\Gamma_{ij}\lambda_i - L_j - \mu &= \sum_{\beta\in\mathcal{J}} C^\beta(x_j,V)\,,\ j=1,\ldots,N\,,\\
\sum_{i=1}^{N}\lambda_i = 1\,, &\qquad \sum_{i=1}^{N} L_i\lambda_i = 0\,,\quad L_i\,,\,\lambda_i \geq 0\,,
\end{aligned}$$

with residual

$$\begin{aligned}
\sigma^2 &= \mu + C^-(V,V) + C^m(V,V) + C^+(V,V)\\
&\quad -\sum_{i=1}^{N}\lambda_i\left(C^-(x_i,V)+C^m(x_i,V)+C^+(x_i,V)\right).
\end{aligned}$$

The proof proceeds from the projection , with metric the expectation of the L_2-distance of $\mathcal{T}$-valued random variables. Further details can be found in (Diamond, 1989).

11.4 CONCLUDING REMARKS

We have surveyed regression analysis with fuzzy data. All methods discussed above fit a fuzzy linear model to the given data, for both crisp input, fuzzy output and fuzzy input-output.

Broadly speaking, the fitting procedures fall into two distinct areas of techniques:

- **Possibilistic regression,** based on ideas of possibilistic distributions. Fitting the data to a model is usually taken to mean that parameters are to be found so that the predicted spread of the model is to be minimized, yet still envelop the data in some sense. This class of methods usually reduce to a linear programming problem, although more general programming is possible in some models.

- **Least squares regression,** based on the notion of a metric distance between fuzzy sets. Fitting the model is to find parameters that minimize the distance between observed output data and the output predicted by the model from the given input data. Such methods thus involve unconstrained or unconstrained optimization and lead to analogues of the familiar Normal Equations.

A number of examples were given for the various types of PR. In particular, the house data were fitted to several models and the results compared.

In similar fashion, several examples illustrate the different least squares methods. Celmins projectile data is processed and compared with these. In the simple case of fuzzy triangular data, fairly straightforward criteria are given for a model to be compatible with the data. When these criteria are abrogated, a model can still be found by introducing a concept of best least squares approximation to Hukuhara differences.

References

J. Kacprzyk and M. Fedrizzi ed. (1992), Fuzzy Regression Analysis, Physica-Verlag, Heidelberg.

L. A. Zadeh (1995), The concept of a linguistic variable and its application to approximate reasoning-I, Inform. Sci. **8**, 199-249.

H. Tanaka, S. Uejima and K. Asai (1980), Fuzzy linear regression model, Int. Congress on Applied Systems Research and Cybernetics, Acapulco, Mexico Vol. VI, 2933-2938.

H. Tanaka, S. Uejima and K. Asai (1982), Linear regression analysis with fuzzy model, IEEE Trans. on Syst., Man, Cybern. **12**, 903-907.

H. Tanaka (1987), Fuzzy data analysis by possibilistic linear models, Fuzzy Sets and Systems **24**, 363-375.

H. Tanaka and J. Watada (1988), Possibilistic linear systems and their application to the linear regression model, Fuzzy Sets and Systems **27**, 275-289.

H. Tanaka, I. Hayashi and J. Watada (1989), Possibilistic linear regression analysis for fuzzy data, European J. Oper. Res. **40**, 389-396.

H. Tanaka and H. Ishibuchi (1991), Identification of possibilistic linear systems by quadratic membership functions, Fuzzy Sets and Systems **41**, 145-160.

H. Tanaka, H. Ishibuchi and S. Yoshikawa (1995), Exponential possibility regression analysis, Fuzzy Sets and Systems **69**, 305-318.

L. A. Zadeh (1977), Fuzzy sets as a basis for a theory of possibility, Fuzzy Sets and Systems **1**, 3-28.

H. Tanaka , I. Hayashi and J. Watada (1987), Possibilistic linear regression analysis based on possibility measure, Second IFSA world Congress, Tokyo, Japan, 317-320.

A. Bardossy (1990), Note on fuzzy regression, Fuzzy Sets and Systems **37**, 65-75.

I. Hayashi and H. Tanaka (1990), The fuzzy GMDH algorithm by possibility models and its application, Fuzzy Sets and Systems **36**, 245-258.

B. Heshmaty and A Kandel (1985), Fuzzy linear regression and its application to forecasting in uncertain environment, Fuzzy Sets and Systems **15**, 159-191.

D. A. Savic and W. Pedrycz (1991), Evaluation of fuzzy linear regression models, Fuzzy Sets and Systems **39**, 51-63.

M. Sakawa and H. Yano (1992), Multiobjective fuzzy linear regression analysis for fuzzy input-output data, Fuzzy Sets and Systems **47**, 173-181.

D. T. Redden and W. H. Woodall (1992), Further examination of fuzzy linear regression, Fuzzy Sets and Systems **79**, 203-211.

H. Moskowitz and K. Kim (1993), On assessing the H value in fuzzy linear regression, Fuzzy Sets and Systems **58**, 303-327.

D. T. Redden and W. H. Woodall (1994), Properties of certain fuzzy linear regression methods, Fuzzy Sets and Systems **64**, 361-375.

G. Peters (1994). Fuzzy linear regression with fuzzy intervals, Fuzzy Sets and Systems **63**, 45-55.

M. Inuiguchi, N. Sakawa and S. Ushiro (1993), Interval regression based on Minkowski's subtraction, Fifth IFSA World Congress, Soul, Korea, 505-508.

M. Kaneyoshi, H. Tanaka, M. Kamei and H. Furuta (190), New system identification Technique using fuzzy regression analysis, Proceedings of First Int. Sym. on Uncertainty Modeling and Analysis, Maryland, USA, 528-533.

P.-T. Chang, E. S. Lee and S. A. Konz (1996), Applying fuzzy linear regression to VDT legibility, Fuzzy Sets and Systems **80**, 197-204.

P. Diamond and P. Kloeden (1994), "Metric Spaces of Fuzzy Sets: Theory and Applications", World Scientific, Singapore.

P. Diamond (1988), Fuzzy least squares, Information Sciences, **46**, 141–157.

P. Diamond (1987), Least squares fitting of several fuzzy variables, in "Analysis of Fuzzy Information", J.C. Bezdek (ed.), CRC Press, Tokyo, 329–331.

A. Celmins (1987), Least squares model fitting to fuzzy vector data, Fuzzy Sets and Systems, **22**, 245–269.

A. Celminsm (1987), Multidimensional least–squares fitting of fuzzy models, Mathl. Modelling, **9**, 669–690.

P. Diamond (1989), Fuzzy Kriging, Fuzzy Sets and Systems, **33**, 315–332.

R. Körner 1995, On the variance of fuzzy random variables, preprint.

M. Fréchet (1948), Les élements aléatoires de natures quelconque dans un espace distancié, Ann. Inst. H. Poincaré, **10**, 215–310.

R. Körner and W. Näther (1996), Linear regression with random fuzzy variables: extended classical estimates, best linear estimates, least squares estimates. Preprint.

IV Reliability, Maintenance and Replacement

12 RELIABILITY

Etienne Kerre

Takehisa Onisawa

Bart Cappelle

Igor Gazdik

Abstract: The chapter presents applications of the fuzzy set theory to the reliability engineering field. It starts by a historical perspective of reliability engineering and some questions in classical reliability. Then, it reviews the probabilistic method of the system reliability analysis from the fuzzy set theoretical point of view. Next, it makes a survey of the probability approach and the non-probabilistic measure approach. Especially, the non-probabilistic measure approach focuses on the introduction of natural language expressions and experts' subjectivity to the system reliability analysis.

12.1 INTRODUCTION

The mathematical theory of reliability is still quite young and has been developed continuously since the thirties. Giving a concise overview of the main achievements and progress made since then is a difficult task, but nevertheless we would like to stress some important improvements since the introduction of the first reliability concepts about sixty years ago.

The growing complexity of machinery and the continuously developing technology lead to the first reliability concepts in the thirties as Barlow (1984) mentions. Grown out of the quality control theory, it may hardly surprise that these concepts were probabilistic. Somewhere in the late thirties, e.g., Weibull introduced a probability distribution to study the breaking strength of materials. The distribution has been named after him later on and is widely used amongst reliability theoreticians and engineers. During the second World War, the teams of Werner Von Braun that were developing Germany's ultimate *Vergeltungswaffe* -the V1 and V2 rockets- studied the reliability behaviour of their rockets by means of some basic reliability concepts.

After the second World War, the US Air Force and the US Department of Defense established the Advisory Group on Reliability of Electronic Equipment (AGREE), a first step towards the mathematical formulation of reliability theory. It was clear that the growing complexity of military equipment forced the engineers to develop more complex and adequate models to study the reliability of their new equipment. Especially life testing problems were studied through other areas like renewal problems and the development of optimal renewal policies were studied since the early forties. The AGREE group introduced amongst others minimal acceptability limits and requirements for reliability tests. During the fifties many reliability related problems were studied by many scientists, especially but not exclusively in the United States. Amongst others we mention missile reliability, the application of the Weibull distribution to reliability problems as an important alternative to the widely applied exponential distributions, life testing of electronic equipment, etc. Reliability theory became more and more popular amongst scientists due to its the growing importance (for example, see Barlow and Proschan (1965, 1975) and Kaufmann et al. (1977)).

The Moore and Shannon papers (1956a and 1956b) about relay network reliability showed that reliable systems could be composed from less reliable components. Their consideration probably led to a major step forward in the study of the reliability behaviour of complex systems: the study of the system behaviour from the component behaviour. Birnbaum, Esary and Saunders (1961) published an important paper on the reliability of coherent structures, and, hence, introduced a general framework to study the reliability of complex systems. Their concepts are still widely used and are still the basic tools to study the reliability behaviour of systems and their components.

The basic idea is that a system is made up of components that assume one of two possible states, *good* and *bad*. The system equally assumes only one of these two possible states. The relation between the system and the component states turns out to be a mapping that must satisfy some boundary conditions and must be increasing. Many research revealed interesting links between coherent structure theory, graph theory and boolean polynomials (for example, see Carvallo (1965)), and many other interesting results like the three module theorem by Birnbaum and Esary (1965). Several important notions were introduced like paths, minimal paths, cuts and minimal cuts and lead to several interesting and useful representations of systems (for example, see Barlow and Proschan (1975), Esary and Proschan (1962) and Kaufmann et al. (1977)). The classes of life distributions were further explored. Barlow and Proschan (1965) collected the main results of reliability theory and, hence, influenced many scientists. In the former Soviet Union a team of probability specialists (Gnedenko et al. (1972)) wrote a very important contribution to the mathematical foundations of reliability, while Barlow and Proschan (1975) presented their second basic work on mathematical reliability theory that is still highly important in nowadays reliability research. Meanwhile the widely-used concept of fault-trees has been introduced to study the reliability of complex systems and phenomena by, for example, Fussell (1973), and many other specialist fields have been developed to cover the main reliability problems. Amongst others we mention special techniques in aircraft industries, space technology, software reliability and the outstanding developments in nuclear technology, for example, by Cohen (1984), Fussell (1984) and Shooman (1984).

12.2 QUESTIONS IN CLASSICAL RELIABILITY AND THEIR POSSIBLE ANSWERS

Though classical reliability covers many problems, since the beginning of reliability theory many basic assumptions have been questioned:

(1) Are systems and their components binary by nature ?

The probabilistic method of the system reliability analysis is based on the assumption that a system state is either good or bad, and that the two states are exclusive each other. As shown in Figure 1, a success state and a failure state are considered as the binary state. These states are also considered to occur at random. The probability of the failure state is defined as unreliability and the probability of the success state is defined as reliability. Based on the assumption that the two states are exclusive, (*Unreliability*)= 1 – (*Reliability*). In practice, however, it may be natural that the system state is a multistate and that the transition from one state to the other is not sharp as shown in Figure 2, that is, the system state is divided into several states not clearly but fuzzily.

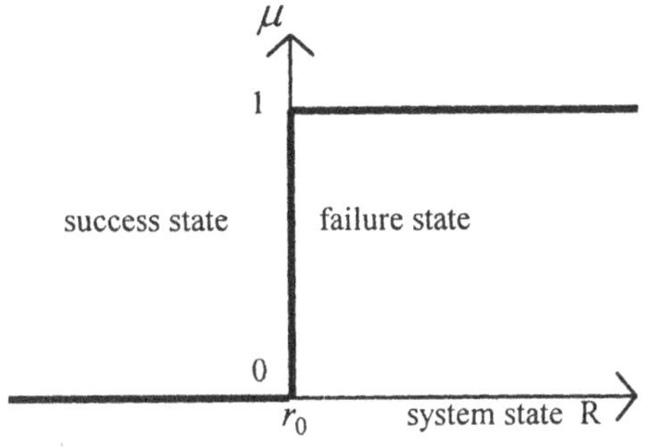

Figure 1 Example of two system states

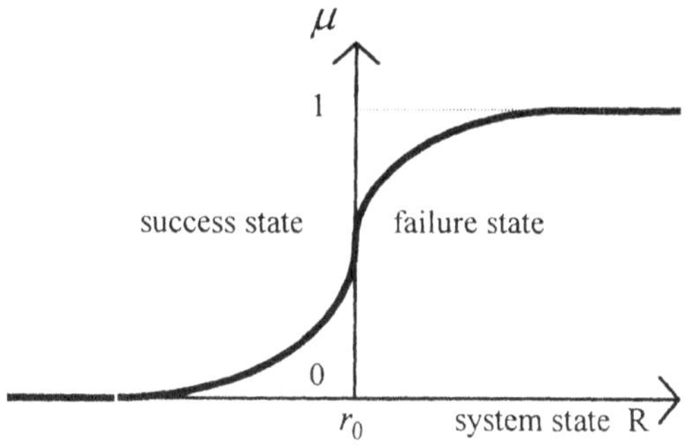

Figure 2 Example of fuzzy system state

(2) What is the nature of a state in mathematical theory of reliability ?

(3) When the binary model is not sufficient to represent the structural reliability of a system, how can we mathematically represent the multistateness of both components and systems ?

Since the introduction of fuzzy set theory the basic assumption about reliability has been questioned more frequently:

(4) Is the reliability of a system and components probabilistic by nature ?

Let us have a closer look at these questions and mention some possible answers. From the early beginning of binary structure function theory, reliability theoreticians have been studying systems that assume a finite number of states (for example, see Caldarola (1980), Dhillon (1980), Elsayed and Zebib (1979), and Murchland (1975)). Two main possibilities are considered:

(1) the system and component are subject to n different types of failure,

(2) the system is subject to some intermediate states.

The former case has been studied quite intensively from the beginning of the sixties, and the classical reliability theory, especially the probabilistic approach to reliability and the related definitions of availability and the probabilistic approach to renewal theory, have been modified to deal with components and systems subject to n different types of failure. Even the study of dual failures can be considered as a special case of components and systems subject to n different types of failures.

From 1975 on, more attention has been devoted to a general multistate approach of the structure of systems and their components. Several scientists consider concepts that lead to a general framework to study the multistate behaviour of multistate systems and their multistate components. Barlow and Wu (1978) introduce a special class of multistate structure functions and are the first to study several important aspects of this kind of systems. The same year El-Neweihi, Proschan and Sethuraman (1978) introduce general concepts to study multistate systems. Their approach covers the Barlow-Wu class completely and they take fundamental ideas about the concepts of minimal path and cuts, generalize to multistate components. Nevertheless both components and systems assume one out of a finite number of states, and the proposed model can not cope with components or systems that are subject to n different types of failure. Moreover, most papers dealt with new concepts about reliability, availability and renewal policies, and, hence, study all kinds of probabilistic aspects of multistate structure functions (for example, see Block and Savits (1982), Griffith (1980), Griffith (1982), and Natvig

(1982)). Baxter (1984) introduces a generalized Barlow-Wu class and he studies systems and components that assume one out of an infinite number of possible states. Baxter et al. (1986, 1986, 1987, 1988) have devoted most of their attention to the probabilistic aspects of this kind of multistate systems.

From 1988 on, Montero, Tejada and Yáñez (1988) introduce lattice theory in structural reliability theory. This innovation lead to a model that is able to study both types of failures, i.e., components and systems subject to n different types of failures and components and systems subject to degrading sates. Their step forward is of major importance to study the structural reliability behaviour of components and systems. Many questions since 1978 have been answered by Montero, Tejada and Yáñez (1988), Montero, Cappelle and Kerre (1995), Cappelle (1991, 1995) and Cappelle and Kerre (1995a) by applying the lattice based approach of multistate structure function theory. Cappelle and Kerre (1995c) also present a practical application of their approach to a subsystem of a nuclear power plant.

In the meanwhile, in the fuzzy state approach the system state is defined as a fuzzy state (event), as shown in Figure 2, a unreliability is defined as the probability of the fuzzy state (event) as Zadeh (1968) proposes. This consideration includes the case where the system state is divided into the two states clearly as shown in Figure 1 (See Cai et al. (1990, 1991a, 1993), for details).

Cai et al. (1991b, 1991c, 1995) considers a fuzzy reliability theory as follows: (1) The PROFUST theory which is based on a fuzzy state and the probability theory, (2) the POSFUST theory which is based on a fuzzy state and the possibility theory, (3) the POSBIST theory which is based on a binary state and the possibility theory, and (4) the PROBIST theory which is based on a binary state and the probability theory. Misra et al. (1981), Cappelle et al. (1991, 1995) and Onisawa and Misra (1993) take the fuzzy theory approach to the multistate consideration of the system state in the reliability analysis, which are expansion of the two fuzzy states (the success state and the failure state) to the multi fuzzy state.

The second question about the nature of a state in structural reliability theory has been answered hardly. Most authors assume that the reader is familiar with the notion state and does not dwell upon this topic. Nevertheless, Gnedenko, Beliaev and Soloviev (1972) discuss some ideas about the notion state though they are not the first since several years before them, Dubes (1963) mentions some similar ideas about the nature of a state of a component and system. Cappelle (1995), however, studies the subject profoundly and presents some new ideas about the nature and the meaning of a state in multistate structure function theory. It turns out that a state depends upon the context, i.e., the conditions that a system or component must fulfill to achieve maximal performance. Therefore, the problem of determining a state of a component or system is quite similar to the evaluation problems presented by De Cooman (1993) in fuzzy set theory.

The answer to the third question is straightforward now. It is clear that though binary structure functions easily can be related to graph theory and Boolean polynomials, multistate structure functions seem to be the only possibility to represent the multistateness of components and systems. Though multivalued logic has been proposed to model the multistateness of both components and systems by Premo (1963), it turns out that as a matter of fact a special kind of multistate structure functions are studied. Therefore, multistate structure function theory will continue to play a major role in structural reliability theory, since a better understanding of the behaviour of systems and the relation between the system state and the components states, must lead to a safer society.

The notion reliability, however, has been defined in a probabilistic way since its very introduction. We explain that this fact is not very surprising since reliability theory has been developed from quality control theory. The reliability of a system and its components has been defined as the probability that the system or component will work during a fixed time interval. Many related concepts have been derived from this basic definition. As we mention, availability theory is based on the probability that a system or component is available during a fixed time interval and renewal policies are developed. The determination of the system reliability from the component reliabilities is one of the fundamental problems in probabilistic reliability theory. There have been many applications of the probabilistic approach to the system reliability analysis. For example, U.S. Atomic Energy Commission (1974) presents WASH-1400, which is famous as risk assessment of nuclear power plants, and Sawin (1963) presents THERP (Technique for Human Error Rate Prediction), which is famous as the model of human reliability analysis.

The probabilistic approach has been introduced in order to assess system reliability rationally, i.e., objectively. However experts' experience and engineering judgements and knowledge also play an important role even in the probabilistic approach of the system reliability analysis. As Svenson (1989) points out, experts' judgements hold a central position in all reliability analyses of complex technical system. Especially they play a much important role in the reliability analysis of a large scale and complex system. Even experts, however, cannot have every knowledge about the large scale and complex system and highly technical development system since the system is composed of many components and their functional relations are much complex. In practice, many unexpected accidents of large scale and complex systems or highly technical development systems have occured, e.g., accidents of nuclear power plants at Three Miles Island in 1979 and at Chernobyl in 1986, an explosion accident of Challenger in 1986, crash accidents of a Japan Air Line jumbo in 1985 and a China Air Line airbus in 1994, an accident of a chemical plant at Bhopal in 1984. These accidents have been evaluated to be hard to occur from the viewpoint of the accident probability since their estimated probabilities are very very small. The evaluation of the analyzed system is based on experts' engineering judgements even if the accident probability is estimated. Their judgements are related to their subjectivity. The probability theory does not necessarily deal with human subjectivity. Although the subjective probability is considered in the probability theory, it follows probability axioms. Human subjectivity does not necessarily follow probability axioms. Zadeh(1965) proposes fuzzy sets to handle human subjectivity which cannot be represented clearly by any other formalism. In the next section the probabilistic approach of the system reliability analysis is reviewed from the fuzzy theory point of view.

12.3 PROBABILISTIC APPROACH OF SYSTEM RELIABILITY ANALYSIS FROM FUZZY THEORY POINT OF VIEW

12.3.1 Failure Probability and Error Probability

The system reliability is usually estimated from each components reliability and system functional relations. System components are hardware, e.g., machinery, instruments, and human ware, e.g., human operators. Hardware reliability and human reliability must be considered in a system reliability analysis since hardware components and human operators have influence on system behavior and several serious failures have been caused by human errors. However there are some problems in the classical reliability models: how to cope with human behaviour and how to estimate the probability of failure of a human action.

Apart from the problems, it is necessary to estimate the failure probability of hardware and the error probability of human ware in order to analyze system reliability quantitatively. Classical reliability models often must deal with extremely small probabilities , e.g., 10^{-7} or 3×10^{-8}. It is to be desired that these probabilities should be estimated from a large amount of data. In practice, however, it is quite obvious that it is almost impossible to determine these probabilities adequately for each component of some complex machinery due to financial and time restrictions. Especially, we have not sufficient data for human errors. Therefore the failure probability and the error probability are usually estimated from the reference data in similar areas based on experts' engineering and experience judgements which are closely related to their subjectivity. Once hardware reliability and human reliability are estimated by numerical values, e.g., the failure probability and the error probability, their reliabilities seem to be estimated objectively. However experts' subjectivity, which is closely related to fuzziness, is inherent in these values.

Hardware reliability and human reliability are affected by many environmental factors, e.g., a set of conditions under which a machine is operated, or they may just be the psychological stress of the operator. Then it is necessary to modify the basic failure probability and the basic error probability according to environmental conditions. The modification is usually performed by multiplying the basic failure probability and the basic error probability by numerical factors. For example, let us consider the influence of psychological stress on human reliability. In Swain and Guttmann (1983) the model as shown in Table 1 is employed. In this model psychological stress is divided into four levels such as *very low stress level*, *optimum stress level*, *moderate high stress level*, and *extremely high stress level*. In this table the stress level, the kind of a task, and experience of human operators are considered as the environmental factors. According to these environmental factors, various numerical factors, by which the basic error probability is multiplied, are selected.

Table 1 Modification of human error probability of experienced personnel and novices under different levels of stress

Stress	*Step-by-Step Procedures*		*Dynamic Interaction*	
Level	*Experienced*	*Novice*	*Experienced*	*Novice*
very low	$P_{SM} \times 2$	$P_{SM} \times 2$	No Account	No Account
optimum	P_{SM}	P_{SM}	P_{DM}	P_{DM}
moderate high	$P_{SM} \times 2$	$P_{SM} \times 4$	$P_{DM} \times 5$	$P_{DM} \times 10$
extremely high	0.25	0.25	0.25	0.25

(1)A novice is defined as a person with less than 6 months on the job in which he/she has been licensed or otherwise qualified.

(2)Probabilities P_{SM} and P_{DM} are basic error probabilities of tasks.

In the modification of the basic failure probability and the basic error probability, the evaluation of environmental factors and the determination of numerical factors are performed by experts based on their engineering and experience judgements. Therefore, estimates of the modified failure probability and error probability depend on experts' subjectivity. Furthermore, for example, in Table 1, stress levels are expressed by linguistic terms, but their meanings cannot be represented by numerical values or formulae definitely. One way how to represent linguistic terms by numerical values can be found in Gazdík (1986).

12.3.2 Dependence

The system reliability analysis is in need of information on system functional relationship since a large scale and complex system, e.g., a nuclear power plant, a jumbo jet plane, is complicated in its functional relation. Therefore,

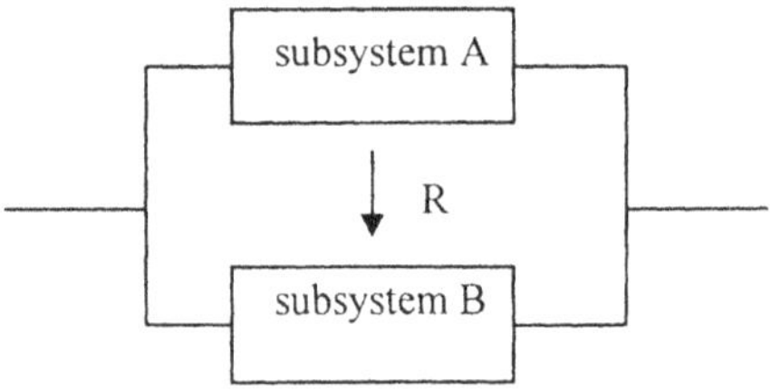

Figure 3 Parallel system with dependence

a redundant system does not always work redundantly in such a system. For example, let us consider the parallel system as shown in Figure 3. When subsystems A and B are independent of each other functionally, the parallel system works as a redundant system. In a large scale and complex system, however, the parallel system does not always work as a redundant system in the sense that the failure of the subsystem A often has an influence on the failure of the subsystem B. This kind of dependence is often observed in a large scale and complex system or in consecutive human tasks.

When the dependence is estimated to exist in a redundant system or in human tasks, the analysis method must deal with the dependence since analysis results sometimes become optimistic without consideration of the dependence. The conditional probability, that is, the failure probability of subsystem B under the condition that subsystem A fails in the example of Figure 3, is usually considered in order to deal with the dependence in the probabilistic analysis method. In the conditional probability method, it is to be desired that the conditional probability should be estimated from lots of immediate data. In practice, however, we usually have not sufficient data to estimate the conditional probability since the data is dependent on an analyzed system or a human task. Information on the dependence is usually given by experts based on their engineering and experience judgements. Their subjectivity is inherent in this kind of information.

In Swain and Guttmann (1983), the model shown in Table 2 is employed in human reliability analysis. They recommend using the conditional probability in the dependence analysis. But when it is difficult to estimate the conditional probability, they recommend employing this model. In this model it is assumed that a human operator performs task B after task A and that errors in both tasks A and B lead to an error in the whole task, and that the error in task A has an influence on the error in task B. Subsystems A and B in Figure 3 are considered as tasks A and B. The dependence level is divided into five levels such as *zero dependence*, *low dependence*, *moderate dependence*, *high dependence* and *complete dependence*. The *zero dependence* means that the error in task A is independent of the error in task B. The error probability in the whole task under the zero dependence is estimated by $P_A \times P_B$, where P_A and P_B are the error probabilities of tasks A and B, respectively. The *complete dependence* means that when a human operator makes an error in task A, the operator always makes an error in task B. The error probability of the whole task under the complete dependence is estimated by $P_A \times 1.0$. In other three dependence levels the error probability in task B, i.e., the conditional probability, is defined according to the dependence level.

Table 2 Conditional probabilities of failure on task B given failure on task A for different levels of dependence

Dependence Level	*Equations Giving Conditional Probability*
zero dependence	P_B
low dependence	$\frac{1+19\times P_B}{20}$
moderate dependence	$\frac{1+6\times P_B}{7}$
high dependence	$\frac{1+P_B}{2}$
complete dependence	1.0

In this model the evaluation of the dependence level on the analyzed task is dependent on experts engineering and experience judgements which are related to their subjectivity. The error probability of task B in the three dependence levels may be also defined based on experts' engineering and experience judgements. The dependence level is expressed by linguistic terms. However the meanings of the linguistic terms cannot be expressed by numerical values or a formula definitely since the borderlines among stress levels are not defined clearly.

12.3.3 Scenario about Sequence of System Failure

It is necessary to consider a scenario about the sequence of a system failure based on system functional relations in the analysis of system reliability. There are two kinds of considerations about the scenario. The first is a bottom-up consideration. As the trigger event the failure of a hardware component or a human error is considered. The scenario from the trigger event to the system failure is constructed. The other is the top-down consideration. First of all, an accident of the system is considered. The scenario from the system failure to component failures or human errors is constructed. In the conventional reliability analysis method the former typical examples are an event tree and failure modes and effects analysis, and the latter is a fault tree. Both scenarios are usually constructed by experts based on their engineering and experience judgements and knowledge, who are familiar with the analyzed system.

The following problems should also be considered in the scenario construction: Even experts cannot have complete knowledge on a large scale and complex system and a highly technical development system such as a nuclear power plant. The scenario cannot be obtained precisely. Uncertainty is included in the scenario. Experts' judgements are used to deal with the uncertainty.

As mentioned before, human reliability must be considered in a system reliability analysis. Concerning human reliability there are also some problems. A human complex task must be broken down into basic tasks in order to estimate the error probability of the complex task by the use of error probabilities of the basic tasks. However even if human tasks are analyzed in detail, human reliability is not estimated precisely since human behavior is complex and unpredictable. To what extent must a human complex task be broken down into basic tasks ? This kind of judgement is also dependent on experts.

12.3.4 System Reliability Evaluation

Let us assume that the probability of a system failure is estimated to be 10^{-6} [1/year] as the result of the system reliability analysis. How do we evaluate reliability of the analyzed system ? The probability of the system failure, e.g., 10^{-6} [1/year], is usually a very very small value from our daily life point of view. The obtained probability of the system failure is usually compared with the probability of a rare event by experts. However experts' subjectivity is inherent in the process of the comparison and the evaluation. The evaluation may be dependent on the analyzed system, for example, whether the analyzed system is a nuclear power plant or a chemical plant. The evaluation may be also dependent on the person who evaluates system reliability, for example, whether the person is concerned with the analyzed system or the person is a neighbor of the system.

What does a very very small probability mean ? Does it imply that the analyzed failure hardly ever happen ? As Dubois and Prade (1980) point out, the low probability of an event does not necessarily mean the low possibility of the event and the high possibility of the event does not neccesarily mean the high probability of the event. This is also found by the facts that failures of many high technical developed systems, e.g., a nuclear power plant, a jumbo jet plane, have happened. The probability of the system failure obtained by the quantitative reliability analysis is not so meaningful as the value, i.e., the probability, shows since experts' subjectivity is inherent in estimation of failure probabilities of system components, error probabilities, and the functional relation of the analyzed system.

12.3.5 Use of Natural Language

As Schmucker (1984) points out, the use of natural language expressions is useful in the system reliability analysis. If the use of probability may be considered as the analysis from the microscopic point of view, the use of natural language may be considered as the analysis from the macroscopic point of view. As a matter of fact, natural language expressions are used in the construction of a fault tree, an event tree, and in the failure modes and effects analysis. In human reliability model Swain and Guttmann (1983) use linguistic terms of stress and dependence levels as shown in Tables 1 and 2. And natural language expressions are appropriate for the evaluation of system reliability since it is possible to communicate system reliability to other people by linguistic terms. As a matter of fact, IAEA and NEA (OECD) (1992) propose the International Nuclear Event Scales (INES) User's Manual, which represents event rating form about nuclear power plant accident by not the probability but linguistic terms such as level 0-*below scale*, level 1-*anomaly*, level 2-*incident*, level 3-*serious incident*, level 4-*accident without significant off-site risk*, level 5-*accident with off-site risk*, level 6-*serious accident*, level 7-*major accident.*

System reliability is discussed with natural language better than with numerical values since natural language plays an important role in our daily communication. A fuzzy set expresses the meaning of linguistic terms better than the numerical value since the meaning of linguistic terms is fuzzy in itself. The fuzzy theory is employed in the system reliability analysis with natural language as one of approaches. In addition some other fuzzy approaches to deal with the problems abovementioned have been proposed (see, for example, Onisawa and Misra (1993), Onisawa and Kacprzyk (1995), Bowels and Pelaez (1995)).

Roughly speaking there are three kinds of applications of the fuzzy theory to the field of the system reliability engineering. The first is based on the consideration that the system state is not restricted to two states definitely, a success state and a failure state, as mentioned before. The second approach uses a fuzzy probability and the third approach defines the non-probabilistic measure and focuses on the introduction of experts' subjectivity and the use of natural language expressions to the system reliability engineering. The latter two approaches are reviewed in the next section.

The introduction of the fuzzy theory has shown that probability is not the only tool to represent vagueness and uncertainty, though some scientists still only accept probability theory to represent uncertainty (see for example French (1990)).

12.4 FUZZY SET THEORY APPROACH

12.4.1 Fuzzy Probability Approach

The fuzzy probability approach is proposed by Noma et al. (1981) and Tanaka et al. (1983), and is employed by many researchers, for example, by Singer (1990), Misra et al. (1990), and Kenarangui (1991). In this approach natural language expressions such as *the failure probability of the component is about* 10^{-3} are available and the meaning of *about* 10^{-3} is expressed by the fuzzy set defined on the unit interval [0,1], which is called a fuzzy probability. The fuzzy probability is defined as a normalized convex fuzzy set. Figure 4 shows an example of a fuzzy probability.

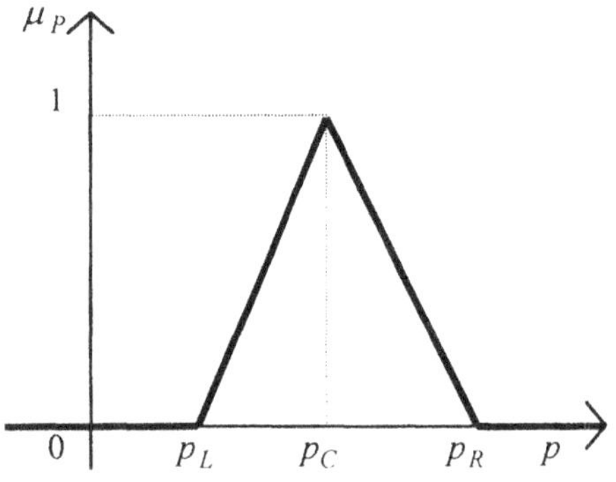

Figure 4 Example of fuzzy probability

The addition, the subtraction and the multiplication of fuzzy probabilities are performed simply by the extension principle (for example, see Gazdík (1996)). Let f be a two-valued function $X \times Y$ to Z, and A and B be fuzzy sets on X and Y, respectively. A fuzzy set C on Z through f is defined by

$$\begin{aligned} C_\alpha &= [c_L(\alpha),\ c_R(\alpha)] \\ &= f(A,\ B)_\alpha \\ &= f(A_\alpha,\ B_\alpha) \end{aligned} \tag{1}$$

where $f(A,\ B)_\alpha$, A_α, B_α and $C_\alpha = [c_L(\alpha),\ c_R(\alpha)]$ are $\alpha-cuts$ of $f(A,\ B)$, A, B and C, respectively, and $\alpha \in [0,\ 1]$. In order to consider the operations of fuzzy probabilities, X, Y and Z are defined as the unit interval [0, 1]. The addition, the subtraction and the multiplication of two fuzzy probabilities P_A and P_B are defined as follows. Let us assume $P_C = P_A * P_B$, where * means the addition, the subtraction or the multiplication. Let $\alpha-cuts$ of P_A, P_B and P_C be $[p_{AL}(\alpha),\ p_{AR}(\alpha)]$, $[p_{BL}(\alpha),\ P_{BR}(\alpha)]$ and $[p_{CL}(\alpha),\ p_{CR}(\alpha)]$, respectively, where $\alpha \in [0,\ 1]$.

Addition

$$(P_C)_\alpha = (P_A + P_B)_\alpha$$
$$= \begin{cases} [p^A{}_L(\alpha), p^A{}_R(\alpha)], & p^A{}_R(\alpha) < 1 \\ [p^A{}_L(\alpha), 1], & p^A{}_L(\alpha) \le 1 \le p^A{}_R(\alpha) \\ 1, & 1 < p^A{}_L(\alpha) \end{cases} \quad (2)$$

where $p^A{}_L(\alpha) = p_{AL}(\alpha) + p_{BL}(\alpha)$ and $p^A{}_R(\alpha) = p_{AR}(\alpha) + p_{BR}(\alpha)$.

Subtraction

$$(P_C)_\alpha = (P_A - P_B)_\alpha$$
$$= \begin{cases} [p^S{}_L(\alpha), p^S{}_R(\alpha)], & 0 < p^S{}_L(\alpha) \\ [0, p^S{}_R(\alpha)], & p^S{}_L(\alpha) \le 0 \le p^S{}_R(\alpha) \\ 0, & p^S{}_R(\alpha) < 0 \end{cases} \quad (3)$$

where $p^S{}_L(\alpha) = p_{AL}(\alpha) - p_{BR}(\alpha)$ and $p^S{}_R(\alpha) = p_{AR}(\alpha) - p_{BL}(\alpha)$.

Multiplication

$$(P_C)_\eta = (P_A \times P_B)_\alpha$$
$$= [p_{AL}(\alpha) \times p_{BL}(\alpha), p_{AR}(\alpha) \times p_{BR}(\alpha)] \quad (4)$$

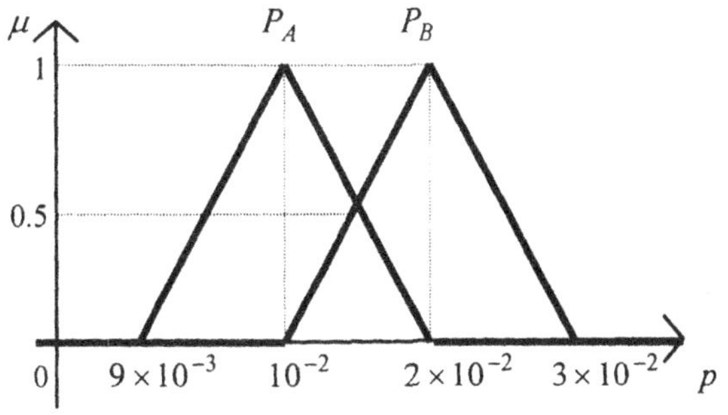

(1)

Figure 5 Addition, subtraction and multiplication of fuzzy probabilities

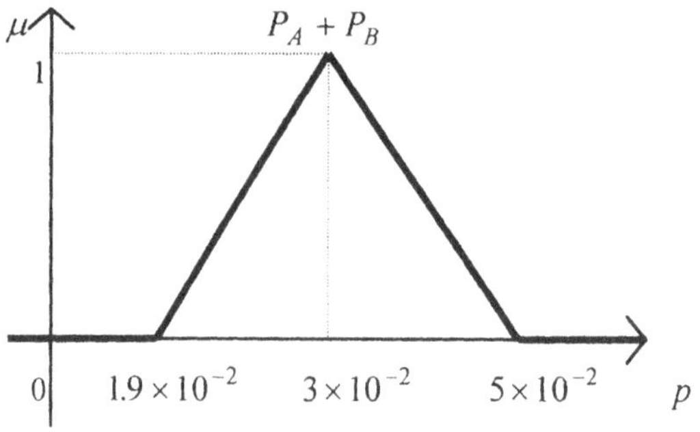

(2)

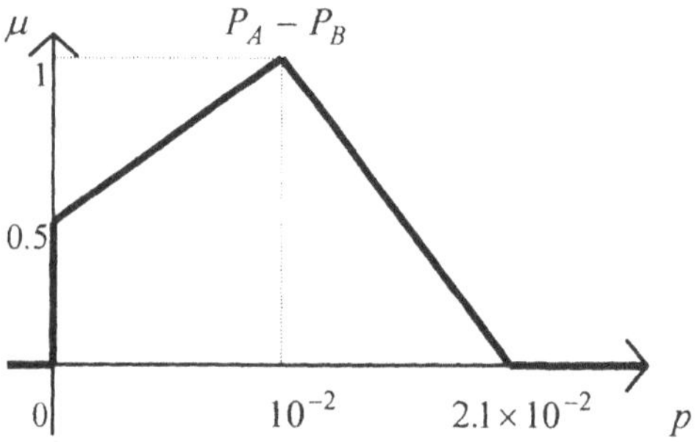

(3)

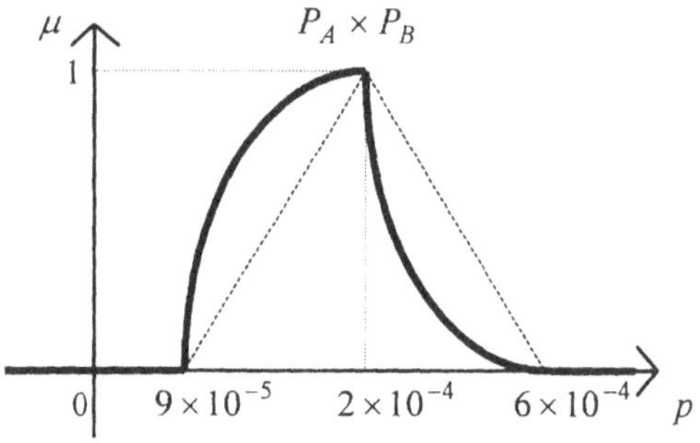

(4)

Figure 5 Addition, subtraction and multiplication of fuzzy probabilities (continued)

Figure 5 shows examples of these operations where two fuzzy probabilities are defined by triangular membership functions as shown in Figure 5(1). In this example, results of the addition and the subtraction are expressed by triangular membership functions as shown in Figures 5(2) and 5(3). However the result of the multiplication is not expressed by the triangular membership function any longer as shown in Figure 5(4). Tanaka et al. (1983) propose the linearization by the use of the membership function $\mu_{P_C}(p)$ as follows:

$$\mu_{P_C}(p) = \begin{cases} 0, & p < p^M{}_L(0) \\ \dfrac{p - p^M{}_L(0)}{p^M{}_L(1) - p^M{}_L(0)}, & p^M{}_L(0) \le p \le p^M{}_L(1) \\ 1, & p^M{}_L(1) < p < p^M{}_R(1) \\ \dfrac{p - p^M{}_R(0)}{p^M{}_R(1) - p^M{}_R(0)}, & p^M{}_R(1) \le p \le p^M{}_R(0) \\ 0, & p^M{}_R(0) < p \end{cases} \tag{5}$$

where $p^M{}_L(0) = p_{AL}(0) \times p_{BL}(0)$, $p^M{}_L(1) = p_{AL}(1) \times p_{BL}(1)$, $p^M{}_R(0) = p_{AR}(0) \times p_{BR}(0)$, $p^M{}_R(1) = p_{AR}(1) \times p_{BR}(1)$, and $(P_A)_0 = [p_{AL}(0), p_{AR}(0)]$, $(P_B)_0 = [p_{BL}(0), p_{BR}(0)]$, $(P_A)_1 = [p_{AL}(1), p_{AR}(1)]$ and $(p_B)_1 = [p_{BL}(1), \ p_{BR}(1)]$ are 0-cuts and 1-cuts of the fuzzy probabilities P_A and P_B, respectively. In this example, $p_{AL}(1) = p_{AR}(1)$ and $p_{BL}(1) = p_{BR}(1)$ and hence $p^M{}_L(1) = p^M{}_R(1)$. The result of the linearization is shown by the dotted line in Figure 5(4).

12.4.2 Fault Tree Analysis by Fuzzy Failure Probability

Basic operations in the fault tree analysis by the use of the non-fuzzy probability are the addition and the multiplication, i.e., the algebraic product and the algebraic sum. Basic operations in the fault tree analysis by the use of the fuzzy failure probability is the algebraic product and the algebraic sum which are extended to fuzzy sets operations. The linearization may be used in the multiplication of fuzzy probabilities for simplicity.

The fault tree analysis by the fuzzy failure probability has the following advantages: The conventional probabilistic methodology is used by the use of the fuzzy probability and the extension principle. It is not necessary to know crisp values of the failure and the error probabilities of basic events in a fault tree. Natural language expressions such as *the probability of the system accident is about* 10^{-4} are also available.

However the analysis does not deal with all the problems mentioned before. Experts' subjectivity towards the analyzed system cannot be necessarily expressed in this approach. And the conditional fuzzy probability must be obtained for the dependence analysis. But it is difficult to estimate the conditional fuzzy probability.

The fault tree analysis by the fuzzy failure probability has the same problems as the conventional analysis method.

12.4.3 Non-Probabilistic Measure Approach

Onisawa (1988,1990,1996) proposes the non-probabilistic measure approach, which focuses on the use of natural language expressions and introduces experts' subjectivity. In this approach natural language expressions of reliability estimates are interpreted as follows:

(1)Comparison results of a few data about reliability with experts' subjective standards are expressed by natural language.

(2)As mentioned before, reliability is affected by many environmental factors. Results of experts' consideration about the factors effects on reliability are expressed as reliability estimates by the use of natural language. In Onisawa (1991) the fuzzy reasoning technique is used in order to evaluate the effects of environmental factors.

In the non-probabilistic measure approach, hardware component reliability, human reliability and system reliability are expressed by natural language in the form of reliability estimate and its fuzziness. Table 3 shows an example of linguistic terms about reliability estimate and Table 4 shows an example of linguistic terms about the fuzziness of reliability estimate.

The fuzzy set with the following membership function is employed in order to express the meanings of linguistic terms.

$$F(x) = \frac{1}{1 + C \times |x - x_0|^m} \tag{6}$$

where $0 \le x, x_0 \le 1$, and x_0 and m are parameters, and C is a constant. The parameter x_0 offers a maximal grade of 1 and the parameter m is related to fuzziness. This fuzzy set is called a subjective measure of reliability in this approach. Figure 6 shows an example of the subjective measure of reliability,

Table 3 Linguistic terms about reliability estimate and parameter x_0

Class	*Expressions of Reliability Estimate*	*Corresponding Parameter* x_0	*Representative Value*
	(System, component hardware and human operator has)		
1	no reliability	-	-
2	very low reliability	0.9 - 1.0	0.95
3	low reliability	0.7 - 0.9	0.8
4	rather low reliability	0.55 - 0.7	0.625
5	standard reliability	0.45 - 0.55	0.5
6	rather high reliability	0.3 - 0.45	0.375
7	high reliability	0.2 - 0.3	0.25
8	quite high reliability	0.1-0.2	0.15
9	extremely high reliability	0.05 - 0.1	0.075
	(Accident, failure and human error is)		
10	next to impossible	0.0 - 0.05	0.025
11	impossible	-	-

Table 4 Linguistic terms about fuzziness of reliability estimate and parameter m

Class	*Expressions of Fuzziness*	*Parameter* m	*Range of Parameter* k
	(Fuzziness is)		
1	low	2.0	$k \leq 3$
2	medium	2.5	$3 < k \leq 5$
3	rather high	3.0	$5 < k \leq 10$
4	high	3.5	$10 < k$

where $C = 20$. The constant term, e.g., $C = 20$, can be replaced by another one, e.g., $C = 10$, in the sense that almost the same analysis results are obtained in both cases (see Onisawa (1994)). In this article, the value of C is assumed to be 20. The subjective measure of reliability expresses reliability of any component or any system. The subjective measure of reliability expressing system reliability is obtained by the fault tree analysis as mentioned later.

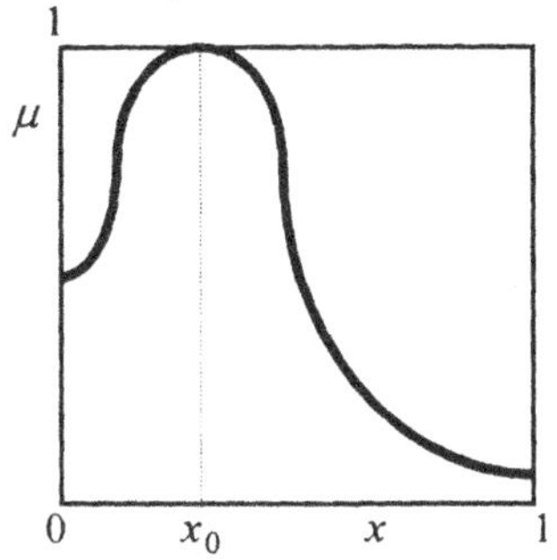

Figure 6 Subjective measure of reliability

The subjective measure of reliability is defined on the unit interval [0, 1] which means subjective evaluation of reliability. The value in the unit interval means the subjective evaluation of reliability. For example, the subjective evaluation of reliability 1 means that a system, a hardware component and a human operator are subjectively estimated to fail. On the other hand 0 means that they are subjectively estimated to work without fail. The midpoint 0.5 means a subjective standard evaluation of reliability, that is, reliability is evaluated to be equal to experts' standards. The value in the unit interval [0, 1] is not probability but subjective evaluation. The subjective measure of reliability is a non-probabilistic reliability measure in the sense that it does not satisfy axiomatic laws of probability such as additivity.

The subjective measure of reliability has the following properties:

(1) The subjective measure of reliability is normal and convex. It is easy to express reliability by simple linguistic terms.

(2) $F(1) \neq 0$ and $F(0) \neq 0$. These mean that even if reliability is estimated to be low, a component hardware does not necessarily fail, or a human operator does not necessarily make an error in a given task, and that even if reliability is estimated to be quite high, a component hardware does not always work without failure or a human operator does not always perform the task without error.

In this approach the parameters x_0 and m correspond to the linguistic terms about reliability estimate and its fuzziness, respectively, as shown in Tables 3 and 4. Especially, the subjective measures of reliability in class 1 and in class 11 are defined as follows:

$$F(x) = \begin{cases} 1, & x = 1 \\ 0, & x \neq 1 \end{cases} \tag{7}$$

$$F(x) = \begin{cases} 0, & x \neq 0 \\ 1, & x = 0 \end{cases} \tag{8}$$

The fuzziness of reliability estimate is not taken into consideration for linguistic terms in classes 1 and 11.

This approach also considers the case where the information on the estimate of reliability is given by numerical value, i.e., the failure probability or the error probability. The failure probability and the error probability are usually given as a triplet, i.e., $[p_L, p_M, p_U]$, where p_M is a recommended value of the failure probability or the error probability, p_L is its lower bound and p_U is its upper bound. The parameter x_0 in the membership function of the subjective measure of reliability (6) is considered as one of subjective reliability indices. The subjective index is usually based on human sensation. Human sensation is said to be proportional to a logarithmic value of a physical quantity. In the reliability field Embrey et al. (1984) and Washio et al. (1992) use the following conversion rules in order to transform the error probability into reliability indices, e.g., a success likelihood x and x', respectively,

$$\log(1-p_M) = a \times x + b \tag{9}$$

$$\log(1-p_M) - \log p_M = a' \times x' + b' \tag{10}$$

where p_M is the error probability, a, b, a' and b' are parameters which are determined based on experts' consideration or by the least square fitting to available reference data. In Embrey et al. (1984) and in Washio et al. (1992) human reliability is discussed by subjective indices obtained by Eqs. (9) and (10) rather than the error probability p_M.

In the non-probabilistic measure approach system reliability is discussed by the subjective index. The logarithmic function is also used in order to transform the recommended value of the failure probability or the error probability to the parameter x_0. This approach considers the following properties in order to obtain the conversion rule $f(p_M)$ since we do not have the reference data to determine parameters a, b, a' and b':

(1) The failure probability or the error probability p_M is compared with the expert's standard probability p_S, where p_S means that the expert considers

subjectively that the failure probability of a hardware component of interest or the error probability of a human operator in a given task should be p_S .

(2) When p_M is equal to p_S, the value of the parameter x_0 is equal to 0.5, which means subjective standard evaluation of reliability.

(3) The value of the parameter x_0 is larger than 0.5 at $p_M > p_S$ and the value of x_0 is smaller than 0.5 at $p_M < p_S$.

(4) The conversion rule is expressed by a monotone increasing function about the failure probability or the error probability.

The following conversion rule is employed considering the above properties.

$$
\begin{aligned}
x_0 &= f(p_M) \\
&= \frac{1}{1+(\frac{\log p_M}{\log p_S})^3}
\end{aligned} \tag{11}
$$

The standard probability p_S is dependent on experts' judgements and the nature of analyzed systems. For example, the p_S value should be very small in the analysis of nuclear power plant reliability. That is, experts' subjectivity is included in the standard probability p_S.

Concerning the parameter m, ratios p_M / p_L and p_U / p_M need to be considered. The parameter m is defined as shown in Table 4, where the parameter k is defined by p_M / p_L or p_U / p_M .

12.4.4 Fuzzy Fault Tree Analysis by Subjective Measure of Reliability

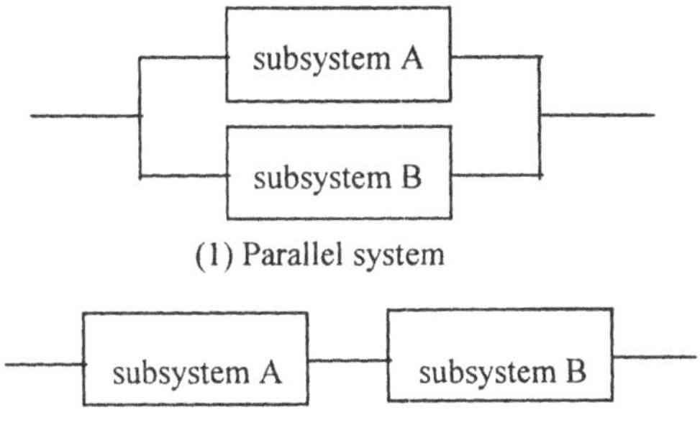

Figure 7 Parallel system and series system

Let us consider a parallel system and a series system depicted in Figure 7. First of all let us consider the parallel system, that is, failures of all subsystems lead to the failure of the system. This process is expressed by the and-gate in a fault tree. The failure probability of the parallel system is not larger than that of each subsystem of the parallel system. A t-norm satisfies this property. The failure probability of the parallel system is obtained by the algebraic product which is one of the possible t-norms. Next, let us consider the series system, that is, the failure of at least one of subsystems leads to the failure of the system. This process is represented by the or-gate in a fault tree. The failure probability of the series system is not smaller than that of each subsystem of the series system. A t-conorm satisfies this property. The failure probability of the series system is obtained by the algebraic sum which is one of the possible t-conorms.

In the subjective measure of reliability approach the following parametrized t-norm and t-conorm are employed as the **and** and the **or** operations of the subjective measures of reliability, respectively. The extension principle is applied to these operations.

$$H(x,\ y)=\frac{1}{1+\left\{\left(\frac{1-x}{x}\right)^{1/nH}+\left(\frac{1-y}{y}\right)^{1/nH}\right\}^{nH}} \tag{12}$$

where $0<x,\ y\leq 1$, $H(0,\ y)=H(x,\ 0)=0$, and nH is a non-negative parameter.

$$G(x,\ y)=\frac{\left\{\left(\frac{x}{1-x}\right)^{nG}+\left(\frac{y}{1-y}\right)^{nG}\right\}^{1/nG}}{1+\left\{\left(\frac{x}{1-x}\right)^{nG}+\left(\frac{y}{1-y}\right)^{nG}\right\}^{1/nG}} \tag{13}$$

where $0\leq x,\ y<1$, $G(1,\ y)=G(x,\ 1)=1$, and nG is a non-negative parameter. The operations H and G are so called Dombi t-norm and Dombi t-conorm, respectively. According to Mizumoto(1989), Dombi t-norm and t-conorm has the following properties: The operation H becomes the drastic product when nH is infinite,

$$H(x,\ y)=\begin{cases} x, & y=1 \\ y, & x=1 \\ 0, & otherwise \end{cases} \tag{14}$$

and the min operation when nH is zero.

$$H(x,\ y)=\min(x,\ y) \tag{15}$$

On the other hand the operation G is the max operation when nG is infinite,

$$G(x,\ y)=\max(x,\ y) \tag{16}$$

and the drastic sum when nG is zero

$$G(x, y) = \begin{cases} x, & y = 0 \\ y, & x = 0 \\ 1, & otherwise \end{cases} \tag{17}$$

The min operation is the greatest t-norm and the drastic product is the smallest t-norm. The drastic sum is the greatest t-conorm and the max operation is the smallest t-conorm. According to Mizumoto (1989), Dombi t-norm is between the min operation and the drastic product, and the Dombi t-conorm is between the drastic sum and the max operation. From the viewpoint of the reliability analysis, this implies that the operations H and G can cover the range of reliability estimate from the most pessimistic estimate through the most optimistic one. The larger the parameters nH and nG are, the more optimistic the reliability estimations are (See Onisawa (1993), for details). So Dombi t-norm and t-conorm are applied to the system reliability analysis using the subjective measure of reliability. The parameters nH and nG can reflect experts' subjectivity towards the analyzed system, that is, whether the system reliability is evaluated optimistically or not. The determination of the parameters nH and nG is mentioned later.

A fault tree is built based on binary logic. However experts' subjectivity is reflected in the ***and*** operation (12) and the ***or*** operation (13). So in the fault tree analysis using subjective measure of reliability approach, the parallel system and the series system are not necessarily dealt in the same way as the fault tree using the probabilistic approach.

12.4.5 Dependence Analysis

Dependence situations are different among analyzed systems. So the dependence level is usually estimated by experts based on their engineering and experience judgements and knowledge, who are familiar with the analyzed system. The natural language expressions are more appropriate for the expressions of the dependence level evaluation than the numerical expressions, e.g., the conditional probability expressions. In the non-probabilistic approach the dependence level evaluation is expressed by natural language in the form of dependence level estimate and its fuzziness. Table 5 shows linguistic terms about the estimate of the dependence level. For the sake of simplicity, the meanings of linguistic terms about the dependenec level are expressed by a fuzzy set with the same membership function as the subjective measure of reliability.

Table 5 Linguistic terms about dependence level estimate

Level	*Expressions of Dependence Level Estimate*
1	complete dependence
2	high dependence
3	moderate dependence
4	low dependence
5	zero dependence

$$R(r) = \frac{1}{1+20\times|r-r_0|^{m_r}} \tag{18}$$

where $0 \le r,\ r_0 \le 1$, and r_0 and m_r are parameters. With regard to natural language expressions of fuzziness of the dependence level estimate and their corresponding numerical values of the parameter m_r, the same expressions and values as shown in Table 4 are used. The determination of the parameter r_0 is mentioned later.

Only a parallel system with the dependence as shown in Figure 3 is considered here. A series system is not dealt with here since the failure of any subsystem in a series system leads to the system failure whether or not the dependency exists. Let us assume that the failure of subsystem A has an influence on that of subsystem B. Let F_A and F_B be subjective measures of reliability of subsystems A and B, respectively, and R be a fuzzy set representing the dependence level. The following two cases are considered in the dependency analysis as far as the dependence level is not complete.

Failure of subsystem A has an influence on failure of subsystem B:
The subjective measure of reliability of the whole system in this case is estimated by F_A *and* R, i.e.,

$$F_B' = H(F_A,\ R), \tag{19}$$

where the extension principle is employed in this and the following operations.

Failure of subsystem A has no influence on failure of subsystem B:
As far as the dependence level is not complete, the failure of subsystem A does not always have an influence on the failure of subsystem B. The portion of the subjective measure of reliability of subsystem A, which is estimated not to have any influence on the failure of subsystem B, is estimated by

$$F_A = G(F_B',\ F_A') \tag{20}$$

where F_A' corresponds to the portion. This consideration implies that F_A *is* F_B' *or* F_A'. The subjective measure of reliability of the whole system in the second case is estimated by

$$F' = H(F_A{}', F_B).\tag{21}$$

The subjective measure of reliability of the whole system considering the above both two cases is estimated by

$$F = G(F_B{}', F').\tag{22}$$

This implies that the failure of this system occurs when the failure of subsystem A has an influence on the failure of subsystem B, or when subsystem A fails and subsystem B fails without influence of the failure of subsystem A.

12.4.6 Natural Language Expressions of the Analysis Results

The analysis results are expressed by the use of natural language in the form of estimate of reliability and its fuzziness. Let F_R be the subjective measure of reliability obtained by the analysis, and F_S be the subjective measure of reliability with the membership function

$$F_S(x) = \frac{1}{1 + 20 \times |x - x_0|^m}\tag{23}$$

where $0 \leq x, x_0 \leq 1$ and $m = 2.0, 2.5, 3.0, 3.5$. Let the $\alpha - cuts$ of F_R and F_S be $(F_R)_\alpha = (x_{1R}(\alpha), x_{2R}(\alpha))$ and $(F_S)_\alpha = (x_{1S}(\alpha), x_{2S}(\alpha))$, respectively. The distance between F_R and F_S is defined as

$$d = \int_0^1 \sqrt{((x_{1R}(\alpha) - x_{1S}(\alpha))^2 + (x_{2R}(\alpha) - x_{2S}(\alpha))^2}\, d\alpha\tag{24}$$

The parameters $x_0 (0 \leq x_0 \leq 1)$ and $m (m = 2.0, 2.5, 3.0, 3.5)$ of the membership function of the subjective measure of reliability F_S are selected so as to minimize the distance. The analysis results are expressed by linguistic terms referring to Tables 3 and 4.

12.4.7 Determination of the Parameters nH and nG

The parameters nH and nG are determined depending on experts' reliability estimates of the parallel system and the series system, respectively, assuming that reliability estimate of each subsystem and its fuzziness are *standard reliability* and *medium*, respectively. For example, when an expert estimates reliability of the parallel system and that of the series system are *quite high reliability* $(x_0 = 0.15)$ and *rather low* $(x_0 = 0.625)$, respectively, these estimates lead to the determination of the parameters $nH = 2.9$ and $nG = 1.0$, respectively. These estimates may be dependent on the analyzed system whether reliability of the system can be evaluated optimistically or not. The parameters nH and nG can reflect experts' subjectivity towards the analyzed system.

12.4.8 Determination of the Parameter r_0

Let us consider the case where the failure of the subsystem A has an influence on the failure of subsystem B shown in Figure 3. Let the reliability estimates of subsystems A and B, and their fuzziness be *standard reliability* and *medium*, respectively. Let x_{0i} be the parameter x_0 of the subjective measure of reliability F_S of the whole system with the dependence level i, and let r_{0i} be the parameter r_0 of R representing the dependence level i($i = 1$:*complete dependence*, $i = 2$:*high dependence*, $i = 3$:*moderate dependence*, $i = 4$:*low dependence*, $i = 5$:*zero dependence*). When the dependence level is estimated to be *complete* $(i = 1)$, the reliability estimate of the whole system is *standard reliability* $(x_{01} = 0.5)$ and its fuzziness is *medium*. When the dependence level is estimated to be *zero* $(i = 5)$, the reliability estimate of the whole system is obtained by the operation H with the determined parameter nH. The parameters $r_{0i}\,(i = 2,\ 3,\ 4)$ are determined so as to satisfy the following:

$$(x_{0i} - x_{05}):(x_{01} - x_{0i}) = \begin{cases} 3:1 & (i = 2) \\ 1:1 & (i = 3) \\ 1:3 & (i = 4) \end{cases} \tag{25}$$

That is, the parameters $r_{0i}\,(i = 2,\ 3,\ 4)$ are determined so that $(x_{01} - x_{05})$ is divided equally. The parameters $r_{0i}\,(i = 2,\ 3,\ 4)$ are dependent on the parameters nH and nG.

12.4.9 Mutual Agreement of the Analysis Results

Many results can be obtained depending on experts' subjectivity. Some of the results can be classified into the same group since the subjective measure of reliability approach expresses the result by natural language expressions. However, there still exists a problem in which all results are not classified into the same group. Onisawa (1995) proposes some of the mutual agreements in the non-probabilistic approach. The intersection and the union of the results are considered here. Let $F_{Rj}\,(j = 1,\ 2,\ \ldots\ ,\ n)$ be the subjective measure of reliability obtained by the $j-th$ expert, where n is the number of experts.

Let F_I and F_U be the intersection and the union of $F_{Rj}\,(j = 1,\ 2,\ \ldots\ ,\ n)$,

$$F_I = \bigcap_j F_{Rj} \tag{26}$$

$$F_U = \bigcup_j F_{Rj} \tag{27}$$

and let F_{NI} and F_{CU} be normalized F_I by the maximal grade of F_I, and the result of the convexity operation on F_U, respectively. The F_{NI} and the F_{CU} are represented by natural language in the form of reliability estimates and their fuzziness by the aforementioned process. The F_{NI} is interpreted as the common part of the reliability evaluation which all experts agree with one another, since the F_{NI} includes only the part of F_{Rj} which all experts agree with. On the other hand the F_{CU} is interpreted as the largest part of the reliability evaluation which all experts agree with, since the F_{CU} includes all parts of F_{Rj}.

Onisawa(1996) develops the system reliability analyzer which introduces experts subjectivity and natural language expressions to the system reliability analysis.

Other fuzzy techniques, e.g., fuzzy measures and fuzzy integrals, are also applied to the field of the reliability engineering (for example, see Onisawa (1986), Washio et al. (1992), Sadatoku et al. (1993)). Many fuzzy approaches are found in this field.

12.5 CONCLUSIONS

First of all, a historical perspective of reliability engineering is performed and some questions in classical reliability are reviewed. Next the probabilistic method of the system reliability analysis is reviewed from the fuzzy theoretical point of view. System reliability evaluation is dependent on experts' subjectivity even in the probabilistic method. As the application of the fuzzy theory to the system reliability analysis this chapter makes a survey of the fuzzy probability approach and the non-probabilistic measure approach. Especially the non-probabilistic measure approach focuses on the introduction of natural language expressions and experts' subjectivity to the system reliability analysis.

Many fuzzy approaches have been found in the field of reliability engineering. However the application of the fuzzy set theory to system reliability engineering does not mean exclusion of the probabilistic approach in this field. It is necessary to apply the fuzzy set theory approach, the probabilistic approach and other soft computing technologies to the field of system reliability engineering.

Human subjectivity, human intuition, human sensitivity, human creativity have been excluded from natural science. As far as we adhere to objectivity, axiomatic laws and formalism in natural science, we cannot hope for the real progress of technologies. The application of the fuzzy set theory and the soft computing technology may break through problems in the field of the system reliability engineering.

References

Barlow, R.E. and Proschan, F.(1965). *Mathematical theory of reliability*, John Wiley and Sons Inc., New York.

Barlow, R.E. and Proschan, F.(1975). *Statistical theory of reliability and life testing*, Holt, Rinehart and Winston, New York.

Barlow, R.E. and Wu, S.A.(1978). Coherent systems with multistate components, *Mathematics in Operations Research*, 3, 275-281.

Barlow, R.E.(1984). Mathematical theory of reliability: A historical perspective, *IEEE Transactions on Reliability* 33, 16-20.

Baxter, L.A. (1984). Continuum structures I, *Journal of Applied Probability*, 21, 802-815.

Baxter, L.A. (1986). Continuum structures II, *Mathematical Proceedings of the Cambridge Philosophical Society*, 99, 331-338.

Baxter, L.A. and Kim, C. (1986). Bounding the stochastic performance of continuum structure functions I, Journal of Applied Probability, 23, 660-669.

Baxter, L.A. and Kim, C. (1987). Bounding the stochastic performance of continuum structure functions II, *Journal of Applied Probability*, 24, 609-618.

Baxter, L.A. (1988). On the theory of cannibalization, *Journal of Mathematical Analysis and Applications*, 136, 290-297.

Birnbaum, Z.W., Esary, J.D. and Saunders, S.C. (1961). Multicomponent systems and structures and their reliability, Technometrics, 3, 55-77.

Birnbaum, Z.W. and Esary, J.D. (1965). Modules of coherent binary systems, *SIAM Journal on Applied Mathematics*, 13, 444-462.

Block, H.W. and Savits, T.H. (1982). A decomposition theorem for multistate structure functions, *Journal of Applied Probability*, 19, 391-402.

Bowels, J.B. and Pelaez, C. (1995). Application of fuzzy logic to reliability engineering, *Proceedings of the IEEE*, 83, 435-449.

Cai, K.Y., Wen, C.Y. (1990). Stree-lighting lamps replacement: A fuzzy viewpoint, *Fuzzy Sets and Systems*, 37, 161-172.

Cai, K.Y., Wen, C.Y. and Zhang, M.L. (1991a). Fuzzy reliability modelling of gracefully degradable computing systems, *Reliability Engineering and System Safety*, 33, 141-157.

Cai, K.Y., Wen, C.Y. and Zhang, M.L. (1991b). Posbist reliability behavior of typical systems with two types of failures, *Fuzzy Sets and Systems*, 43, 17-32.

Cai, K.Y., Wen, C.Y. and Zhang, M.L. (1991c). Fuzzy variables as a basis for theory of fuzzy reliability in the possibility context, *Fuzzy Sets and Systems*, 42, 145-172.

Cai, K.Y., Wen, C.Y. and Zhang, M.L. (1993). Fuzzy states as a basis for a theory of fuzzy reliability, *Microelectronics and Reliability*, 33, 2253-2263.

Cai, K.Y., Wen, C.Y. and Zhang, M.L. (1995). Coherent systems in profust reliability theory, in *Reliability and Safety Analyses under Fuzziness*, Onisawa, T. and Kacprzyk, J. eds., Physica-Verlag, A Springer-Verlag Company, Heidelberg, 81-94.

Caldarola, L. (1980). Fault-tree analysis with multistate components, *in Synthesis and Analysis for Safety and Reliability Studies*, Apostolakis, G, Garriba, S. and Volta, G. eds., Plenum Press, New York, 199-248.

Cappelle, B. (1991). Multistate structure functions and possibility theory: An alternative approach to reliability, in *Introduction to the Basic Principles of Fuzzy Set Theory and Some of its Applications*, Kerre, E.E. ed., Communication and Cognition, Gent, 252-293.

Cappelle, B. and Kerre, E.E. (1994a). Possibilistic and necessitic reliability functions: Fundamental concepts and theorems to represent non-probabilistic uncertainty in reliability theory, in *Uncertainty Modelling and Analysis: Theory and Applications*, Ayyub, B.M. and Gupta, M.M. eds., Elsevier Science B.V., North Holland, 131-144.

Cappelle, B. and Kerre, E.E. (1994b). A general possibilistic framework for reliability theory, *Proceedings of the Information Processing and management of Uncertainty in Knowledge-Based Systems Conference-IPMU 5*, Bouchon-Meunier, B. and Yager, R. eds., 25-35.

Cappelle, B. (1995). On the notion state in multistate structure function theory, in *Fuzzy Set Theory and Advanced Mathematical Applications*, Ruan, D. ed., Kluwer Academic Publishers, Boston, 201-221.

Cappelle, B. and Kerre, E.E. (1995a). Issue in possibilistic reliability theory, in *Reliability and Safety Analyses under Fuzziness*, Onisawa, T. and Kacprzyk, J. eds., Physica-Verlag, A Springer-Verlag Company, Heidelberg, 61-80.

Cappelle, B. and Kerre, E.E. (1995b). Computer assisted reliability analysis: An application of possibilistic reliability theory to a subsystem of a nuclear power plant, *Fuzzy Sets and Systems*, 74, 103-113.

Cappelle, B. and Kerre, E.E. (1995c). An algorithm to compute possibilistic reliability, *Proceedings of the Third of International Symposium on Uncertainty Modelling and Analysis and the Annual Conference of the North American Fuzzy Infromation Processing Society*.

Carvallo, M. (1965). *Principles et applications de l'analyse booléenne*, Gauthier-Villars, Paris.

Cohen, H. (1984). Space reliability technology: A historical perspective, *IEEE Transactions on Reliability*, 33, 36-40.

de Cooman, G. (1993). *Evaluation Sets and mappings: An order theoretic approach to vagueness and uncertainty*, Doctoral Dissertation, Universiteit Gent(in Dutch).

Dhillon, B.S. (1980). A system with two kinds of 3-state elements, *IEEE Transactions on Reliability*, 29, 345.

Dubes, R.C. (1963). Two algorithms for computing reliability, *IEEE Transactions on Reliability*, 12, 55-63.

Dubois, A. and Prade, H. (1980). *Fuzzy sets and systems: Theory and applications*, Academic Press, New York.

El-Neweihi, E., Proschan, F. and Sethuraman, J. (1978). Multistate coherent systems, *Journal of Applied Probability*, 15, 675-688.

Elsayed, E.A. and Zebib, A. (1979). A repairable multistate device, *IEEE, Transactions on Reliability*, 28, 81-82.

Embrey, D.E. (1984). SLIM-MAUD(Success Likelihood Index Methodology Multi-Attribute Utility Decomposition): An approach to assessing human error probabilities using structured expert judgement, *NUREG/CR*-3518.

Esary, J.D. and Proschan, F. (1962). The reliability of coherent systems, in *Redundancy Techniques for Computing Systems*, Wilcox, R.H. and Mann, C.W. eds., Spartan Books, Washington D.C., 47-61.

French, S. (1990). On being enumerate, *The University of Leeds Review*, 33, 119-134.

Fussell, J.B. (1973). Fault-tree analysis: Concepts and techniques, *Proceedings of the NATO Conference on Reliability*, NATO Advanced Study Institute on Generic Techniques of System Reliability, Liverpool.

Fussell, J.B. (1984). Nuclear power system reliability: A historical perspective, *IEEE Transactions on Reliability*, 33, 41-47.

Gazdík, I. (1986). RELSHELL, An expert system shell for fault diagnosis, in *Fault Diagnosis, Reliability and related Knowledge-Based Approaches*, Tzafestas, Singh, Schmidt, eds., D Reidel.

Gazdík, I. (1996). Zadeh's extension principle in design reliability, *Fuzzy Sets and Systems*, 83, 169-178.

Gnedenko, B., Beliaev, Y. and Soloviev, A. (1972). *Méthodes mathématiques en théorie de la fiabilité*, Mir, Moscow.

Griffith, W.S. (1980). Multistate reliability models, *Journal of Applied Probability*, 17, 735-744.

Griffith, W.S. (1982). A multistate availability model: System performance and component importance, *IEEE Transactions on Reliability*, 31, 97-98.

IAEA and NEA (OECD) (1992). INES: *The international nuclear event scale user's manual*, revised and extended edition.

Kaufmann, A. Cyouokko, D. and Cryon, R. (1977). *Mathematical models for the study of the reliability of systems*, Academic Press, New York.

Kenarangui, R. (1991). Event-tree analysis by fuzzy probability, *IEEE Trans. on Reliability*, 40, 120-124.

Misra, K.B. and Sharma, A. (1981). Performance index to quantify reliability using fuzzy subset theory, *Microelectronics and Reliability*, 21, 543-549.

Misra, K.B. and Weber, G.G. (1990). Use of fuzzy set theory for level-1 studies in probabilistic risk assessment, *Fuzzy Sets and Systems*, 37, 139-160.

Mizumoto, M. (1989). Pictorial representations of fuzzy connectives, Part I: Cases of t-norms, t-conorms, and averaging operators, *Fuzzy Sets and Systems*, 31, 217-242.

Montero, J. Tejada, J. and Yáñez, J. (1988). General structure functions, *Proceedings of Workshop on Knowledge-Based Systems and Models of Logical Reasoning*, Egypt.

Montero, J., Cappelle, B. and Kerre, E.E. (1995). The usefulness of complete lattices in reliability theory, in *Reliability and Safety Analyses under Fuzziness*, Onisawa, T. and Kacprzyk, J. eds., Physica-Verlag, A Springer-Verlag Company, Heidelberg, 95-110.

Moore, E.F. and Shannon, C.E. (1956a). Reliable circuits using less reliable relays I, *Journal of the Franklin Institute*, 262, 191-208.

Moore, E.F. and Shannon, C.E. (1956b). Reliable circuits using less reliable relays II, *Journal of the Franklin Institute*, 262, 281-297.

Murchland, J.D. (1975). Fundamental concepts and relations for reliability analysis of multi state systems in *Reliability and Fault Tree Analysis*, Barlow, R.E, Fussell, J.B. and Singpurwalla, N.D. eds., SIAM, Philadelphia, 581-618.

Natvig, B. (1982). Two suggestions of how to define a multistate coherent system, *Advances in Applied Probability*, 14, 434-455.

Noma, K. Tanaka, H. and Asai, K. (1981). On fault tree analysis with fuzzy probability, *The Japanese Journal of Ergonomics*, 17, 291-297(in Japanese).

Onisawa, T. (1986). Performance shaping factors modelling using the error possibility and fuzzy integrals, *The Japanese Journal of Ergonomics*, 22, 81-89(in Japanese).

Onisawa, T. (1988). An approach to human reliability in man-machine systems using error possibility, *Fuzzy Sets and Systems*, 27. 87-103.

Onisawa, T. (1990). An application of fuzzy concepts to modelling of reliability analysis, *Fuzzy Sets and Systems*, 37, 269-286.

Onisawa, T. (1991). Fuzzy reliability assessment considering the influence of many factors on reliability, *International Journal of Approximate Reasoning*, 5, 265-280.

Onisawa, T. (1993). On fuzzy sets operations in system reliability analysis with natural language, *Journal of Japan Society for Fuzzy Theory and Systems*, 5, 43-54(in Japanese).

Onisawa, T. and Misra, K.B. (1993). Use of fuzzy sets theory(Part-II: Applications), in *New Trends in System Reliability Evaluation*, Misra, K.B. ed., Elsevier, The Netherlands, 551-586.

Onisawa, T. (1994). A few remarks on membership function in fuzzy reliability analysis, *Proc. of 3rd International Conference on Fuzzy Logic, Neural Nets and Soft Computing*, Iizuka, Japan, 635-636.

Onisawa, T. and Kacprzyk, J. eds. (1995). *Reliability and Safety Analyses under Fuzziness*, Physica-Verlag, A Springer-Verlag Company, Heidelberg.

Onisawa, T. (1995). Subjective system reliability analysis and mutual agreements of its results, in *Fuzzy Logic and Its Applications to Engineering, Information Sciences, and Intelligent Systems*, Bien, Z. and Min, K.C. eds., Kluwer Academic Publishers, The Netherlands, 265-274.

Onisawa, T. (1996). Subjective analysis of system reliability and its analyzer, *Fuzzy Sets and Systems*, 83, 249-269.

Premo, A.F. (1963). The use of Boolean algebra and a truth table in the formulation of a mathematical model of success, *IEEE Transactions on Reliability*, 12, 45-49.

Sadatoku, H., Nagamachi, M., Matsubara, Y. and Onisawa, T. (1993). An analysis of human-error factors using fuzzy integral-measure model and natural language, *The Japanese Journal of Ergonomics*, 29, 289-297(in Japanese).

Schmucker, K.J. (1984). *Fuzzy sets, natural language computations, and risk analysis*, Computer Press, Rockville, M.D.

Shooman, M.L. (1984). Software reliability: A historical perspective, *IEEE Transactions on Reliability*, 33, 48-55.

Singer, D. (1990). A fuzzy set approach to fault and reliability analysis, *Fuzzy Sets and Systems*, 34, 145-155.

Svenson, O. (1989). On expert judgement in safety analyses in the process industries, *Reliability Engineering and System Safety*, 25, 219-256.

Swain, A.D.(1963). *A method for performing a human factors reliability analysis*, Report SCR-685, Sandia Corporation.

Swain, A.D. and Guttmann, H.E. (1983). Handbook of human reliability analysis with emphasis on nuclear power plant applications, *NUREG/CR*-1278.

Tanaka, H., Fan, L.T., Lui, F.S. and Toguchi, K. (1983). Fault tree analysis by fuzzy probability, *IEEE Trans. on Reliability*, 32, 453-457.

U.S. Atomic Energy Commission (1974). *Reactor safety study: An assessment of accident risks in U.S. commercial nuclear power plants*, WASH-1400.

Washio, T., Takahashi, H. and Kitamura, M. (1992). A method for supporting decision-making on plant operation based on human reliability analysis with fuzzy integral, *Proc. of the 2nd Int. Conf. on Fuzzy Logic and Neural Networks*, Iizuka, 2, 841-845.

Zadeh, L.A. (1965). Fuzzy sets, *Information and Control*, 8, 338-353.

Zadeh, L.A. (1968). Probability measures of fuzzy events, *J. Math. Analysis and Appl.*, 23, 421-427.

13 MAINTENANCE AND REPLACEMENT MODELS UNDER A FUZZY FRAMEWORK

Augustine O. Esogbue
Warren E. Hearnes II

Abstract: Uncertainty is present in virtually all replacement decisions due to unknown future events, such as revenue streams, maintenance costs, and inflation. Fuzzy sets provide a mathematical framework for explicitly incorporating imprecision into the decision-making model, especially when the system involves human subjectivity. This chapter illustrates the use of fuzzy sets and possibility theory to explicitly model uncertainty in replacement decisions via fuzzy variables and fuzzy numbers. In particular, a fuzzy set approach to *economic life of an asset* calculation as well as a finite horizon single asset replacement problem with multiple challengers is discussed. Because the use of triangular fuzzy numbers provides a compromise between computational efficiency and realistic modeling of the uncertainty, this discussion emphasizes fuzzy numbers. The algorithms used to determine the optimal replacement policy incorporate fuzzy arithmetic, dynamic programming with fuzzy rewards, the vertex method, and various ranking methods for fuzzy numbers. A brief history of replacement analysis, current conventional techniques, the basic concepts of fuzzy sets and possibility theory, and the advantages of the fuzzy generalization are also discussed.

13.1 ECONOMIC DECISION ANALYSIS

Economic decision analysis is a useful tool, offering individuals and organizations the techniques to model economic decision-making problems, such as maintenance and replacement decisions, and determine an optimal decision.

However, the accuracy of the model determines the validity of the conclusion. In many cases, the assumption of certainty in many models is made not so much for validity but the need to obtain simpler and more readily solvable formulations. Essentially, the tradeoff is between an *inaccurate but solvable* model and a *more accurate but potentially unsolvable* one. In most real-world systems, however, there are elements of uncertainty in the process or its parameters, which may lack precise definition or precise measurement, especially when the system involves human subjectivity.

When developing a model of a system with uncertainty, the decision-maker can either *ignore* the uncertainty, *implicitly acknowledge* it, or *explicitly model* it. Ignoring the uncertainty usually results in a deterministic model of the process with precise values for all parameters. Implicitly acknowledging the uncertainty may still result in a deterministic model in which sensitivity analysis or discount factors can be used to get an idea of how this uncertainty affects the outcome. Lastly, the decision-maker can explicitly model the uncertainty using specific paradigms such as interval analysis, possibility theory, probability theory, or evidence theory (Behrens and Choobineh, 1989).

The proper paradigm depends on the nature of the uncertainty. When the probabilities are specified for the outcomes, then the theory of Von Neumann and Morgenstern (Von Neumann and Morgenstern, 1944) provides the tools necessary to determine the optimal decision. However, in many cases these probabilities are neither defined nor directly attainable. Under these circumstances, other theories are needed. The most common choice is the use of subjective probability distributions and the theory of choice due to Savage (Savage, 1954). However, considerable debate on the use of subjective probabilities exists and is well documented in the literature (Bezdek, 1994, Dubois and Prade, 1994, Klir, 1994, Kosko, 1994, Laviolette and Seaman, 1994). From a psychological standpoint, the methods used to elicit these subjective probabilities and the validity of the subjective probabilities themselves have been the focus of research led by Tversky and Kahneman (Tversky and Kahneman, 1971, Tversky and Kahneman, 1981, Tversky and Kahneman, 1974). Their studies show that the heuristics employed to assess probabilities and predict values can sometimes lead to "severe and systematic errors" (Tversky and Kahneman, 1974).

Because humans do not think naturally in probabilistic terms, they tend to find the notions of fuzzy sets and their linguistic based approaches more user-friendly and appealing. We may view fuzzy set theory as a generalization of classical set theory since it provides us with a mathematical tool for describing sets that have no sharp transition from membership to nonmembership. Membership in a fuzzy set is defined by a generalized version of the classical indicator function called a membership function. Fuzzy sets allow the definition of vague or imprecise concepts such as "*approximately* 1000". This theory has been developed and successfully applied to numerous areas such as control, decision-making, engineering, and medicine. Its application to economic analysis is natural due to the uncertainty inherent in many financial and investment decisions. As noted earlier, it provides a precise mathematical language to model uncertainty due to vagueness and imprecision in events or statements describing a system. More information on fuzzy set theory, particularly fundamental concepts such as fuzzy numbers, which are invoked in our presentation, can be found in (Zadeh, 1978, Zimmerman, 1991).

13.2 REPLACEMENT ANALYSIS

One of the most practical and topical areas of engineering economics is replacement analysis. Mathematical models and analysis methods are used to determine the sequence of replacement decisions that provides a required service for a specified time horizon in an optimal manner. Maintenance and replacement decisions are assumed to occur on a periodic basis. The decision-maker chooses from various options, such as to keep, overhaul, or perform preventive maintenance on the existing asset or replace it with a new/used asset. Any sequence of decisions is called a *replacement policy* and any sequence that optimizes some performance measure such as net present value or annual equivalent cost is an *optimal replacement policy.*

In replacement analysis the *economic life of an asset* determines the replacement cycle that gives the minimum annual equivalent cost (MAEC) of operating a single asset over an infinite horizon (Sharp-Bette and Park, 1990). Dynamic programming (DP), with discounting to put more emphasis on short-term income, is an acknowledged tool for the determination of the optimal replacement policy for more general replacement models (Gluss, 1972). Using the DP approach some of the restrictive assumptions of the economic life method can be relaxed and still produce a computationally feasible solution algorithm. Early pioneers in the use of dynamic programming for finite horizon equipment replacement problems were Bellman and Wagner. Bellman (Bellman, 1955, Bellman and Dreyfus, 1962) was the first to formulate the replacement problem as a dynamic program. Optimal replacement policies were proposed first for the case with no technological change and then assuming some type of technological improvement. Bellman formulated a discounted DP version of the *economic life of an asset* model and determined analytically the optimal age T to replace the asset. In the more challenging technological improvement version, the revenue, the upkeep costs, and the replacement costs are assumed to be functions of the date the asset is installed as well as its age with respect to installation. Wagner (Wagner, 1975) formulated the replacement problem as a network and solved for the shortest path, which corresponded to the minimum outlay. Terborgh (Terborgh, 1949) included linear obsolescence in his formulation while Alchian (Alchian, 1952) allowed operating revenues and operating costs to increase linearly with time. Oakford (Oakford, 1970) modeled increasing revenues and data. Dreyfus (Dreyfus and Law, 1977) modeled technological change in revenue, maintenance, and replacement costs using bounded exponential functions. The Dynamic Replacement Economy Decision Model (DREDM) developed in (Oakford et al., 1984) is a generalization of Wagner's dynamic programming model that allows for multiple challengers and time-varying parameters.

Replacement models of great interest and relevance to this research are those that model uncertainty. Dreyfus and Law (Dreyfus and Law, 1977) treat the replacement problem where determinism yields to stochasticity. Their model includes the probability of a catastrophic failure in the asset being used as well as an uncertain net operating cost that is modeled by another probability distribution. The DP algorithm determines the minimum expected cost for the process. The Stochastic Replacement Economy Decision Model (SRE) presented by Lohmann (Lohmann, 1986) is a stochastic generalization of the DREDM. The assumption that the cash flows and relationships are known with certainty was relaxed and the component cash flows are modeled as triangular probability distributions based on the decision-maker's subjective judgment. The solution for this model is generated through Monte-Carlo simulation, which determines the probability that each asset is the optimal choice at time 0 as well as the probability distribution of the optimal net present value of the policy.

A tacit assumption implicit in the foregoing models is that uncertainty in the replacement decision can be fully modeled either deterministically or stochastically. This is not, however, always the case. Limiting replacement models to these two approaches either ignores the uncertainty or assumes that all uncertainty is probabilistic in nature and that the probabilistic information is fully known. Categorically classifying all uncertainty as randomness may not be reasonable or adequate.

In recent years the debate concerning the use of nonprobabilistic uncertainty, and specifically fuzzy sets, has surfaced in the area of economic analysis (Behrens and Choobineh, 1989, Buckley, 1987, Buckley, 1992, Chiu and Park, 1994, Choobineh and Behrens, 1992, Gupta, 1993, Hearnes, 1995, Ward, 1985). The replacement decisions made at each time period are based not only on the current cash flows but also on projected future cash flows of all possible assets (Lohmann, 1986). Therefore, uncertainty in these cash flows can have a pronounced effect on the optimal replacement policy. A fuzzy set theoretic approach, as described in the sequel, may lead to more informed replacement decisions when the assumptions for a probabilistic approach are not met. For the deterministic case, a fuzzy set theoretic approach using fuzzy numbers is equivalent to multivariable sensitivity analysis and immediately provides both the deterministic optimal value and the possible range due to uncertainty.

13.3 FUZZY CONCEPTS IN CASH FLOW ANALYSIS

Cash flows, the basic variable in replacement decisions, are used by managers and financial analysts to measure the streams of money going into and flowing out of a particular organization's operation (Sharp-Bette and Park, 1990). Traditionally, cash flows are treated as either deterministic or stochastic. However, as shown in simulation studies (Oakford et al., 1981), uncertain information in estimating these cash flows can limit the value of the analysis. Errors in deterministic cash flow estimations can skew the results of the analysis. Similarly, the use of subjective probability distributions is of concern since they generally cannot be verified, and the required historical information for generating frequency-based probability distributions is not always available.

The fundamental types of uncertainty, *nonspecificity*, *fuzziness*, and *strife*, are examined by Klir and Yuan (Klir and Yuan, 1995). Uncertainty occurs in replacement and maintenance decisions in various ways. Of particular interest are nonspecificity and fuzziness which may factor into the estimates of the aggregate cash flows, the purchase prices/salvage values, the minimum attractive rate of return (MARR), or the physical lifetimes of the assets. This is especially true when these variables are based on the estimates provided by experts via the use of such natural language statements as "*approximately* \$1000". Through the use of fuzzy variables one can represent this vagueness and imprecision. In this presentation, however, these vague quantities will be represented using triangular fuzzy numbers (TFNs).

Several basic concepts of fuzzy sets applicable to the foregoing forms of uncertainty are *fuzzy sets*, *α-level sets*, *convexity*, and *triangular fuzzy numbers*. It is assumed that the reader is familiar with these concepts. If not, the reader is referred to (Zadeh, 1978, Zimmerman, 1991) for a detailed review. Fuzzy numbers are fuzzy sets defined on the set of real numbers, generally used to represent vague expressions, such as "*about* 20" or "*approximately* 1000", used often in the description of uncertain economic decision systems. A special type of fuzzy number is the L-R fuzzy number introduced by Dubois and Prade (Dubois and Prade, 1978). These types of fuzzy numbers simplify the arithmetic operations considerably. Triangular fuzzy numbers are a special type of L-R fuzzy number that are used in the models developed in this research. A *triangular fuzzy number* (TFN) is a fuzzy number $M_T = [l, m, r]$ defined by the membership function

$$\mu_{M_T}(x) = \begin{cases} (x-l)/(m-l), & \text{for } x \in (l, m] \\ (r-x)/(r-m), & \text{for } x \in (m, r) \\ 0, & \text{otherwise} \end{cases} \tag{1}$$

where m is the modal value and l and r are the left and right limits of the support for the fuzzy number, respectively.

13.3.1 Nonprobabilistic Methods in Cash Flow Analysis

Research into fuzzy versions of cash flows began with Ward (Ward, 1985), defining them as trapezoidal fuzzy numbers and solving a fuzzy present worth problem. Buckley (Buckley, 1987) used fuzzy numbers to develop fuzzy net present value (NPV) and fuzzy net future value (NFV) with fuzzy interest rates over a period of n years where n may also be fuzzy set. Buckley developed fuzzy equivalents to continuous interest payments, the effective rate of interest, and regular annuities as well. Restricting the fuzzy cash flows to positive fuzzy numbers allows the multiplication operation to be distributive over addition. The fuzzy number of time periods produces nonlinearities that make computations more complex. Li Calzi (Li Calzi, 1990) provided an axiomatic development for the fuzzy extension of financial mathematics with a desire to maintain consistency in the calculations. He examined two classes of fuzzy quantities, compact fuzzy intervals and invertible fuzzy intervals, and proved general theorems regarding consistency.

Two of the most recent and practical applications of nonprobabilistic uncertainty to economic analysis are given in (Chiu and Park, 1994, Choobineh and Behrens, 1992). Choobineh and Behrens (Choobineh and Behrens, 1992) call attention to the use of intervals and possibility theory in economic analysis. The weak distributivity of interval arithmetic is noted, but a technique called the vertex method (Dong and Shah, 1987) is utilized to bypass this problem in interval and fuzzy arithmetic. Their approach to modeling cash flows as fuzzy intervals is similar to Ward's. Chiu and Park (Chiu and Park, 1994) use fuzzy numbers in cash flow analysis and provide a good survey of the major methods for ranking mutually exclusive fuzzy projects. The cash flows are modeled as triangular fuzzy numbers and the linear approximation to the product of two triangular fuzzy numbers is investigated. The present worth of a fuzzy project is also examined. Their resultant formulation of a fuzzy present worth is

$$PW = \sum_{t=0}^{n} \left\{ (P_t) \left(\prod_{s=0}^{t} (1 + R_s) \right)^{-1} \right\} \tag{2}$$

where P_t is a positive or negative TFN representing the cash flow at the end of time t, n is the number of evaluation periods, and R_s is the nonnegative TFN representing the discount rate at the end of time s. Extending these ideas to replacement analysis, Hearnes (Hearnes, 1995) formulated fuzzy versions of the economic life of an asset model and the finite single asset replacement problem. This work is used as a point of departure for the discussions that follow.

13.3.2 Fuzzy Arithmetic and Interval Analysis

Fuzzy numbers represent vague notions of precise quantities. It is essential to be able to perform algebraic operations on them. The arithmetic operations of addition, subtraction, multiplication, and division developed by Dubois and Prade (Dubois and Prade, 1978) and particularly in (Dubois and Prade, 1987) are particularly useful when modeling and analyzing cash flows. The algebraic operations for TFNs are specifically reviewed. The choice of TFNs is in part due to their simplified algebraic operations. However, the set of TFNs is not closed under the operations of multiplication and division. The effect of using a linear (TFN) approximation, which is studied thoroughly in (Kaufmann and Gupta, 1988), is not significant for multiplication, inverse, and division operations. The TFN approximation to the multiplication operation is used in the following models for computational simplicity. The approximation error was also empirically determined to be not significant for the example problems in the sequel.

Fuzzy numbers are a family of nested intervals (Choobineh and Behrens, 1992) which correspond to levels of "confidence" by the decision-maker and therefore are closely related to interval analysis. However, like interval arithmetic, the multiplication operation for fuzzy numbers is only weakly distributive over addition. This presents a problem when modeling with fuzzy numbers since the outcome can depend on the form of the equation used. There are some special cases, however, where the multiplication operation is distributive over addition. If M_1, M_2, M_3 are fuzzy numbers, the multiplication operation is distributive over addition (i.e., $M_1 \cdot (M_2 + M_3) = M_1 \cdot M_2 + M_1 \cdot M_3$) when (Dubois and Prade, 1987):

1. M_1 is a real number—i.e., scalar multiplication is distributive over addition.

2. M_2 and M_3 are both positive or both negative.

3. $M_2 = -M_2$ and $M_3 = -M_3$.

When conditions (1) – (3) are not met, a procedure called the *vertex method* (Dong and Shah, 1987) preserves the distributivity of multiplication over addition. The price, however, is an exponential increase in the number of computations.

13.3.3 Some Sources of Uncertainty in Replacement Analysis

Several aspects of replacement analysis contain imprecision and vagueness that warrant further discussion. We postulate that some of these variables, such as the physical lifetime of an asset, aggregate cash flow estimates, MARR, and inflation significantly impact the optimal replacement policy.

Physical Lifetime of an Asset. The physical lifetime of some assets may not be known with certainty, yet this is tacitly assumed and treated as deterministic in many models. In some situations, if enough information is known, then it is appropriate to treat it probabilistically. However, in cases where the asset is a new technology or a new model, this information is not generally available and such an approach must be considered suspect. In this case, it is instructive to treat the uncertainty in this variable through the use of fuzzy DP models but particularly those that allow stochastic or fuzzy termination times (Buckley, 1987, Kacprzyk, 1978). For example, for an older asset the historical information about failures may be known and a probability distribution for failure can be derived. However, for an asset with new technology estimate of physical lifetime may be a fuzzy set such as "*about* 5 years" or "*more or less* 10 years". In each of these cases, the decision space has an uncertain boundary that affects the overall decision policy. Stochastic and fuzzy DP (Esogbue and Kacprzyk, 1996) provide methods for dealing with this type of uncertainty. Similarly, there may be uncertainty in the actual horizon N of the project which may be either stochastic or fuzzy. For example, the project duration of asset life may be determined if the state of the asset reaches some imprecisely definable point.

Aggregate Cash Flows. Another source of uncertainty in replacement and maintenance decisions is the estimation of the aggregate cash flow for each time period. In previous models, aggregate cash flows were treated as either deterministic or stochastic variables, but errors in these can lead to skewed analysis (Oakford et al., 1981). Subjective probability distributions generally cannot be verified, while the required historical information for generating frequency-based probability distributions is not generally available. In these cases, it may be more appropriate to define the aggregate cash flows as possibility distributions based on a decision-maker's opinion or expert judgment.

Minimum Attractive Rate of Return. The minimum attractive rate of return (MARR), which is usually used for project evaluation and comparison, is also another variable which may realistically possess forms of uncertainty (Sharp-Bette and Park, 1990). However, in classical approaches this either is not addressed or is erroneously assumed to be well known or deterministic. The selection of the proper MARR plays an important role in the outcome of the maintenance and replacement decisions. There are a number of ways to determine a corporation's MARR, such as the use of the Delphi method involving its directors or some chosen mathematical formula. However, the MARR can be investment or management dependent. Because of the uncertainties characteristic of investment and management decision processes, it is inevitable that any MARR thus determined is imprecise or fraught with uncertainties. Variation in the MARR and its effect on the optimal policy are vital pieces of information to decision-makers. These may be better modeled as a fuzzy variable or fuzzy number.

Inflation. A number of engineering economic studies discuss the incorporation of inflation and inflation rate in their models. It is tacitly assumed or conceded that the measurement of this variable is precise. This, however, is not the case. We know that there is a considerable degree of uncertainty due to the way that it is measured. For example, the Consumer Price Index presently used by the government of the United States of America is now under review due to the concern expressed by certain economists. It is argued that the "basket" of goods and services it uses may not accurately reflect the *true* inflation (see, for example, (Samuelson, 1996)). The lack of specificity or precision involved in the measurement of the inflation rate may necessitate the injection of fuzzy modeling such as the use of fuzzy numbers to represent it.

13.4 ECONOMIC LIFE OF AN ASSET MODEL

In some replacement decisions, an asset is required for a long period of time. In these cases, an infinite horizon can be assumed and the decision variable becomes the life of the asset. This is commonly called the *economic life of an asset*. The chosen replacement cycle is the cycle corresponding to the minimum annual equivalent (AE) cost of owning and operating the asset (Sharp-Bette and Park, 1990). An infinite sequence of replacements and stationary cash flows (with respect to installation time) is assumed.

The general deterministic n-period replacement cycle gives the following AE cost:

$$AE_n(i) = (P - S_n)(A/P, i, n) + S_n i + (A/P, i, n) \sum_{n'=1}^{n} (C_{n'}(P/F, i, n')) \quad (3)$$

where

$$\begin{aligned} i &\equiv \text{minimum attractive rate of return (MARR)} \\ P &\equiv \text{initial purchase price} \\ S_n &\equiv \text{salvage value at end of period } n \\ C_{n'} &\equiv \text{aggregate cash outflow at end of period } n' \\ (A/P, i, n) &\equiv \text{capital recovery factor} \\ (P/F, i, n) &\equiv \text{present worth factor} \end{aligned}$$

Two significant factors determining the optimal replacement cycle are the *aggregate cash flows* at each time period and the MARR. The MARR is set by the organization and is considered a crisp (deterministic) number in the model. The future aggregate cash flows and salvage values, however, are a source of considerable uncertainty and are modeled as triangular fuzzy numbers. These parameters are represented as fuzzy versions of their original counterparts by $\tilde{C}_{n'}$ and $\tilde{S}_n$, respectively. The decision-maker determines a best, worst, and most likely estimate for each. This method of elicitation is quick and has been used previously in replacement analysis (Lohmann, 1986). These estimates (l, m, and r, respectively) are transformed into a TFN defined as $[l, m, r]$.

Generalizing to consider fuzzy cash flows and salvage values, one obtains the Possibilistic Economic Life of an Asset Model (PELAM) (Hearnes, 1995). The fuzzy economic life of an asset is defined as the replacement cycle, n, corresponding to the minimum *fuzzy AE* (FAE) of all possible replacement cycles. The traditional model in Equation 3 is manipulated into a *proper representation*—i.e., all fuzzy numbers appear only once in the equation:

$$FAE_n(i) = P(A/P, i, n) + \tilde{S}_n \left(i - (A/P, i, n)\right) + (A/P, i, n) \sum_{n'=1}^{n} \left(\tilde{C}_{n'}(P/F, i, n')\right) \quad (4)$$

where

$$\begin{aligned} \tilde{S}_n &\equiv \text{TFN representing the "salvage value at end of period } n\text{"} \\ \tilde{C}_{n'} &\equiv \text{TFN representing the "aggregate cash flow at end of period } n'\text{"} \end{aligned}$$

The operations used in PELAM are scalar multiplication, addition, and subtraction. Therefore, the use of TFNs to model the cash flows gives TFNs as a result. However, if the MARR is also modeled as a fuzzy number, the result is not a TFN and the linear approximation to TFN multiplication and the vertex method must be used.

An Example Problem

An asset to perform service A is required by XYZ corporation indefinitely. Asset B can be purchased for \$50 (all dollar amounts are in thousands) and has a physical life of 5 years. The aggregate cash flows (operating costs - operating revenues) for each year of the life of asset B are \$3, \$4, \$6, \$10, and \$12, respectively, for $n = 1, \ldots, 5$. If the asset is sold at the end of the year, its salvage value is \$35, \$30, \$27, \$23, and \$20, respectively, for $n = 1, \ldots, 5$. Assume a MARR of 10%. Determine the economic life of Asset B.

The assumption of certainty in future cash flows is unrealistic, except in some cases such as when the asset is covered by a service contract. Likewise, future salvage values are dependent on the state of the equipment at that time, possible technological breakthroughs that have occurred, and numerous other uncertain events. The deterministic data is treated as the "most likely" estimates. The local expert or decision-maker provides additional information, namely the "best" and "worst" estimates:

Being pessimistic, the decision-maker believes that the cash flows (operating costs - revenues) might be much higher than the "most likely" estimates, and the salvage values might be much lower. Therefore the high estimates for the cash flows are \$5, \$7, \$10, \$15, and \$18, respectively, for $n = 1, \ldots, 5$. The low estimates remain near the "most likely" estimates—\$2, \$3, \$5, \$8, and \$10. The salvage values high estimates are \$38, \$32, \$29, \$27, and \$25. The low estimates are \$32, \$24, \$21, \$18, and \$15.

Table 1 Results of PELAM for Example Problem.

Replacement Cycle (Years)	Fuzzy *AE* Cost of Replacement Cycle
1	$[19.00, 23.00, 28.00]$
2	$[16.05, 18.00, 23.33]$
3	$[14.58, 16.19, 20.94]$
4	$[14.22, 16.30, 20.75]$
5	$[14.30, 16.46, 21.09]$

Table 1 gives the FAE costs for each replacement cycle, $n = 1, \ldots, 5$. Of these, the "minimum" must be chosen. Comparing alternatives described by fuzzy numbers requires a ranking method. All ranking methods reported in the literature suffer from some pathological examples where the result is counterintuitive (Bortolan and Degani, 1985, Chiu and Park, 1994). The rankings of selected methods are given in Table 2. Note that the rankings are not all in agreement as shown in Table 2. However, several of them do agree with each other (Chiu and Park, Choobineh and Behrens, and Kaufmann and Gupta). All three of these methods use a ranking function based on the l, m, and r values of the TFN. In the remainder of this paper, the Kaufmann and Gupta method is used as the preferred ranking function.

The Kaufmann and Gupta method is a hierarchical test. It can be described as follows: Let $M_1 = [l_1, m_1, r_1]$ and $M_2 = [l_2, m_2, r_2]$ be two different TFNs.

Table 2 Ranking of FAE Costs in Example Problem by Various Methods.

Ranking Method	Ranking
Adamo, $\alpha = 0.9$	$FAE_3 \sim FAE_4 \sim FAE_5 < FAE_2 < FAE_1$
Chang	$FAE_3 < FAE_4 < FAE_5 < FAE_2 < FAE_1$
Chiu and Park, $w = 0.3$	$FAE_4 < FAE_3 < FAE_5 < FAE_2 < FAE_1$
Choobineh and Behrens	$FAE_4 < FAE_3 < FAE_5 < FAE_2 < FAE_1$
Kaufmann and Gupta	$FAE_4 < FAE_3 < FAE_5 < FAE_2 < FAE_1$
Traditional model	$AE_3 < AE_4 < AE_5 < AE_2 < AE_1$

TEST 1: Compare the *ordinary numbers*: If $0.25(l_1 + 2m_1 + r_1) < 0.25(l_2 + 2m_2 + r_2)$ then $M_1 < M_2$ else if $0.25(l_1 + 2m_1 + r_1) > 0.25(l_2 + 2m_2 + r_2)$ then $M_1 > M_2$, else go to TEST 2.

TEST 2: Compare the *modal values*: If $m_1 < m_2$ then $M_1 < M_2$ else if $m_1 > m_2$ then $M_1 > M_2$ else go to TEST 3.

TEST 3: Compare the *divergence*: If $(r_1 - l_1) < (r_2 - l_2)$ then $M_1 < M_2$ else $M_1 > M_2$.

This method is usually used because (1) it is relatively easy to compute and (2) it always chooses a maximum when the two TFNs are not equal. The latter property is especially important in models based on dynamic programming where a unique optimal value at each stage is desired.

The Optimal Replacement Cycle for the Example Problem

We now show the determination of the optimal replacement policy for the example problem. Using the Kaufmann and Gupta ranking method for reasons discussed above, the optimal replacement cycle for Asset B is 4 years with a fuzzy annual equivalent cost of $[14.22, 16.30, 20.75]$. The modal values of the TFNs for each replacement cycle correspond to the *deterministic* AE costs from the traditional economic life of an asset model. Therefore, the traditional optimal replacement cycle is immediately available from the fuzzy solution, and the fuzzy solution using fuzzy numbers is equivalent to performing sensitivity analysis on all uncertain variables. It is also interesting to note that using the traditional model the optimal replacement cycle is 3 years, which is different from the 4 years determined by PELAM. Additionally, PELAM determines the optimal replacement cycle based on the decision-maker's estimates of the uncertainty and therefore provides a more informative answer than the traditional model.

13.5 DYNAMIC PROGRAMMING FORMULATION OF ECONOMIC LIFE OF AN ASSET

Bellman formulated a discounted DP approach to determining the optimal replacement cycle of an asset (Bellman and Dreyfus, 1962). This approach allows for the modeling of *technological change*, unlike the standard *economic life of an asset* model. Technological change implies that the operating revenues and costs, as well as purchase and salvage values, change over time due to technological improvements. An excellent example is the personal computer. As technology improves it affects the purchase price and salvage values of existing computers. The revenues and costs may also be affected by the faster processors.

In the traditional model, at each time period $n = 0, 1, 2, \ldots$, the choice is made to either keep the existing asset or purchase a new one. The following model is an adaptation of Bellman's model (Bellman and Dreyfus, 1962) to the variables and data used in the economic life of an asset model in Section 13.4. This model also uses the year-end convention for operating cash flows. The functional equation is defined:

$f(t) \equiv$ total discounted return (TDR) over an infinite horizon starting with a machine of age t and using an optimal policy.

Thus

$$f(t) = \max \begin{bmatrix} PURCHASE: & S_t - P + \alpha\,(f(1) - C_1) \\ KEEP: & \alpha\,(f(t+1) - C_{t+1}) \end{bmatrix} \tag{5}$$

where

$$\begin{aligned} \alpha &\equiv \text{discounting rate} = (1 + MARR)^{-1} \\ P &\equiv \text{initial purchase price} \\ S_n &\equiv \text{salvage value at the end of period } n \\ C_n &\equiv \text{aggregate cash outflow at end of period } n \end{aligned}$$

Noting that an optimal policy has the form of keeping a new machine until it is T years old and then replace it with a new machine results in an analytical solution of the optimal life cycle

$$T = \operatorname{argmax}_{N=1,2,\ldots} \left\{ \begin{array}{ll} \frac{(S_1 - P - C_1)}{1-\alpha} & \text{if } N = 1 \\ \frac{\alpha^{N-1}(S_N - P) - \alpha^N C_1 - \sum_{n=1}^{N-1} \alpha^{n-1} C_{n+1}}{1-\alpha^N} & \text{otherwise} \end{array} \right\} \tag{6}$$

with the optimal total discounted return of a new machine being

$$f(0) = -C_1 + \alpha \left[\max_{N=1,2,\ldots} \left\{ \begin{array}{ll} \frac{(S_1 - P - C_1)}{1-\alpha} & \text{if } N = 1 \\ \frac{\alpha^{N-1}(S_N - P) - \alpha^N C_1 - \sum_{n=1}^{N-1} \alpha^{n-1} C_{n+1}}{1-\alpha^N} & \text{otherwise} \end{array} \right\} \right] \tag{7}$$

Generalizing this model, the discount factor α is related to the MARR and remains a crisp number. The future aggregate cash flows and salvage values, however, are a source of considerable uncertainty. These are modeled as TFNs and represented as fuzzy versions of their original counterparts by the tilde. Generalizing to consider fuzzy cash flows and salvage values gives the following:

$$T = \text{argmax}_{N=1,2,\ldots} \left\{ \begin{array}{ll} \frac{(\tilde{S}_1 - P - \tilde{C}_1)}{1-\alpha} & \text{if } N = 1 \\ \frac{\alpha^{N-1}(\tilde{S}_N - P) - \alpha^N \tilde{C}_1 - \sum_{n=1}^{N-1} \alpha^{n-1} \tilde{C}_{n+1}}{1-\alpha^N} & \text{otherwise} \end{array} \right\} \quad (8)$$

with the optimal total discounted return of a new machine being

$$f(0) = -\tilde{C}_1 + \alpha \left[\max_{N=1,2,\ldots} \left\{ \begin{array}{ll} \frac{(\tilde{S}_1 - P - \tilde{C}_1)}{1-\alpha} & \text{if } N = 1 \\ \frac{\alpha^{N-1}(\tilde{S}_N - P) - \alpha^N \tilde{C}_1 - \sum_{n=1}^{N-1} \alpha^{n-1} \tilde{C}_{n+1}}{1-\alpha^N} & \text{otherwise} \end{array} \right\} \right] \quad (9)$$

An Example Problem

Using the same data as in the example problem in Section 13.4 above, the total discounted returns for the DP formulation are given in Table 3 below.

Table 3 Results of Possibilistic Version of Bellman's Model for the Example Problem.

Replacement Cycle (Years)	Fuzzy TDR of Replacement Cycle
1	[−230.00, −180.00, −140.00]
2	[−186.67, −131.91, −111.90]
3	[−164.71, −115.02, −98.34]
4	[−164.99, −117.60, −95.90]
5	[−170.01, −120.45, −97.71]

The Optimal Replacement Cycle for the Example Problem

In this example the optimal replacement cycle is 3 years for the fuzzy version under all five ranking methods. By using fuzzy numbers that have a modal value equivalent to the deterministic value of each variable, the solution to the traditional formulation of Bellman's model are the modal values of the respective fuzzy TDRs. Thus the traditional model also has an optimal replacement cycle of 3 years.

13.6 THE GENERAL SINGLE ASSET REPLACEMENT PROBLEM

When a service is required for only a finite period or the aggregate cash flows are nonstationary with respect to installation time, a more general approach than PELAM is needed. The general single asset replacement problem is widely studied in the literature (Bean et al., 1985, Dreyfus and Law, 1977, Lohmann,

1986, Oakford et al., 1984). It may be defined as follows:

$$\begin{aligned} a_n &\equiv \text{asset in use at time period } n \\ N &\equiv \text{number of time periods that service is required} \\ A_n &\equiv \text{number of challenging assets at time period } n \end{aligned}$$

The time periods $n = 0, 1, \ldots, N$ represent the periodic replacement decisions. If N is finite, the problem is a finite horizon replacement problem. Otherwise, the problem is an infinite horizon problem. The existing asset is known as the *defender* and can only be placed into service at period 0. The A_0 assets available for replacement at time 0 are known as *current challengers*, and the A_n assets available at future periods are known as *future challengers*. For each period n in the lifetime of each asset, there are three component cash flows describing the *installation cost and/or salvage value*, *operating revenues*, and *operating costs* at period n. The component cash flows of the future challengers are related to the corresponding component cash flow of a current challenger by a scalar function $f(a, n, C)$ where n is the time period in which asset a is installed and $C \in \{1, 2, 3\}$ represents the respective component cash flow. This function allows for the modeling of inflation, technological improvements, and other time-dependent effects on cash flows. For example, to model a crisp 3% inflation rate, then $f(a, n, C) \equiv 1.03^n$ for all a, C. Other more complicated variations can be defined to model a wide range of factors. The component cash flows and relation functions are either known with certainty or estimated by the decision-maker and may also be a fuzzy variable. The problem is to find the sequence(s) of keep/replace decisions that maximize net present value (NPV).

The Possibilistic Model for Single Asset Replacement via dynamic programming (PMSAR) (Hearnes, 1995) is a generalization of SREDM in (Oakford et al., 1984) to allow for fuzzy parameters such as aggregate cash flows, inflation, or technological change. SREDM used the "best", "worst", and "most likely" estimates of the parameters, as in PERT analysis, to create triangular probability distributions. Using those probability distributions, Monte-Carlo simulation provided estimates of the probability of each asset being the optimal choice at period 0. Under such conditions of estimating the distributions, it is arguable that a possibility theory approach is more appropriate. Like PELAM, PMSAR uses TFNs for cash flows. However, PMSAR does not assume an infinite horizon and allows for replacement by competing assets. There is also the possibility of having uncertainty in technological improvements, inflation, or other aspects of the future challengers, which is also modeled as a TFN through a relation function f. The solution technique is a forward dynamic program that uses the Kaufmann and Gupta ranking method to determine the optimal decision and functional equation value at each time period.

We now present a fuzzy analog of the SREDM model. For ease of exposition, we define the following variables of the model:

1. Let k be the state of the system, $k = 1, 2, \ldots, N$, which represents the number of periods of required service.

2. There are two decision variables: (1) n, the period to place an asset into service and (2) a, the asset to place into service.

3. The immediate reward $r(a, n, k)$ is the FPV generated by placing asset a into service at period n and keeping it in service until period k. Defining L_a as the physical lifetime, in time periods, of asset a gives:

$$r(a,n,k) = \begin{cases} \sum_{C=1}^{3} FPV(a,n,C,i,k-n) & \text{if } k-n \leq L_a \\ -M & \text{otherwise} \end{cases} \tag{10}$$

where $M >> 0$ is some sufficiently large number and

$FPV(a, n, C, i, k - n) \equiv$ Fuzzy Present Value of the *installation cost and/or salvage value*, *operating revenues*, and *operating costs* for $C = 1, 2, 3$, respectively, of placing asset a into service at period n for the remaining $k - n$ time periods using an MARR of i.

4. The transition function $\tau(a, n, k)$ for placing asset a into service at time n for the remaining $k - n$ time periods is $\tau(a, n, k) = n$.

Define the FPV of the functional for this process as the value obtained using an optimal replacement policy from state 0 to state k. Invoking Bellman's Principle of Optimality results in the following functional equation of dynamic programming:

$$FPV^{\star}(k) = \max_{n,a}\{r(a,n,k) + FPV^{\star}(\tau(a,n,k))\} \tag{11}$$

where

$$n \in \{0, 1, \ldots, k-1\} \text{ and } a \in \begin{cases} \{0, 1, \ldots, A_0\} & \text{if } n = 0 \\ \{1, 2, \ldots, A_n\} & \text{if } n > 0 \end{cases}$$

and boundary condition $FPV^{\star}(0) = [0, 0, 0]$. The max operation is performed through the Kaufmann and Gupta ranking method on TFNs.

Relating the parameters of future challengers to the parameters of current challengers via a fuzzy relation function $f(a, n, C)$ is a desirable feature since there may exist considerable uncertainty of the nonprobabilistic nature in future events. This addition is not without its price, however. The model requires multiplication of two fuzzy numbers, which is only weakly distributive and is not closed over TFNs. This problem may be readily circumvented by the adroit use of the vertex method (Dong and Shah, 1987) and a TFN approximation to the product of two TFNs (Kaufmann and Gupta, 1988).

An Example Problem

Let us now consider an adaptation of a replacement problem discussed by Lohmann (Oakford et al., 1984) in which we specifically incorporate fuzzy uncertainty and use it as a vehicle for illustrating our point.

Three current challengers, $A_n = 1, 2, 3$ for $n = 1, \ldots, 15$, can replace the defender, $a = 0$. The time horizon of $N = 15$ years is established. Each challenger has a physical lifetime of 5 years. The defender has a remaining life of 3 years. For capital transfers, the cash flow at period 0 is the purchase cost and the cash flows at periods $n > 0$ are salvage values. The component cash flows for the most likely estimates plus or minus a percentage are listed in Table 4. Assume a MARR of 10%. Suppose the tax rate is assumed to be 50% and MACRS depreciation tables for a 7-year recovery period are utilized in the determination of depreciation tax shield.

Table 4 Data for Example Problem.

a	n	$C = 1$	$C = 2$	$C = 3$
0	0	5423	0	0
	1	4194 ±5%	-15791 ±5%	19200 ±5%
	2	3355 ±5%	-17685 ±5%	19200 ±5%
	3	2684 ±10%	-19808 ±10%	19200 ±10%
1	0	20000	0	0
	1	16100 ±5%	-8000 ±5%	19200 ±5%
	2	13200 ±5%	-8960 ±5%	19200 ±5%
	3	10840 ±10%	-10035 ±10%	19200 ±10%
	4	8192 ±10%	-11239 ±10%	19200 ±10%
	5	6554 ±10%	-12588 ±10%	19200 ±10%
2	0	21000	0	0
	1	16800 ±5%	-7500 ±5%	19200 ±5%
	2	13440 ±5%	-8400 ±5%	19200 ±5%
	3	10252 ±10%	-9408 ±10%	19200 ±10%
	4	8602 ±10%	-10537 ±10%	19200 ±10%
	5	6881 ±10%	-11801 ±10%	19200 ±10%
3	0	22000	0	0
	1	17600 ±25%	-7250 ±25%	19200 ±5%
	2	14080 ±25%	-8120 ±25%	19200 ±5%
	3	11264 ±25%	-9094 ±25%	19200 ±10%
	4	9011 ±25%	-10186 ±25%	19200 ±10%
	5	7209 ±25%	-11408 ±25%	19200 ±10%

The optimal sequence of decisions is determined via forward dynamic programming with rewards modeled as fuzzy numbers. The functional equation for the first decision stage is computed as follows:

$$\begin{aligned}
FPV(0,0,1,0.1,1) &= \text{- Purchase + PV(Salvage)} \\
&\quad + \text{PV(Tax shield from depreciation)} \\
&\quad - \text{PV(Tax on gain at sale)} \\
&= -[5423,5423,5423] + [3984.3,4194,4403.7]\cdot(1.1)^{-1} \\
&\quad +[5423,5423,5423]\cdot 0.5 \cdot 0.1429 \\
&\quad -([3984.3,4194,4403.7] \\
&\quad -[5423,5423,5423]\cdot(1-0.1429)) \\
&= [-1146.95,-1051.64,-956.32] \\
FPV(0,0,2,0.1,1) &= \text{PV(After-tax costs)} \\
&= [-7536.61,-7177.73,-6818.84] \\
FPV(0,0,3,0.1,1) &= \text{PV(After-tax revenues)} \\
&= [8290.91,8727.27,9163.64]
\end{aligned}$$

Note that the salvage and purchase values and functional relation (if used) appear more than once in the FPV calculations for $C = 1$, therefore the vertex method is used to generate the correct FPV. This entails computing deterministically the problem for the modal (most likely values) as well as an additional 2^3 times for all the possible combinations of "best" and "worst" estimates for these parameters. This gives the following for $N = 1$:

$$\begin{aligned}
FPV^\star(1) &= \max_{n=0,a\in\{0,1,2,3\}}\left\{\sum_{C=1}^{3} FPV(a,0,C,0.1,1) + FPV^\star(0)\right\} \\
&= \max\left\{\begin{array}{ll}
[-392.66,497.91,1388.48]+[0,0,0] & \text{for } a=0, n=0 \\
[515.91,1500.00,2484.09]+[0,0,0] & \text{for } a=1, n=0 \\
[511.36,1500,2488.64]+[0,0,0] & \text{for } a=2, n=0 \\
[-228.41,1431.82,3092.04]+[0,0,0] & \text{for } a=3, n=0
\end{array}\right\} \\
&= [515.91,1500.00,2484.09] \text{ for } a=1.
\end{aligned} \tag{12}$$

The same calculations are performed for the remaining decision stages, up to $k = 15$, recording both the functional equation value and the optimal decision for each stage as in Table 5.

The Optimal Replacement Policy for the Example Problem

As earlier we show the determination of the optimal replacement problem for the model example under fuzziness. Solving the functional equation given in Equation 11 for $N = 15$ results in the optimal replacement policy:

$$(3,0)(3,5)(3,10) \text{ with } FPV^\star(15) = [829.63,15480.30,30131.10].$$

This translates into buying asset 3 at time 0, again at time 5, and again at time 10. In this solution, the modal values of each $FPV^\star(k)$ represent the $NPV^\star(k)$ of the deterministic model, while the lower and upper values indicate the overall uncertainty in the decision. The uncertainty signified by the width of the base of each $FPV^\star(k)$ also is equivalent to the range of possible values determined via multivariable sensitivity analysis. Thus a fuzzy model immediately provides

Table 5 Dynamic Programming Results for Basic Example Problem.

N	Optimal Policy (asset, time installed)	$FPV^\star(N)$
1	(1,0)	[515.91,1500.00,2484.09]
2	(1,0)	[1686.70,3159.42,4632.15]
3	(1,0)	[2054.50,4759.95,7465.39]
4	(2,0)	[2799.62,6360.66,9921.71]
5	(3,0)	[413.48,7715.23,15017.00]
6	(3,0) (2,5)	[731.00,8646.61,16562.20]
7	(3,0) (1,5)	[1460.79,9676.98,17893.20]
8	(2,0) (2,4)	[4711.79,10705.10,16698.40]
9	(3,0) (2,5)	[2151.82,11664.70,21177.60]
10	(3,0) (3,5)	[670.22,12505.80,24341.40]
11	(3,0) (3,5) (2,10)	[867.37,13084.10,25300.80]
12	(3,0) (3,5) (1,10)	[1320.51,13723.90,26127.20]
13	(3,0) (2,5) (2,9)	[3339.13,14362.20,25385.40]
14	(3,0) (3,5) (2,10)	[1749.59,14958.10,28166.60]
15	(3,0) (3,5) (3,10)	[829.63,15480.30,30131.10]

both the traditional deterministic NPV for this policy as well as the range of values the NPV may take due to uncertainty in the parameters.

Other uncertain factors may be introduced via the relation function $f(a, n, C)$. If, for example, a moderate inflation increase of $[1\%, 2\%, 3\%]$ per year were expected while the other parameters remained the same, the optimal replacement policy becomes:

$$(3,0)(2,5)(2,10) \text{ with } FPV^\star(15) = [-969.74, 16558.10, 34313.60].$$

Contrast this with the traditional stochastic model, SREDM, where the solution is derived using Monte-Carlo simulation. A large number of realizations of the uncertain (and assumed random) variables are generated and each set of them is solved deterministically (Oakford et al., 1984). From this large sample, the probability that each alternative current asset is the optimal first choice can be estimated (see Figure 1) as well as the corresponding cumulative probability distributions of (1) the economic life of each current asset, (2) the NPV of the optimal sequence of challengers for a finite horizon (see Figure 2, and (3) the equivalent finite horizon time for infinite horizon problems. We note that the probability distributions generated by SREDM are *subjectively* interpreted to determine the optimal *current* decision and no information regarding future decisions is available.

13.7 INSPECTION AND REPLACEMENT MODELS

Another problem of great interest and relevance is inspection and replacement under uncertainty. In real-life problems, it is more common to have a com-

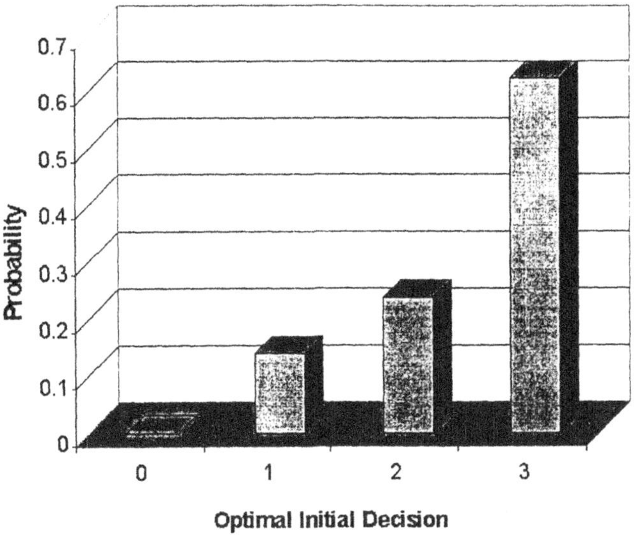

Figure 1 Monte-Carlo simulation results for the example problem depicting the probability distribution for the optimal first choice.

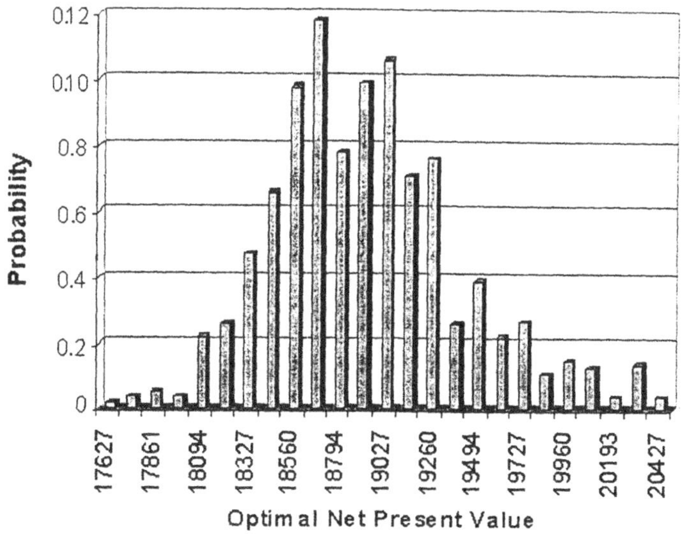

Figure 2 Monte-Carlo simulation results for the example problem depicting the probability distribution for the optimal NPV.

bination of inspection and replacement, with elements of uncertainty in each. Dreyfus and Law (Dreyfus and Law, 1977) treat this situation stochastically. There do not exist any analogous models involving either a possibilistic or fuzzy approach to this problem. For ease of presentation, let us examine the stochastic model using it as the leitmotif for discussing the role of fuzziness in this more realistic situation.

In the inspection and replacement model, parts are inspected at the beginning of a time period and fail at the end of a time period. Parts that are inspected and judged to be in working order are assumed to work just as well as a new part and are given an age of 0. Because of the series configuration, any part failing in the system causes the entire system to fail. The probability that a part j is good at the beginning of a period given that it was also good at the beginning of the previous period is p_j. For inspection of a single part j it takes c_j time periods. Inspection of multiple parts $j_1, \ldots, j_m$ saves time and takes $c_{j_1 \cdots j_m} < c_{j_1} + \cdots + c_{j_m}$. Similarly, replacement takes r_j and $r_{j_1 \cdots j_m} < r_{j_1} + \cdots + r_{j_m}$ time periods. Let the parts have initial ages $t_0(j_1), \ldots, t_0(j_m)$. The objective is to maximize the expected number of periods that the system remains in working order in the next N periods.

As an example, let $m = 2$. Several DP functional equations are needed to clarify the model:

$f_n(g_1, g_2)$ = Maximum expected number of periods from period n through N that the system is working, given that period n starts with parts 1 and 2 last known to be in good order g_1 and g_2 periods previously.

For convenience let

$K_n(g_1)$ = Maximum expected number of periods from period n through N that the system is working, given that period n starts with parts 1 last known to be in good order g_1 and part 2 has just been inspected and found to be in a failed condition

$L_n(g_2)$ = Maximum expected number of periods from period n through N that the system is working, given that period n starts with parts 2 last known to be in good order g_2 and part 1 has just been inspected and found to be in a failed condition

M_n = Maximum expected number of periods from period n through N that the system is working, given that period n starts with parts 1 and 2 just inspected and found to be in a failed condition

The possible actions are to leave the system alone (LA), inspect any combination of parts (I_1, I_2, I_{12}), or replace any combination of parts (R_1, R_2, R_{12}).

This gives

$$f_n(g_1,g_2) = \max \begin{bmatrix} LA: & p_1^{g_1} p_2^{g_2} + f_{n+1}(g_1+1, g_2+1) \\ I_1: & p_1^{g_1} f_{n+c_1}(0,g_2) + (1-p_1^{g_1}) L_{n+c_1}(g_2) \\ I_2: & p_2^{g_2} f_{n+c_2}(g_1,0) + (1-p_2^{g_2}) K_{n+c_2}(g_1) \\ I_{12}: & p_1^{g_1} p_2^{g_2} f_{n+c_{12}}(0,0) + (1-p_1^{g_1}) p_2^{g_2} L_{n+c_{12}}(g_2) \\ & +p_1^{g_1}(1-p_2^{g_2}) K_{n+c_{12}}(g_1) + (1-p_1^{g_1})(1-p_2^{g_2}) M_{n+c_{12}} \\ R_1: & f_{n+r_1}(0,g_2) \\ R_2: & f_{n+r_2}(g_1,0) \\ R_{12}: & f_{n+r_{12}}(0,0) \end{bmatrix}$$

$$K_n(g_1) = \max \begin{bmatrix} I_1: & p_1^{g_1} K_{n+c_1}(0) + (1-p_1^{g_1}) f_{n+c_1+r_{12}}(0,0) \\ R_2: & f_{n+r_2}(g_1,0) \\ R_{12}: & f_{n+r_{12}}(0,0) \end{bmatrix}$$

$$L_n(g_2) = \max \begin{bmatrix} I_1: & p_2^{g_2} L_{n+c_2}(0) + (1-p_2^{g_2}) f_{n+c_2+r_{12}}(0,0) \\ R_2: & f_{n+r_1}(0,g_2) \\ R_{12}: & f_{n+r_{12}}(0,0) \end{bmatrix}$$

and

$$M_n = f_{n+r_{12}}(0,0)$$

with boundary conditions for $n > N$ given as

$$f_n(g_1,g_2) = K_n(g_1) = L_n(g_2) = M_n = 0 \quad \forall g_1, g_2$$

There are several questions that can be raised regarding the assumptions of this model. First, the inspections are assumed to be perfect in that no misdiagnoses are made. This is rarely the case in the real world. Varying degrees of state imperfections can be represented via fuzzy modeling. Second, the "working" state is rarely as good as the original new part and therefore the inspection should tell to what degree the machine is working. This again can be more realistically modeled as a fuzzy variable with the membership function being a surrogate for the extent to which the machine is in a "working" order. Finally, the probability distributions used may not be known or of a familiar type and therefore may need to be replaced by possibility distributions. This model also suffers from the "curse of dimensionality" in that the state and action spaces grow exponentially with the number of parts m in the system.

13.8 MAINTENANCE DECISIONS

Not all replacement decisions are limited to simply choosing to "keep" or "replace" the existing asset. Fortunately, the DP formulation is sufficiently general to allow for other options. For example, a third option may be the "purchase of a used machine". Bellman describes the DP formulation of the replacement model with the option to purchase a used machine as well as that of a model with an "overhaul" option (Bellman and Dreyfus, 1962). The overhaul option can be done either in a general manner, where the cash flows and other variables are functions of both the installation time and the overhaul time, or by allowing the overhauled asset (t years old) to maintain the characteristics of a younger asset ($t' < t$ years old).

Bellman's general model of overhauling machines is formulated in (Bellman and Dreyfus, 1962). We briefly describe it below. To model overhauls, another dimension is added to the problem. Let the functional equation be:

$$f_N(t_1, t_2) \quad = \quad \text{value at year } N \text{ of the overall discounted return from a machine of age } t_1, \text{ last overhauled at age } t_2, \text{ where an optimal replacement policy is employed for the remainder of the process.} \tag{13}$$

Then the recurrence relation becomes:

$$f_N(t_1, t_2) = \max \begin{bmatrix} R: & -C_N(0,0) + S_N(t_1, t_2) - P + \alpha f_{N+1}(1,0) \\ K: & -C_N(t_1, t_2) + \alpha f_{N+1}(t_1 + 1, t_2) \\ O: & -C_N(t_1, t_2) - O_N(t_1, t_2) + \alpha f_{N+1}(t_1 + 1, t_1) \end{bmatrix} \tag{14}$$

where $O_N(t_1, t_2)$ is the cost of overhauling in year N a machine of age t_1 last overhauled at age t_2. Using fuzzy numbers to model the aggregate cash flows, overhaul costs, and salvage values increases the number of calculations by a factor of 3. Thus the general complexity of the fuzzy version remains unchanged from the traditional model.

Modeling an overhaul option in PMSAR can be done either by defining functions that relate an overhauled asset's aggregate cash flows and salvage value to its installation time and overhaul time or defining challenging assets that represent the costs and characteristics of an overhauled machine. By defining a function, an extra dimension is added to the model as above, while defining "overhaul" challengers keeps the solution technique the same as in the original PMSAR.

We return to the example problem for the PMSAR above to illustrate a maintenance option with an imprecise or vague effect. Suppose that the maintenance option is modeled as a "ghost" asset which is defined as follows: The purchase cost P' of the "ghost" asset a' is a function of the cost of the maintenance option on the asset a currently in place and the number of years n' since a was installed.

$$P' = ([0.15, 0.20, 0.25] \cdot P_a)(1.03)^{n'}$$

where P_a is the purchase price of asset a. This particular function states that the base maintenance cost is a fuzzy number that is [15%, 20%, 25%] of the original cost of the asset with an increase of 3% per year since installation. The result of the maintenance is also fuzzy, a [6%, 10%, 12%] reduction in the original operating costs for the following year. The component cash flows corresponding to this maintenance challenger a' are a function of the time n' that the maintenance occurs as well as the original asset and its installation time. Let us denote performing maintenance on the existing asset at period n as (M, n). The resulting optimal maintenance and replacement policy for $N = 15$

then becomes:

$$(M,0)(M,1)$$
$$(1,2)(M,3)(M,4)(M,5)(M,6)$$
$$(1,7)(M,8)(M,9)(M,10)(M,11)$$
$$(2,12)(M,13)(M,14)$$

with an optimal fuzzy present value $FPV^{*}(15) = [25709.00, 40749.90, 54887.80]$. This translates to performing maintenance on the *current defender* at periods 0 and 1, purchasing asset 1 at period 2 and performing maintenance on this asset each period until asset 1 is purchased again in time 7. The maintenance at each period continues on asset 1 until asset 2 is purchased at period 12. Maintenance on asset 2 then performed at each remaining period. The fuzzy present value with this maintenance option, in this example, has risen significantly.

13.9 SUMMARY

Probability theory has been used as the traditional approach for modeling uncertainty in economic analysis. This is acceptable only to the extent that uncertainty is satisfactorily equated with randomness. However, there exist other types of uncertainty which are especially relevant to economic decision analysis. Thus, there is a role to be played by nonprobabilistic uncertainty as shown in this effort. Many approaches have been shown to be possible. A brief survey of replacement analysis, focusing on the use of nonprobabilistic uncertainty, is given. The use of triangular fuzzy numbers provides a compromise between computational efficiency and realistic modeling of the uncertainty. Thus, this discussion emphasizes fuzzy numbers. In the extension to the *economic life of an asset* model, the uncertainty in the parameters is explicitly modeled. By only a three-fold increase in the number of computations, the optimal choice based on the decision-maker's best estimates of these parameters is easily obtained. The traditional deterministic models are a special case of this new possibilistic model. In effect, PELAM performs multivariable sensitivity analysis on all the uncertain parameters concurrently and incorporates this uncertainty into the determination of the optimal decision. Bellman's DP model is also extended to use fuzzy numbers with similar benefits. The benefits for PMSAR are virtually the same, except there is a greater increase in the number of computations due to the vertex method. Contrast this with the large number of repetitions that Monte-Carlo simulation requires, as well as the subjective interpretation of the results, and this disadvantage is not too severe.

When probability distributions are not known, or when a stochastic model is too difficult to solve, fuzzy sets and possibility theory offer an efficient alternative in replacement analysis. There are a number of benefits for modeling the uncertainty in the replacement problem via fuzzy numbers. We outline a few of them:

- The use of fuzzy uncertainty may be more appropriate when modeling systems with human subjectivity. The only existing technique in replacement analysis that modeled general uncertainty in the replacement decision was a Monte-Carlo simulation method (Lohmann, 1986).

- Creating a triangular distribution from the best, worst, and most likely estimates of an expert is more appropriate for possibility theory than probability theory due to the lack of probabilistic information.

- The results of each model can be easily plotted to provide a graphical representation of the effects of the uncertain parameters and can provide this information without a significant increase in the computational complexity. The algorithms are easily implemented on a personal computer.

- The use of TFNs retains the triangular distribution throughout the solution algorithm and conveniently shows the best, worst, and most likely NPVs of each possible decision.

- The use of fuzzy numbers corresponds to performing sensitivity analysis on all uncertain parameters simultaneously, as well as identifying the answer to the deterministic version of the problem which corresponds to the modal value of each fuzzy number.

The models detailed here are limited to extensions of standard economic decision-making techniques. They take a cost approach to determining the optimal replacement policy. Fuzzy sets allow other approaches, such as a fuzzy control approach using IF-THEN rules. For instance, a maintenance or replacement decision may depend on several variables such as *operating revenues*, *operating costs*, *inflation rate*, *technological age* (asset's status as a current technology in the industry), *chronological age*, and numerous others. While no precise mathematical models of these more realistic situations, fuzzy sets defined on these linguistic variables could generate a fuzzy rulebase consisting of rules such as the following two examples:

IF *operating revenues* are **MEDIUM** AND *operating costs* are **HIGH**, THEN *preventive maintenance priority* is **MEDIUM**.
IF *inflation rate* is **HIGH** AND *technological age* is **CUTTING EDGE** AND *chronological age* is **YOUNG**, THEN *replacement priority* is **LOW**.

This approach develops a maintenance and replacement policy based on priorities which differs from the traditional NPV approach. These can be readily solved to yield maintenance and replacement policies that are in many cases intuitive and human-friendly. There are, of course, some limitations such as those induced by fuzzification and defuzzification via standard tools. These can however be ameliorated by recourse to optimal defuzzification strategies as discussed by Esogbue and Song (Esogbue and Song, 1995, Esogbue and Song, 1996, Esogbue et al., 1996). This method could also utilize some of the advances in membership function generation such as reinforcement learning and neural networks as well as other new techniques in fuzzy rulebase control.

References

Alchian, A. (1952). Economic replacement policy. Technical Report Publication R-224, The RAND Corporation, Santa Monica, CA.

Bean, J., Lohmann, J., and Smith, R. (1985). A dynamic infinite horizon replacement economy decision model. *The Engineering Economist*, 30(2):99-120.

Behrens, A. and Choobineh, F. (1989). Can economic uncertainty always be described by randomness? In *Proceedings of the 1989 International Industrial Engineering Conference*, pages 116–120.

Bellman, R. (1955). Equipment replacement policy. *SIAM Journal Applied Mathematics*, 3:133–136.

Bellman, R. and Dreyfus, S. (1962). *Applied Dynamic Programming*. Princeton University Press, Princeton, NJ.

Bezdek, J. (1994). Fuzziness vs. probability - again (!?). *IEEE Transactions on Fuzzy Systems*, 2(1):1–3.

Bortolan, G. and Degani, R. (1985). A review of some methods for ranking fuzzy subsets. *Fuzzy Sets and Systems*, 15:1–19.

Buckley, J. (1987). The fuzzy mathematics of finance. *Fuzzy Sets and Systems*, 21:257–273.

Buckley, J. (1992). Solving fuzzy equations in economics and finance. *Fuzzy Sets and Systems*, 48:289–296.

Chiu, C. and Park, C. (1994). Fuzzy cash flow analysis using present worth criterion. *The Engineering Economist*, 39(2):113–138.

Choobineh, F. and Behrens, A. (1992). Use of intervals and possibility distributions in economic analysis. *Journal of the Operational Research Society*, 43(9):907-918.

Dong, W. and Shah, H. (1987). Vertex method for computing functions of fuzzy variables. *Fuzzy Sets and Systems*, 14:65–78.

Dreyfus, S. and Law, A. (1977). *The Art and Theory of Dynamic Programming*. Academic Press, New York.

Dubois, D. and Prade, H. (1978). Operations on fuzzy numbers. *International Journal of Systems Science*, 9:612–626.

Dubois, D. and Prade, H. (1987). Fuzzy numbers: An overview. In Bezdek, J., editor, *Analysis of Fuzzy Information*, volume 1, pages 3–39. CRC Press, Boca Raton, FL.

Dubois, D. and Prade, H. (1994). Fuzzy sets–a convenient fiction for modeling vagueness and possibility. *IEEE Transactions on Fuzzy Systems*, 2(1):16–21.

Esogbue, A. and Kacprzyk, J. (1996). Fuzzy dynamic programming: A survey of main developments and applications. *Archives of Control Sciences*, 5(1-2):39–59.

Esogbue, A. and Song, Q. (1995). Optimal defuzzification and applications. In *Proceedings of the International Joint Conference on Information Science, Fourth Annual Conference on Fuzzy Theory & Technology*, pages 182–185, Wrightsville Beach, NC.

Esogbue, A. and Song, Q. (Submitted 1996). Defuzzification filters: Design and applications. *IEEE Transactions on Fuzzy Sets*.

Esogbue, A., Song, Q., and Hearnes, W. (1996). Defuzzification filters and applications to power system stabilization problems. In *Proceedings of the 1996*

IEEE International Conference on Systems, Man and Cybernetics Information, Intelligence and Systems, Beijing, China.

Gluss, B. (1972). *An Elementary Introduction to Dynamic Programming: A State Equation Approach.* Allyn and Bacon, Inc.

Gupta, C. (1993). A note on the transformation of possibilistic information into probabilistic information for investment decisions. *Fuzzy Sets and Systems*, 56:175–182.

Hearnes, W. (1995). Modeling with possibilistic uncertainty in the single asset replacement problem. In *Proceedings of the International Joint Conference on Information Science, Fourth Annual Conference on Fuzzy Theory & Technology*, pages 552–555, Wrightsville Beach, NC.

Kacprzyk, J. (1978). Decision-making in a fuzzy environment with fuzzy termination time. *Fuzzy Sets and Systems*, 1:169–179.

Kaufmann, A. and Gupta, M. (1988). *Fuzzy Mathematical Models in Engineering and Management Science.* Elsevier Science Publishers, B.V.

Klir, G. (1994). On the alleged superiority of probabilistic representation of uncertainty. *IEEE Transactions on Fuzzy Systems*, 2(1):27–31.

Klir, G. and Yuan, B. (1995). *Fuzzy Sets and Fuzzy Logic: Theory and Applications.* Prentice Hall PTR, Upper River Saddle, NJ.

Kosko, B. (1994). The probability monopoly. *IEEE Transactions on Fuzzy Systems*, 2(1):32–33.

Laviolette, M. and Seaman, J. J. (1994). The efficacy of fuzzy representations of uncertainty. *IEEE Transactions on Fuzzy Systems*, 2(1):4–15.

Li Calzi, M. (1990). Towards a general setting for the fuzzy mathematics of finance. *Fuzzy Sets and Systems*, 35:265–280.

Lohmann, J. (1986). A stochastic replacement economic decision model. *IIE Transactions*, 18:182–194.

Oakford, R. (1970). *Capital Budgeting: A Quantitative Evanluation of Investment Alternatives.* The Ronald Press Co., New York.

Oakford, R., Lohmann, J., and Salazar, A. (1984). A dynamic replacement economy decision model. *IIE Transactions*, 16:65–72.

Oakford, R., Salazaar, A., and DiGiulio, H. (1981). The long term effectiveness of expected net present value maximization in an environment of incomplete and uncertain information. *AIIE Transactions*, 13:265–276.

Samuelson, R. (1996). Imperfect vision: A flawed CPI makes it harder for us to read the past and plan for the future. *Newsweek*, 55.

Savage, L. (1954). *The Foundations of Statistics.* Wiley, New York.

Sharp-Bette, G. and Park, C. (1990). *Advanced Engineering Economics.* John Wiley and Sons, Inc.

Terborgh, G. (1949). *Dynamic Equipment Policy.* Machinery and Applied Products Institute, Washington, DC.

Tversky, A. and Kahneman, D. (1971). Belief in the law of small numbers. *Psychological Bulletin*, 2:105–110.

Tversky, A. and Kahneman, D. (1974). Judgement under uncertainty: Heuristics and biases. *Science*, 185:1124–1131.

Tversky, A. and Kahneman, D. (1981). The framing of decisions and the psychology of choice. *Science*, 211:3–8.

Von Neumann, J. and Morgenstern, O. (1944). *Theory of Games and Economic Behavior*. Princeton University Press, Princeton, NJ.

Wagner, H. (1975). *Principles of Operations Research*. Prentice-Hall.

Ward, T. (1985). Discounted fuzzy cash flow analysis. In *Proceedings of the 1985 Fall Industrial Engineering Conference*, pages 476–481.

Zadeh, L. (1978). Fuzzy sets as a basis for a theory of possibility. *IEEE Transactions on Systems, Man, and Cybernetics*, 1:3–28.

Zimmerman, H. (1991). *Fuzzy Set Theory and Its Applications*. Kluwer Academic Press, Boston, MA, second edition.

Index

Made in the USA
Las Vegas, NV
28 October 2024

10602825R00267